MW01027461

Lecture Notes in Physics

Springer
Berlin
Heidelberg
New York
Barcelona
Hong Kong
London
Milan
Paris
Singapore
Tokyo

Editorial Policy

The series *Lecture Notes in Physics* (LNP), founded in 1969, reports new developments in physics research and teaching -- quickly, informally but with a high quality. Manuscripts to be considered for publication are topical volumes consisting of a limited number of contributions, carefully edited and closely related to each other. Each contribution should contain at least partly original and previously unpublished material, be written in a clear, pedagogical style and aimed at a broader readership, especially graduate students and nonspecialist researchers wishing to familiarize themselves with the topic concerned. For this reason, traditional proceedings cannot be considered for this series though volumes to appear in this series are often based on material presented at conferences, workshops and schools (in exceptional cases the original papers and/or those not included in the printed book may be added on an accompanying CD ROM, together with the abstracts of posters and other material suitable for publication, e.g. large tables, colour pictures, program codes, etc.).

Acceptance

A project can only be accepted tentatively for publication, by both the editorial board and the publisher, following thorough examination of the material submitted. The book proposal sent to the publisher should consist at least of a preliminary table of contents outlining the structure of the book together with abstracts of all contributions to be included.
Final acceptance is issued by the series editor in charge, in consultation with the publisher, only after receiving the complete manuscript. Final acceptance, possibly requiring minor corrections, usually follows the tentative acceptance unless the final manuscript differs significantly from expectations (project outline). In particular, the series editors are entitled to reject individual contributions if they do not meet the high quality standards of this series. The final manuscript must be camera-ready, and should include both an informative introduction and a sufficiently detailed subject index.

Contractual Aspects

Publication in LNP is free of charge. There is no formal contract, no royalties are paid, and no bulk orders are required, although special discounts are offered in this case. The volume editors receive jointly 30 free copies for their personal use and are entitled, as are the contributing authors, to purchase Springer books at a reduced rate. The publisher secures the copyright for each volume. As a rule, no reprints of individual contributions can be supplied.

Manuscript Submission

The manuscript in its final and approved version must be submitted in camera-ready form. The corresponding electronic source files are also required for the production process, in particular the online version. Technical assistance in compiling the final manuscript can be provided by the publisher's production editor(s), especially with regard to the publisher's own Latex macro package which has been specially designed for this series.

Online Version/ LNP Homepage

LNP homepage (list of available titles, aims and scope, editorial contacts etc.):
http://www.springer.de/phys/books/lnp/

LNP online (abstracts, full-texts, subscriptions etc.):
http://link.springer.de/series/lnp/

J. Trampetić J. Wess (Eds.)

Particle Physics in the New Millennium

Proceedings of the 8th Adriatic Meeting

Editors

Josip Trampetić
Rudjer Boskovic Institute
Theoretical Physics Division
P.O.Box 180
10 002 Zagreb
Croatia

Julius Wess
Sektion Physik der Ludwig-Maximilians-Universität
Theresienstr. 37
80333 München
and
Max-Planck-Institut für Physik
(Werner-Heisenberg-Institut)
Föhringer Ring 6
80805 München
Germany

Cover Picture: (see Fig.1 contribution by D. Denegri in this volume)

Library of Congress Cataloging-in-Publication Data.
Cataloging-in-Publication Data applied for

Bibliographic information published by Die Deutsche Bibliothek Die Deutsche Bibliothek lists this publication in the Deutsche Nationalbibliografie; detailed bibliographic data is available in the Internet at <http://dnb.ddb.de>

ISSN 0075-8450
ISBN 3-540-00711-3 Springer-Verlag Berlin Heidelberg New York

Springer-Verlag Berlin Heidelberg New York
a member of BertelsmannSpringer Science+Business Media GmbH

http://www.springer.de

Printed in Germany

Typesetting: Camera-ready by the authors/editor
Camera-data conversion by Steingraeber Satztechnik GmbH Heidelberg
Cover design: *design & production*, Heidelberg

Printed on acid-free paper
54/3111 – 5 4 3 2 1 SPIN 10970041

Preface

The traditional purpose of the Adriatic Meeting is to present most advanced scientific research conducted by the lecturers who take part in the development of their fields and, in addition, to provide a school-like atmosphere for young scientists.

Dubrovnik, as a geographical centre of this region of Europe, provided a most adequate location for this conference. Having very agreeable surroundings, the conference site nevertheless gave a focus for very strong scientific interaction.

The subjects chosen for the 8th meeting, in September 2001, were gauge theories, particle phenomenology, string theories and cosmology.

We were able to bring together a very good cross section of outstanding scientists who gave extraorinarily good presentations. Certainely one reason for this success is that most of us feel obliged to help the scientific life in South East Europe return to its former level. However, there are very exciting new scientific developments as well.

Part of the meeting was dominated by neutrino physics which has just seen exciting progress by establishing neutrino masses experimentally. This was discussed within neutrino masses and grand unified theories (GUTs). General aspects of neutrino physics and CP violation, neutrino mixing and the bayron asymmetry were presented along the same lines.

On the theoretical side the idea of the construction of gauge theories on non-commutative spaces and their phenomenological implications is accepted worldwide within the particle physics community.

Both status of CP violation and heavy meason decay were discussed at a moment when the recent experimental results became available. A remarkable achievement of heavy quark physics in general and CP violation within the framework of heavy quark effective field theory (HQEFT) was made on the basis of nonperturbative QCD.

For an outlook on Large Hadron Collider (LHC) and the physics to be done there, the meeting took place at just the right time. We are especially proud that at this 8th Adriatic Meeting there were a number of representatives from huge experimental collaborations ranging from ATLAS and CMS at CERN to CLEO at Cornel. It is a pleasure to note here the excellent contact that was established with particle physicists from South America, in particular from Brazil.

Cosmology with its new estimate of dark matter was another subject of huge interest.

Finally, there was a number of extremely interesting presentations concerning the theoretical and experimental problems in: SUSY, magnetic monopoles in QCD, the perturbative QCD approach, hot matter in QCD and physics beyond the standard model at new accelerators.

All of this gave an impressive overview of the present activities and the progress in those areas of physics represented at the meeting. At the same time it created an active atmosphere which drew many of the young scientists into these fields.

We would like to thank young members of Theory Division, Rudjer Bošković Institute for help during the Conference: G. Duplančić, H. Nikolić and H. Štefančić. For a substantial help during the organization of the Conference we would especially like to thank K. Passek-Kumerički. We would like to thank L. Jonke for great help in preparing these Lectures Notes.

Muenchen,
March 2002

Josip Trampetić
Julius Wess

Table of Contents

Part V Diverse Topics in Theoretical Physics

List of Contributors

Guido ALTARELLI
CERN, Geneva

Gilvan A. ALVES
Centro Brasileiro de Pesquisas Fisicas, Rio de Janeiro

Neven BILIĆ
Rudjer Bošković Institute, Zagreb

Loriano BONORA
SISSA, Trieste

Wilfried BUCHMULLER
DESY, Hamburg

Daniel DENEGRI
CERN, Geneva

N. G. DESHPANDE
University of Oregon, Eugene

John ELLIS
CERN, Geneva

Yitzhak FRISHMAN
Weizmann Institute, Rehovot

Harald GROSSE
University of Vienna

Roman JACKIW
MIT, Cambridge

Konrad KLEINKNECHT
Johannes-Gutenberg Universitaet, Mainz

Antonio MASIERO
SISSA, Trieste

Holger B. NIELSEN
Niels Bohr Institute, Copenhagen

Sandip PAKVASA
University of Hawaii, Honolulu

Yoram ROZEN
Technion, Haifa

Paolo SALUCCI
SISSA, Trieste

A. I. SANDA
Nagoya University

Helmut SATZ
University of Bielefeld

Peter SCHUPP
University of Munich

Dmitry SHIRKOV
Bogoliubov Lab at JINR, Dubna

Nikolaos G. STEFANIS
Ruhr-University Bochum

Claudio VERZEGNASSI
University of Trieste

Roland WALDI
University of Rostock

Julius WESS
University of Munich

Valentine ZAKHAROV
Max-Planck Institute, Munich

Part I

Neutrino Physics, Cosmology

Models of Neutrino Masses and Mixing

Guido Altarelli

Theory Division, CERN, CH-1211 Geneva 23, Switzerland

Abstract. We briefly review models of neutrino masses and mixings. In view of the existing experimental ambiguities many possibilities are still open. After an overview of the main alternative options we focus on the most constrained class of models based on three widely split light neutrinos within SUSY Grand Unification

1 Introduction

At present there are many alternative models of neutrino masses. This variety is in part due to the considerable existing experimental ambiguities. The most crucial questions to be clarified by experiment are whether the LSND signal will be confirmed or will be excluded and which solar neutrino solution will eventually be established. If LSND is right we need four light neutrinos, if not we can do with only the three known ones. Other differences are due to less direct physical questions like the possible cosmological relevance of neutrinos as hot dark matter. If neutrinos are an important fraction of the cosmological density, say $\Omega_\nu \sim 0.1$, then the average neutrino mass must be considerably heavier than the splittings that are indicated by the observed atmospheric and solar oscillation frequencies. For example, for three light neutrinos, only models with almost degenerate neutrinos, with common mass $|m_\nu| \approx 1\ eV$, are compatible with a large hot dark matter component. On the contrary hierarchical three neutrino models have the largest neutrino mass fixed by $m \approx \sqrt{\Delta m^2_{atm}} \approx 0.05\ eV$. In most models the smallness of neutrino masses is related to the fact that $\nu's$ are completely neutral (i.e. they carry no charge which is exactly conserved), they are Majorana particles and their masses are inversely proportional to the large scale where the lepton number L conservation is violated. Majorana masses can arise from the see-saw mechanism, in which case there is some relation with the Dirac masses, or from higher dimension non renormalisable operators which come from a different sector of the lagrangian density than other fermion mass terms.

In my lecture first I will briefly summarise the main categories of neutrino mass models and give my personal views on them. Then, I will argue in favour of the most constrained set of models, where there are only three widely split neutrinos, with masses dominated by the see-saw mechanism and inversely proportional to a large mass close to the Grand Unification scale M_{GUT}. In this framework neutrino masses are a probe into the physics of GUT's and one can aim at a comprehensive discussion of all fermion masses. This is for example possible in models based on $SU(5) \bigotimes U(1)_{flavour}$ or on $SO(10)$ (we always consider

SUSY GUT's). This will also lead us to consider the status of GUT models in view of the experimental bounds on p decay, which are now very severe also for SUSY models, and of well known naturality problems, like the doublet-triplet splitting problem. So we will discuss "realistic" as opposed to minimal models, including a description of the pattern of all fermion masses. We will also mention some recent ideas on a radically different concept of SUSY $SU(5)$ where the symmetry is valid in 5 dimensions but is broken by compactification and not by some Higgs system in the 24 or larger representation. In this version of $SU(5)$ the doublet-triplet splitting problem is solved elegantly and p decay can naturally be suppressed or even forbidden by the compactification mechanism.

This review is in part based on work that I have done over the recent months with Ferruccio Feruglio and Isabella Masina [1], [2], [3], [4], [5], [6], [7].

2 Neutrino Masses and Lepton Number Violation

Neutrino oscillations imply neutrino masses which in turn demand either the existence of right-handed neutrinos (Dirac masses) or lepton number L violation (Majorana masses) or both. Given that neutrino masses are certainly extremely small, it is really difficult from the theory point of view to avoid the conclusion that L must be violated. In fact, it is only in terms of lepton number violation that the smallness of neutrino masses can be explained as inversely proportional to the very large scale where L is violated, of order M_{GUT} or even M_{Planck}.

Once we accept L violation we gain an elegant explanation for the smallness of neutrino masses which turn out to be inversely proportional to the large scale where lepton number is violated. If L is not conserved, even in the absence of ν_R, Majorana masses can be generated for neutrinos by dimension five operators of the form

$$O_5 = \frac{L_i^T \lambda_{ij} L_j HH}{M} \tag{1}$$

with H being the ordinary Higgs doublet, λ a matrix in flavour space and M a large scale of mass, of order M_{GUT} or M_{Planck}. Neutrino masses generated by O_5 are of the order $m_\nu \approx v^2/M$ for $\lambda_{ij} \approx$ O(1), where $v \sim$ O(100 GeV) is the vacuum expectation value of the ordinary Higgs.

We consider that the existence of ν_R is quite plausible because all GUT groups larger than SU(5) require them. In particular the fact that ν_R completes the representation 16 of SO(10): 16=$\bar{5}$+10+1, so that all fermions of each family are contained in a single representation of the unifying group, is too impressive not to be significant. At least as a classification group SO(10) must be of some relevance. Thus in the following we assume that there are both ν_R and lepton number violation. With these assumptions the see-saw mechanism [8] is possible which leads to:

$$m_\nu = m_D^T M^{-1} m_D \tag{2}$$

That is, the light neutrino masses are quadratic in the Dirac masses and inversely proportional to the large Majorana mass. Note that for $m_\nu \approx \sqrt{\Delta m^2_{atm}} \approx$

0.05 eV and $m_\nu \approx m_D^2/M$ with $m_D \approx v \approx 200\ GeV$ we find $M \approx 10^{15}\ GeV$ which indeed is an impressive indication for M_{GUT}.

If additional non renormalisable terms from O_5 are comparatively non negligible, they should simply be added. After elimination of the heavy right-handed fields, at the level of the effective low energy theory, the two types of terms are equivalent. In particular they have identical transformation properties under a chiral change of basis in flavour space. The difference is, however, that in the see-saw mechanism, the Dirac matrix m_D is presumably related to ordinary fermion masses because they are both generated by the Higgs mechanism and both must obey GUT-induced constraints. Thus if we assume the see-saw mechanism more constraints are implied. In particular we are led to the natural hypothesis that m_D has a largely dominant third family eigenvalue in analogy to m_t, m_b and m_τ which are by far the largest masses among u quarks, d quarks and charged leptons. Once we accept that m_D is hierarchical it is very difficult to imagine that the effective light neutrino matrix, generated by the see-saw mechanism, could have eigenvalues very close in absolute value.

3 Four Neutrino Models

The LSND signal has not been confirmed by KARMEN. It will be soon double-checked by MiniBoone. Perhaps it will fade away. But if an oscillation with $\Delta m^2 \approx 1\ eV^2$ is confirmed then, in presence of three distinct frequencies for LSND, atmospheric and solar neutrino oscillations, at least four light neutrinos are needed. Since LEP has limited to three the number of "active" neutrinos (that is with weak interactions, or equivalently with non vanishing weak isospin, the only possible gauge charge of neutrinos) the additional light neutrino(s) must be "sterile", i.e. with vanishing weak isospin. Note that ν_R that appears in the see-saw mechanism, if it exists, is a sterile neutrino, but a heavy one.

A typical pattern of masses that works for 4-ν models consists of two pairs of neutrinos [9], the separation between the two pairs, of order 1 eV, corresponding to the LSND frequency. The upper doublet would be almost degerate at $|m|$ of order 1 eV being only split by (the mass difference corresponding to) the atmospheric ν frequency, while the lower doublet is split by the solar ν frequency. This mass configuration can be compatible with an important fraction of hot dark matter in the universe. A complication is that the data appear to be incompatible with pure 2-ν oscillations for $\nu_e - \nu_s$ oscillations for solar neutrinos and for $\nu_\mu - \nu_s$ oscillations for atmospheric neutrinos (with ν_s being a sterile neutrino). There are however viable alternatives. One possibility is obtained by using the large freedom allowed by the presence of 6 mixing angles in the most general 4-ν mixing matrix. If 4 angles are significantly different from zero, one can go beyond pure 2-ν oscillations and, for example, for solar neutrino oscillations ν_e can transform into a mixture of $\nu_a + \nu_s$, where ν_a is an active neutrino, itself a superposition of ν_μ and ν_τ [9]. A different alternative is to have many interfering sterile neutrinos: this is the case in the interesting class of models with extra dimensions, where a whole tower of Kaluza-Klein neutrinos is introduced. This picture of sterile neutrinos from extra dimensions is exciting and we now discuss it in some detail.

The context is theories with large extra dimensions. Gravity propagates in all dimensions (bulk), while SM particles live on a 4-dim brane. As well known [10], this can make the fundamental scale of gravity m_s much smaller than the Planck mass M_P. In fact, for $d = n + 4$, if R is the compactification radius we have a geometrical volume factor that suppresses gravity so that: $(m_s R)^n = (M_P/m_s)^2$ and, as a result, m_s can be as small as $\sim 1\ TeV$. For neutrino phenomenology we need a really large extra dimension with $1/R \lesssim 0.01\ eV$ plus $n-1$ smaller ones with $1/\rho \gtrsim 1\ TeV$. Then we define m_5 by $m_5 R = (M_P/m_s)^2$, or $m_5 = m_s(m_s\rho)^{n-1}$. In string theories of gravity there are always scalar fields associated with gravity and their SUSY fermionic partners (dilatini, modulini). These are particles that propagate in the bulk, have no gauge interactions and can well play the role of sterile neutrinos. The models based on this framework [11] have some good features that make them very appealing at first sight. They provide a "physical" picture for ν_s. There is a KK tower of recurrences of ν_s:

$$\nu_s(x,y) = \frac{1}{\sqrt{R}} \sum_n \nu_s^{(n)}(x) \cos\frac{ny}{R} \tag{3}$$

with $m_{\nu_s} = n/R$. The tower mixes with the ordinary light active neutrinos in the lepton doublet L:

$$L_{mix} = h\frac{m_s}{M_P} L\nu_s^{(n)} H \tag{4}$$

where H is the Higgs doublet field. Note that the geometrical factor m_s/M_P, which automatically suppresses the Yukawa coupling h, arises naturally from the fact that the sterile neutrino tower lives in the bulk. Note in passing that ν_s mixings must be small due to existing limits from weak processes, supernovae and nucleosynthesis, so that the preferred solution for 4-ν models is MSW-(small angle). The interference among a few KK states makes the spectrum compatible with solar data:

$$P(\nu_e \to X) = \sum_n \frac{m_e^2}{M_e^2 + \frac{n^2}{R^2}} \tag{5}$$

provided that $1/R \sim 10^{-2} - 10^{-3}\ eV$ or $R \sim 10^{-3} - 10^{-2}\ cm$, that is a really large extra dimension barely compatible with existing limits [12].

In spite of its good properties there are problems with this picture, in my opinion. The first property that I do not like of models with large extra dimensions is that the connection with GUT's is lost. In particular the elegant explanation of the smallness of neutrino masses in terms of the large scale where the L conservation is violated in general evaporates. Since $m_s \sim 1\ TeV$ is relatively small, what forbids on the brane an operator of the form $\frac{1}{m_s} L_i^T \lambda_{ij} L_j HH$ which would lead to by far too large ν masses? One must assume L conservation on the brane and that it is only broken by some Majorana masses of sterile ν's in the bulk, which I find somewhat ad hoc. Another problem is that we would expect gravity to know nothing about flavour, but here we would need right-handed partners for ν_e, ν_μ and ν_τ. Also a single large extra dimension has problems, because it implies [13] a linear evolution of the gauge couplings with energy from 0.01 eV to $m_s \sim 1\ TeV$. But more large extra dimensions lead to

$$P(\nu_e \to X) = \sum_n \frac{m_e^2}{M_e^2 + \frac{n^2}{R^2}} = \int dn n^{d-1} \frac{m_e^2}{M_e^2 + \frac{n^2}{R^2}} \tag{6}$$

For $d > 2$ the KK recurrences do not decouple fast enough (the divergence of the integral is only cut off at m_s) and the mixing becomes very large. Perhaps a compromise at $d = 2$ is possible.

In conclusion the models with large extra dimension are interesting because they are speculative and fascinating but the more conventional framework still appears more plausible at closer inspection.

4 Three Neutrino Models

We now assume that the LSND signal will not be confirmed, that there are only two distinct neutrino oscillation frequencies, the atmospheric and the solar frequencies, which can be reproduced with the known three light neutrino species (for reviews of three neutrino models see [4], [14] where a rather complete set of references can be found). The two frequencies, are parametrised in terms of the ν mass eigenvalues by

$$\Delta_{sun} \propto m_2^2 - m_1^2, \qquad \Delta_{atm} \propto m_3^2 - m_{1,2}^2 \tag{7}$$

The numbering 1,2,3 corresponds to our definition of the frequencies and in principle may not coincide with the family index although this will be the case in the models that we favour. Given the observed frequencies and our notation in (7), there are three possible patterns of mass eigenvalues:

$$\begin{aligned} &\text{Degenerate} &&: |m_1| \sim |m_2| \sim |m_3| \\ &\text{Inverted hierarchy} &&: |m_1| \sim |m_2| \gg |m_3| \\ &\text{Hierarchical} &&: |m_3| \gg |m_{2,1}| \end{aligned} \tag{8}$$

We now discuss pro's and con's of the different cases and argue in favour of the hierarchical option.

4.1 Degenerate Neutrinos

At first sight the degenerate case is the most appealing: the observation of nearly maximal atmospheric neutrino mixing and the possibility that also the solar mixing is large (at present the MSW-(large angle) solution of the solar neutrino oscillations appears favoured by the data) suggests that all ν masses are nearly degenerate. Moreover, the common value of $|m_\nu|$ could be compatible with a large fraction of hot dark matter in the universe for $|m_\nu| \sim 1 - 2\ eV$. In this case, however, the existing limits on the absence of neutrino-less double beta decay ($0\nu\beta\beta$) imply [15] double maximal mixing (bimixing) for solar and atmospheric neutrinos. In fact the quantity which is bound by experiments is the 11 entry of the ν mass matrix, which is given by [4]:

$$m_{ee} = m_1 cos^2\theta_{12} + m_2 sin^2\theta_{12} \lesssim 0.3 - 0.5\ eV \tag{9}$$

To satisfy this constraint one needs $m_1 = -m_2$ (recall that the sign of fermion masses can be changed by a phase redefinition) and $cos^2\theta_{12} \sim sin^2\theta_{12}$ to a good accuracy (in fact we need $sin^2 2\theta_{12} > 0.96$ in order that $|cos 2\theta_{12}| = |cos^2\theta_{12} - sin^2\theta_{12}| < 0.2$). Of course this strong constraint can be relaxed if the common mass is below the hot dark matter maximum. It is true in any case that a signal of $0\nu\beta\beta$ near the present limit (like a large relic density of hot dark matter) would be an indication for nearly degenerate ν's.

In general, for naturalness reasons, the splittings cannot be too small with respect to the common mass, unless there is a protective symmetry [16]. This is because the wide mass differences of fermion masses, in particular charged lepton masses, would tend to create neutrino mass splittings via renormalization group running effects even starting from degenerate masses at a large scale. For example, the vacuum oscillation solution for solar neutrino oscillations would imply $\Delta m/m \sim 10^{-9} - 10^{-11}$ which is difficult to obtain. In this respect the MSW-(large angle) solution would be favoured, but, if we insist that $|m_\nu| \sim 1 - 2\ eV$, it is not clear that the mixing angle is sufficiently maximal.

It is clear that in the degenerate case the most likely origin of ν masses is from dim-5 operators $O_5 = L_i^T \lambda_{ij} L_j HH/M$ and not from the see-saw mechanism $m_\nu = m_D^T M^{-1} m_D$. In fact we expect the ν-Dirac mass m_D to be hierarchical like for all other fermions and a conspiracy to reinstaure a nearly perfect degeneracy between m_D and M, which arise from completely different physics, looks very unplausible. Thus in degenerate models, in general, there is no direct relation with Dirac masses of quarks and leptons and the possibility of a simultaneous description of all fermion masses within a grand unified theory is more remote [17].

4.2 Inverted Hierarchy

The inverted hierarchy configuration $|m_1| \sim |m_2| \gg |m_3|$ consists of two levels m_1 and m_2 with small splitting $\Delta m^2_{12} = \Delta m^2_{sun}$ and a common mass given by $m^2_{1,2} \sim \Delta m^2_{atm} \sim 2.5 \cdot 10^{-3}\ eV^2$ (no large hot dark matter component in this case). One particularly interesting example of this sort [18], which leads to double maximal mixing, is obtained with the phase choice $m_1 = -m_2$ so that, approximately:

$$m_{diag} = M[1, -1, 0] \tag{10}$$

The effective light neutrino mass matrix

$$m_\nu = U m_{diag} U^T \tag{11}$$

which corresponds to the mixing matrix of double maximal mixing $c = s = 1/\sqrt{2}$:

$$U_{fi} = \begin{bmatrix} c & -s & 0 \\ s/\sqrt{2} & c/\sqrt{2} & -1/\sqrt{2} \\ s/\sqrt{2} & c/\sqrt{2} & +1/\sqrt{2} \end{bmatrix} \quad . \tag{12}$$

is given by:

$$m_\nu = \frac{M}{\sqrt{2}} \begin{bmatrix} 0 & 1 & 1 \\ 1 & 0 & 0 \\ 1 & 0 & 0 \end{bmatrix} \quad . \tag{13}$$

The structure of m_ν can be reproduced by imposing a flavour symmetry $L_e - L_\mu - L_\tau$ starting from $O_5 = L_i^T \lambda_{ij} L_j HH/M$. The $1-2$ degeneracy remains stable under radiative corrections. The preferred solar solutions are vacuum oscillations or the LOW solution. The MSW-(large angle) could be also compatible if the mixing angle is large enough. The required dominance of O_5 leads to the same comments as the degenerate models of the previous section.

4.3 Hierarchical

We now discuss the class of models which we consider of particular interest because this is the most constrained framework which allows a comprehensive combined study of all fermion masses in GUT's. We assume three widely split ν's and the existence of a right-handed neutrino for each generation, as required to complete a 16-dim representation of $SO(10)$ for each generation. We then assume dominance of the see-saw mechanism $m_\nu = m_D^T M^{-1} m_D$. We know that the third-generation eigenvalue of the Dirac mass matrices of up and down quarks and of charged leptons is systematically the largest one. It is natural to imagine that this property will also be true for the Dirac mass of ν's: $diag[m_D] \sim [0, 0, m_{D3}]$. After see-saw we expect m_ν to be even more hierarchical being quadratic in m_D (barring fine-tuned compensations between m_D and M). The amount of hierarchy, $m_3^2/m_2^2 = \Delta m^2_{atm}/\Delta m^2_{sun}$, depends on which solar neutrino solution is adopted: the hierarchy is maximal for vacuum oscillations and LOW solutions, is moderate for MSW in general and could become quite mild for the upper Δm^2 domain of the MSW-(large angle) solution. A possible difficulty is that one is used to expect that large splittings correspond to small mixings because normally only close-by states are strongly mixed. In a 2 by 2 matrix context the requirement of large splitting and large mixings leads to a condition of vanishing determinant. For example the matrix

$$m \propto \begin{bmatrix} x^2 & x \\ x & 1 \end{bmatrix} \quad . \tag{14}$$

has eigenvalues 0 and $1 + x^2$ and for x of $0(1)$ the mixing is large. Thus in the limit of neglecting small mass terms of order $m_{1,2}$ the demands of large atmospheric neutrino mixing and dominance of m_3 translate into the condition that the 2 by 2 subdeterminant 23 of the 3 by 3 mixing matrix approximately vanishes. The problem is to show that this vanishing can be arranged in a natural way without fine tuning. Once near maximal atmospheric neutrino mixing is reproduced the solar neutrino mixing can be arranged to be either small of large without difficulty by implementing suitable relations among the small mass terms.

It is not difficult to imagine mechanisms that naturally lead to the approximate vanishing of the 23 sub-determinant. For example [18], [19], assume that one ν_R is particularly light and coupled to μ and τ. In a 2 by 2 simplified context if we have

$$M \propto \begin{bmatrix} \epsilon & 0 \\ 0 & 1 \end{bmatrix}; \qquad M^{-1} \approx \begin{bmatrix} 1/\epsilon & 0 \\ 0 & 0 \end{bmatrix} \tag{15}$$

then for a generic m_D we find

$$m_\nu \;=\; m_D^T M^{-1} M_D \sim \begin{bmatrix} a & c \\ b & d \end{bmatrix} \begin{bmatrix} 1/\epsilon & 0 \\ 0 & 0 \end{bmatrix} \begin{bmatrix} a & b \\ c & d \end{bmatrix} \;=\; \frac{1}{\epsilon} \begin{bmatrix} a^2 & ac \\ ac & c^2 \end{bmatrix} \tag{16}$$

A different possibility that we find attractive is that, in the limit of neglecting terms of order $m_{1,2}$ and, in the basis where charged leptons are diagonal, the Dirac matrix m_D, defined by $\bar{R} m_D L$, takes the approximate form:

$$m_D \propto \begin{bmatrix} 0 & 0 & 0 \\ 0 & 0 & 0 \\ 0 & x & 1 \end{bmatrix} \quad . \tag{17}$$

This matrix has the property that for a generic Majorana matrix M one finds:

$$m_\nu = m_D^T M^{-1} m_D \propto \begin{bmatrix} 0 & 0 & 0 \\ 0 & x^2 & x \\ 0 & x & 1 \end{bmatrix} \quad . \tag{18}$$

The only condition on M^{-1} is that the 33 entry is non zero. But when the approximately vanishing matrix elements are replaced by small terms, one must also assume that no new $o(1)$ terms are generated in m_ν by a compensation between small terms in m_D and large terms in M. It is important for the following discussion to observe that m_D given by (17) under a change of basis transforms as $m'_D -> V^\dagger m_D U$ where V and U rotate the right and left fields respectively. It is easy to check that in order to make m_D diagonal we need large left mixings (i.e. large off diagonal terms in the matrix that rotates left-handed fields). Thus the question is how to reconcile large left-handed mixings in the leptonic sector with the observed near diagonal form of V_{CKM}, the quark mixing matrix. Strictly speaking, since $V_{CKM} = U_u^\dagger U_d$, the individual matrices U_u and U_d need not be near diagonal, but V_{CKM} does, while the analogue for leptons apparently cannot be near diagonal. However nothing forbids for quarks that, in the basis where m_u is diagonal, the d quark matrix has large non diagonal terms that can be rotated away by a pure right-handed rotation. We suggest that this is so and that in some way right-handed mixings for quarks correspond to left-handed mixings for leptons.

In the context of (Susy) SU(5) there is a very attractive hint of how the present mechanism can be realized. In the $\bar{5}$ of SU(5) the d^c singlet appears together with the lepton doublet (ν, e). The (u, d) doublet and e^c belong to the 10 and ν^c to the 1 and similarly for the other families. As a consequence, in the simplest model with mass terms arising from only Higgs pentaplets, the Dirac matrix of down quarks is the transpose of the charged lepton matrix: $m^d_D = (m^l_D)^T$. Thus, indeed, a large mixing for right-handed down quarks corresponds to a large left-handed mixing for charged leptons. At leading order we may have:

$$m_d = (m_l)^T = \begin{bmatrix} 0 & 0 & 0 \\ 0 & 0 & x \\ 0 & 0 & 1 \end{bmatrix} v_d \tag{19}$$

In the same simplest approximation with 5 or $\bar{5}$ Higgs, the up quark mass matrix is symmetric, so that left and right mixing matrices are equal in this case. Then small mixings for up quarks and small left-handed mixings for down quarks are sufficient to guarantee small V_{CKM} mixing angles even for large d quark right-handed mixings. If these small mixings are neglected, we expect:

$$m_u = \begin{bmatrix} 0 & 0 & 0 \\ 0 & 0 & 0 \\ 0 & 0 & 1 \end{bmatrix} v_u \tag{20}$$

When the charged lepton matrix is diagonalized the large left-handed mixing of the charged leptons is transferred to the neutrinos. Note that in SU(5) we can diagonalize the u mass matrix by a rotation of the fields in the 10, the Majorana matrix M by a rotation of the 1 and the effective light neutrino matrix m_ν by a rotation of the $\bar{5}$. In this basis the d quark mass matrix fixes V_{CKM} and the charged lepton mass matrix fixes neutrino mixings. It is well known that a model where the down and the charged lepton matrices are exactly the transpose of one another cannot be exactly true because of the e/d and μ/s mass ratios. It is also known that one remedy to this problem is to add some Higgs component in the 45 representation of SU(5) [20]. A different kind of solution [21] will be described later. But the symmetry under transposition can still be a good guideline if we are only interested in the order of magnitude of the matrix entries and not in their exact values. Similarly, the Dirac neutrino mass matrix m_D is the same as the up quark mass matrix in the very crude model where the Higgs pentaplets come from a pure 10 representation of SO(10): $m_D = m_u$. For m_D the dominance of the third family eigenvalue as well as a near diagonal form could be an order of magnitude remnant of this broken symmetry. Thus, neglecting small terms, the neutrino Dirac matrix in the basis where charged leptons are diagonal could be directly obtained in the form of (17).

5 Simple Examples with Horizontal Abelian Charges

We discuss here some explicit examples of the mechanism under discussion in the framework of a unified Susy $SU(5)$ theory with an additional $U(1)_F$ flavour symmetry [22]. If, for a given interaction vertex, the $U(1)_F$ charges do not add to zero, the vertex is forbidden in the symmetry limit. But the symmetry is spontaneously broken by the vev v_f of a number of "flavon" fields with non vanishing charge. Then a forbidden coupling is rescued but is suppressed by powers of the small parameters v_f/M with the exponent larger for larger charge mismatch. We expect $v_f \gtrsim M_{GUT}$ and $M \lesssim M_P$. Here we discuss some aspects of the description of fermion masses in these models. In the following sections we will consider how to imbed these concepts within more complete and realistic $SU(5)$ models. We will also discuss the need and the options to go beyond minimal models.

In these models the known generations of quarks and leptons are contained in triplets Ψ^a_{10} and $\Psi^a_{\bar{5}}$, $(a = 1, 2, 3)$ transforming as 10 and $\bar{5}$ of $SU(5)$, respectively. Three more $SU(5)$ singlets Ψ^a_1 describe the right-handed neutrinos. In SUSY models we have two Higgs multiplets, which transform as 5 and $\bar{5}$ in the minimal model. We first assume that they have the same charge. The simplest models are obtained by allowing all the third generation masses already in the symmetric limit. This is realised by taking vanishing charges for the Higgses and for the third generation components Ψ^3_{10}, $\Psi^3_{\bar{5}}$ and Ψ^3_1. We can arrange the unit of charge in such a way that the Cabibbo angle, which we consider as the typical hierachy parameter of fermion masses and mixings, is obtained when the suppression exponent is unity. Remember that the Cabibbo angle is not too small, $\lambda \sim 0.22$ and that in $U(1)_F$ models all mass matrix elements are of the form of a power of a suppression factor times a number of order unity, so that only their order of suppression is defined. As a consequence, in practice, we can limit ourselves to integral charges in our units, for simplicity (for example, $\sqrt{\lambda} \sim 1/2$ is already almost unsuppressed).

After these preliminaries let's first try a simplest model with all charges being non negative and containing one single flavon of negative charge. For example, we could take [23] (see also [24])

$$\Psi_{10} \sim (4, 2, 0) \tag{21}$$

$$\Psi_{\bar{5}} \sim (2, 0, 0) \tag{22}$$

$$\Psi_1 \sim (4, 2, 0) \tag{23}$$

In this case a typical mass matrix has the form

$$m = \begin{bmatrix} y_{11}\lambda^{q_1+q'_1} & y_{12}\lambda^{q_1+q'_2} & y_{13}\lambda^{q_1+q'_3} \\ y_{21}\lambda^{q_2+q'_1} & y_{22}\lambda^{q_2+q'_2} & y_{23}\lambda^{q_2+q'_3} \\ y_{31}\lambda^{q_3+q'_1} & y_{32}\lambda^{q_3+q'_2} & y_{33}\lambda^{q_3+q'_3} \end{bmatrix} v \tag{24}$$

where all the yij are of order 1 and q_i and q'_i are the charges of 10,10 for m_u, of $\bar{5}$,10 for m_d or m_l^T, of 1,$\bar{5}$ for m_D (the Dirac ν mass), and of 1,1 for M, the RR Majorana ν mass. Note the two vanishing charges in $\Psi_{\bar{5}}$. They are essential for this mechanism: for example they imply that the 32, 33 matrix elements of m_D are of order 1. It is important to observe that m can be written as:

$$m = \lambda^q y \lambda^{q'} \tag{25}$$

where $\lambda_q = diag[\lambda_{q1}, \lambda_{q2}, \lambda_{q3}]$ and y is the y_{ij} matrix. As a consequence when we start from the Dirac ν matrix: $m_D = \lambda^{q_1} y_D \lambda^{q_{\bar{5}}}$ and the RR Majorana matrix $M = \lambda^{q_1} y_M \lambda^{q_1}$ and write down the see-saw expression for $m_\nu = m_D^T M^{-1} m_D$, we find that the dependence on the q_1 charges drops out and only that from $q_{\bar{5}}$ remains. On the one hand this is good because it corresponds to the fact that the effective light neutrino Majorana mass matrix $m_\nu \sim L^T L$ can be written in terms of $q_{\bar{5}}$ only. In particular the 22,23,32,33 matrix elements of m_ν are of order 1, which implies large mixings in the 23 sector. On the other hand the sub determinant 23 is not suppressed in this case, so that the splitting between the 2 and 3 light neutrino masses is in general small. In spite of the fact that m_D is, in first approximation, of the form in (17) the strong correlations between m_D and M implied by the simple charge structure of the model destroy the vanishing of the 23 sub determinant that would be guaranteed for generic M. Models of this sort have been proposed in the literature [23], [24]. The hierarchy between m_2 and m_3 is considered accidental and better be moderate. The preferred solar solution in this case is MSW-(small angle) because if m_1 is suppressed the solar mixing angle is typically small.

Models with natural large 23 splittings are obtained if we allow negative charges and, at the same time, either introduce flavons of opposite charges or stipulate that matrix elements with overall negative charge are put to zero. We now discuss a model of this sort [3]. We assign to the fermion fields the set of F-charges given by:

$$\Psi_{10} \sim (3,2,0) \tag{26}$$
$$\Psi_{\bar{5}} \sim (3,0,0) \tag{27}$$
$$\Psi_1 \sim (1,-1,0) \tag{28}$$

We consider the Yukawa coupling allowed by $U(1)_F$-neutral Higgs multiplets φ_5 and $\varphi_{\bar{5}}$ in the 5 and $\bar{5}$ $SU(5)$ representations and by a pair θ and $\bar{\theta}$ of $SU(5)$ singlets with $F = 1$ and $F = -1$, respectively.

In the quark sector we obtain :

$$m_u = (m_u)^T = \begin{bmatrix} \lambda^6 & \lambda^5 & \lambda^3 \\ \lambda^5 & \lambda^4 & \lambda^2 \\ \lambda^3 & \lambda^2 & 1 \end{bmatrix} v_u \ , \qquad m_d = \begin{bmatrix} \lambda^6 & \lambda^5 & \lambda^3 \\ \lambda^3 & \lambda^2 & 1 \\ \lambda^3 & \lambda^2 & 1 \end{bmatrix} v_d \ , \tag{29}$$

from which we get for the eigenvalues the order-of-magnitude relations:

$$\begin{aligned} m_u : m_c : m_t &= \lambda^6 : \lambda^4 : 1 \\ m_d : m_s : m_b &= \lambda^6 : \lambda^2 : 1 \end{aligned} \tag{30}$$

and

$$V_{us} \sim \lambda \ , \qquad V_{ub} \sim \lambda^3 \ , \qquad V_{cb} \sim \lambda^2 \ . \tag{31}$$

Here $v_u \equiv \langle \varphi_5 \rangle$, $v_d \equiv \langle \varphi_{\bar{5}} \rangle$ and λ, arising from the $\bar{\theta}$ vev, is, as above, of the order of the Cabibbo angle. For non-negative F-charges, the elements of the quark mixing matrix V_{CKM} depend only on the charge differences of the left-handed quark doublet [22]. Up to a constant shift, this defines the choice in (26). Equal F-charges for $\Psi_{\bar{5}}^{2,3}$ (see (27)) are then required to fit m_b and m_s. We will comment on the lightest quark masses later on.

At this level, the mass matrix for the charged leptons is the transpose of m_d:

$$m_l = (m_d)^T \tag{32}$$

and we find:

$$m_e : m_\mu : m_\tau = \lambda^6 : \lambda^2 : 1 \tag{33}$$

The O(1) off-diagonal entry of m_l gives rise to a large left-handed mixing in the 23 block which corresponds to a large right-handed mixing in the d mass matrix. In the neutrino sector, the Dirac and Majorana mass matrices are given by:

$$m_D = \begin{bmatrix} \lambda^4 & \lambda & \lambda \\ \lambda^2 & \lambda' & \lambda' \\ \lambda^3 & 1 & 1 \end{bmatrix} v_u \quad , \qquad M = \begin{bmatrix} \lambda^2 & 1 & \lambda \\ 1 & \lambda'^2 & \lambda' \\ \lambda & \lambda' & 1 \end{bmatrix} \bar{M} \quad , \tag{34}$$

where λ' is related to θ and $\bar{M}$ denotes the large mass scale associated to the right-handed neutrinos: $\bar{M} \gg v_{u,d}$.

After diagonalization of the charged lepton sector and after integrating out the heavy right-handed neutrinos we obtain the following neutrino mass matrix in the low-energy effective theory:

$$m_\nu = \begin{bmatrix} \lambda^6 & \lambda^3 & \lambda^3 \\ \lambda^3 & 1 & 1 \\ \lambda^3 & 1 & 1 \end{bmatrix} \frac{v_u^2}{\bar{M}} \tag{35}$$

where we have taken $\lambda \sim \lambda'$. The O(1) elements in the 23 block are produced by combining the large left-handed mixing induced by the charged lepton sector and the large left-handed mixing in m_D. A crucial property of m_ν is that, as a result of the sea-saw mechanism and of the specific $U(1)_F$ charge assignment, the determinant of the 23 block is *automatically* of $O(\lambda^2)$ (for this the presence of negative charge values, leading to the presence of both λ and λ' is essential [2]).

It is easy to verify that the eigenvalues of m_ν satisfy the relations:

$$m_1 : m_2 : m_3 = \lambda^4 : \lambda^2 : 1 \quad . \tag{36}$$

The atmospheric neutrino oscillations require $m_3^2 \sim 10^{-3}$ eV^2. From (35), taking $v_u \sim 250$ GeV, the mass scale $\bar{M}$ of the heavy Majorana neutrinos turns out to be close to the unification scale, $\bar{M} \sim 10^{15}$ GeV. The squared mass difference between the lightest states is of $O(\lambda^4)$ m_3^2, appropriate to the MSW solution to the solar neutrino problem. Finally, beyond the large mixing in the 23 sector, m_ν provides a mixing angle $s \sim (\lambda/2)$ in the 12 sector, close to the range preferred by the small angle MSW solution. In general U_{e3} is non-vanishing, of $O(\lambda^3)$.

In general, the charge assignment under $U(1)_F$ allows for non-canonical kinetic terms that represent an additional source of mixing. Such terms are allowed by the underlying flavour symmetry and it would be unnatural to tune them to the canonical form. The results quoted up to now remain unchanged after including the effects related to the most general kinetic terms, via appropriate rotations and rescaling in the flavour space.

Obviously, the order of magnitude description offered by this model is not intended to account for all the details of fermion masses. Even neglecting the parameters associated with the CP violating observables, some of the relevant observables are somewhat marginally reproduced. For instance we obtain $m_u/m_t \sim \lambda^6$ which is perhaps too large. However we find it remarkable that in such a simple scheme most of the 12 independent fermion masses and the 6 mixing angles turn out to have the correct order of magnitude. Notice also that this model prefers large values of $\tan\beta \equiv v_u/v_d$. This is a consequence of the equality $F(\Psi_{10}^3) = F(\Psi_{\bar{5}}^3)$ (see (26) and (27)). In this case the Yukawa couplings of top and bottom quarks are expected to be of the same order of magnitude, while the large m_t/m_b ratio is attributed to $v_u \gg v_d$ (there may be factors O(1) modifying these considerations, of course). Alternatively, to keep $\tan\beta$ small, one could suppress m_b/m_t by adopting different F-charges for the $\Psi_{\bar{5}}^3$ and Ψ_{10}^3 or for the 5 and $\bar{5}$ Higgs, as we will see in the next section.

A common problem of all $SU(5)$ unified theories based on a minimal higgs structure is represented by the relation $m_l = (m_d)^T$ that, while leading to the successful $m_b = m_\tau$ boundary condition at the GUT scale, provides the wrong prediction $m_d/m_s = m_e/m_\mu$ (which, however, is an acceptable order of magnitude equality). We can easily overcome this problem and improve the picture [21] by introducing an additional supermultiplet $\bar{\theta}_{24}$ transforming in the adjoint representation of $SU(5)$ and possessing a negative $U(1)_F$ charge, $-n$ $(n > 0)$. Under these conditions, a positive F-charge f carried by the matrix elements $\Psi_{10}^a \Psi_{\bar{5}}^b$ can be compensated in several different ways by monomials of the kind $(\bar{\theta})^p(\bar{\theta}_{24})^q$, with $p + nq = f$. Each of these possibilities represents an independent contribution to the down quark and charged lepton mass matrices, occurring with an unknown coefficient of O(1). Moreover the product $(\bar{\theta}_{24})^q \varphi_{\bar{5}}$ contains both the $\bar{5}$ and the $\overline{45}$ $SU(5)$ representations, allowing for a differentiation between the down quarks and the charged leptons. The only, welcome, exceptions are given by the O(1) entries that do not require any compensation and, at the leading order, remain the same for charged leptons and down quarks. This preserves the good $m_b = m_\tau$ prediction. Since a perturbation of O(1) in the subleading matrix elements is sufficient to cure the bad $m_d/m_s = m_e/m_\mu$ relation, we can safely assume that $\langle\bar{\theta}_{24}\rangle/M_P \sim \lambda^n$, to preserve the correct order-of-magnitude predictions in the remaining sectors.

A general problem common to all models dealing with flavour is that of recovering the correct vacuum structure by minimizing the effective potential of the theory. It may be noticed that the presence of two multiplets θ and $\bar{\theta}$ with opposite F charges could hardly be reconciled, without adding extra structure to the model, with a large common VEV for these fields, due to possible analytic terms of the kind $(\theta\bar{\theta})^n$ in the superpotential. We find therefore instructive to explore the consequences of allowing only the negatively charged $\bar{\theta}$ field in the theory.

It can be immediately recognized that, while the quark mass matrices of (29) are unchanged, in the neutrino sector the Dirac and Majorana matrices get modified into:

$$m_D = \begin{bmatrix} \lambda^4 & \lambda & \lambda \\ \lambda^2 & 0 & 0 \\ \lambda^3 & 1 & 1 \end{bmatrix} v_u \quad , \qquad M = \begin{bmatrix} \lambda^2 & 1 & \lambda \\ 1 & 0 & 0 \\ \lambda & 0 & 1 \end{bmatrix} \bar{M} \quad . \tag{37}$$

The zeros are due to the analytic property of the superpotential that makes impossible to form the corresponding F invariant by using $\bar{\theta}$ alone. These zeros should not be taken literally, as they will be eventually filled by small terms coming, for instance, from the diagonalization of the charged lepton mass matrix and from the transformation that put the kinetic terms into canonical form. It is however interesting to work out, in first approximation, the case of exactly zero entries in m_D and M, when forbidden by F.

The neutrino mass matrix obtained via see-saw from m_D and M has the same pattern as the one displayed in (35). A closer inspection reveals that the determinant of the 23 block is identically zero, independently from λ. This leads to the following pattern of masses:

$$m_1 : m_2 : m_3 = \lambda^3 : \lambda^3 : 1 \quad , \qquad m_1^2 - m_2^2 = \mathrm{O}(\lambda^9) \quad . \tag{38}$$

Moreover the mixing in the 12 sector is almost maximal:

$$\frac{s}{c} = \frac{\pi}{4} + \mathrm{O}(\lambda^3) \quad . \tag{39}$$

For $\lambda \sim 0.2$, both the squared mass difference $(m_1^2 - m_2^2)/m_3^2$ and $\sin^2 2\theta_{sun}$ are remarkably close to the values required by the vacuum oscillation solution to the solar neutrino problem. This property remains reasonably stable against the perturbations induced by small terms (of order λ^5) replacing the zeros, coming from the diagonalization of the charged lepton sector and by the transformations that render the kinetic terms canonical. We find quite interesting that also the just-so solution, requiring an intriguingly small mass difference and a bimaximal mixing, can be reproduced, at least at the level of order of magnitudes, in the context of a "minimal" model of flavour compatible with supersymmetric SU(5). In this case the role played by supersymmetry is essential, a non-supersymmetric model with $\bar{\theta}$ alone not being distinguishable from the version with both θ and $\bar{\theta}$, as far as low-energy flavour properties are concerned.

6 From Minimal to Realistic SUSY SU(5)

In this section, following the lines of a recent study [6], we address the question whether the smallest SUSY SU(5) symmetry group can still be considered as a basis for a realistic GUT model. The minimal model has large fine tuning problems (e.g. the doublet-triplet splitting problem) and phenomenological problems from the new improved limits on proton decay [25]. Also, analyses of particular aspects of GUT's often leave aside the problem of embedding the sector under discussion into a consistent whole. So the problem arises of going beyond minimal toy models by formulating sufficiently realistic, not unnecessarily complicated, relatively complete models that can serve as benchmarks to be compared with experiment. More appropriately, instead of "realistic" we should say "not grossly unrealistic" because it is clear that many important details cannot be sufficiently controlled and assumptions must be made. The model we aim at should not rely on large fine tunings and must lead to an acceptable phenomenology. This includes coupling unification with an acceptable value of $\alpha_s(m_Z)$, given α and $sin^2\theta_W$ at m_Z, compatibility with the bounds on proton decay, agreement with the observed fermion mass spectrum, also considering neutrino masses and mixings and so on. The success or failure of the programme of constructing realistic models can decide whether or not a stage of gauge unification is a likely possibility.

We indeed have presented in [6] an explicit example of a "realistic" SU(5) model, which uses a $U(1)_F$ symmetry as a crucial ingredient. In this model the doublet-triplet splitting problem is solved by the missing partner mechanism [26] stabilised by the flavour symmetry against the occurrence of doublet mass lifting due to non renormalisable operators. Relatively large representations (50, $\bar{50}$, 75) have to be introduced for this purpose. A good effect of this proliferation of states is that the value of $\alpha_s(m_Z)$ obtained from coupling unification in the next to the leading order perturbative approximation receives important negative corrections from threshold effects near the GUT scale arising from mass splittings inside the 75. As a result, the central value changes from $\alpha_s(m_Z) \approx 0.129$ in minimal SUSY SU(5) down to $\alpha_s(m_Z) \approx 0.116$, in better agreement with observation. At the same time, an increase of the effective mass that mediates proton decay by a factor of typically 20-30 is obtained to optimize the value of $\alpha_s(m_Z)$. So finally the value of the strong coupling is in better agreement with the experimental value and the proton decay rate is smaller by a factor 400-1000 than in the minimal model (in addition the rigid relation of the minimal model between mass terms and proton decay amplitudes is released, so that the rate can further be reduced) . The presence of these large representations also has the consequence that the asymptotic freedom of SU(5) is spoiled and the associated gauge coupling becomes non perturbative below M_P. We argue that this property far from being unacceptable can actually be useful to obtain better results for fermion masses and proton decay. The same $U(1)_F$ flavour symmetry that stabilizes the missing partner mechanism explains the hierarchical structure of fermion masses. In the neutrino sector, mass matrices similar to those discussed in the previous section are obtained. In the present particular version maximal mixing also for solar neutrinos is preferred.

While we refer to the original paper for a complete discussion, here we only summarise the fermion mass sector of the model, which is of relevance for neutrinos. At variance with the previous models we adopt in this case different $U(1)_F$

charges for the Higgs field $H \sim 5$ and $\bar{H} \sim \bar{5}$:

$$F(H) = -2 \text{ and } \mathrm{F}(\bar{\mathrm{H}}) = 1, \tag{40}$$

For matter fields

$$\begin{aligned} F(\Psi_{10}) &= (4,3,1) \\ F(\Psi_{\bar{5}}) &= (5,2,2) \\ F(\Psi_1) &= (1,-1,0) \end{aligned} \tag{41}$$

The Yukawa mass matrices are, in first approximation, of the form:

$$m_u = \begin{bmatrix} \lambda^6 & \lambda^5 & \lambda^3 \\ \lambda^5 & \lambda^4 & \lambda^2 \\ \lambda^3 & \lambda^2 & 1 \end{bmatrix} v_u/\sqrt{2} \quad , \tag{42}$$

$$m_d = \begin{bmatrix} \lambda^6 & \lambda^5 & \lambda^3 \\ \lambda^3 & \lambda^2 & 1 \\ \lambda^3 & \lambda^2 & 1 \end{bmatrix} v_d \lambda^4/\sqrt{2} \quad = m_l^T, \tag{43}$$

$$m_\nu = \begin{bmatrix} \lambda^4 & \lambda & \lambda \\ \lambda^2 & 0 & 0 \\ \lambda^3 & 1 & 1 \end{bmatrix} v_u/\sqrt{2} \quad , \tag{44}$$

$$m_{maj} = \begin{bmatrix} \lambda^2 & 1 & \lambda \\ 1 & 0 & 0 \\ \lambda & 0 & 1 \end{bmatrix} M \quad , \tag{45}$$

For a correct first approximation of the observed spectrum we need $\lambda \approx \lambda_C \approx 0.22$, λ_C being the Cabibbo angle. These mass matrices closely match those of the previous section, with two important special features. First, we have here that $tan\beta = v_u/v_d \approx m_t/m_b \lambda^4$, which is small. The factor λ^4 is obtained as a consequence of the Higgs and matter fields charges F, while previously the H and $\bar{H}$ charges were taken as zero. We recall that a value of $\tan\beta$ near 1 is an advantage for suppressing proton decay. Of course the limits from LEP that indicate that $tan\beta \gtrsim 2-3$ must be and can be easily taken into account. Second, the zero entries in the mass matrices of the neutrino sector occur because the negatively F-charged flavon fields have no counterpart with positive F-charge in this model. Neglected small effects could partially fill up the zeroes. As already explained these zeroes lead to near maximal mixing also for solar neutrinos. A problematic aspect of this zeroth order approximation to the mass matrices is the relation $m_d = m_l^T$. The necessary corrective terms can arise from the neglected higher order terms from non renormalisable operators with the insertion of n factors of the 75, which break the transposition relation between m_d and m_l. With reasonable values of the coefficients of order 1 we obtain double nearly maximal mixing and $\theta_{13} \sim 0.05$. The preferred solar solutions are LOW or vacuum oscillations.

7 SU(5) Unification in Extra Dimensions

Recently it has been observed that the GUT gauge symmetry could be actually realized in 5 (or more) space-time dimensions and broken down to the the Standard Model (SM) by compactification [1]. In particular a model with N=2 Supersymmetry (SUSY) and gauge SU(5) in 5 dimensions has been proposed [28] where the GUT symmetry is broken by compactification on $S^1/(Z_2 \times Z_2')$ down to a N=1 SUSY-extended version of the SM on a 4-dimensional brane. In this model many good properties of GUT's, like coupling unification and charge quantization are maintained while some unsatisfactory properties of the conventional breaking mechanism, like doublet-triplet splitting, are avoided. In a recent paper of ours [7] we have elaborated further on this class of models. We differ from [28] (and also from the later reference [29]) in the form of the interactions on the 4-dimensional brane. As a consequence we not only avoid the problem of the doublet-triplet splitting but also directly suppress or even forbid proton decay, since the conventional higgsino and gauge boson exchange amplitudes are absent, as a consequence of $Z_2 \times Z_2'$ parity assignments on matter fields on the brane. Most good predictions of SUSY SU(5) are thus maintained without unnatural fine tunings as needed in the minimal model. We find that the relations among fermion masses implied by the minimal model, for example $m_b = m_\tau$ at M_{GUT} are preserved in our version of the model, although the Yukawa interactions are not fully SU(5) symmetric. The mechanism that forbids proton decay still allows Majorana mass terms for neutrinos so that the good potentiality of SU(5) for the description of neutrino masses and mixing is preserved. This class of models offers a new perspective on how the GUT symmetry and symmetry-breaking could be realized.

8 SO(10) Models

Models based on $SO(10)$ times a flavour symmetry are more difficult to construct because a whole generation is contained in the 16, so that, for example for $U(1)_F$, one would have the same value of the charge for all quarks and leptons of each generation, which is too rigid. But the mechanism discussed sofar, based on asymmetric mass matrices, can be embedded in an $SO(10)$ grand-unified theory in a rather economic way [30], [14]. The 33 entries of the fermion mass matrices can be obtained through the coupling $\mathbf{16}_3\mathbf{16}_3\mathbf{10}_H$ among the fermions in the third generation, $\mathbf{16}_3$, and a Higgs tenplet $\mathbf{10}_H$. The two independent VEVs of the tenplet v_u and v_d give mass, respectively, to t/ν_τ and b/τ. The keypoint to obtain an asymmetric texture is the introduction of an operator of the kind $\mathbf{16}_2\mathbf{16}_H\mathbf{16}_3\mathbf{16}'_H$. This operator is thought to arise by integrating out an heavy $\mathbf{10}$ that couples both to $\mathbf{16}_2\mathbf{16}_H$ and to $\mathbf{16}_3\mathbf{16}'_H$. If the $\mathbf{16}_H$ develops a VEV breaking $SO(10)$ down to $SU(5)$ at a large scale, then, in terms of $SU(5)$ representations, we get an effective coupling of the kind $\bar{\mathbf{5}}_2\mathbf{10}_3\bar{\mathbf{5}}_H$,

[1] Grand unified supersymmetric models in six dimensions, with the grand unified scale related to the compactification scale were also proposed by Fayet [27].

with a coefficient that can be of order one. This coupling contributes to the 23 entry of the down quark mass matrix and to the 32 entry of the charged lepton mass matrix, realizing the desired asymmetry. To distinguish the lepton and quark sectors one can further introduce an operator of the form $\mathbf{16}_i\mathbf{16}_j\mathbf{10}_H\mathbf{45}_H$, $(i,j=2,3)$, with the VEV of the $\mathbf{45}_H$ pointing in the $B-L$ direction. Additional operators, still of the type $\mathbf{16}_i\mathbf{16}_j\mathbf{16}_H\mathbf{16}'_H$ can contribute to the matrix elements of the first generation. The mass matrices look like:

$$m_u = \begin{bmatrix} 0 & 0 & 0 \\ 0 & 0 & \epsilon/3 \\ 0 & -\epsilon/3 & 1 \end{bmatrix} v_u \ , \qquad m_d = \begin{bmatrix} 0 & \delta & \delta' \\ \delta & 0 & \sigma+\epsilon/3 \\ \delta' & -\epsilon/3 & 1 \end{bmatrix} v_d \ , \tag{46}$$

$$m_D = \begin{bmatrix} 0 & 0 & 0 \\ 0 & 0 & -\epsilon \\ 0 & \epsilon & 1 \end{bmatrix} v_u \ , \qquad m_e = \begin{bmatrix} 0 & \delta & \delta' \\ \delta & 0 & -\epsilon \\ \delta' & \sigma+\epsilon & 1 \end{bmatrix} v_d \ . \tag{47}$$

They provide a good fit of the available data in the quarks and the charged lepton sector in terms of 5 parameters (one of which is complex). In the neutrino sector one obtains a large θ_{23} mixing angle, $\sin^2 2\theta_{12} \sim 6.6 \cdot 10^{-3}$ eV^2 and θ_{13} of the same order of θ_{12}. Mass squared differences are sensitive to the details of the Majorana mass matrix.

Looking at models with three light neutrinos only, i.e. no sterile neutrinos, from a more general point of view, we stress that in the above models the atmospheric neutrino mixing is considered large, in the sense of being of order one in some zeroth order approximation. In other words it corresponds to off diagonal matrix elements of the same order of the diagonal ones, although the mixing is not exactly maximal. The idea that all fermion mixings are small and induced by the observed smallness of the non diagonal V_{CKM} matrix elements is then abandoned. An alternative is to argue that perhaps what appears to be large is not that large after all. The typical small parameter that appears in the mass matrices is $\lambda \sim \sqrt{m_d/m_s} \sim \sqrt{m_\mu/m_\tau} \sim 0.20-0.25$. This small parameter is not so small that it cannot become large due to some peculiar accidental enhancement: either a coefficient of order 3, or an exponent of the mass ratio which is less than 1/2 (due for example to a suitable charge assignment), or the addition in phase of an angle from the diagonalization of charged leptons and an angle from neutrino mixing. One may like this strategy of producing a large mixing by stretching small ones if, for example, he/she likes symmetric mass matrices, as from left-right symmetry at the GUT scale. In left-right symmetric models smallness of left mixings implies that also right-handed mixings are small, so that all mixings tend to be small. Clearly this set of models [31] tend to favour moderate hierarchies and a single maximal mixing, so that the SA-MSW solution of solar neutrinos is preferred.

9 Conclusion

By now there are rather convincing experimental indications for neutrino oscillations. If so, then neutrinos have non zero masses. As a consequence, the phenomenology of neutrino masses and mixings is brought to the forefront. This is a very interesting subject in many respects. It is a window on the physics of GUTs in that the extreme smallness of neutrino masses can only be explained in a natural way if lepton number is violated. Then neutrino masses are inversely proportional to the large scale where lepton number is violated. Also, the pattern of neutrino masses and mixings can provide new clues on the long standing problem of quark and lepton mass matrices. The actual value of neutrino masses is important for cosmology as neutrinos are candidates for hot dark matter: nearly degenerate neutrinos with a common mass around 1- 2 eV would significantly contribute to the matter density in the universe.

While the existence of oscillations appears to be on a solid ground, many important experimental ambiguities remain. For solar neutrinos it is not yet clear which of the solutions, MSW-SA, MSW-LA, LOW and VO, is true, and the possibility also remains of different solutions if not all of the experimental input is correct (for example, energy independent solutions are resurrected if the Homestake result is modified). Finally a confirmation of the LSND alleged signal is necessary, in order to know if 3 light neutrinos are sufficient or additional sterile neutrinos must be introduced. We argued in favour of models with 3 widely split neutrinos. Reconciling large splittings with large mixing(s) requires some natural mechanism to implement a vanishing determinant condition. This can be obtained in the see-saw mechanism if one light right-handed neutrino is dominant, or a suitable texture of the Dirac matrix is imposed by an underlying symmetry. In a GUT context, the existence of right-handed neutrinos indicates SO(10) at least as a classification group. The symmetry group at M_{GUT} could be either (Susy) SU(5) or SO(10) or a larger group. We have presented a class of natural models where large right-handed mixings for quarks are transformed into large left-handed mixings for leptons by the approximate transposition relation $m_d = m_e^T$ which is approximately realised in SU(5) models. We have shown that these models can be naturally implemented by simple assignments of $U(1)_F$ horizontal charges.

In conclusion the fact that some neutrino mixing angles are large, while surprising at the start, was eventually found to be well be compatible, without any major change, with our picture of quark and lepton masses within GUTs. In fact, it provides us with new important clues that can become sharper when the experimental picture will be further clarified.

Acknowledgements

I am grateful to Josip Trampetic for inviting me to this exceptionally interesting School and for his beautiful organisation and splendid hospitality.

References

1. G. Altarelli and F. Feruglio, Phys. Lett. B **439**, 112(1998).
2. G. Altarelli and F. Feruglio, JHEP **11**, 21 (1998).

3. G. Altarelli and F. Feruglio, Phys. Lett. B **451**, 388 (1999).
4. G. Altarelli and F. Feruglio, Phys. Rep. **320**, 295 (1999).
5. G. Altarelli, F. Feruglio and I. Masina, Phys. Lett. B **472**, 382 (2000).
6. G. Altarelli, F. Feruglio and I. Masina, JHEP **11**, 040 (2000).
7. G. Altarelli and F. Feruglio, hep-ph/0102301.
8. M. Gell-Mann, P. Ramond and R. Slansky in Supergravity, ed. P. van Nieuwenhuizen and D. Z. Freedman, North-Holland, Amsterdam, 1979, p.315; T. Yanagida, in Proceedings of the Workshop on the unified theory and the baryon number in the universe, ed. O. Sawada and A. Sugamoto, KEK report No. 79-18, Tsukuba, Japan, 1979. See also R. Mohapatra and G. Senjanovic, Phys. Rev. Lett. **44**, 912 (1980).
9. See, for example, M. C. Gonzalez-Garcia and C. Pena-Garay, hep-ph/0011245; G. L. Fogli, E. Lisi and A. Marrone, Phys. Rev. D**63**, 053008 (2001); S. M. Bilenkii, C. Giunti, W. Grimus and T. Schwetz, Phys. Rev. D **60**, 0073007 (1999).
10. P. Horava and E. Witten, Nuc. Phys. B **475**, 94 (1996); N. Arkani-Hamed, S. Dimopoulos and G. Dvali, Phys. Lett. B **429**, 263 (1998); I. Antoniadis, N. Arkani-Hamed, S. Dimopoulos and G. Dvali, Phys. Lett. B **436**, 257 (1998).
11. For an immersion into this subject, see, for example, the recent paper by A. Lukas, P. Ramond, A. Romanino and G. Ross, hep-ph/0011295 and references therein.
12. C.D. Hoyle et al, Phys. Rev. Lett. 86(2001)1418 (hep-ph/0011014).
13. See, for example, I. Antoniadis and K. Benakli, hep-ph/0007226.
14. S. M. Barr and I. Dorsner, hep-ph/0003058.
15. F. Vissani, hep-ph/9708483; H. Georgi and S.L. Glashow, hep-ph/9808293.
16. J. Ellis and S. Lola, hep-ph/9904279; J.A. Casas et al, hep-ph/9904395, hep-ph/9905381, hep-ph/9906281; R. Barbieri, G.G. Ross and A. Strumia, hep-ph/9906470; E. Ma, hep-ph/9907400; K.R.S. Balaji et al, hep-ph/0001310 and hep-ph/0002177.
17. Examples of degenerate models are described in A. Ioannisian, J. W. F. Valle, Phys. Lett. B **332**, 93 (1994); M. Fukugita, M. Tanimoto, T. Yanagida, Phys. Rev. D **57**, 4429 (1998) M. Tanimoto, hep-ph/9807283 and hep-ph/9807517; H. Fritzsch, Z. Xing, hep-ph/9808272; R. N. Mohapatra, S. Nussinov, hep-ph/9808301 and hep-ph/9809415; M. Fukugita, M. Tanimoto, T. Yanagida, hep-ph/9809554; Yue-Liang Wu, hep-ph/9810491; J. I. Silva-Marcos, hep-ph/9811381; C. Wetterich, hep-ph/9812426; S.K. Kang and C.S. Kim, hep-ph/9811379.
18. R. Barbieri, L. J. Hall, D. Smith, A. Strumia and N. Weiner, hep/ph 9807235.
19. S. F. King, Phys. Lett. B **439**, 350 (1998) and hep-ph/9904210; S. Davidson and S. F. King, Phys. Lett. B **445**, 191 (1998); Q. Shafi and Z. Tavartkiladze, Phys. Lett B **451**, 129 (1999).
20. H. Georgi and C. Jarlskog, Phys. Lett. B **86**, 297 (1979).
21. J. Ellis and M. K. Gaillard, Phys. Lett. B **88**, 315 (1979).
22. C. Froggatt and H. B. Nielsen, Nucl. Phys. B **147**, 277 (1979).
23. W. Buchmuller and T. Yanagida, hep-ph/9810308.
24. P. Binetruy, S. Lavignac, S. Petcov and P. Ramond, Nucl. Phys. B **496**, 3 (1997); N. Irges, S. Lavignac, P. Ramond, Phys. Rev. D **58**, 5003 (1998); Y. Grossman, Y. Nir, Y. Shadmi, hep-ph/9808355.
25. Y. Hayato et al, (SuperKamiokande Collab.), Phys. Rev. Lett. **83**, 1529 (1999).
26. A. Masiero et al, Phys. Lett. B **115**, 380 (1982); B Grinstein, Nucl. Phys. B **206**, 387 (1982); Z. Berezhiani and Z. Tavartkiladze, Phys. Lett. B **409**, 220 (1997).
27. P. Fayet, Phys. Lett. B **146**, 41 (1984).
28. Y. Kawamura, hep-ph/0012125.

29. L. Hall and Y. Nomura, hep-ph/0103125.
30. C. H. Albright and S. M. Barr, Phys. Rev. D **58**, 013002 (1998); hep-ph/9901318; hep-ph/0002155; hep-ph/0003251; C. H. Albright, K. S. Babu and S. M. Barr, Phys. Rev. Lett. **81**, 1167 (1998).
31. See, for example, S. Lola and G. G. Ross, hep-ph/9902283; K. Babu, J. Pati and F. Wilczek, hep-ph/9912538.

Dark Matter in the Galaxy

Neven Bilić[1,2], Gary B. Tupper[2], and Raoul D. Viollier[2]

[1] Rudjer Bošković Institute, P.O. Box 180, 10002 Zagreb, Croatia
[2] Institute of Theoretical Physics and Astrophysics, Department of Physics, University of Cape Town, Private Bag, Rondebosch 7701, South Africa

Abstract. After a brief introduction to standard cosmology and the dark matter problem in the Universe, we consider a self-gravitating noninteracting fermion gas at nonzero temperature as a model for the dark matter halo of the Galaxy. This fermion gas model is then shown to imply the existence of a supermassive compact dark object at the Galactic center.

1 Introduction

At some stage of the evolution of the Universe, primordial density fluctuations must have become gravitationally unstable forming dense clumps of dark matter (DM) that have survived until today in the form of galactic halos. In the recent past, galactic halos were successfully modeled as a self-gravitating isothermal gas of particles of arbitrary mass, the density of which scales asymptotically as r^{-2}, yielding flat rotation curves [1]. The aim of this paper is to describe the halo of our Galaxy in terms of a self-gravitating fermion gas in hydrostatic and thermal equilibrium at finite temperature.

Self-gravitating weakly interacting fermionic matter has been exploited in a wide range of astrophysical phenomena. Originally, self-gravitating degenerate neutrino stars were suggested as a model for quasars [1], and later neutrino matter was used as a model for dark matter in galactic halos and clusters of galaxies, with a neutrino mass in the $\sim$ eV range [3]. Recently, degenerate superstars, composed of weakly interacting fermions in the ~ 10 keV range, were suggested [3–7] as an alternative to the supermassive black holes that are believed to exist at the centers of galaxies. It was shown [5] that such degenerate fermion stars could explain the whole range of supermassive compact dark objects which have been observed so far, with masses ranging from 10^6 to $3 \times 10^9 M_\odot$, merely assuming that a weakly interacting quasistable fermion of mass $m \simeq 15$ keV exists in Nature. Most recently, it has been pointed out that a weakly interacting dark matter particle in the mass range $1 \lesssim m/\mathrm{keV} \lesssim 5$ could solve the problem of the excessive structure generated on subgalactic scales in N-body and hydrodynamical simulations of structure formation in this Universe [9].

Of course, it is well known that the interval 1-15 keV lies squarely in the cosmologically forbidden mass range for stable active neutrinos ν [12]. However, for an initial lepton asymmetry of $\sim 10^{-3}$, a sterile neutrino ν_s of mass $m_s \sim 10$ keV may be resonantly produced in the early Universe with near closure density, i.e., $\Omega \simeq 1$ [11]. The resulting energy spectrum is not thermal but rather cut off so that it approximates a degenerate Fermi gas. In this mass range, sterile neutrinos are also constrained by astrophysical bounds on the radiative $\nu_s \to \nu\gamma$ decay [14]. However, the allowed parameter space includes $m_s \simeq 15$ keV, contributing $\Omega_d \simeq 0.3$ to the critical density, as favored by the BOOMERANG data [15]. As an alternative possibility, the ~ 15 keV sterile neutrino could be replaced by the axino [14] or the gravitino [15,16] in soft supersymmetry breaking scenarios.

As the supermassive compact dark objects at the galactic centers are well described by a degenerate gas of fermions, it is tempting to explore the possibility that one could describe both the supermassive compact dark objects and their galactic halos in a unified way in terms of a fermion gas at finite temperature. We will show that this is indeed the case, and that the observed dark matter distribution in the Galactic halo is consistent with the existence of a supermassive compact dark object at the center of the Galaxy which has about the right mass and size, and is in thermal and hydrostatic equilibrium with the halo.

2 Standard Cosmology

Standard cosmology provides a successful description of the evolution of the Universe from a fraction of a second after the creation until today. A short review on the standard model of cosmology is given in [17]. For our purpose, it is sufficient to state the basic underlying principles. Standard cosmology is based on the following three theoretical assumptions:

1. Cosmological Principle. The cosmological principle asserts that the Universe is homogeneous and isotropic on large scales. The most general metric satisfying the cosmological principle is the Friedmann-Robertson-Walker metric [18]

$$ds^2 = dt^2 - a(t)^2 \left[\frac{dr^2}{1 - kr^2} - r^2(d\theta^2 + \sin\theta d\phi^2) \right] , \tag{1}$$

where the curvature constant k takes on the values 1, 0, or -1, for a closed, flat, or open universe, respectively. The time-dependent quantity $a(t)$ is the scale factor of the expansion conveniently normalized to unity at present time, i.e., $a(t_0) = 1$. In other words, a is the radius of the Universe measured in units of its current radius.

2. General Relativity. Gravity is described by Einstein's general theory of relativity governed by the *equivalence principle* and Einstein's field equations.

3. Perfect Fluid. Matter is approximated by a homogeneous perfect fluid. The energy-momentum tensor then takes a simple form

$$T_{\mu\nu} = (\rho + p)u_\mu u_\nu - g_{\mu\nu} , \tag{2}$$

where ρ and p are the density and the pressure of the fluid, respectively.

With these assumptions, the set of Einstein's equations reduces to the Friedmann-Robertson-Walker (FRW) equations

$$H(t)^2 \equiv \left(\frac{\dot{a}}{a} \right)^2 = \frac{8\pi\rho}{3} - \frac{k}{a^2} + \frac{\Lambda}{3} , \tag{3}$$

$$\frac{\ddot{a}}{a} = \frac{\Lambda}{3} - 4\pi(\rho + 3p), \tag{4}$$

where the natural system of units $\hbar = c = G = 1$ is assumed. The first FRW equation describes the expansion of the Universe. The quantity $H(t)$ is the Hubble "constant" and Λ is the cosmological constant. We define the critical density as $\rho_{\rm cr} = 3H_0^2/8\pi$ and the ratio of the density to the critical density is denoted by $\Omega = \rho/\rho_{\rm cr}$. The precise present value of the Hubble constant is not known, but the widely accepted value is $H_0 = 100h_0\,{\rm kms}^{-1}{\rm Mpc}^{-1}$, with a dimensionless parameter h_0 between 0.4 and 1. Dividing by $H(t_0)^2$, (3) at $t = t_0$ can be conveniently written as a sum rule, i.e.,

$$\Omega_0 - k/(H_0 a_0)^2 + \Omega_\Lambda = 1, \tag{5}$$

where $\Omega_\Lambda = \Lambda/(3H_0^2)$ is the vacuum energy contribution to the critical density today. Observational evidence favors a flat universe today, i.e., $k = 0$. Thus, (5) becomes

$$\Omega_0 + \Omega_\Lambda = 1. \tag{6}$$

The second FRW equation (4) describes the acceleration of the expansion. The expansion will accelerate or decelerate, depending on whether the vacuum energy dominates the matter or vice versa. However, even if the cosmological term vanishes, the expansion could accelerate if the dominant component of the DM obeys a peculiar equation of state such that the pressure is negative and higher than one third of the density. One popular example is the scalar field model called *quintessence* [19]. Another scenario is based on a fluid obeying the Chaplygin gas equation of state $p \propto -1/\rho$ [20,21], which has been intensively investigated for its solubility in 1+1-dimensional space-time, for its supersymmetric extension and connection to d-branes [22].

The observational evidence in support of standard cosmology may be summarized in four empirical pillars [23]: Hubble's law, cosmic background radiation (CBR), anisotropy of CBR, and abundance patterns of light elements. However, despite an overwhelming observational support, a number of problems remain unsolved:

- What caused the Big Bang and the expansion? Did the Universe begin with more than 3+1 dimensions?
- Why is the cosmological constant Λ about 50-120 orders of magnitude smaller than the value expected from quantum field theory?
- What caused the initial baryon-antibaryon asymmetry that led to the absence of antimatter today?
- Why is the Universe so smooth on large scales as evidenced by CBR?
- What caused the primordial density fluctuations that provided the seeds for structure formation?
- What does nonbaryonic DM consist of?

A solution to these problems most probably goes beyond the standard model of cosmology and certainly beyond the standard model of particle physics.

3 Dark Matter

Here, we briefly discuss the DM problem and possible DM candidates. A more detailed analysis may be found in a number of recent review articles [24].

DM has to be introduced because of the following facts:

- Astronomical observations, such as the flattness of the rotation curves of spiral galaxies and the peculiar motion of galaxies within clusters, strongly indicate that
$$\Omega_0 \equiv \frac{\rho_{\rm matt}}{\rho_{\rm cr}} \gtrsim 0.3 \,. \tag{7}$$
- Consistency with the Big Bang nucleosynthesis implies for baryonic matter
$$0.008 h_0^{-2} < \Omega_B < 0.024 h_0^{-2}. \tag{8}$$
- Astronomical observations yield a small relative density of luminous matter
$$\Omega_{\rm lum} = 0.003 h_0^{-1}. \tag{9}$$

From these facts we conclude that

- About 99% of matter is **dark**.
- About 60%-90% of DM is **nonbaryonic**.
- At least 75% of baryonic matter is **dark**.

Whereas baryonic DM is most likely in the form of relatively standard astrophysical objects, e.g., cold hydrogen clouds or compact objects such as neutron stars, brown dwarfs, MACHOs, and even black holes, the nature of nonbaryonic DM is unknown and still a subject of speculations. Candidates within the standard model are practically excluded and those beyond the standard model have not yet been detected in particle physics experiments. Nevertheless, cosmological and astrophysical observations tell us what these, yet undetected, particles could or could not be.

The different DM scenarios are conveniently classified as *hot, warm,* and *cold* DM [25], depending on the thermal velocities of DM particles in the early Universe.

Hot DM refers to low-mass neutral particles that are still relativistic when galaxy-size masses ($\sim 10^{12} M_\odot$) are first encompassed within the horizon. Hence, fluctuations on galaxy scales are wiped out by the "free streaming" of the dark matter. Standard examples of hot DM are neutrinos and majorons. They are still in thermal equilibrium after the QCD deconfinement transition, which took place at $T_{\rm QCD} \simeq 150$ MeV. Hot DM particles have a cosmological number density comparable with that of microwave background photons, which implies an upper bound to their mass of a few tens of eV.

Warm DM particles are just becoming nonrelativistic when galaxy-size masses enter the horizon. Warm DM particles interact much more weakly than neutrinos. They decouple (i.e., their mean free path first exceeds the horizon size) at

$T \gg T_{\rm QCD}$. As a consequence, their number is expected to be roughly an order of magnitude lower and their mass an order of magnitude larger, than hot DM particles. Examples of warm DM are $\sim$ keV sterile neutrinos, axinos [14], or gravitinos in soft supersymmetry breaking scenarios [15,16]. There has been renewed interest in the standard model neutrino as a candidate for warm DM [26].

Cold DM particles are already nonrelativistic when even globular cluster masses ($\sim 10^6 M_\odot$) enter the horizon. Hence, their free streaming is of no cosmological importance. In other words, all cosmologically relevant fluctuations survive in a universe dominated by cold DM. The two main particle candidates for cold dark matter are the lowest supersymmetric *weakly interacting massive particles* (WIMPs) and the *axion*.

One of the central issues in dark-matter modeling is the problem of structure formation on subgalactic scales. The combination of cold DM and a small cosmological constant (ΛCDM) seems to be in good agreement with many observational constraints. However, N-body and hydrodynamical simulations of galaxy formation evidence that ΛCDM overpredicts structure on small scales [9]. In addition to that, high-resolution simulations generally find a dark matter profile with a central cusp $\rho \propto r^{-1.5}$ for galactic halos [27,28] which seems to contradict the observations.

Clustering on small scales could be suppressed by an upper limit to the phase-space density of DM particles owing either to degeneracy pressure if they are fermions or to a repulsive interaction if they are bosons. The fermion mass would have to lie in the range $0.1 \lesssim m \lesssim 10$ keV. A specific scenario invoking keV mass fermions is the *cool-dark-matter* proposal [14] for which candidates exist in shadow-world models and the axino.

4 Galactic Halo

We now discuss the properties of the halo of our Galaxy assuming that it consists of a self-gravitating gas of keV mass fermions in hydrostatic and thermal equilibrium at finite temperature. The Milky Way Galaxy consists of five major components which are nested within each other [29]. A spheroidal **halo** with modest concentations of stars and about 170 globular clusters extends out to a radius of perhaps 200 kpc. Within a radius of $\sim$ 25 kpc, the halo contains stars and open clusters that are concentrated into two essentially coplanar disks: the **thin disk** and the **thick disk**. In their innermost part the disks merge with a spheroidal **bulge**, the central concentration of luminous matter in the Galaxy. Finally, a compact dark object with a mass of $M_{\rm c} \simeq 2.6 \times 10^6 M_\odot$ is located in the vicinity of the enigmatic radio source Sgr A* at the Galactic center [11], within a radius of 18 mpc.

Here, we demonstrate that by extending the Thomas-Fermi theory to nonzero temperature it is possible to explain, within the same model, both the Galactic halo and the compact dark object at the Galactic center. Extending the Thomas-Fermi theory to finite temperature [18,17,33], it has been shown that, at some critical temperature T_c, weakly interacting massive fermionic matter with a total mass below the Oppenheimer-Volkoff limit [34] undergoes a first-order gravitational phase transition from a diffuse to a clustered state, i.e., a nearly degenerate fermion star. However, during this first-order phase transition a large amount of latent heat must be released in order to substantially decrease the entropy of the initial diffuse configuration. In the absence of a mechanism which would make such a release possible, the system would remain in a thermodynamic quasistable supercooled state close to the point of gravothermal collapse. The Fermi gas will be caught in the supercooled state even if the total mass of the gas exceeds the Oppenheimer-Volkoff limit as a stable condensed state does not exist in this case.

The formation of a quasistable supercooled state may be understood as a process similar to that of violent relaxation, which was introduced to describe rapid virialization of stars of different mass in globular clusters [35,36]. Through the gravitational collapse of an overdense fluctuation at about one Gyr after the Big Bang, part of gravitational energy transforms into the kinetic energy of random motion of small-scale density fluctuations. The resulting virialized cloud will thus be well approximated by a gravitationally stable thermalized halo. In order to estimate the mass-to-temperature ratio, we assume that an overdense cloud of mass M stops expanding at the time t_{m}, reaching its maximal radius R_{m} and the average density $\rho_{\mathrm{m}} = 3M/(4\pi R_{\mathrm{m}}^3)$. The total energy per particle is just the gravitational energy

$$E = -\frac{3}{5}\frac{M}{R_{\mathrm{m}}} \,. \tag{10}$$

From the spherical model of nonlinear collapse [37] it follows

$$\rho_{\mathrm{m}} = \frac{9\pi^2}{16}\bar{\rho}(t_{\mathrm{m}}) = \frac{9\pi^2}{16}\Omega_d \rho_{\mathrm{cr}}(1+z_{\mathrm{m}})^3, \tag{11}$$

where $\bar{\rho}(t_{\mathrm{m}})$ is the background density at the time t_{m} or the cosmological redshift z_{m}. We approximate the virialized cloud by a singular isothermal sphere [36] of the mass of the Galaxy M and radius R. The singular isothermal sphere is characterized by a constant circular velocity $\Theta = (2T/m)^{1/2}$ and the density profile $\rho(r) = \Theta^2/(4\pi r^2)$. Its total energy per particle is the sum of gravitational and thermal energies, i.e.,

$$E = -\frac{1}{4}\frac{M}{R} = -\frac{1}{4}\Theta^2. \tag{12}$$

Combining (10), (11), and (12), we find

$$\Theta^2 = \frac{6\pi}{5}(6\Omega_d \rho_{\mathrm{cr}} M^2)^{1/3}(1+z_{\mathrm{m}}). \tag{13}$$

Taking $\Omega_{\mathrm{d}} = 0.3$, $M = 2\times 10^{12} M_{\odot}$, $z_{\mathrm{m}} = 4$, and $h_0 = 0.65$, we find $\Theta \simeq 220\,\mathrm{km\,s^{-1}}$, which corresponds to the mass-temperature ratio $m/T \simeq 4\times 10^6$.

Next, we briefly discuss the general-relativistic Thomas-Fermi theory [33] for a self-gravitating gas, consisting of N fermions of mass m in equilibrium at a temperature T, enclosed in a sphere of radius R. We denote by p, ρ, and n the pressure, energy density, and particle number density of the gas, respectively. The metric generated by the mass distribution is static, spherically symmetric, and asymptotically flat, i.e.,

$$ds^2 = \xi^2 dt^2 - (1 - 2\mathcal{M}/r)^{-1} dr^2 - r^2(d\theta^2 + \sin\theta d\phi^2). \tag{14}$$

For numerical convenience, we introduce the parameter

$$\alpha = \frac{\mu}{T} \tag{15}$$

and the substitution

$$\xi = \frac{\mu}{m}(\varphi + 1)^{-1/2}, \tag{16}$$

where μ is the chemical potential associated with the conserved particle number N. Using this, the equation of state for a self-gravitating ideal gas may be represented in a parametric form [38]:

$$n = \frac{1}{\pi^2} \int_0^\infty dy \frac{y^2}{1 + \exp\{[(y^2+1)^{1/2}/(\varphi+1)^{1/2} - 1]\alpha\}}, \tag{17}$$

$$\rho = \frac{1}{\pi^2} \int_0^\infty dy \frac{y^2(y^2+1)^{1/2}}{1 + \exp\{[(y^2+1)^{1/2}/(\varphi+1)^{1/2} - 1]\alpha\}}, \tag{18}$$

$$p = \frac{1}{3\pi^2} \int_0^\infty dy \frac{y^4(y^2+1)^{-1/2}}{1 + \exp\{[(y^2+1)^{1/2}/(\varphi+1)^{1/2} - 1]\alpha\}}. \tag{19}$$

We have chosen appropriate length and mass scales a and b, respectively, such that

$$a = b = \sqrt{\frac{2}{g}} \frac{1}{m^2}, \tag{20}$$

or, restoring $\hbar$, c, and G, we have

$$a = \sqrt{\frac{2}{g}} \frac{\hbar M_{\rm Pl}}{cm^2} = 1.0798 \times 10^{10} \sqrt{\frac{2}{g}} \left(\frac{15\text{keV}}{m}\right)^2 \text{km}, \tag{21}$$

$$b = \sqrt{\frac{2}{g}} \frac{M_{\rm Pl}^3}{m^2} = 0.7251 \times 10^{10} \sqrt{\frac{2}{g}} \left(\frac{15\text{keV}}{m}\right)^2 M_\odot \,. \tag{22}$$

Here, $M_{\rm Pl} = \sqrt{\hbar c/G}$ denotes the Planck mass and g the combined spin-degeneracy factor of neutral fermions and antifermions, i.e., g=2 or 4 for Majorana or Dirac fermions, respectively. In this way, the fermion mass, the degeneracy factor, and the chemical potential are eliminated from the equation of state.

Einstein's field equations for the metric (14) are given by

$$\frac{d\varphi}{dr} = -2(\varphi + 1)\frac{\mathcal{M} + 4\pi r^3 p}{r(r - 2\mathcal{M})}, \tag{23}$$

$$\frac{d\mathcal{M}}{dr} = 4\pi r^2 \rho. \tag{24}$$

To these two equations we add

$$\frac{d\mathcal{N}}{dr} = 4\pi r^2 (1 - 2\mathcal{M}/r)^{-1/2} n, \tag{25}$$

imposing the particle-number constraint as a condition at the boundary

$$\mathcal{N}(R) = N. \tag{26}$$

Equations (23)-(25) should be integrated using the boundary conditions at the origin:

$$\varphi(0) = \varphi_0 > -1\,; \qquad \mathcal{M}(0) = 0\,; \qquad \mathcal{N}(0) = 0. \tag{27}$$

It is useful to introduce the degeneracy parameter $\eta = \alpha\varphi/2$, which, in the Newtonian limit, approaches $\eta_{\rm nr} = (\mu_{\rm nr} - V)/T$. Here, we have introduced the nonrelativistic chemical potential $\mu_{\rm nr} = \mu - m$, with $\mu_{\rm nr} \ll m$, and the approximation $\xi = e^{V/m} \simeq 1 + V/m$, with V being the Newtonian potential. As φ is monotonously decreasing with increasing r, the strongest degeneracy is obtained at the center with $\eta_0 = \alpha\varphi_0/2$. The parameter η_0, uniquely related to the central density and pressure, will eventually be fixed by the requirement (26). For $r \geq R$, the function φ yields the usual empty-space Schwarzschild solution

$$\varphi(r) = \frac{\mu^2}{m^2}\left(1 - \frac{2M}{r}\right)^{-1} - 1\,, \tag{28}$$

with

$$M = \mathcal{M}(R) = \int_0^R dr 4\pi r^2 \rho(r). \tag{29}$$

Given the temperature T, the set of self-consistency equations (17)-(25), with the boundary conditions (26)-(29) defines the general-relativistic Thomas-Fermi equation.

The numerical procedure is now straightforward. For a fixed, arbitrarily chosen α, we first integrate (23) and (24) numerically on the interval $[0, R]$ and find solutions for various central values η_0. Integrating (25) simultaneously, we obtain $\mathcal{N}(R)$ as a function of η_0. We then select the value of η_0 for which $\mathcal{N}(R) = N$. The chemical potential μ corresponding to this particular solution is given by (28). If we now eliminate μ using (15), we finally get the parametric dependence on temperature through α.

The quantities N, T, and, R are free parameters in our model and their range and choice are dictated by physics. At $T = 0$ the number of fermions N is restricted by the OV limit $N_{\rm OV} = 2.89 \times 10^9 \sqrt{2/g}(15\,{\rm keV}/m)^2 M_\odot/m$. However, at nonzero temperature, stable solutions exist with $N > N_{\rm OV}$, depending on temperature and radius. In the following, N is required to be of the order $2 \times 10^{12} M_\odot/m$, so that for any m, the total mass is close to the estimated mass of the halo [39]. As we have demonstrated, the expected particle mass-temperature ratio of the halo is given by $\alpha \simeq m/T = 4 \times 10^4$. The halo radius R is in principle unlimited; in practice, however, it should not exceed half the average intergalactic distance. It is known that an isothermal configuration has no natural boundary, in contrast to the degenerate case of zero temperature, where for given N (up to the OV limit) the radius R is naturally fixed by the condition of vanishing pressure and density. At nonzero temperature, with R being unbounded, our gas would occupy the entire space, and fixing N would make p and ρ vanish everywhere. Conversely, if we do not fix N and integrate the equations on the interval $[0, \infty)$, both M and N will diverge at infinity for $T > 0$. Thus, one is forced to introduce a cutoff. In an isothermal model of a similar kind [40], the cutoff was set at the radius R, where the energy density was by about six orders of magnitude smaller than the central value. Our choice of $R = 200$ kpc is based on the estimated size of the Galactic halo. The only remaining free parameters of our model are the fermion mass m and the degeneracy factor g, which always appear in the combination $m^4 g$. We fix these parameters at $m = 15$ keV and $g = 2$, and justify this choice *a posteriori*.

We now present the results of the calculations for fixed particle number and temperatures near the point of gravothermal collapse. In Fig. 1 the energy per particle defined as $E = M/N - m$ is plotted as a function of temperature for fixed $N = 2 \times 10^{12} M_\odot/m$. The plot looks very much like that of a canonical Maxwell-Boltzmann ensemble [36], with one important difference: in the Maxwell-Boltzmann case, the curve continues to spiral inwards *ad infinitum* approaching the point of the singular isothermal sphere, that is characterized by an infinite central density. In the Fermi-Dirac case, the spiral consists of two, almost identical curves. The inwards winding of the spiral begins for

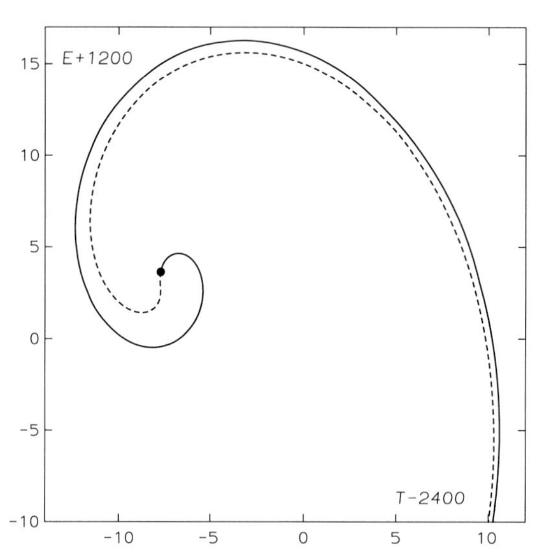

Fig. 1. Energy (shifted by $12 \times 10^{-8} m$) versus temperature (shifted by $-24 \times 10^{-8} m$), both in units of $10^{-10} m$, for fixed $N = 2 \times 10^{12} M_\odot/m$.

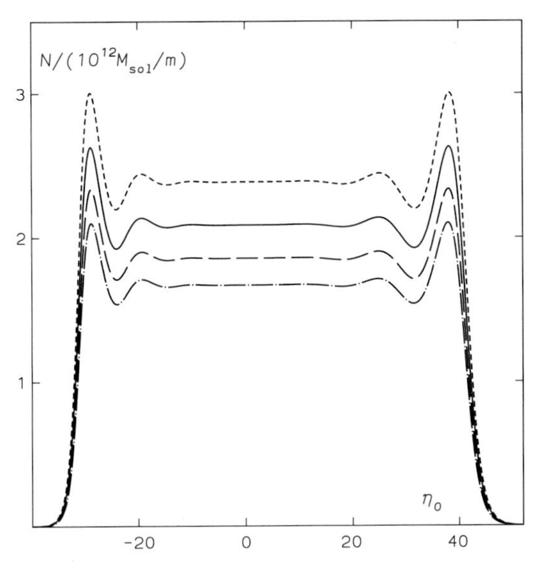

Fig. 2. Number of particles versus central degeneracy parameter for $m/T = 4 \times 10^6$ (solid), 3.5×10^6 (short dashs), 4.5×10^6 (long dashs), and 5×10^6 (dot-dashed line).

some negative central degeneracy and stops at the point $T = 2.3923 \times 10^{-7}m$, $E = -1.1964 \times 10^{-7}m$, where η_0 becomes zero. This part of the curve, which basically depicts the behavior of a nondegenerate gas, we call the *Maxwell-Boltzmann branch.* By increasing the central degeneracy parameter further to positive values, the spiral begins to unwind outwards very close to the inwards winding curve. The outwards winding curve will eventually depart from the Maxwell-Boltzmann branch for temperatures $T \gtrsim 10^{-3}m$. Further increase of the central degeneracy parameter brings us to a region where general-relativistic effects become important. The curve will exhibit another spiral for temperatures and energies of the order of a few $10^{-3}m$ approaching the limiting temperature $T_\infty = 2.4 \times 10^{-3}m$ and energy $E_\infty = 3.6 \times 10^{-3}m$, with both the central degeneracy parameter and the central density approaching infinite values. It is remarkable that gravitationally stable configurations with arbitrary large central degeneracy parameters exist at finite temperature even though the total mass exceeds the OV limit by several orders of magnitude.

The results of the numerical integration of (23) and (24), without restricting N, are presented in Fig. 2, where we plot the particle number N as a function of the central degeneracy parameter η_0 for several values of α close to 4×10^6. For fixed N, there is a range of α, where the Thomas-Fermi equation has multiple solutions. For example, for $N = 2 \times 10^{12}$ and $\alpha = 4 \times 10^6$ six solutions are found, which are denoted by (1), (2), (3), (3'), (2'), and (1'), corresponding to the values η_0 = -30.528, -25.354, -22.390, 29.284, 33.380, and 40.479, respectively. In Fig. 3 we plot the corresponding density profiles. For negative central value η_0, for which the degeneracy parameter is negative everywhere, the system behaves basically as a Maxwell-Boltzmann isothermal sphere. Positive values of the cen-

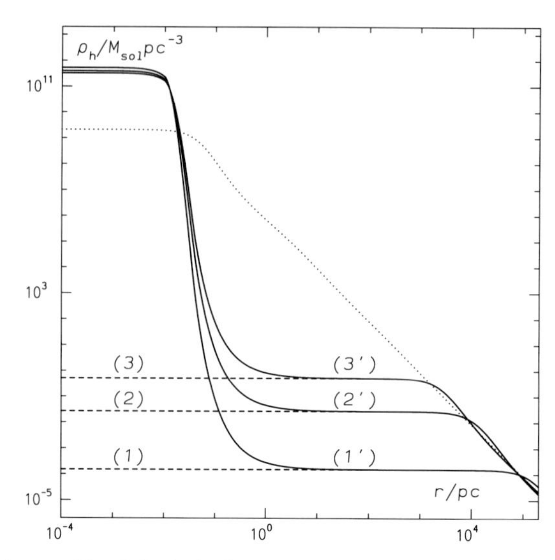

Fig. 3. The density profile of the halo for a central degeneracy parameter $\eta_0 = 0$ (dotted line) and for the six η_0-values discussed in the text. Configurations with negative η_0 ((1)-(3)) are depicted by the dashed and those with positive η_0 ((1')-(3')) by the solid line.

tral degeneracy parameter η_0 are characterized by a pronounced central core of mass of about $2.5 \times 10^6 M_\odot$ within a radius of about 20 mpc. The presence of this core is obviously due to the degeneracy pressure of the Fermi-Dirac statistics. A similar structure was obtained in collisionless stellar systems modeled as a nonrelativistic Fermi gas [41].

Figure 3 shows two important features. First, a galactic halo at a given temperature T may or may not have a central core, depending on whether the central degeneracy parameter η_0 is positive or negative. As the potential is nearly harmonic up to about 1 to 10 kpc for negative η_0, this may favor the formation of a barred galaxy. Second, the closer to zero η_0 is, the smaller the radius is at which the r^{-2} asymptotic behavior of the density begins. The flattening of the Galactic rotation curve begins in the range $1 \lesssim r/\mathrm{kpc} \lesssim 10$, hence the solution (3') most likely describes the Galaxy's halo. This may be verified by calculating the rotational curves in our model. We know already from our estimate (13) that our model yields the correct asymptotic circular velocity of 220 km/s. In order to make a more realistic comparison with the observed Galactic rotation curve, we must include two additional matter components: the bulge and the disk. The bulge is modeled as a spherically symmetric matter distribution of the form [42]

$$\rho_{\mathrm{b}}(s) = \frac{e^{-hs}}{2s^3} \int_0^\infty du \frac{e^{-hsu}}{[(u+1)^8 - 1]^{1/2}}, \tag{30}$$

where $s = (r/r_0)^{1/4}$, r_0 is the effective radius of the bulge and h is a parameter. We adopt $r_0 = 2.67$ kpc and h yielding the bulge mass $M_{\mathrm{b}} = 1.5 \times 10^{10} M_\odot$ [43].

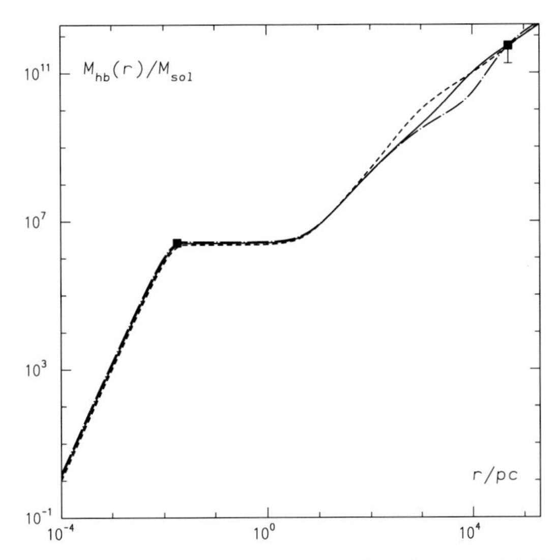

Fig. 4. Enclosed mass of halo plus bulge versus radius for $\eta_0 = 24$ (dashed), 28 (solid), and 32 (dot-dashed line).

In Fig. 4 the mass of halo and bulge enclosed within a given radius is plotted for various η_0. Here, the gravitational backreaction of the bulge on the fermionic halo has been taken into account. The data points, indicated by squares, are the mass $M_c = 2.6 \times 10^6 M_\odot$ within 18 mpc, estimated from the motion of the stars near Sgr A* [44], and the mass $M_{50} = 5.4^{+0.2}_{-3.6} \times 10^{11}$ within 50 kpc, estimated from the motions of satellite galaxies and globular clusters [39]. Variation of the central degeneracy parameter η_0 between 24 and 32 does not change the essential halo features.

In Fig. 5 we plot the circular velocity components of the halo, the bulge, and the disk. The contribution of the disk is modeled as [45]

$$\Theta_{\mathrm{d}}(r)^2 = \Theta_{\mathrm{d}}(r_{\mathrm{o}})^2 \frac{1.97(r/r_{\mathrm{o}})^{1.22}}{[(r/r_{\mathrm{o}})^2 + 0.78^2]^{1.43}} , \tag{31}$$

where we have taken $r_{\mathrm{o}} = 13.5$ kpc and $\Theta_{\mathrm{d}} = 100$ km/s. Here it is assumed for simplicity that the disk does not influence the mass distribution of the bulge and the halo. Choosing the central degeneracy $\eta_0 = 28$ for the halo, the data by Merrifield and Olling [46] are reasonably well fitted.

We now turn to the discussion of our choice of the fermion mass $m = 15$ keV for the degeneracy factor $g = 2$. To that end, we investigate how the mass of the central object, i.e., the mass M_c within 18 mpc, depends on m in the interval 5 to 25 keV, for various η_0. We find that $m \simeq 15$ keV always gives the maximal value of M_c ranging between 1.7 and 2.3 $\times 10^6 M_\odot$ for η_0 between 20 and 28. Hence, with $m \simeq 15$ keV we get the value closest to the mass of the central object M_c estimated from the motion of the stars near Sgr A* [44].

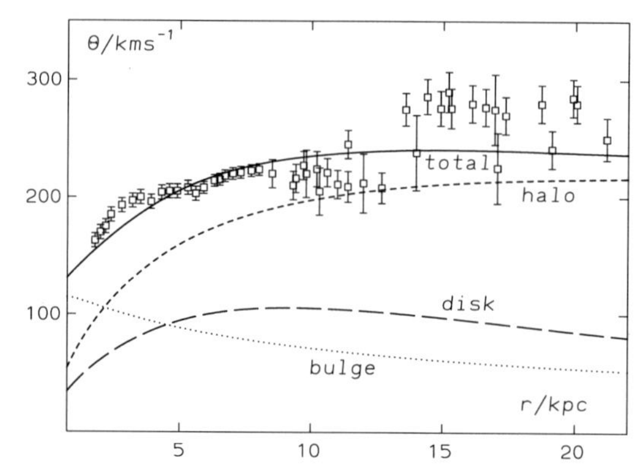

Fig. 5. Fit to the rotation curve of the Galaxy. The data points are from [46] for $R_0 = 8.5$ kpc and $\Theta_0 = 220$ km/s.

5 Conclusions

In summary, using the Thomas-Fermi theory, we have shown that a weakly interacting fermionic gas at finite temperature yields a mass distribution that successfully describes both the center and the halo of the Galaxy. For a fermion mass $m \simeq 15$ keV, a reasonable fit to the rotation curve is achieved with the temperature $T = 3.75$ meV and the central degeneracy parameter $\eta_0 = 28$. With the same parameters, we obtain the mass $M_{50} = 5.04 \times 10^{11} M_\odot$ and $M_{200} = 2.04 \times 10^{12} M_\odot$ within 50 and 200 kpc, respectively. These values agree quite well with the mass estimates based on the motions of satellite galaxies and globular clusters [39]. Moreover, the mass of $M_c \simeq 2.27 \times 10^6 M_\odot$, enclosed within 18 mpc, agrees reasonably well with the observations of motion of stars near the compact dark object at the center of the Galaxy.

Acknowledgement

We thank P. Salucci for valuable discussions and comments. We are grateful to M.R. Merrifield and R.P. Olling for sending us the Galactic rotation curve. This research is in part supported by the Foundation of Fundamental Research (FFR) grant number PHY99-01241 and the Research Committee of the University of Cape Town. The work of N.B. is supported in part by the Ministry of Science and Technology of the Republic of Croatia under Contract No. 0098002.

References

1. S. Cole and C. Lacey, MNRAS **281**, 716 (1996).
2. M.A. Markov, Phys. Lett. **10**, 122 (1964).
3. G. Marx and A.S. Szalay, in *Neutrino '72*, **1**, 191 (Technoinform, Budapest, 1972); R. Cowsik and J. McClelland, ApJ **180**, 7 (1973); R. Ruffini, Lett. Nuovo Cim. **29**, 161 (1980).

4. R.D. Viollier, D. Trautmann, and G.B. Tupper, Phys. Lett. B **306**, 79 (1993); R.D. Viollier, Prog. Part. Nucl. Phys. **32**, 51 (1994).
5. N. Bilić, D. Tsiklauri, and R.D. Viollier, Prog. Part. Nucl. Phys. **40**, 17 (1998).
6. N. Bilić, F. Munyaneza, and R.D. Viollier, Phys. Rev. D **59**, 024003 (1999).
7. D. Tsiklauri, and R.D. Viollier, Astropart. Phys. **12**, 199 (1999); F. Munyaneza and R.D. Viollier, astro-ph/9907318.
8. F. Munyaneza, D. Tsiklauri, and R.D. Viollier, ApJ **509**, L105 (1998); *ibid.* **526**, 744 (1999); F. Munyaneza and R.D. Viollier, ApJ **563**, 0000 (2001).
9. P. Bode, J.P. Ostriker, and N. Turok, ApJ **556**, 93 (2001).
10. E.W. Kolb and M.S. Turner, *The Early Universe*, (Addison-Wesley, San Francisco, 1989).
11. X. Shi and G.M. Fuller, Phys. Rev. Lett. **82**, 2832 (1999); K. Abazajian, G.M. Fuller, and M. Patel, Phys. Rev. D **64**, 023501 (2001), astro-ph/0101524; G.B. Tupper, R.J. Lindebaum, and R.D. Viollier, Mod. Phys. Lett. A **15**, 1221 (2000).
12. M. Drees and D. Wright, hep-ph/0006274.
13. P. de Bernardis *et al.*, Nature **404**, 955 (2000).
14. T. Goto and M. Yamaguchi, Phys. Lett. B **276**, 123 (1992); L. Covi , J.E. Kim, and L. Roszkowski, Phys. Rev. Lett. **82**, 4180 (1999); L. Covi , H.-B. Kim, J.E. Kim, and L. Roszkowski, hep-ph/0101009.
15. M. Dine and A.E. Nelson, Phys. Rev. D **48**, 1277 (1993); M. Dine, A.E. Nelson, and Y. Shirman, Phys. Rev. D **51**, 1362 (1995); M. Dine, A.E. Nelson, Y. Nir, and Y. Shirman, Phys. Rev. D **53**, 2658 (1996); D.H. Lyth, Phys. Lett. B **488**, 417 (2000).
16. H. Murayama, Phys. Rev. Lett. **79**, 18 (1997); S. Dimopoulos *et al.*, Nucl. Phys. B **510**, 12 (1998); E.A. Baltz and H. Murayama, astro-ph/0108172.
17. E.W. Kolb and M.S. Turner, in *Review of Particle Physics*, Eur. Phys. J. C **15**, 1 (2000).
18. S. Weinberg, "Gravitation and Cosmology" (Wiley, New York, 1972).
19. C. Wetterich, Nucl. Phys. B **302**, 668 (1988); P.J.E. Peebles and B. Ratra, ApJ **325**, L17 (1988); R. Caldwell, R. Dave, and P.J. Steinhardt, Phys. Rev. Lett. **80**, 1589 (1998); D. Huterer and M.S. Turner, Phys. Rev. D **60**, 081301 (1999).
20. A. Kamenshchik, U. Moschella, and V. Pasquier, Phys. Lett. B **511**, 265 (2001).
21. N. Bilić, G.B. Tupper, and R.D. Viollier, Phys. Lett. B **535**, 17 (2002).
22. Y. Bergner and R. Jackiw, Phys. Lett. A **284**, 146 (2001); for a review see R. Jackiw, physics/0010042.
23. M.S. Turner, Phys. World **9**, No 9, 31 (1996).
24. A.D. Dolgov, hep-ph/9910532; M. Srednicki, in *Review of Particle Physics*, Eur. Phys. J. C **15**, 1 (2000); L. Bergstrom, Rep. Prog. Phys. **63**, 793 (2000), hep-ph/0002126.
25. J.R. Primack, in *Formation of Structure in the Univerese*, eds. A. Dekel and J.P. Ostriker (Cambridge University Press, 1999), astro-ph/9707285.
26. G.F. Giudice *e al.*, Phys. Rev. D **64**, 043512 (2001).
27. A. Klypin et al., ApJ **554**, 903 (2001).
28. B. Moore et al., Phys. Rev. D in press, astro-ph/0106217.
29. R. Buser, Science **287**, 69 (2000).
30. R. Mahadevan, Nature **394**, 651 (1998).
31. W. Thirring, Z. Physik **235**, 339 (1970); P. Hertel, H. Narnhofer, and W. Thirring, Comm. Math. Phys. **28**, 159 (1972); J. Messer, J. Math. Phys. **22**, 2910 (1981).
32. N. Bilić and R.D. Viollier, Phys. Lett. B **408**, 75 (1997).
33. N. Bilić and R.D. Viollier, Gen. Rel. Grav. **31**, 1105 (1999); Eur. Phys. J. C **11**, 173 (1999).

34. J.R. Oppenheimer and G.M. Volkoff, Phys. Rev. **55**, 374 (1939).
35. D. Lynden-Bell, MNRAS **136**, 101 (1967).
36. J. Binney and S. Tremaine, *Galactic Dynamics* (Princeton University Press, Princeton, New Jersey, 1987), and references cited therein.
37. T. Padmanabhan, *Structure formation in the Universe* (Cambridge University Press, Cambridge, 1993).
38. J. Ehlers, in *Relativity, Astrophysics and Cosmology*, edited by W. Israel (D. Reidel Publishing Company, Dordrecht/Boston 1973)
39. M.I. Wilkinson and N.W. Evans, MNRAS **310**, 645 (1999).
40. W.Y. Chau, K. Lake, and J. Stone, ApJ **281**, 560 (1984).
41. P.-H. Chavanis and J. Sommeria, MNRAS **296**, 569 (1998).
42. P.J. Young, ApJ **81**, 807 (1976); G. de Vaucouleurs and W.D. Pence, ApJ **83**, 1163 (1978).
43. P.D. Sackett, ApJ **483**, 103 (1997).
44. A. Eckart and R. Genzel, MNRAS **284**, 576 (1997); A.M. Ghez, B.L. Klein, M. Morris, and E.E. Becklin, ApJ **509**, 678 (1998).
45. M. Persic, P. Salucci, and F. Stell, MNRAS **281**, 27 (1986).
46. R.P. Olling and M.R. Merrifield, MNRAS **311**, 361 (2000).

Neutrino Masses in GUTs and Baryon Asymmetry

Wilfried Buchmüller

Deutsches Elektronen-Synchrotron DESY, 22603 Hamburg, Germany

Abstract. We discuss the implications of large neutrino mixings for grand unified theories based on the seesaw mechanism. In SU(5) GUTs large mixings can be accomodated by means of $U(1)_F$ flavour symmetries. In these models the heavy Majorana neutrinos are essentially decoupled from low energy neutrino physics. On the contrary in SO(10) GUTs large neutrino mixings severely constrain the mass spectrum of the heavy Majorana neutrinos. This leads to predictions for a variety of observables in neutrino physics as well as for the cosmological baryon asymmetry.

1 Status of Neutrino Mixing

Recent results from the Sudbury Neutrino Observatory[1] and from the Super-Kamiokande experiment[2] provide further evidence for neutrino oscillations as the solution of the solar neutrino problem. Neutrino oscillations can also account for the atmospheric neutrino anomaly[3,4]. It is remarkable that a consistent picture can be obtained with just three neutrinos, ν_e, ν_μ and ν_τ, undergoing 'nearest neighbour' oscillations, $\nu_e \leftrightarrow \nu_\mu$ and $\nu_\mu \leftrightarrow \nu_\tau$.

For massive neutrinos a mixing matrix U appears in the leptonic charged current,

$$\mathcal{L}_{CC} = -\frac{g}{\sqrt{2}} \sum_{\alpha,i} \overline{e}_\alpha \gamma^\mu (1-\gamma_5) U_{\alpha,i} \nu_i \, W_\mu^- \; + \ldots \, , \tag{1}$$

where e_α and ν_i are mass eigenstates. In the case of three neutrinos, one for each generation, U is a unitary matrix.

The experimental results on the ν_e deficit in the solar neutrino flux favour the LMA or LOW solutions[5] of the MSW conversion with large mixing angle. A large mixing also fits the atmospheric neutrino oscillations. As a result, the leptonic mixing matrix $U_{\alpha i}$ appears to be very different from the familiar CKM quark mixing matrix $V_{\alpha i}$. The emerging pattern is rather simple[6],

$$U = \begin{pmatrix} * & * & \diamond \\ * & * & * \\ * & * & * \end{pmatrix} . \tag{2}$$

Here the '$*$' denotes matrix elements whose value is consistent with the range $0.5 \ldots 0.8$, whereas for the matrix element '$\diamond$' only an upper bound exits, $|U_{e3}| < 0.16$. The neutrino masses may be hierarchical or quasi-degenerate. Note, however, that a possible hierarchy has to be much weaker than the known mass hierarchy of quarks and charged leptons.

Several interesting phenomenological schemes have been suggested, such as 'bi-maximal' or 'democratic' mixing, which describe the pattern (2) rather well[7]. Is is unclear, however, how these schemes are related to a more fundamental theory. We shall therefore focus on the question how large neutrino mixings can be obtained in a grand unified theory based on the gauge groups SU(5) or SO(10). In both cases we shall rely on the seesaw mechanism which naturally explains the smallness of light Majorana neutrino masses m_ν by the largeness of right-handed neutrino masses M[8],

$$m_\nu \simeq -m_D \frac{1}{M} m_D^T \;, \tag{3}$$

where m_D is the Dirac neutrino mass matrix. In unified theories m_D is related to the quark and charged lepton mass matrices. Since they have a large hierarchy, the almost non-hierarchical structure of the leptonic mixing matrix is very surprising and requires some explanation. In the following we shall discuss two qualitatively different examples based on the GUT groups SU(5) and SO(10), respectively, which illustrate present attempts to solve the puzzle of the large neutrino mixings.

2 Models with SU(5)

In the simplest GUT based on the gauge group SU(5)[9] quarks and leptons are grouped into the multiplets $\mathbf{10} = (q_L, u_R^c, e_R^c)$, $\mathbf{5}^* = (d_R^c, l_L)$ and $\mathbf{1} = \nu_R$. Hence, unlike the gauge fields, quarks and leptons are not unified in a single irreducible representation. In particular, the right-handed neutrinos are gauge singlets and can therefore have Majorana masses not generated by spontaneous symmetry breaking. In addition one has three Yukawa interactions, which couple the fermions to the Higgs fields $H_1(\mathbf{5})$ and $H_2(\mathbf{5}^*)$,

$$\mathcal{L} = h_{uij} \mathbf{10}_i \mathbf{10}_j H_1(\mathbf{5}) + h_{dij} \mathbf{5}^*{}_i \mathbf{10}_j H_2(\mathbf{5}^*) + h_{\nu ij} \mathbf{5}^*{}_i \mathbf{1}_j H_1(\mathbf{5}) + M_{ij} \mathbf{1}_i \mathbf{1}_j \;. \tag{4}$$

The mass matrices of up-quarks, down-quarks, charged leptons and the Dirac neutrino mass matrix are given by $m_u = h_u v_1$, $m_d = h_d v_2$, $m_e = m_d$ and $m_D = h_\nu v_1$, respectively, with $v_1 = \langle H \rangle_1$ and $v_2 = \langle H \rangle_2$. The Majorana masses M are independent of the Higgs mechanism and can therefore be much larger than the electroweak scale v.

An attractive framework to explain the observed mass hierarchies of quarks and charged leptons is the Froggatt-Nielsen mechanism[10] based on a spontaneously broken $\mathrm{U}(1)_F$ generation symmetry. The Yukawa couplings are assumed to arise from non-renormalizable interactions after a gauge singlet field $\mathcal{F}$ acquires a vacuum expectation value,

$$h_{ij} = g_{ij} \left(\frac{\langle \mathcal{F} \rangle}{\Lambda} \right)^{Q_i + Q_j} \;. \tag{5}$$

Here g_{ij} are couplings $\mathcal{O}(1)$ and Q_i are the $\mathrm{U}(1)_F$ charges of the various fermions, with $Q_{\mathcal{F}} = -1$. The interaction scale Λ is usually chosen to be very large, $\Lambda > \Lambda_{GUT}$.

Table 1. *Lopsided $U(1)_F$ charges of $SU(5)$ multiplets. From [14].*

ψ_i	$\mathbf{10}_3$	$\mathbf{10}_2$	$\mathbf{10}_1$	$\mathbf{5}^*_3$	$\mathbf{5}^*_2$	$\mathbf{5}^*_1$	$\mathbf{1}_3$	$\mathbf{1}_2$	$\mathbf{1}_1$
Q_i	0	1	2	a	a	$a+1$	b	c	d

The symmetry group SU(5)×U(1)$_F$ has been considered by a number of authors. Particularly interesting is the case with a 'lopsided' family structure where the chiral U(1)$_F$ charges are different for the **5***-plets and the **10**-plets of the same family[11–13]. Note, that such lopsided charge assignments are not consistent with the embedding into a higher-dimensional gauge group, like SO(10)×U(1)$_F$ or E$_6$×U(1)$_F$. An example of phenomenologically allowed lopsided charges Q_i is given in Table 1.

This charge assignement determines the structure of the Yukawa matrices, e.g.,

$$h_e = h_d \sim \begin{pmatrix} \epsilon^3 & \epsilon^2 & \epsilon^2 \\ \epsilon^2 & \epsilon & \epsilon \\ \epsilon & 1 & 1 \end{pmatrix} , \tag{6}$$

where the parameter $\epsilon = \langle \mathcal{F} \rangle / \Lambda$ controls the flavour mixing, and coefficients $\mathcal{O}(1)$ are unknown. The corresponding mass hierarchies for up-quarks, down-quarks and charged leptons are

$$m_t : m_c : m_u \simeq 1 : \epsilon^2 : \epsilon^4 , \tag{7}$$

$$m_b : m_s : m_d = m_\tau : m_\mu : m_e \simeq 1 : \epsilon : \epsilon^3 . \tag{8}$$

The differences between the observed down-quark mass hierarchy and the charged lepton mass hierarchy can be accounted for by introducing additional Higgs fields[15]. From a fit to the running quark and lepton masses at the GUT scale one obtains for the flavour mixing parameter $\epsilon \simeq 0.06$.

The light neutrino mass matrix is obtained from the seesaw formula,

$$m_\nu = -m_D \frac{1}{M} m_D^T \sim \epsilon^{2a} \begin{pmatrix} \epsilon^2 & \epsilon & \epsilon \\ \epsilon & 1 & 1 \\ \epsilon & 1 & 1 \end{pmatrix} . \tag{9}$$

Note, that the structure of this matrix is determined by the U(1)$_F$ charges of the **5***-plets only. It is independent of the U(1)$_F$ charges of the right-handed neutrinos.

Since all elements of the 2-3 submatrix of (9) are $\mathcal{O}(1)$, one naturally obtains a large $\nu_\mu - \nu_\tau$ mixing angle Θ_{23}[11,12]. At first sight one may expect that $\Theta_{12} = \mathcal{O}(\epsilon)$, which would correspond to the SMA solution of the MSW conversion. However, one can also have a large mixing angle Θ_{12} if the determinent of the 2-3 submatrix of m_ν is $\mathcal{O}(\epsilon)$[16]. Choosing the coefficients $\mathcal{O}(1)$ randomly, in the spirit of 'flavour anarchy'[17], the SMA and the LMA solutions are about equally probable for $\epsilon \simeq 0.1$[18]. The corresponding neutrino masses are consistent with $m_2 \sim 5 \times 10^{-3}$ eV and $m_3 \sim 5 \times 10^{-2}$ eV. We conclude that the neutrino mass matrix (9) naturally yields a large angle Θ_{23}, with Θ_{12} large or small. In order to have maximal mixings the coefficients $\mathcal{O}(1)$ have to obey special relations.

The model can also explain the cosmological baryon asymmetry via leptogenesis[19] for an appropriate choice of the parameters in table 1[14]. The mass of the heaviest Majorana neutrino is

$$M_3 \sim \epsilon^{2(a+b)} \frac{v_1^2}{\overline{m}_\nu} \sim \epsilon^{2(a+b)}\, 10^{15}\ \mathrm{GeV}\ , \tag{10}$$

where $\overline{m}_\nu = \sqrt{m_2 m_3} \sim 10^{-2}$ eV. The special choice $a = b = 0,\ c = 1,\ d = 2$ yields the scenario of [20] where $B - L$ is broken at the GUT scale.

For the CP asymmetry in the decays of the heavy neutrinos N_1,

$$\varepsilon_1 = \frac{\Gamma(N_1 \to l\, H_2) - \Gamma(N_1 \to l^c\, H_2^c)}{\Gamma(N_1 \to l\, H_2) + \Gamma(N_1 \to l^c\, H_2^c)}\ , \tag{11}$$

one has in the case $M_1 < M_{2,3}$,

$$\varepsilon_1 \simeq -\frac{3}{16\pi}\frac{M_1}{(h_\nu^\dagger h_\nu)_{11}} \mathrm{Im}\left(h_\nu^\dagger h_\nu \frac{1}{M} h_\nu^T h_\nu^*\right)_{11} \sim \frac{3}{16\pi}\epsilon^{2(a+d)}\ . \tag{12}$$

Successful baryogenesis requires $a + d = 2$. With $\epsilon \sim 0.1$ the corresponding CP asymmetry is $\varepsilon_1 \sim 10^{-6}$. The baryogenesis temperature is then $T_B \sim M_1 \sim \epsilon^4 M_3 \sim 10^{10}$ GeV. The effective neutrino mass which controls the out-of-equilibrium condition of the decaying heavy Majorana neutrino is given by $\widetilde{m}_1 = (m_D^\dagger m_D)_{11}/M_1 \sim 10^{-2}$ eV.

Thermal leptogenesis leads to the baryon asymmetry[21]

$$Y_B = \frac{n_B - n_{\overline{B}}}{s} = \kappa c_S \frac{\varepsilon_1}{g_*}\ , \tag{13}$$

where n_B and s are baryon number and entropy densities, respectively; $g_* \sim 100$ is the number of degrees of freedom in the plasma of the early universe and $c_S = \mathcal{O}(1)$ is the conversion factor from lepton asymmetry to baryon asymmetry due to sphaleron processes. Washout processes are accounted for by $\kappa < 1$, which can be computed by solving the full Boltzmann equations [22,23]. The resulting baryon asymmetry then reads

$$Y_B \sim \kappa\ 10^{-8}\ . \tag{14}$$

With $\kappa \sim 0.1 \ldots 0.01$ this is indeed the correct order of magnitude in accord with observation, $Y_B \simeq (0.6 - 1) \times 10^{-10}$.

The magnitude for the generated baryon asymmetry depends crucially on the parameters ε_1, $\widetilde{m}_1$ and M_1. In the models with SU(5)×U(1)$_F$ symmetry low energy neutrino physics is essentially decoupled from the heavy Majorana neutrinos and does not constrain the value of M_1. Hence, successful baryogenesis is consistent with the SU(5)×U(1)$_F$ symmetry, but it cannot be considered a generic prediction. This is different in unified theories with larger gauge groups.

3 Models with SO(10)

The simplest grand unified theory which unifies one generation of quarks and leptons including the right-handed neutrino in a single irreducible representation is based on the gauge group SO(10)[24]. The quark and lepton mass matrices are obtained from the couplings of the fermion multiplet $\mathbf{16} = (q_L, u^c_R, e^c_R, d^c_R, l_L, \nu_R)$ to the Higgs multiplets $H_1(\mathbf{10})$, $H_2(\mathbf{10})$ and $\Phi(\mathbf{126})$,

$$\mathcal{L} = h_{uij}\mathbf{16}_i\mathbf{16}_j H_1(\mathbf{10}) + h_{dij}\mathbf{16}_i\mathbf{16}_j H_2(\mathbf{10}) + h_{Nij}\mathbf{16}_i\mathbf{16}_j \Phi(\mathbf{126}) \,. \tag{15}$$

Here we have assumed that the two Higgs doublets of the standard model are contained in the two[1] ten-plets H_1 and H_1, respectively. This yields the quark mass matrices $m_u = h_u v_1$, $m_d = h_d v_2$, with $v_1 = \langle H\rangle_1$ and $v_2 = \langle H\rangle_2$, and the lepton mass matrices

$$m_D = m_u \,, \qquad m_e = m_d \,. \tag{16}$$

Contrary to SU(5) GUTs, the Dirac neutrino and the up-quark mass matrices are now related. Note, that all matrices are symmetric. The Majorana mass matrix $M = h_N \langle\Phi\rangle$ is also generated by spontaneous symmetry breaking and a priori independent of m_u and m_d.

With $m_D = m_u$ the seesaw mass relation becomes

$$m_\nu \simeq -m_u \frac{1}{M} m_u^T \,. \tag{17}$$

The large neutrino mixings now appear very puzzling, since the quark mass matrices are hierarchical and the quark mixings are small. It turns out, however, that because of the known properties of the up-quark mass matrix this puzzle can be resolved provided the heavy neutrino masses also obey a specific hierarchy. This then leads to predictions for a number of observables in neutrino physics including the cosmological baryon asymmetry. In the following we shall describe these implications of large neutrino mixings in SO(10) GUTs following ref.[27]. The role of the heavy neutrino mass hierarchy for the light neutrino mixings has previously been discussed in different contexts[25].

From the phenomenology of weak decays we know that the quark matrices have approximately the form[28,29],

$$m_{u,d} \propto \begin{pmatrix} 0 & \epsilon^3 e^{i\phi} & 0 \\ \epsilon^3 e^{i\phi} & \rho\epsilon^2 & \eta\epsilon^2 \\ 0 & \eta\epsilon^2 & e^{i\psi} \end{pmatrix} . \tag{18}$$

Here $\epsilon \ll 1$ is the parameter which determines the flavour mixing, and

$$\rho = |\rho| e^{i\alpha} \,, \qquad \eta = |\eta| e^{i\beta} \,, \tag{19}$$

are complex parameters $\mathcal{O}(1)$. We have chosen a 'hierarchical' basis, where off-diagonal matrix elements are small compared to the product of the corresponding eigenvalues, $|m_{ij}|^2 \leq \mathcal{O}(|m_i m_j|)$. In contrast to the usual assumption of

[1] Note, that this is unavoidable in models with SO(10) breaking by orbifold compactification[26].

hermitian mass matrices[28,29], SO(10) invariance dictates the matrices to be symmetric. All parameters may take different values for up– and down–quarks. Typical choices for ϵ are $\epsilon_u = 0.07$, $\epsilon_d = 0.21$[29]. The agreement with data can be improved by adding in the 1-3 element a term $\mathcal{O}(\epsilon^4)$[30,31] which, however, is not important for the following analysis. Data also imply one product of phases to be 'maximal', i.e., $\Delta = \phi_u - \alpha_u - \phi_d + \alpha_d \simeq \pi/2$.

We do not know the structure of the Majorana mass matrix $M = h_N \langle \Phi \rangle$. However, in models with family symmetries it should be similar to the quark mass matrices, i.e., the structure should be independent of the Higgs field. In this case, one expects

$$M = \begin{pmatrix} 0 & M_{12} & 0 \\ M_{12} & M_{22} & M_{23} \\ 0 & M_{23} & M_{33} \end{pmatrix} , \tag{20}$$

with $M_{12} \ll M_{22} \sim M_{23} \ll M_{33}$. M is diagonalized by a unitary matrix, $U^{(N)\dagger} M U^{(N)*} = \mathrm{diag}(M_1, M_2, M_3)$. Using the seesaw formula one can now evaluate the light neutrino mass matrix. Since the choice of the Majorana matrix m_N fixes a basis for the right-handed neutrinos the allowed phase redefinitions of the Dirac mass matrix m_D are restricted. In (18) the phases of all matrix elements have therefore been kept.

The ν_μ-ν_τ mixing angle is known to be large. This leads us to require $m_{\nu_{i,j}} = \mathcal{O}(1)$ for $i,j = 2,3$. It is remarkable that this determines the hierarchy of the heavy Majorana mass matrix to be[2]

$$M_{12} : M_{22} : M_{33} = \epsilon^5 : \epsilon^4 : 1 . \tag{21}$$

With $M_{33} \simeq M_3$, $M_{22} = \sigma\epsilon^4 M_3$, $M_{23} = \zeta\epsilon^4 M_3 \sim M_{22}$ and $M_{12} = \epsilon^5 M_3$, one obtains for masses and mixings to order $\mathcal{O}(\epsilon^4)$,

$$M_1 \simeq -\frac{\epsilon^6}{\sigma} M_3 , \quad M_2 \simeq \sigma\epsilon^4 M_3 , \tag{22}$$

$$U_{12}^{(N)} = -U_{21}^{(N)} = \frac{\epsilon}{\sigma} , \quad U_{23}^{(N)} = \mathcal{O}(\epsilon^4) , \quad U_{13}^{(N)} = 0 . \tag{23}$$

Note, that σ can always be chosen real whereas ζ is in general complex. This yields for the light neutrino mass matrix

$$m_\nu = -\begin{pmatrix} 0 & \epsilon e^{2i\phi} & 0 \\ \epsilon e^{2i\phi} & -\sigma e^{2i\phi} + 2\rho e^{i\phi} & \eta e^{i\phi} \\ 0 & \eta e^{i\phi} & e^{2i\psi} \end{pmatrix} \frac{v_1^2}{M_3} . \tag{24}$$

The complex parameter ζ does not enter because of the hierarchy. The matrix (24) has the same structure as the mass matrix (9) in the $SU(5) \times U(1)_F$ model, except for additional texture zeroes. Since, as required, all elements of the 2-3

[2] We also note that this result is independent of the zeroes in the mass matrix (18) if its 1-3 element is smaller than ϵ^3, as required by data.

submatrix are $\mathcal{O}(1)$, the mixing angle Θ_{23} is naturally large. A large mixing angle Θ_{12} can again occur in case of a small determinant of the 2-3 submatrix,

$$(-\sigma + 2\rho e^{-i\phi})e^{2i\psi} - \eta^2 \equiv \partial e^{2i\gamma} = \mathcal{O}(\epsilon) . \tag{25}$$

Such a condition can be fullfilled without fine tuning if $\sigma, \rho, \eta = \mathcal{O}(1)$. It implies relations between the moduli as well as the phases of ρ and η. In the special case of a somewhat smaller mass of the second heavy neutrino, i.e., $|\sigma| < |\rho|$, the condition (25) becomes

$$\psi - \beta \simeq \frac{1}{2}(\phi - \alpha) , \quad |\eta|^2 \simeq 2|\rho| . \tag{26}$$

The mass matrix m_ν can again be diagonalized by a unitary matrix, $U^{(\nu)\dagger} \cdot m_\nu U^{(\nu)*} = \mathrm{diag}(m_1, m_2, m_3)$. A straightforward calculation yields ($s_{ij} = \sin\Theta_{ij}$, $c_{ij} = \cos\Theta_{ij}$, $\xi = \epsilon/(1+|\eta|^2)$),

$$U^{(\nu)} = \begin{pmatrix} c_{12}e^{i(\phi-\beta+\psi-\gamma)} & s_{12}e^{i(\phi-\beta+\psi-\gamma)} & \xi s_{23}e^{i(\phi-\beta+\psi)} \\ -c_{23}s_{12}e^{i(\phi+\beta-\psi+\gamma)} & c_{23}c_{12}e^{i(\phi+\beta-\psi+\gamma)} & s_{23}e^{i(\phi+\beta-\psi)} \\ s_{23}s_{12}e^{i(\gamma+\psi)} & -s_{23}c_{12}e^{i(\gamma+\psi)} & c_{23}e^{i\psi} \end{pmatrix} , \tag{27}$$

with the mixing angles,

$$\tan 2\Theta_{23} \simeq \frac{2|\eta|}{1-|\eta|^2} , \quad \tan 2\Theta_{12} \simeq 2\sqrt{1+|\eta|^2}\frac{\epsilon}{\partial} . \tag{28}$$

Note, that the 1-3 element of the mixing matrix is small, $U_{13}^{(\nu)} = \mathcal{O}(\epsilon)$. The masses of the light neutrinos are

$$m_1 \simeq -\tan^2\Theta_{12}\, m_2 , \quad m_2 \simeq \frac{\epsilon}{(1+|\eta|^2)^{3/2}}\cot\Theta_{12}\, m_3 , \quad m_3 \simeq (1+|\eta|^2)\,\frac{v_1^2}{M_3} . \tag{29}$$

This corresponds to the weak hierarchy,

$$m_1 : m_2 : m_3 = \epsilon : \epsilon : 1 , \tag{30}$$

with $m_2^2 \sim m_1^2 \sim \mathrm{d}m_{21}^2 = m_2^2 - m_1^2 \sim \epsilon^2$. Since $\epsilon \sim 0.1$, this pattern is consistent with the LMA solution of the solar neutrino problem, but not with the LOW solution.

The large ν_μ-ν_τ mixing has been obtained as consequence of the required very large mass hierarchy (22) of the heavy Majorana neutrinos. The large ν_e-ν_μ mixing follows from the particular values of parameters $\mathcal{O}(1)$. Hence, one expects two large mixing angles, but single maximal or bi-maximal mixing would require fine tuning. On the other hand, one definite prediction is the occurence of exactly one small matrix element, $U_{13}^{(\nu)} = \mathcal{O}(\epsilon)$. Note, that the obtained pattern of neutrino mixings is independent of the off-diagonal elements of the mass matrix M. For instance, replacing the texture (20) by a diagonal matrix, $M = \mathrm{diag}(M_1, M_2, M_3)$, leads to the same pattern of neutrino mixings.

In order to calculate various observables in neutrino physics we need the leptonic mixing matrix

$$U = U^{(e)\dagger} U^{(\nu)} , \tag{31}$$

where $U^{(e)}$ is the charged lepton mixing matrix. In our framework we expect $U^{(e)} \simeq V^{(d)}$, and also $V = V^{(u)\dagger} V^{(d)} \simeq V^{(d)}$ for the CKM matrix since $\epsilon_u < \epsilon_d$. This yields for the leptonic mixing matrix

$$U \simeq V^{\dagger} U^{(\nu)} . \tag{32}$$

To leading order in the Cabibbo angle $\lambda \simeq 0.2$ we only need the off-diagonal elements $V_{12}^{(d)} = \overline{\lambda} = -V_{21}^{(d)*}$. Since the matrix m_d is complex, the Cabibbo angle is modified by phases, $\overline{\lambda} = \lambda \exp\{i(\phi_d - \alpha_d)\}$. The resulting leptonic mixing matrix is indeed of the wanted form (2) with all matrix elements $\mathcal{O}(1)$, except U_{13},

$$U_{13} = \xi s_{23} e^{i(\phi-\beta+\psi)} - \overline{\lambda} s_{23} e^{i(\phi+\beta-\psi)} = \mathcal{O}(\lambda, \epsilon) \sim 0.1 , \tag{33}$$

which is close to the experimental limit.

Let us now consider the CP violation in neutrino oscillations. Observable effects are controlled by the Jarlskog parameter J_l [32] ($\widetilde{\epsilon}_{ij} = \sum_{k=1}^{3} \epsilon_{ijk}$)

$$\mathrm{Im}\{U_{\alpha i} U_{\beta j} U^*_{\alpha j} U^*_{\beta i}\} = \widetilde{\epsilon}_{\alpha\beta} \widetilde{\epsilon}_{ij} J_l , \tag{34}$$

for which one finds

$$J_l \simeq \lambda s_{12} c_{12} c_{23} s_{23}^2 \sin\left(2(\beta - \psi + \gamma) + \phi_d - \alpha_d\right) . \tag{35}$$

In the case of a small mass difference $\mathrm{d}m_{12}^2$ the CP asymmetry $P(\nu_\mu \to \nu_e) - P(\overline{\nu}_\mu \to \overline{\nu}_e)$ is proportinal to ∂ (cf. (25)). Hence, the dependence of J_l on the angle γ is not surprising.

For large mixing, $c_{ij} \simeq s_{ij} \simeq 1/\sqrt{2}$, and in the special case (26) one obtains from the SO(10) phase relation $\phi - \alpha = \phi_u - \alpha_u$ and $\phi_u - \alpha_u - \phi_d + \alpha_d = \Delta \simeq \pi/2$,

$$J_l \simeq \frac{\lambda}{4\sqrt{2}} \sin\left(-\frac{\pi}{2} + 2\gamma\right) . \tag{36}$$

For small γ this corresponds to maximal CP violation, but without a deeper understanding of the fermion mass matrices this case is not singled out. Due to the large neutrino mixing angles, J_l is much bigger than the Jarlskog parameter in the quark sector, $J_q = \mathcal{O}(\lambda^6) \sim 10^{-5}$, which may lead to observable effects at future neutrino factories[33].

According to the seesaw mechanism neutrinos are Majorana fermions. This can be directly tested in neutrinoless double β-decay. The decay amplitude is proportional to the complex mass

$$\begin{aligned}\langle m \rangle &= \sum_i U_{ei}^2 m_i = -(U U^{(\nu)\dagger} m_\nu U^{(\nu)*} U^T)_{ee} \simeq -(V^{(d)\dagger} m_\nu V^{(d)*})_{ee} \\ &= -\frac{1}{1+|\eta|^2} \left(\lambda^2 |\eta|^2 e^{2i(\phi_d - \alpha_d + \beta + \phi - \psi)} - 2\lambda\epsilon e^{i(\phi_d - \alpha_d + 2\phi)}\right) m_3 . \end{aligned} \tag{37}$$

With $m_3 \simeq \sqrt{\mathrm{d}m^2_{atm}} \simeq 5 \times 10^{-2}$ eV this yields $\langle m \rangle \sim 10^{-3}$ eV, more than two orders of magnitude below the present experimental upper bound[34].

Finally, consider again the baryon asymmetry which should eventually be related to the CP violation in neutrino oscillations and quark mixing. This possibility has recently been discussed also in other contexts[35,36]. In the special case[3] (26) one obtains for the CP asymmetry,

$$\varepsilon_1 \simeq \frac{3}{16\pi}\epsilon^6 \frac{|\eta|^2}{\sigma} \frac{(1+|\rho|)^2}{|\eta|^2+|\rho|^2} \sin(\phi_u - \alpha_u)\ . \tag{38}$$

As expected ϵ_1 depends only on phases of the up-quark matrix and not on the combination of up− and down−quark phases d which appears in the CKM matrix. In addition, the parameter σ enters. Hence, the baryon asymmetry is not completely determined by properties of the quark matrices and the CP violation in the neutrino sector.

Numerically, with $\epsilon \sim 0.1$ one has $\varepsilon_1 \sim 10^{-7}$, $|M_1| \simeq (\epsilon^6/|\sigma|)(1+|\eta|^2)v_1^2/m_3 \sim 10^9$ GeV and $\widetilde{m}_1 \sim (|\eta|^2+|\rho|^2)/(\sigma(1+|\eta|^2))m_3 \sim 10^{-2}$ eV. The baryon asymmetry is then given by

$$Y_B \sim -\kappa\ \mathrm{sign}(\sigma)\ \sin(\phi_u - \alpha_u) \times 10^{-9}\ . \tag{39}$$

The parameters ε_1, M_1 and $\widetilde{m}_1$ are rather similar to those considered in the previous section. Hence, a solution of the Boltzmann equations can be expected to yield again a baryon asymmetry in accord with observation.

4 Conclusions

Large neutrino mixings, together with the known small quark mixings, have important implications for the structure of GUTs. In SU(5) models this difference between the lepton and quark sectors can be explained by lopsided $U(1)_F$ family symmetries. In these models the heavy Majorana neutrino masses are not constrained by low energy physics, i.e., light neutrino masses and mixings. Successful leptogenesis then depends on the choice of the heavy neutrino masses and is not a generic prediction of the theory.

In SO(10) models the implications of large neutrino mixings are much more stringent because of the connection between Dirac neutrino and up-quark mass matrices. It is remarkable that the requirement of large neutrino mixings determines the relative magnitude of the heavy Majorana neutrino masses in terms of the known quark mass hierarchy. This leads to predictions for neutrino mixings and masses, CP violation in neutrino oscillations and neutrinoless double β-decay. The predicted order of magnitude for the baryon asymmetry is in accord with observation. It would be very interesting to relate directly the CP violation in the quark sector and in neutrino oscillations to the baryon asymmetry. This, however, will require a deeper unterstanding of the quark and lepton mass matrices.

[3] For the discussion of the general case, see ref.[27].

Acknowledgements

I would like to thank Michael Plümacher, Daniel Wyler and Tsutomu Yanagida for an enjoyable collaboration on the topic of this lecture, and I am grateful to the organisers of the 8th Adriatic Meeting for the warm hospitality in Dubrovnik.

References

1. SNO Collaboration, Q. R. Ahmad et al., nucl-ex/0106015.
2. Super-Kamiokande Collaboration, S. Fukuda et al., Phys. Rev. Lett. **86**, 5651 (2001).
3. Super-Kamiokande Collaboration, S. Fukuda et al., Phys. Rev. Lett. **81**, 1562 (1998).
4. K2K Collaboration, S. H. Ahn et al., Phys. Lett. B **511**, 178 (2001).
5. M. C. Gonzalez-Garcia, P. C. de Holanda, C. Peña-Garay, J. W. F. Valle, Nucl. Phys. B **573**, 3 (2000);
   J. N. Bahcall, M. C. Gonzalez-Garcia, C. Peña-Garay, JHEP, **0108**, 2001, 014;
   G. L. Fogli, E. Lisi, D. Montanino, A. Palazzo, Phys. Rev. D **64**, 093007 (2001).
6. M. Fukugita, M. Tanimoto, Phys. Lett. B **515**, 30 (2001).
7. For a review and references, see
   G. Altarelli, F. Feruglio, Phys. Rep. C **320**, 295 (1999);
   H. Fritzsch, Z. Xing, Prog. Part. Nucl. Phys. **45**,1 (2000).
8. T. Yanagida, in *Workshop on Unified Theories*, KEK report 79-18 (1979) p. 95;
   M. Gell-Mann, P. Ramond, R. Slansky, in *Supergravity* (North Holland, Amsterdam, 1979) eds. P. van Nieuwenhuizen, D. Freedman, p. 315
9. H. Georgi, S. L. Glashow, Phys. Rev. Lett. **32**, 438 (1974).
10. C. D. Froggatt, H. B. Nielsen, Nucl. Phys. B **147**, 277 (1979).
11. J. Sato, T. Yanagida, Phys. Lett. B **430**, 127 (1998).
12. N. Irges, S. Lavignac, P. Ramond, Phys. Rev. D **58**, 035003 (1998).
13. J. Bijnens, C. Wetterich, Nucl. Phys. B **292**, 443 (1987).
14. W. Buchmüller, T. Yanagida, Phys. Lett. B **445**, 399 (1999).
15. H. Georgi, C. Jarlskog, Phys. Lett. B **86**, 297 (1979).
16. F. Vissani, JHEP **9811** (1998) 25.
17. L. Hall, H. Murayama, N. Weiner, Phys. Rev. Lett. **84**, 2572 (2000).
18. J. Sato, T. Yanagida, Phys. Lett. B **493**, 356 (2000).
19. M. Fukugita, T. Yanagida, Phys. Lett. B **174**, 45 (1986).
20. W. Buchmüller, M. Plümacher, Phys. Lett. B **389**, 73 (1996).
21. For a review and references, see
    W. Buchmüller, M. Plümacher, Int. J. Mod. Phys. **A 15**, 5086 (2000); hep-ph/0007176.
22. M. A. Luty, Phys. Rev. D **45**, 455 (1992).
23. M. Plümacher, Z. Phys. **C 74** (1997) 549; Nucl. Phys. B **530**, 207 (1998).
24. H. Georgi, in *Particles and Fields*, ed. C. E. Carlson (AIP, NY, 1975) p. 575;
    H. Fritzsch, P. Minkowski, Ann. of Phys. **93**, 193 (1975).
25. A. Yu. Smirnov, Phys. Rev. D **48**, 3264 (1993);
    G. Altarelli, F. Feruglio, I. Masina, Phys. Lett. B **472**, 382 (2000);
    D. Falcone, Phys. Lett. B **479**, 1 (2000);
    C. Albright and S. Barr, hep-ph/0104294.
26. T. Asaka, W. Buchmüller, L. Covi, hep-ph/0108021;
    L. Hall, Y. Nomura, T. Okui, D. Smith, hep-ph/0108071.

27. W. Buchmüller, D. Wyler, Phys. Lett. B **521**, 291 (2001).
28. H. Fritzsch, Z. Xing, ref.[7]
29. R. Rosenfeld, J. L. Rosner, Phys. Lett. B **516**, 408 (2001).
30. G. Branco, D. Emmanuel-Costa, R. Gonzalez Felipe, Phys. Lett. B **483**, 87 (2000).
31. R. G. Roberts, A. Romanino, G. G. Ross, L. Velasco-Sevilla, Nucl. Phys. B **615**, 358 (2001).
32. C. Jarlskog, Phys. Rev. Lett. **55**, 1039 (1985).
33. A. Blondel et al., Nucl. Instr. and Meth. **A 451** (2000) 102;
C. Albright et. al., hep-ex/0008064.
34. Heidelberg-Moscow Collaboration, L. Baudis et al., Phys. Rev. Lett. **83**, 41 (1999).
35. A. S. Joshipura, E. A. Paschos, W. Rodejohann, JHEP **0108**, 029 (2001).
36. G. C. Branco, T. Morozumi, B. M. Nobre, M. N. Rebelo, hep-ph/0107164.

Neutrinos in the New Century

Sandip Pakvasa

Department of Physics and Astronomy, University of Hawaii, Honolulu, HI 96822, USA

Abstract. I review the current status of neutrino physics and summarize future prospects.

1 Introduction

With the explosion of new data, new experiments coming on line, new facilities being planned and dreamed about, it would seem that we are entering the era of the neutrino century. Here I will review recent progress in neutrino physics and look ahead to what we may expect with new neutrino detectors, proposals and plans for neutrino factories, neutrino telescopes, and the ties between neutrinos and cosmology.

2 Neutrino Counting

Since 1989, from SLC and LEP data on the invisible width of Z, we have known that the number of light, active neutrinos is 3 to an increasing accuracy (2.994 $\pm$ 0.012)[1]. These results put no restriction on the number of sterile $I_w = 0$ neutrinos, however.

In the standard hot big bang nucleosynthesis[2], the neutron number gets frozen at a temperature T^* when the inverse reactions $e^- p \to n\nu_e$ and $\bar{\nu}_e p \to e^+ n$ get suppressed and the n/p ratio is fixed at exp $(-(m_n - m_p))/T^*$. The temperature T* is governed by the reaction rate Γ_W being equal to the expansion rate Γ, yielding $T^* \sim 0.66$ MeV. The fractional number of helium-4 nuclei is

$$x = N(^4He)/(p-n) = \frac{n/2}{p-n} = 1/12 \tag{1}$$

and the mass fraction is

$$Y = \frac{M(^4He)}{M(^1H_1) + M(^4He)} = \frac{4x}{1+4x} = 1/4 \tag{2}$$

A more detailed treatment leads to Y=0.24 + 0.04. If the number of "neutrinos" is increased and thus the number of degrees of freedom, then expansion rate increases and thus T* also rises. A higher T^* leads to a higher n and higher Y; roughly $\delta Y \sim 0.02 \delta N_\nu$.

From the observed helium-4 abundance, as well as those of other light elements such as ^2D and ^3Li, a bound on the number of light neutrinos of about 5 can be derived[3]. This bound applies to sterile neutrinos which mix with active neutrinos as well as to other light degrees of freedom.

3 Neutrino Mass

From the study of the end point of Tritium beta decay, the limit on mass of ν_e is[1]

$$m_{\bar{\nu_e}} < 3\ eV \tag{3}$$

There are proposals for new experiments[4] which may make it possible to lower this to about 0.3 eV. The direct bounds on ν_μ and ν_τ are much weaker: 190 KeV for ν_μ and 18 MeV for ν_τ. However, if ν_e, ν_μ and ν_τ all mix with $\delta m^{2'} s$ suggested by the atmospheric and solar neutrino data then all of them will satisfy the above bound[5].

For Majorana neutrinos, there are strong bounds from the lack of neutrinoless double beta decay:

$$Z \to (Z+2) + e^- + e^- \tag{4}$$

The current bound from ^{76}Ge is[6]

$$\langle m_{\nu_e}^M \rangle = \Sigma_i U_{ei}^2 m_i \leq 0.6\ eV \tag{5}$$

with a likely uncertainty, due to the nuclear matrix element, of about of factor of 2 to 3. Proposals for future large detectors can push these limits to 0.02 eV[7]. This gets to an interesting range since the Super-K atmospheric data suggest that at least one neutrino has a mass greater than 0.07 eV.

4 Atmospheric Neutrinos

The strongest evidence for neutrino oscillations comes from atmospheric neutrino data. The L and E dependence of the $\nu_\mu' s$ is given by

$$P_{\mu\mu} = 1 - A_{\mu\mu} sin^2(\delta m^2 L/4E) \tag{6}$$

where $A_{\mu\mu} = sin^2 2\theta$ for the two flavor case. From the Super-K data[8], we have $A_{\mu\mu} \approx 1$ and $\delta m^2 \approx 5.10^{-3} eV^2$. From the study of matter effects and neutral current events the favored scenario is $\nu_\tau - \nu_\tau$ oscillations[9].

Explanations which do not involve conventional oscillations have been attempted. Amongst the various possibilities, the following fail to account [10] for all the data: (i) massless neutrinos with FCNC; (ii) massless neutrinos with violation of equivalence principle or Lorentz invariance; (iii) decay with mixing and very large δm^2. In all these three cases, the reason that the inclusion of high energy upcoming muon events makes the fits poorer is very simple. The upcoming muons come from much higher energy $\nu_\mu' s$ and although there is some suppression, it is less than what is observed for lower energy events at the same L (zenith angle). This is in accordance with expectations from conventional oscillations. The energy dependence in the above three scenarios is different and fails to account for the data. In the FCNC case there is no energy dependence and so the high energy $\nu_\mu' s$ should have been equally depleted, in the FV Gravity (or Lorentz invariance violation) at high energies the oscillations should average out to give uniform 50% suppression and in this decay scenario due to time dilation the decay is suppressed and there is hardly any depletion of $\nu_\mu' s$.

• Decay

Consider the following decay possibility which does work [11]. The three states ν_μ, ν_τ, ν_s (where ν_s is a sterile neutrino) are related to the mass eigenstates ν_2, ν_3, ν_4 by the approximate mixing matrix

$$\begin{pmatrix} \nu_\mu \\ \nu_\tau \\ \nu_s \end{pmatrix} = \begin{pmatrix} \cos\theta & \sin\theta & 0 \\ -\sin\theta & \cos\theta & 0 \\ 0 & 0 & 1 \end{pmatrix} \begin{pmatrix} \nu_2 \\ \nu_3 \\ \nu_4 \end{pmatrix} \tag{7}$$

and the decay is $\nu_2 \to \bar{\nu}_4 + J$. The electron neutrino, which we identify with ν_1, cannot mix very much with the other three because of the more stringent bounds on its couplings, and thus our preferred solution for solar neutrinos would be small angle matter oscillations.

In this case the δm^2_{23} in (6) is not related to the δm^2_{24} in the decay, and can be very small, say $< 10^{-4}\,\text{eV}^2$ (to ensure that oscillations play no role in the atmospheric neutrinos). Then $P(\nu_\mu \to \nu_\mu)$ is given:

$$P_{\mu\mu} = (sin^2\theta + \cos^2\theta exp(-\alpha L/2E))^2 \tag{8}$$

The decay model of (8) above gives a very good fit to the Super-K data[9] with a minimum $\chi^2 = 33.7$ (32 d.o.f.) for the choice of parameters

$$\tau_\nu/m_\nu = 63\ \text{km/GeV},\ \cos^2\theta = 0.30 \tag{9}$$

and normalization $\beta = 1.17$.

The best fits of the two models (viz. $\nu_\mu - \nu_\tau$ oscillations and decay) are of comparable quality. The reason for the similarity of the results obtained in the two models can be understood by comparing the survival probability $P(\nu_\mu \to \nu_\mu)$ of muon neutrinos as a function of L/E_ν for the two models using the best fit parameters.

In the case of the neutrino decay model the probability $P(\nu_\mu \to \nu_\mu)$ monotonically decreases from unity to an asymptotic value $\sin^4\theta \simeq 0.49$. In the case of oscillations the probability has a sinusoidal behaviour in L/E_ν. The two functional forms seem very different; however, taking into account the resolution in L/E_ν, the two forms are hardly distinguishable. In fact, in the large L/E_ν region, the oscillations are averaged out and the survival probability there can be well approximated with 0.5 (for maximal mixing). In the region of small L/E_ν both probabilities approach unity. In the region L/E_ν around 400 km/GeV, where the probability for the neutrino oscillation model has the first minimum, the two curves are most easily distinguishable, at least in principle. This oscillation dip, which is the real smoking gun for oscillations can be seen in Long Baseline experiments such as MINOS[12] and MONOLITH[13].

• Decoherence

There are several different possibilities that can give rise to decoherence of the neutrino beam. An obvious one is violation of quantum mechanics, others are unknown (flavor specific) new interactions with environment etc. Quantum gravity effects are also expected to lead to effective decoherence.

The density matrix describing the neutrinos no longer satisfies the usual equation of motion but rather is modified to

$$\dot{\rho} = -i[H, \rho] + D(\rho) \tag{10}$$

Imposing reasonable conditions on $D(\rho)$ it was shown by Lisi et al.[14] that the ν_μ survival probability $P_{\mu\mu}$ has the form:

$$P_{\mu\mu} = cos^2 2\theta + sin^2 2\theta\ e^{-\gamma L} cos \left(\frac{\delta m^2 L}{2E} \right). \tag{11}$$

where γ is the decoherence parameter. If δm^2 is very small ($\delta m^2 L/2E \ll 1$), this reduces to

$$P_{\mu\mu} = cos^2 2\theta\ + sin^2\ 2\theta\ e^{-\gamma L} \tag{12}$$

If $\gamma = \alpha/E$ with α constant, then an excellent fit to the Super-K data can be obtained with $\theta = \pi/4$ and $\alpha \sim 7.10^{-3}$ GeV/Km. (If γ is a constant, no fit is possible and γ can be bounded by 10^{-22} GeV). The fits to Super-K data are shown in [14]. and they are as good as the decay or $\nu_\mu - \nu_\tau$ oscillations. The shape of $P_{\mu\mu}$ as a function of L/E is very similar to the decay case.

• Large Extra Dimensions

Recently the possibility that SM singlets propagate in extra dimensions with relatively large radii has received some attention. In addition to the graviton, right handed neutrino is an obvious candidate to propagate in some extra dimensions. The smallness of neutrino mass (for a Dirac neutrino) can be linked to this property of the right handed singlet neutrino. The implications for neutrino masses and oscillations in various scenarios have been discussed extensively. I focus on one particularly interesting possibility for atmospheric neutrinos raised by Barbieri et al [15]. The survival probability $P_{\mu\mu}$ is given by

$$P_{\mu\mu}(L) = \mid \Sigma_{i=1}^{3} V_{\alpha i} V_{\alpha i}^{*} A_i(L)\ \mid^2 . \tag{13}$$

$$A_i(L) = \Sigma_{n=0}^{\infty} {U_{on}}^{(i)^2}\ exp(i{\lambda_n}^{(i)^2} L/2ER^2) \tag{14}$$

where n runs over the tower of Kaluza-Klein states, ${\lambda^{(i)}}_n/R^2$ are the eigenvalues of the mass-squared matrix and $U_{on}^{(i)} (\approx 1/\pi^2\xi^2)$ are the matrix elements of the diagonalizing unitary matrix.

An excellent fit to the atmospheric neutrino data can be obtained with the following choice of parameters:

$$\xi_3 = m_3 R \sim 3, 1/R \sim 10^{-3} eV, V_{\mu 3}^2 \approx 0.4. \tag{15}$$

The fit to Super-K data is shown in [15] and obviously it is as good as oscillations. This case corresponds to ν_μ oscillating into ν_τ and a large number(about 25) of Kaluza-Klein states. Because of the mixing with a large number of closely spaced states, the dip in oscillations gets washed out and $P_{\mu\mu}$ looks very much like the decay model.

5 Solar Neutrinos

The results of the four solar neutrino experiments[16] are summarized below;

Detector	R
Homestake	0.33 ± 0.03
GNO	0.51 ± 0.08
SAGE	0.58 ± 0.07
Super-Kamiokande	0.451 ± 0.016

here R is the ratio of the observed rate compared to the rate expected using the SSM fluxes from BP2000[17]. There is energy dependence in the suppression which is driven mainly by the Homestake data. The large data sample of Super K has not shown any smoking gun evidence for oscillations yet: (i) no day night effect; (ii) no spectrum distortion; and (iii) no seasonal dependence. This allows one to rule out large parts of $\delta m^2 - sin^2 2\theta$ parameter space and the allowed regions have become rather small: (i) the most favored region is LMA with $\delta m^2 \sim (0.5-2)10^{-4}\ eV^2$ and $sin^2 2\theta \sim 0.7-0.9$; (ii) SMA with $\delta m^2 \sim (0.5-1)10^{-5}\ eV^2$ and $sin^2 2\theta \sim (0.2-1)10^{-2}$ and (iii) vacuum with $\delta m^2 \sim (0.6-1)10^{-10}\ eV^2$ and $sin^2 2\theta \sim 1$. When the SSM fluxes are relaxed, other solutions, in particular with δm^2 as large as 10^{-3} and large angles, are also allowed[18].

Alternative explanations in terms of FCNC or violation of equivalence principle are not yet ruled out[19]. The possibility of solar neutrinos decaying to account for the deficit is a very old idea[20]. With the current data, decay (even with mixing) is an unlikely scenario[21].

SNO[22] which will measure the reaction $\nu_e D \to e^- pp$ via charged current and $\nu_e D \to \nu np$ via neutral current has stated taking data. Even the cc reaction rate and spectrum measurement will be very useful in restricting the range of solutions. The neutral current will be a clear test of flavor conversion or a lack of it. The first results from the cc reaction were announced by SNO in June [23]. Already they provide very strong constraints. In particular, they establish the following (i) the neutrino fluxes from the Standard Solar Model are more or less current, (ii) the Homestate ^{37}Cl results are current, (iii) the conversion of $\nu_e' s$ into other $\nu' s$ does take place, (iv) the conversion is not dominantly into sterile neutrinos, and (v) confirm that the LMA is the favored solution.

In the near future BOREXINO[24] and KAMLAND[25] should be starting to take data. KAMLAND will be initially observing $\bar{\nu}_e' s$ from nuclear reactors in Japan. The L/E happens to be just right for it to confirm or rule out the LMA solution for solar neutrinos. Borexino (and eventually KAMLAND as well) can detect the ^7Be neutrino line and will be especially sensitive to seasonal variations expected in the vacuum solution. These detectors are also able to measure possible conversion of ν_e to $\bar{\nu_e}$ via transition magnetic dipole moment and large flavor mixing. Furthermore, by detecting $\bar{\nu}_e$ from the interior of the earth, which are supposed to come from decay of U and Th, one can distinguished between different geophysical profiles for U/Th distribution.

Next generation of solar neutrino detectors include LENS[26] which promises to measure pp neutrinos in real time via the low threshold reaction $\nu_e + Yb^{176} \rightarrow e^- + Lu^{176*}$ with a gamma ray tag.

6 LSND

The LSND detector[27] has observed $\bar{\nu}_e$ events in a $\bar{\nu_\mu}$ beam from μ^+ decay $\mu^+ \rightarrow e^+\nu_e\bar{\nu}_\mu$. The event rate corresponds to a fractional rate of 2.10^{-3}. The KARMEN detector[28] does not see the effect and when the LSND data is interpreted in terms of $\nu_\mu - \nu_e$ oscillations, a small amount of parameter space is left corresponding to $\delta m^2 \sim (0.4 - \ 2)\ eV^2$ and $sin^2 2\theta \sim 10^{-2}$ to 10^{-3}. In this case the solar neutrino oscillations have to be due to $\nu_e - \nu_{sterile}$ mixing and at least four neutrinos are needed. The Mini-BOONE experiment[29] under construction at Fermilab will test this possibility.

Now this could be accounted for without oscillations provided that the conventional decay mode $\mu^+ \rightarrow e^+\nu_e\bar{\nu}_\mu$ is accompanied by the rare mode $\mu^+ \rightarrow e^+\bar{\nu}_e X$ at a level of a branching fraction of 2.10^{-3}.

Assuming X to be a single particle, what can X be? It is straightforward to rule out X as being (i) ν_μ (too large a rate for Muonium-Antimuonium transition rate), (ii) ν_e (too large a rate of FCNC decays of Z such as $Z \rightarrow \mu\bar{e} + \mu\bar{e}$) and (iii) ν_τ (too large a rate for FCNC decays of τ such as $\tau \rightarrow \mu ee$). The remaining possibilities for X are $\bar{\nu}_\alpha$ or $\nu_{sterile}$. Models with these features can be constructed[30].

Experimental tests to distinguish this rare decay possibility from the conventional oscillation explanation are easy to state. In the rare decay case, the rate is constant and shows no dependence on L or E. Furthermore, in this case, the LSND results can be correct and Mini-Boone will still not observe any effect since they use $\nu_\mu' s$ from π decay and not μ-decay!

7 Three Flavor Mixing

If we ignore LSND results then three flavors can account for all the neutrino data. Atmospheric neutrino results suggest large $\nu_\mu - \nu_\tau$ mixing and LMA and vacuum solutions for solar neutrinos suggest large $\nu_e - \nu_x$ mixing. A natural candidate for the MNS (Maki-Nakagawa-Sakata)[31] matrix is the class of nearly bi-maximal[32] matrices:

$$U_{BM} = \begin{pmatrix} 1/\sqrt{2} & -1/\sqrt{2} & \epsilon \\ 1/2 & 1/2 & -1/\sqrt{2} \\ 1/2 & 1/2 & 1/\sqrt{2} \end{pmatrix} \tag{16}$$

Many questions remain even if this form is approximately valid:

(i) how large is U_{e3} or ϵ? Is it exactly zero? (ii) is the matrix real or is there a CPV phase? If so, how large is it?

We need to measure all the elements including the phase and all the δm^2 more accurately.

8 Theory Issues

Theory has several tasks: (i) to explain the smallness of neutrino masses, (ii) to predict the Dirac or Majorana nature of the mass, (iii) to account for (predict?) the mass-mixing pattern. Are the masses nearly degenerate or hierarchical?

The traditional explanation for the smallness has been the so-called see-saw mechanism, which suggests

$$m_\nu = m_D^2/M \tag{17}$$

with $m_D \approx$ quark-lepton mass and M a heavy scale, which can be M_{GUT} in Grand Unified theories or $1/R$ where $R =$ the size of an extra dimension. Although large mixing suggests near-degeneracy, most model-building leads to hierarchical spectra.

The bottom line is that although there are many models, none are compelling or have the smell of truth!

9 Neutrino Factories

Muon storage rings could provide intense neutrino beams ($\sim 10^{19} - 10^{21}$ per year) that would yield thousands of charged-current neutrino interactions in a reasonably sized detector (10-50 kt) anywhere on Earth[33]. These neutrino factories would have pure neutrino beams ($\nu_e, \bar{\nu}_\mu$ from stored μ^+ and $\bar{\nu}_e, \nu_\mu$ from stored μ^-) with 50% ν_e or $\bar{\nu}_e$ components. Detection of wrong-sign muons (the muons with opposite sign to the charge current form the beam muon neutrino) would signal $\nu_e \to \nu_\mu$ factory with 2×10^{20} muons a year and a 10kt detector to resolve the issues raised above.

The precision attainable in δm_{32}^2 and $sin^2 2\theta_{23}$ parameters through ν_μ survival measurements based on an $E_\mu = 30\ GeV$ storage ring and a baseline of L=2800 km. This accuracy in measuring $sin^2 2\theta_{23}$ would differentiate the bimaximal model prediction of $\sin^2 2\theta_{23} = 1$ from others.

There would be hundreds of tau-lepton events per year from $\nu_\mu \to \nu_\tau$ oscillations; however, the signal of $\nu_e \to \nu_\tau$ would be difficult with the 2×10^{20} luminosity. The wrong-sign muon event rates are approximately proportional to $sin^2 2\theta_{13}$ and with $sin^2 2\theta_{13} = 0.04$ for a 10 kt detector at L=2900 km with an intensity of 2×10^{20} neutrinos, could be a few hundred.

The observation of $\nu_e \to \nu_\mu$ appearance oscillations at baselines long enough to have significant matter effects will allow a determination of the sign of δm_{32}^2 and thus determine the pattern of the masses.

In optimizing E_μ and L for long-baseline experiments to find the sign of δm^2_{32}, L=732 km is too short (matter effects are small) and L=7332 km is too far (event rates are low). The sensitivity to determine the sign of δm^2_{32} improves linearly with E_μ. There is a tradeoff between energy, detector size and muon beam intensity. The $\nu_e \to \nu_\mu$ sensitivity on $\sin^2 2\theta_{23}$ for a 10kt detector can be as good as 10^{-3}.

It is found that a baseline of 30000 km is ideal for detecting CP violation and if we are fortunate enough to have the CPV phase not too small ($> 20^0$) then it can be measured.

More recently there has been a discussion[34] of upgraded neutrino beams (superbeams) requiring mega-watt proton drivers. With a multi-ton liquid argon detector it may be possible to measure $sin^2 2\theta_{13}$ down to a level of 0.01 or better, although it is unlikely that an unambiguous signal for CP violation can be established. Superbeams with massive detectors would be a useful tool enroute to a neutrino factory, which would provide a further order of magnitude improvement in sensitivity.

10 Cosmology and Neutrino Mass

In the standard hot big bang model[2], the effective neutrino temperature now is

$$T_\nu = (4/11)^{1/3} T_\gamma \approx 1.9^0 K \tag{18}$$

The neutrino number density is

$$n_\nu + n_{\bar{\nu}} = 3/4(4/11) n_\gamma \approx 110/cc \tag{19}$$

per flavor.

Furthermore, if N (≤ 3) families have mass m_ν then the total neutrino energy density is

$$\rho_\nu = N m_\nu (110)\ eV/cc \tag{20}$$

and $\Omega_\nu \approx \left(\frac{NM_\nu}{47}\right) eV/cc$ for a Hubble constant of 70 km/sec/Mpc.

For example, for 3 degenerate $\nu's$ of mass $\sim 0(eV), \Omega_\nu \sim 0.06$. From the Super-K results on atmospheric neutrinos, $\Omega_\nu \geq 0.0015$. Progressively stronger constraints on neutrino masses have been derived starting from the conservative bound $\Omega_\nu < 1$ which yields $Nm_\nu < 47eV$ or $m_\nu < 16eV$ for 3 degenerate neutrinos.

From the recent observations of red shifts of type 1A Supernova and of CMB anisotropy (BOOMERANG and MAXIMA) a consensus about Ω is emerging [35]; namely $\Omega_\Lambda \sim 0.7, \Omega_{DM} \sim 0.25$ and $\Omega_b \sim 0.04$ to 0.06. From these, one may bound $\Omega_\nu < \Omega_{DM}$ leading to total mass in neutrinos bounded by about 4 eV. Furthermore, the Sloan Digital Sky Survey (SDSS) [36] should measure the power spectrum of Large Scale Structure to a 1% accuracy. The presence of massive neutrinos tends to suppress the power at "smaller" scales and a bound can be obtained [37]:

$$m_\nu \cong 0.33 \left(\frac{10\Omega_m}{N}\right)^{0.8} eV. \tag{21}$$

In the near future with the MAP and PLANCK data, neutrino masses as small as a fraction of eV (~ 0.25 eV) can be probed. This compares very favorably with Laboratory experiments which hope to probe masses at a level of 0.3 eV (or less) in Tritium beta-decay and 0.05 eV in neutrino-less double beta decay (for Majorana neutrinos).

11 Neutrinos from AGN's and Neutrino Telescopes

I make two optimistic assumptions. The first one is that distant neutrino sources (e.g. AGN's and GRB's) exist; and furthermore with detectable fluxes at high energies (upto and beyond PeV). The second one is that in the not too far future, very large volume, well instrumented detectors of sizes of order of KM3 and beyond will exist and be operating; and furthermore will have (a) reasonably good energy resolution, (b) good angular resolution ($\sim 1^0$?) and (c) low energy threshold ($\sim 50\ GeV$?).

For AGN's the expectations are that they emit high energy $\nu's$; the total flux probably overtakes atmospheric ν-flux by $E_\nu \sim O(TeV)$ and the most likely flavor mix is $\nu_\mu : \nu_e : \nu_\tau \approx 2 : 1 : 0$.

• ν_τ Signature

For a ν_τ of energy above 2 PeV there is a characteristic "double bang" signature[38]. When ν_τ interacts via charged current there is a hadronic shower (of energy E_1) with about 10^{11} photons emitted; then the τ travels about 90m (for $E_\tau \sim 1.8 PeV$) and when it decays (either to e's or hadrons with 80 % probability) there is again a cascade (of energy E_2) with 2.10^{11} photons emitted in Cerenkov light. The τ track is minimum ionising and may emit $10^6 - 10^7$ photons; even if it is not resolvable, one can connect the two showers by speed of light and reconstruct the event.

The backgrounds (after appropriate cuts) are very small. Hence such "double bang" events represent $\nu_\mu \to \nu_\tau$ oscillations or ν_τ-emission at the source and in any case are extremely interesting. For signal events due to ν_τ, one expects $E_2/E_1 > 2$ on the average, and hence a cut of $E_2/E_1 > 1$ removes many backgrounds; another cut on the distance D between the two bangs of $D > 50m$ eliminates most of the punch-thru backgrounds.

• Sensitivity to Oscillations

The sensitivity to oscillation parameters depends on several factors. If individual AGN's can be identified in $\nu_\tau' s$ (say upto 100 Mpc or more) then δm^2 upto $\geq 10^{-16} eV^2$ and mixing angles upto $sin^2 2\theta \gtrsim 0.05$ can be probed. On the other hand, if the current indications from atmospheric neutrino results are established as due to flavor oscillations, then the oscillating term in the conversion formula:

$$P_{\alpha\beta} = sin^2 2\theta sin^2 \left(\frac{\delta m^2 L}{4E} \right) \tag{22}$$

averages to 1/2 and one can only confirm the expected value of mixing.

To proceed further let us assume: (i) initial fluxes are $\nu_\mu : \nu_e : \nu_\tau \approx 2 : 1 : 0$; (ii) $\# \nu = \# \nu$ (although this is not essential); (iii) all $\delta m^2 \gg 10^{-16} eV^2$, i.e. $\langle sin^2\ (\delta m^2 L/4E)\rangle \approx 1/2$; (iv) matter effects negligible at production (e.g. $N_{e-} = N_{e+}$) and no significant matter effects en-route (this is valid for δm^2 of current interest $\sim 10^{-2} - 10^{-6} eV^2$)[39];(v) atmospheric ν-anomaly caused by $\nu_\mu - \nu_\tau$ oscillations with $\delta m^2 \sim 5.10^{-3} eV^2$ and $\sin^2 2\theta \geq 0.6$. In this case we expect $\nu_\mu : \nu_e : \nu_\tau \approx 1:1:1$ at earth.

If should be stressed that this result viz. $\nu_\mu : \nu_e : \nu_\tau \cong$ 1:1:1 depends crucially on the large $\nu_\mu - \nu_\tau$ mixing and is relatively insensitive to the mixing of ν_e[40]. For example, the wide variety of neutrino mixing matrices currently under consideration: (i) Bi-maximal mixing, (ii)Tri-maximal mixing, (iii) SMA (small angle MSW), (iv) Fritzsch-Xing mixing; all lead to the same result as long as the initial flavor mix is $\nu_\mu : \nu_e : \nu_\tau = 2:1:\epsilon$. Of course, if a source emits a universal flavor mix i.e $\nu_\mu : \nu_e : \nu_\tau = 1:1:1$, it remains unchanged by oscillations.

Extra bonuses from observing the double-bang events are (i) the use of the zenith angle distribution to measure ν_τ cross section via attenuation and (ii) use of enormous light collection and good timing to get good vertex resolution and determine ν_τ direction to within a degree or so. Proposals to account for the highest energy cosmic rays include some[41] in which the neutrino cross-sections are enhanced at very high energies. Because of unitarity constraints, in the 2-20 PeV range they can increase by almost an order of magnitude. Such scenarios can be possibly probed.

12 Cosmology with Neutrinos

We know from supernova studies that there are several effects of neutrino masses and mixings on the observation of neutrino bursts. A pulse spreads in time due to dispersion of velocities (from non-zero mass); a pulse separates into several pulses due to a neutrino of a given flavor being a mixture of different mass eigenstates and the original flavor composition can change due to mixing and oscillations. One can apply these considerations to neutrino pulses from sources which are at cosmological distances. Then the effects come to depend on cosmological parameters.

For example, the time difference [42] between two mass eigen-states which left at the same time is given by

$$\Delta t \approx z/H \left[1 - \frac{(3+q_0)}{2} z....\right] \frac{1}{2} \left[\frac{m_1^2}{E_1^2} - \frac{m_2^2}{E_2^2}\right] \tag{23}$$

where E_i are the energies observed at earth and z, H and q have the usual meanings. The spreading of a pulse of a given mass neutrino is given by [42]

$$\Delta t \approx z/H \left[1 - \frac{(3+q_0)}{2} z.....\right] \frac{1}{2} m^2 \left\{\frac{1}{E_1^2} - \frac{1}{E_2^2}\right\} \tag{24}$$

Finally, the conversion probability for an emitted flavor α to become β at detection is given by

$$\Delta t \approx z/H \left[1 - \frac{(3+q_0)}{2} z....\right] \frac{1}{2} \left[\frac{m_1^2}{E_1^2} - \frac{m_2^2}{E_2^2}\right] \tag{25}$$

where E_i are the energies observed at earth and z, H and q have the usual meanings. The spreading of a pulse of a given mass neutrino is given by [42]

$$\Delta t \approx z/H \left[1 - \frac{(3+q_0)}{2} z.....\right] \frac{1}{2} m^2 \left\{ \frac{1}{E_1^2} - \frac{1}{E_2^2} \right\} \tag{26}$$

Finally, the conversion probability for an emitted flavor α to become β at detection is given by

$$P_{\alpha\beta} = sin^2 2\theta \; sin^2 \phi/2 \tag{27}$$

where the phase ϕ is [43]

$$\phi \cong z/H \left[1 - \frac{(3+q_0)}{2} z....\right] \frac{\delta m^2}{2E} \tag{28}$$

The basic flight time factors are rather small, for eV neutrino masses and GeV energies, $\Delta t \sim 50$ milliseconds at 1000 MPc. These time spreads and separation may be shorter than the times involved in the production process thus making them difficult to observe.

As for flavor conversion, emitted $\nu_\mu' s$ can get converted into $\nu_\tau' s$ and thus produce a significant incoming flux of $\nu_\tau' s$ (which is essentially absent initially in most neutrino production scenarios). With the flavor mix of the incoming beam determined as discussed above, $P_{\alpha\beta}$ and hence the phase ϕ can be deduced by comparing to expected initial relative fluxes. Provided the phase $\phi/2$ is not too large (and $\sin^2 \phi/2$ does not average to 1/2) one has sensitivity to the parameters z, q_0, and H.

With such measurements of Δt and ϕ, one can potentially measure these cosmological parameters. This would be the first time that the red-shift or other cosmological parameters are measured for anything other than light. There is another advantage of using neutrinos. This is the fact that the initial flavor mixing only depends on microphysics and so the comparison is free from problems such as evolution or worries about standard candles etc.

13 Detecting Relic Neutrinos

In the standard hot Big Bang Model[2], the effective temperature today of relic neutrinos is $1.9^0 K$; the number density per flavor is 110/cc (adding $\nu' s$ and $\bar{\nu'} s$); the average momentum is $5.2.10^{-4}$ eV/c. The current density is $\sim 3.10^{12} cm^{-2} s^{-1}$ for massless neutrinos and $5.10^9 cm^{-2} s^{-1}$for a neutrino mass of $5.10^9 cm^{-2} s^{-1}$. The ν_α-scattering cross section (at very low energies) for Dirac neutrinos for allowed transitions goes as

$$\sigma_\alpha \sim \frac{a_\alpha^2 \; G_F^2 \; m_\nu^2}{\pi} \tag{29}$$

where $a_\alpha = (3Z - A)$ for $\alpha = e$ and $a_\alpha = (A - Z)$ for $\alpha = \mu$ or τ. Many early proposals to detect relic neutrinos by reflection or coherent effects turned out to be incorrect. There are three methods which some day may prove to be practical.

The first is a 1975 proposal due to Stodolsky[44]. The idea needs neutrino asymmetry i.e. excess of ν (or $\bar{\nu}$) over $\bar{\nu}$ (or ν) in order to work. Then a polarized electron moving in a background of CMB neutrinos can change its polarization due to the axial vector parity violating interaction. The effective interaction goes as

$$H_{eff} \sim \frac{2G_F}{\sqrt{2}} \underline{\sigma}_e . \underline{v} \; n_\nu \tag{30}$$

With $v \sim 300 \; kms^{-1}$ and $n_\nu \sim 10^7/cc$ this correspond to an energy of about $10^{-33}eV$ and leads to a rotation of the polarization of about 0.02" in a year. Can such small spin rotations can be detected? Certainly not at present, but technology may someday allow this.

The second method is one suggested by Zeldovich and collaborators [45]. The idea is to take advantage of momentum transfer in neutrino-nucleus scattering. Consider an object made up of small spheres of radius $a \approx \lambda$ (neutrino wavelength) packed loosely with pore sizes also of the same size. (to avoid destructive interference). If the number of atoms in the target is N_A, then the effective coherent cross-section is

$$\sigma = \sigma_\alpha \; N_A^2 \tag{31}$$

where σ_α is as given in (10). Assuming total reflection, momentum transfer is

$$\Delta p \cong 2m_\nu \; v_\nu \tag{32}$$

and the force $f = j_\nu \sigma \Delta p$ is given by

$$f = 2n_\nu \sigma_\alpha \; N_A^2 m_\nu \; v_\nu^2 \tag{33}$$

The most optimistic estimates are obtained by assuming some clustering ($n_\nu \sim 10^7/cc$), $m_\nu \sim 0(eV), v_\nu \sim 10^7 cms^{-1}, \rho \sim 20gm/cc$; leading to

$$a = \frac{f}{m} = \frac{f}{N_A m_N} \sim 10^{-23}(a_\alpha/A)^2 \; cm.s^{-2} \tag{34}$$

Such accelerations are at least ten orders of magnitude removed from current sensibility and possible detection remains far in future [46]. Incidentally, this is based on having Dirac neutrinos; for Majorana neutrinos, one would need spin alignment in a macroscopic sample.

The third possibility is the one proposed by Weiler in 1984 [47]. The basic idea is as follows. If neutrinos have masses in the eV range and there are sources of very high energy neutrinos at large distances, then the H.E. ν can annihilate on the C.B.R. $\bar{\nu}$ and make a Z^0 on-shell at resonance creating an absorption dip in the neutrino spectrum. The threshold for Z production would be at $E \sim m_Z^2/2m_\nu$ which is about 4.10^{21} eV for an eV neutrino mass. This seemed like an unlikely possibility, since it required large neutrino fluxes at very high energies to see the neutrino spectrum and then the absorption dip. But all this changed dramatically recently with the clear signal of cosmic rays beyond the GZK cut-off [48]. The GZK cut-off is the energy at which cosmic ray protons pass the threshold for pion production off the CMB photons. This is at an energy $E \sim m_\pi mp/E_\gamma \sim 6.10^{19}$ eV. Above this energy, the mean free path of protons is less than 10 Mpc and hence these protons have to be "local". The flux should then decrease dramatically since we believe the cosmic rays are not produced locally. Recently, what used to be hints of the cosmic ray signal extending beyond this cut-off, has become a clear signal [49]. The events are most likely due to primary protons. Then an explanation is called for. One intriguing proposal [50] is that these events are nothing but a signal for the Z's produced by the $\nu\bar{\nu} \to Z$ process with the protons coming from the subsequent Z decay! Of course, the original problem of needing sources of high energy neutrinos remains. If this explanation is valid, we have already seen (indirect) evidence for the existence of relic neutrinos. In principle, this proposal can be tested: (i) the events should point back at the neutrino sources; (ii) there is an eventual cut-off when the energy reaches the threshold energy for Z production, $E \sim 4.10^{21} \left(\frac{eV}{m_\nu}\right)$ eV; (iii) γ/p ratio should be large near threshold and (iv) the large ν-flux should be eventually seen directly in large ν-telescopes. There is the bonus that the cut-off energy also measures the mass of the relic neutrinos! Thus neutrino telescopes can give existence proof of relic neutrinos as well as measure their mass.

14 SETI with Neutrinos

Yes, I do mean[51] search for extra-terrestrial intelligence. There is a school of thought that holds that this search in futile, pointless, and bound to yield null results. However, as has been pointed out, absence of evidence is not necessarily evidence of absence.

Imagine that an advanced civilization exists in the galaxy, with many outposts. It will need to maintain time standards over a long base line. In turn, this will require (i) stable clocks of high precision, (ii) fast processes for transmitting and receiving time markers and (iii) a form of radiation which will faithfully carry timing data over long distances.

The need for clock synchronization stems from the fact that standard clocks have to exchange timing data to remain synchronized. This is in order to correct for general relativistic effects which depend on positions and motions of nearby massive objects. Furthermore, the presence of chaos in many body systems means that such corrections cannot be calculated indefinitely from initial data alone, so that the synchronization has to be done repeatedly.

The requirements of a mobile, spread out civilization would suggest the use of isotropic synchronization signals. Other arguments suggest the same thing. Hence, even though it raises the required energy budget, this is the most likely scenario.

The fastest known process is the Z^0 decay with a lifetime of $2.5 \times 10^{-25}s$. It also produces neutrinos of 45.6 GeV, satisfying the requirements of radiation which can carry information intact thru many obstacles.

If an advanced civilization is using this process to send timing signals, a neutrino telescope can detect some neutrino events at the energy of 45.6 GeV. If the source is a few kpc from us, then a KM3 water/ice ĉ detector will detect a few events per year (all flavors in equal numbers).

The ETI would have to overcome many technical problems to implement such a scheme. We have addressed some of them elsewhere[51]. The power requirements to give a few events per year in a KM3 detector at a distance of few kpc are huge; approximately the solar luminosity $\sim 10^{45} eV/sec$. This, of course, is *their* problem and we have to imagine that they have solved it. Is it possible that a technology radiating such huge amounts of power within a few kpc has escaped our detection? We speculate that this would correspond to a "Dyson shell." Dyson had suggested that if an advanced civilization surrounds a star with a shell of material and uses heat engines to extract power, then the system would appear as an infra-red source. Since the IRAS data include over 50,000 IR sources, some of these indeed could well be "Dyson shells".

These synchronization neutrino signals at E = $M_Z/2$ are extremely distinctive in that they are not expected to occur naturally and are therefore unlikely to be mistaken for anything else. In view of the spectacular nature of the timing signal and the enormous implications of its detection, we believe it is surely worth keeping watch for it.

15 Conclusion

It is obvious that this new century will bring many new and exciting rewards in neutrino physics, astrophysics and cosmology. We have to be prepared to receive them by building the right detectors as well as utilizing a wide variety of neutrino beams both artificial as well as natural.

Acknowledgements

I thank the organizers (especially Josip Trampetic) for the excellent hospitality and the stimulating atmosphere of the workshop. This work was supported in part by the U.S. Department of Energy.

References

1. Particle Data Group, D.E. Groom et al;, Eur. Phys. J. C **15**, 1 (2000).
2. E. Kolb and M. Turner, *The Early Universe*, Addison-Wesley, N.Y. (1990).
3. S. Burles, K.M. Nollett and M. Turner; astro-ph/0008495.
4. The KATRIN Collaboration, V. Aseev et al., see http://www.ikl.fzk.de/tritium/.
5. V. Barger, T. J. Weiler and K. Whisnant, Phys. Lett. B **442**, 255 (1998).
6. L. Baudis, et al, Phys. Rev. Lett. **83**, (41) (1999).
7. H. V. Klapdor-Kleingrothaus, J. Helling and M. Hinsch, J. Phys. G **24**, 483 (1998).
8. Y. Fukuda, et al., Phys. Rev. Lett. **81**, 1562 (1998).
9. S. Fukuda, et al., Phys. Rev. Lett. **85**, 3999 (2000).

10. P. Lipari and M. Lusignoli, Phys. Rev. D **60**, 013003 (1999). G. Fogli, E. Lisi and A. Marrone, Phys. Rev. **D59**, 117303 (1999); S. Choubey and S. Goswami, Astropart. Phys. **14**, 67 (2000).
11. V. Barger, et al., Phys. Lett. B **462**, 109 (1999).
12. MINOS Collaboration, NAMI-L-375 (1988).
13. MONOLITH, M. Aglietta et al., CERN/SPSC 98-20.
14. E. Lisi, A. Marrone and D. Montanino, Phys. Rev. Lett. **85**, 1166 (2000).
15. R. Barbieri, P. Creminella, and A. Strumia, Nucl. Phys. B **575**, 61 (2000).
16. B.T. Cleveland et al. Nucl. Phys. B (Proc. Supp), **38**, 47 (1995); M. Altmann et al; hep-ex10006034; J. Abdurashitoev et al., Phys. Rev. Lett. **77**, 4700 (1998); Y. Fukuda et al., Phys. Rev. Lett. **82** 2430 (1999).
17. J.N. Bahcall, M.H. Pinsonneault and S. Basu, astro-ph/0010346.
18. G.L. Fogli, E. Lisi and A. Palazzv, hep-ph/0105080; S. Choubey, et al, hep-ph/0103318.
19. A.M. Gago, H. Nunokawa and R. Zukanovich-Funchal, Phys. Rev. Lett. **84**, 4035 (2000). D. Majumdar; A. Raychaudhuri and A. Sil, Phys. Rev. D **63**, 0730014 (2001).
20. S. Pakvasa and K. Tennakone, Phys. Rev. Lett. **28**, 1415 (1972); J. N. Bahcall, N. Cabibbo and A. Yahil, Phys. Rev. Lett. **28**, 316 (1972).
21. A. Acker and S. Pakvasa, Phys. Lett. B **320**, 320 (1994); S. Choubey, S. Goswami and D. Majumdar, Phys. Lett. B **484**, 73 (2000).
22. http://www.sno.phys.queensu.ca/.
23. Q.R. Ahmad et al; Phys. Rev. Lett.**87**, 071301 (2001).
24. http://Almine.mi.infn.it/.
25. http://www.awa.tohoku.ac.jp/html/KamLAND/index.html.
26. R.S. Raghavan, Phys. Rev. Lett. **78**, 361 (1997); R. S. Raghavan et al., ibid **85**, 4442 (2000).
27. Final Neutrino Oscillation Results from LSND, W.C. Louis for the (LSND) Collaboration, Particle Physics and Cosmology, San Juan, May 2000, p. 93.
28. The KARMEN Collaboration, J. Kleinfeller et al., Nucl. Phys. Proc. Suppl., **87**, 281 (2000).
29. http://www.neutrino.lanl.gov/Boone/.
30. K.S. Babu and S. Pakvasa (in preparation).
31. Z. Maki, M. Nakagawa and S. Sakata, Progr. Theoret. Phys., **28**, 870 (1962).
32. V. Barger et al., Phys. Lett. B **437**, 107 (1998); A. Baltz, A. Goldhaber and M. Goldhaber, Phys. Rev. Lett. **81**, 5730 (1998); D. Ahluwalia, Mod. Phys. Lett., A **13**, 2249 (1998).
33. C. Albright et al., hep-ex/0008064; A. Cervera et al., Nucl. Phys. B **579**, 17 (2000); A. Bueno *et al*; hep-ph/9905240.
34. B. Richter, hep-ph/0008222; V. Barger et al., hep-ph/0012017.
35. See for example C.H. Lineweaver, astro-ph/0011448.
36. http://www.sdss.nasa.gov/.
37. W. Hu, D.J. Eisenstein and M. Tegmark, Phys. Rev. Lett. **280**, 5255 (1998).
38. J.G. Learned and S. Pakvasa, Astropart. Phys. **3**, 267 (1995).
39. See for example, C. Lunardini and A. Smirnov, hep-ph/0012056.
40. This was first observed in J.G. Learned and S. Pakvasa, Ref. 38; S. Pakvasa, *Beyond the Desert 1999, Proc. of the Second International Conference on Particle Physics Beyond the Standard Model*, Castle Ringberg, Germany, 6-12 June 1999, Ed. H.V. Klapdor-Kleingrothaus and I.V. Krivosheina, IOP Publishing, Bristol (2000) p. 247; hep-ph/9910246. It has since been re-discovered by a number of

authors: L. Bento, P. Keranen and J. Maalampi, Phys. Lett. B **476**, 205 (2000); H. Athar, M. Jezabek, and O. Yasuda, Phys. Rev. D **62**, 103007 (2000); D. Ahluwalia, hep-ph/00104316.

41. G. Domokos and S. Kovesi-Domokos, Phys. Rev. Lett. **82**, 1366 (1999); J. Bordes et al; Astropart. Phys. **8**, 135 (1998); P. Jain et al; Phys. Lett. B **484**, 267 (2000).
42. L. Stodolsky, Phys. Lett. B **473**, 61 (2000).
43. T.J. Weiler, W. Simmons, J.G. Learned, and S. Pakvasa, hep-ph/9411432.
44. L. Stodolsky, Phys. Rev. Lett. **34**, 110 (1975); B.A. Campbell and P.J. O'Donnell, Phys. Rev. D **26**, 1487 (1982) generalized the result to include neutral currents and massive neutrinos.
45. B.F. Shvartsman et al, JETP Lett. **36**, 277 (1982).
46. Several ideas have been proposed to render this effect detectable: P. F. Smith and J.D. Lewin, Phys. Rep. **187**, 203 (1990); C. Hagmann, hep-ph/9905258.
47. T.J. Weiler, Ap.J. **285**, 495 (1984).
48. K. Greisen, Phys. Rev. Lett. **16**, 748 (1966); G. Zatsepin and V. Kuzmin, JETP Lett. **4**, 78 (1966).
49. M. Takeda et al, Phys. Rev. Lett. **81**, 1163 (1998) 1163; M. Are et al, Phys. Rev. Lett. **85**, 224 (2000).
50. T.J. Weiler Astropart. Phys. **11**, 303 (1999); D. Fargion, S. Mele and J. Salis, Astrophys. J. **517**, 725 (1999).
51. J.G. Learned, S. Pakvasa, W.A. Simmons and X. Tata, Q.J.R. Astro. Soc. **35**, 321 (1994).

The Intriguing Distribution of Dark Matter in Galaxies

Paolo Salucci and Annamaria Borriello

International School for Advanced Studies SISSA-ISAS – Trieste, Italy

Abstract. We review the most recent evidence for the amazing properties of the density distribution of the dark matter around spiral galaxies. Their rotation curves, coadded according to the galaxy luminosity, conform to an Universal profile which can be represented as the sum of an exponential thin disk term plus a spherical halo term with a flat density core. From dwarfs to giants, these halos feature a constant density region of size r_0 and core density ρ_0 related by $\rho_0 = 4.5 \times 10^{-2}(r_0/\mathrm{kpc})^{-2/3}\mathrm{M}_{\odot}\mathrm{pc}^{-3}$. At the highest masses ρ_0 decreases exponentially with r_0, revealing a lack of objects with disk masses $> 10^{11} M_{\odot}$ and central densities $> 1.5 \times 10^{-2}$ $(r_0/\mathrm{kpc})^{-3} M_{\odot}\mathrm{pc}^{-3}$ implying a *maximum* mass of $\approx 2 \times 10^{12} M_{\odot}$ for a dark halo hosting a stellar disk. The fine structure of dark matter halos is obtained from the kinematics of a number of suitable low–luminosity disk galaxies. The halo circular velocity increases linearly with radius out to the edge of the stellar disk, implying a constant dark halo density over the entire disk region. The properties of halos around normal spirals provide substantial evidence of a discrepancy between the mass distributions predicted in the Cold Dark Matter scenario and those actually detected around galaxies.

1 Introduction

Rotation curves (RC's) of disk galaxies are the best probe for dark matter (DM) on galactic scale. Notwithstanding the impressive amount of knowledge gathered in the past 20 years, only very recently we start to shed light to crucial aspects of the mass *distribution* of dark halos, including their radial density profile, and its claimed universality. On a cosmological side, high–resolution N–body simulations have shown that cold dark matter (CDM) halos achieve a specific equilibrium density profile [13 hereafter NFW, 5, 8, 12, 9] characterized by one free parameter, e.g. the halo mass. In the inner region the DM halos density profiles show some scatter around an average profile which is characterized by a power–law cusp $\rho \sim r^{-\gamma}$, with $\gamma = 1 - 1.5$ [13, 12, 2]. In detail, the DM density profile is:

$$\rho_{\mathrm{NFW}}(r) = \frac{\rho_s}{(r/r_s)(1+r/r_s)^2} \tag{1}$$

where r_s and ρ_s are respectively the characteristic inner radius and density. Let us define r_{vir} as the radius within which the mean density is Δ_{vir} times the mean universal density ρ_m at the halo formation redshift, and the associated virial mass M_{vir} and velocity $V_{\mathrm{vir}} \equiv GM_{\mathrm{vir}}/r_{\mathrm{vir}}$. Hereafter we assume the ΛCDM scenario, with $\Omega_m = 0.3$, $\Omega_\Lambda = 0.7$ and $h = 0.75$, so that $\Delta_{\mathrm{vir}} \simeq 340$ at $z \simeq 0$. By assuming the concentration parameter as $c \equiv r_{\rm vir}/r_s$ the halo circular velocity $V_{\rm NFW}(r)$ takes the form [2]:

$$V^2_{\rm NFW}(r) = V^2_{vir} \frac{c}{A(c)} \frac{A(x)}{x} \tag{2}$$

where $x \equiv r/r_s$ and $A(x) \equiv \ln(1+{\rm x}) - {\rm x}/(1+{\rm x})$. As the relation between V_{vir} and r_{vir} is fully specified by the background cosmology, the independent parameters characterizing the model reduce from three to two (c and r_s). Let us stress that a high density $\Omega_m = 1$ model, with a concentration parameter $c > 12$, is definitely unable to account for the observed galaxy kinematics [11].

So far, due to the limited number of suitable RC's and to the serious uncertainties in deriving the actual amount of luminous matter inside the inner regions of spirals, it has been difficult to investigate the internal structure of dark halos. These difficulties have been overcome by means of:

i) a specific investigation of the Universal Rotation Curve [16], built by coadding 1000 RC's, in which we adopt a general halo mass distribution:

$$V^2_{h,URC}(x) = V^2_{opt}\ (1-\beta)\ (1+a^2)\ \frac{x^2}{(x^2+a^2)} \tag{3}$$

with $x{\equiv}r/r_{opt}$, a the halo core radius in units of r_{opt} and $\beta{\equiv}(V_{d,URC}(r_{opt})/V_{opt})^2$. It is important to remark that, out to r_{opt}, this mass model is *neutral* with respect to the halo profile. Indeed, by varying β and a, we can efficiently reproduce the maximum–disk, the solid–body, the no–halo, the all–halo, the CDM and the core-less–halo models. For instance, CDM halos with concentration parameter $c = 5$ and $r_s = r_{opt}$ are well fit by (3) with $a \simeq 0.33$

ii) a number of suitably selected individual RC's [1], whose mass decomposition has been made adopting the cored Burkert–Borriello–Salucci (BBS) halo profile (see below).

2 Dark Matter Properties from the Universal Rotation Curve

The observational framework is the following: *a*) the mass in spirals is distributed according to the Inner Baryon Dominance (IBD) regime [16]: there is a characteristic transition radius $r_{IBD} \simeq 2r_d(V_{opt}/220\ {\rm km/s})^{1.2}$ (r_d is the disk scale–length and $V_{opt} \equiv V(r_{opt})$) according which, for $r \leq r_{IBD}$, the luminous matter totally accounts for the gravitating mass, whereas, for $r > r_{IBD}$, the dark matter shows dynamically up and *rapidly* becomes the dominant component [20, 18, 1]. Then, although dark halo might extend down to the galaxy centers, it is only for $r > r_{IBD}$ that they give a non–negligible contribution to the circular velocity. *b*) DM is distributed in a different way with respect to any of the various baryonic components [16, 6], and *c*) HI contribution to the circular velocity at $r < r_{opt}$, is negligible [e.g. 17].

2.1 Mass Modeling

Persic, Salucci and Stel [16] have derived from $\sim$ 15000 velocity measurements of $\sim$ 1000 RC's the synthetic rotation velocities of spirals $V_{syn}(\frac{r}{r_{opt}}; \frac{L_I}{L_*})$, sorted by luminosity (Fig. 1, with L_I the I–band luminosity $(L_I/L_* = 10^{-(M_I+21.9)/5})$. Remarkably, individual RC's have a negligible variance with respect to their corresponding synthetic curves: spirals sweep a very narrow locus in the RC-profile/amplitude/luminosity space. In addition, kinematical properties of spirals do significantly change with galaxy luminosity [e.g. 16], then it is natural to relate their mass distribution with this quantity. The whole set of synthetic RC's define the Universal Rotation Curve (URC), composed by the sum of two terms: *a*) an exponential thin disk with circular velocity (see [16]):

$$V_{d,URC}^2(x) = 1.28\ \beta\ V_{opt}^2\ x^2\ (I_0K_0 - I_1K_1)|_{1.6x} \tag{4}$$

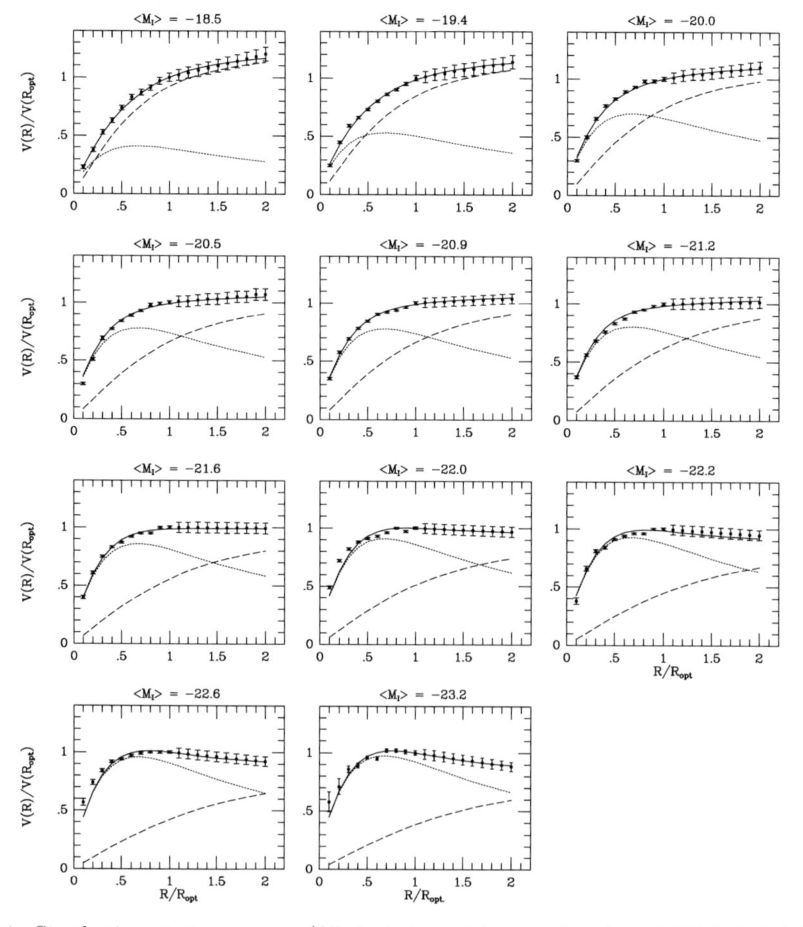

Fig. 1. Synthetic rotation curves (filled circles with error bars) and URC (solid line) with the separate dark/luminous contributions (dotted line: disks; dashed line: halos).

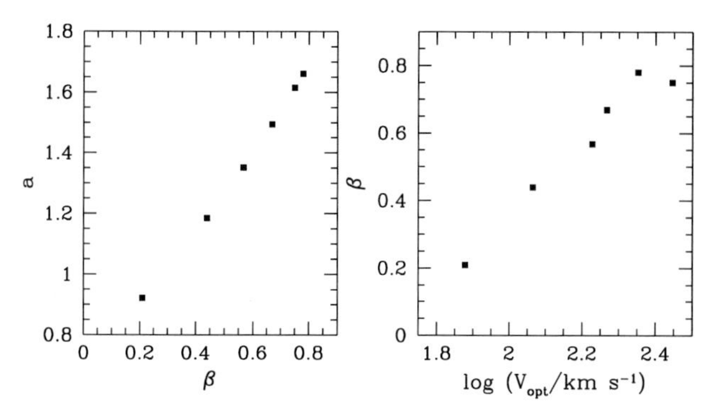

Fig. 2. a *vs.* β and β *vs.* V_{opt}.

and a spherical halo, whose velocity contribution is given by (3). At high luminosities, the contribution from a bulge component has also been considered.

The data (i.e. the synthetic curves V_{syn}) select the actual model out of this family, by setting $V^2_{URC}(x) = V^2_{h,URC}(x, \beta, a) + V^2_{d,URC}(x, \beta)$ with a and β as free parameters. An extremely good fit occurs for $a \simeq 1.5(L_I/L_*)$ [16] or, equivalently, for $a = a(\beta)$ and $\beta = \beta(\log V_{opt})$ as plotted in Fig. 2. With these values the URC reproduces the data $V_{syn}(r)$ up to their *rms* (i.e. within 2%). Moreover, at fixed luminosity the σ fitting uncertainties in a and β are lesser than 20%. The emerging picture is: *i)* smaller objects have more fractional amount of dark matter (inside $r_{\rm opt}$: $M_*/M_{\rm vir} \simeq 0.2\ (M_*/2 \times 10^{11} M_\odot)^{0.75}$ [20]), *ii)* dark mass increses with radius much more that linearly.

2.2 Halo Density Profiles

The above evidence calls for a quite specific DM density profile; we adopt the BBS halo mass distribution [3, 4, 1]:

$$\rho_{\rm BBS}(r) = \frac{\rho_0\ r_0^3}{(r + r_0)(r^2 + r_0^2)} \tag{5}$$

where ρ_0 and r_0 are free parameters which represent the central DM density and the core radius. Of course, for $r_0 \ll r_d$, we recover a cuspy profile. Within spherical symmetry, the mass distribution is given by:

$$M_{\rm BBS}(r) = 4\ M_0\ \{\ln(1 + {\rm r/r_0}) - \arctan({\rm r/r_0}) + 0.5\ln[1 + ({\rm r/r_0})^2]\} \tag{6}$$

$$M_0 \simeq 1.6\ \rho_0\ r_0^3 \tag{7}$$

with M_0 the dark mass within the core. The halo contribution to the circular velocity is then: $V^2_{\rm BBS}(r) = G\ M_{\rm BBS}(r)/r$. Although the dark matter "core" parameters r_0, ρ_0 and M_0 are in principle independent, the observations reveal a quite strong correlation among them [e.g. 19]. Then, dark halos may be an 1–parameter family, completely specified by e.g. their core mass M_0. When we

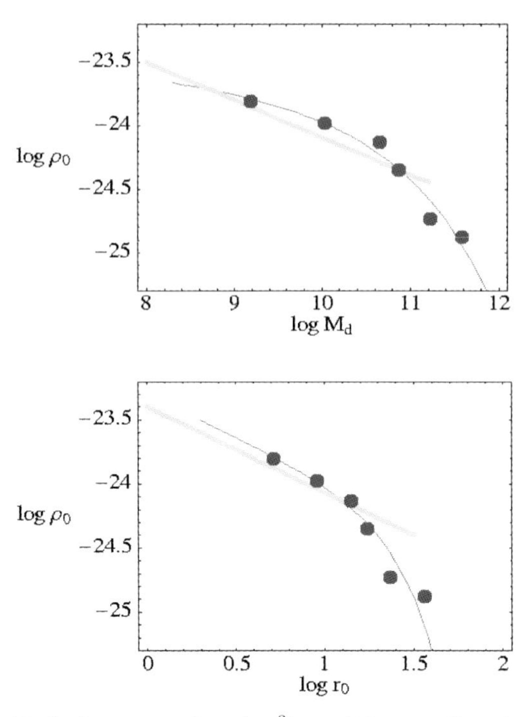

Fig. 3. *up)* Central halo density ρ_0 (in g/cm^3) *vs.* disk mass (in solar units) for normal spirals (*filled circles*); *bottom)* central density *vs.* core radii (in kpc) for normal spirals. The straight lines are from [3], whereas the curved lines are the best fits used in §4.

test the disk+BBS velocities with ρ_0 and r_0 left as free parameters, we find that, at any luminosity and out to $\sim 6\ r_d$, the model is indistinguishable from data (i.e. $V_{syn}(r)$). More specifically, we reproduce the synthetic rotation curves at the level of their *rms*. The values of r_0 and ρ_0 derived in this way agree with the extrapolation at high masses of the scaling law $\rho \propto r_0^{-2/3}$ [3] established for objects with much smaller core radii r_0 and stellar masses (see Fig. 3). Let us notice that the core radii are pretty large ($r_0 \gg r_d$): ever-rising halo RC's cannot be excluded by the data. Moreover, spirals lie on the extrapolation of the disk–mass *vs.* central halo density relationship $\rho_0 \propto M_d^{-1/3}$ found for dwarf galaxies [3], to indicate that the densest halos harbor the least massive disks (see Fig. 3).

The curvature in ρ_0 *vs.* r_0 at the highest masses/lowest densities can be linked to the existence of an *upper mass limit* in $M_{\rm vir}$ which is evident by the sudden decline of the baryonic mass function of disk galaxies at $M_d^{max} = 2\times 10^{11} M_\odot$ [20]. In fact, such a limit implies a maximum halo mass of $M_{\rm vir}^{max} \sim \Omega_0/\Omega_b\ M_d^{max}$. Then, for (6) and (7), $M_{\rm vir} = \eta\ M_0$, with $\eta \simeq 12$ for $(\Omega_0,\ \Omega_b\ ,z) = (0.3, 0.03, 3)$, and the limiting halo mass implies a lack of objects with $\rho_0 > 4 \times 10^{-25}$ g/cm^3 and $r_0 > 30$ kpc, as is evident in Fig. 3. On the other side, the observed deficit of objects with $M_d \sim M_d^{max}$ and $\rho_0 > 4 \times 10^{-25}$ g/cm^3, suggests that, at this mass scale, the total–to–baryonic density ratio nears the cosmological value $\Omega_0/\Omega_b \simeq 10$.

2.3 Testing CDM with the URC

The BBS density profile reproduces in synthetic RC's the DM halo contributions, at least out to two optical radii. This is in contradiction with CDM halo properties according to which the velocity dispersion σ of the dark matter particles decreases towards the center to reach $\sigma \to 0$ for $r \to 0$. Dark halos therefore, are not kinematically cold structures, but "warm" regions of sizes $r_0 \propto \rho_0^{-1.5}$ which, by the way, turn up quite large: $r_0 \sim 4 - 7\ r_d$. Then, the boundary of the core region is well beyond the region of the stellar disk and there is not evidence that dark halos converge to a $\rho \sim r^{-2}$ (or a steeper) regime, as dictated by CDM predictions.

3 Dark Matter Properties from Individual Rotation Curves

Although deriving halo densities from individual RC's is certainly complicated, the belief according to which one always gets ambiguous halo mass modeling [e.g. 22] is incorrect. In fact, this is true only for rotation curves of low spatial resolution, i.e. with less than ~ 3 measures per exponential disk length–scale occurring in most HI RC's. In this case, since the parameters of the galaxy structure are very sensitive to the *shape* of the rotation curve in the region $0 < r < r_d$, there are no sufficient data to constrain models.

In the case of high–quality *optical* RC's tens of independent measurements in the critical region make possible to infer the halo mass distribution. Moreover, since the dark component can be better traced when the disk contributes to the dynamics in a modest way, a convenient strategy leads to investigate DM–dominated objects, like dwarf and low surface brightness (LSB) galaxies. It is well known that for the latter [e.g. 7, 11, 3, 4, 9, 10, 21] the results are far from being definitive in that they are *1)* affected by a quite low spatial resolution and *2)* uncertain, due to the limited amount of available kinematical data [e.g. 23].

Since most of the properties of cosmological halos are claimed universal, an useful strategy is to investigate a number of high–quality *optical* rotation curves of *low luminosity* late–type spirals, with I–band absolute magnitudes $-21.4 < M_I < -20.0$ and $100 < V_{opt} < 170$ km s^{-1}. Objects in this luminosity/velocity range are DM dominated [e.g. 20] but their RC's, measured at an angular resolution of $2''$, have an excellent spatial resolution of $\sim 100(D/10\ \mathrm{Mpc})$ pc and $n_{data} \sim r_{opt}/w$ independent measurements. For nearby galaxies: $w \ll r_d$ and $n_{data} > 25$. Moreover, we select RC's of bulge–less systems, so that the stellar disk is the only baryonic component for $r \lesssim r_d$.

In detail, we extract the best 9 rotation curves, from the 'excellent' subsample of 80 rotation curves of [15], which are all suitable for an accurate mass modeling. In fact, these RC's trace properly the gravitational potential in that: *1)* data extend at least out to the optical radius, *2)* they are smooth and symmetric, *3)* they have small *rms*, *4)* they have high spatial resolution and a homogeneous radial data coverage, i.e. about $30-100$ data points homogeneously distributed with radius and between the two arms. The 9 extracted galaxies are of low luminosity ($5 \times 10^9 L_\odot < L_I < 2 \times 10^{10} L_\odot$; $100 < V_{opt} < 170$ km s^{-1}) and their I–band surface luminosity profiles are (almost) perfect radial exponential. These two last criteria, not indispensable to perform the *mass* decomposition, help inferring the dark halo *density* distribution. Each RC has $7-15$ velocity points inside r_{opt}, each one being the average of $2-6$ independent data. The RC's spatial resolution is better than $1/20\ r_{opt}$, the velocity *rms* is about 3% and the RC's logarithmic derivative is generally known within about 0.05.

3.1 Halo Density Profiles

We model the mass distribution as the sum of two components: a stellar disk and a spherical dark halo. By assuming centrifugal equilibrium under the action of the gravitational potential, the observed circular velocity can be split into the two components: $V^2(r) = V_d^2(r) + V_h^2(r)$. By selection, the objects are bulge–less and stars are distributed like an exponential thin disk. Light traces the mass via an assumed radially constant mass–to–light ratio.

We neglect the gas contribution $V_{gas}(r)$ since in normal spirals it is usually modest within the optical region [17, Fig. 4.13]: $\beta_{gas} \equiv (V_{\rm gas}^2/V^2)_{r_{opt}} \sim 0.1$. Furthermore, high resolution HI observations show that in low luminosity spirals: $V_{gas}(r) \simeq 0$ for $r < r_d$ and $V_{gas}(r) \simeq (20 \pm 5)(r - r_d)/2r_d$ for $r_d \leq r \leq 3r_d$. Thus, in the optical region: *i)* $V_{gas}^2(r) \ll V^2(r)$ and *ii)* $d[V^2(r) - V_{gas}^2(r)]/dr \gtrsim 0$. This last condition implies that by including V_{gas} the halo velocity profiles would result *steeper* and then the core radius in the halo density even *larger*. Incidentally, this is not the case for dwarfs and LSB's: most of their kinematics is affected by the HI disk gravitational pull in such a way that neglecting it could bias the determination of the DM density. The circular velocity profile of the disk is given by (4) and the DM halo will have the form given by (3). Since we normalize (at r_{opt}) the velocity model $(V_d^2 + V_h^2)^{1/2}$ to the observed rotation speed V_{opt}, β enters explicitly in the halo velocity model and this reduces the free parameters of the mass model to two.

For each galaxy, we determine the values of the parameters β and a by means of a χ^2–minimization fit to the observed rotation curves:

$$V_{model}^2(r; \beta, a) = V_d^2(r; \beta) + V_h^2(r; \beta, a) \qquad (8)$$

A central role in discriminating among the different mass decompositions is played by the derivative of the velocity field dV/dr. It has been shown [e.g. 14] that by taking into account the logarithmic gradient of the circular velocity field defined as: $\nabla(r) \equiv \frac{d \log V(r)}{d \log r}$ one can retrieve the crucial information stored in the shape of the rotation curve. Then, we set the χ^2's as the sum of those evaluated on velocities and on logarithmic gradients: $\chi_V^2 = \sum_{i=1}^{n_V} \frac{V_i - V_{model}(r_i; \beta, a)}{\delta V_i}$

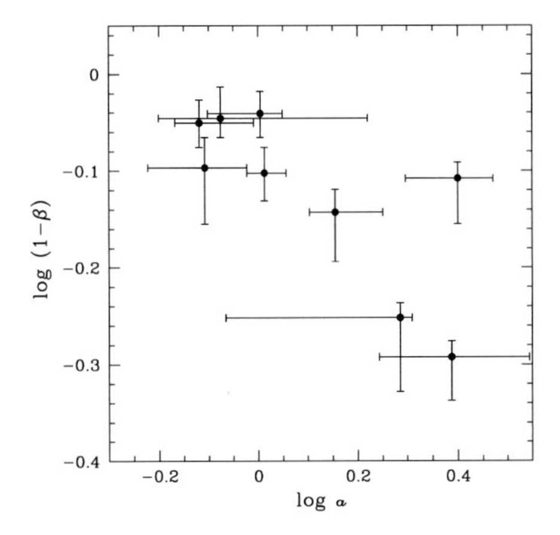

Fig. 4. Halo parameters (a is in units of r_{opt}) for the individual RC's. Notice that for this sample we can derive in model–independent way $a \gtrsim 1$, in disagreement with CDM predictions.

and $\chi^2_\nabla = \sum_{i=1}^{n_\nabla} \frac{\nabla(r_i) - \nabla_{model}(r_i;\beta,a)}{\delta\nabla_i}$, with $\nabla_{model}(r_i, \beta, a)$ given from the above equations. The parameters of the mass models are finally obtained by minimizing the quantity $\chi^2_{tot} \equiv \chi^2_V + \chi^2_\nabla$.

The best–fit models parameters are shown in Fig. 4. The disk–contribution β and the halo core radius a span a range from 0.1 to 0.5 and from 0.8 to 2.5, respectively. They are pretty well constrained in a small and continuous region of the (a, β) space. The derived mass models are shown in Fig. 5, alongside with the separate disk and halo contributions. We also get a "lowest" and a "highest" halo velocity curve (dashed lines in figure) by subtracting from $V(r)$ the maximum and the minimum disk contributions $V_d(r)$ obtained by substituting in (4) the parameter β with $\beta_{best} + \delta\beta$ and $\beta_{best} - \delta\beta$, respectively. It is obvious that the halo curves are steadily increasing, almost linearly, out to the last data point. In each object the uniqueness of the resulting halo velocity model can be realized by the fact that the maximum–disk and minimum–disk models almost coincide. Remarkably, we find that the size of the halo density core is always greater than the disk characteristic scale–length r_d and it can extend beyond the disk edge (and the region investigated).

3.2 Testing CDM

In Fig. 5 we show how the halo velocity profiles of the nine galaxies rise almost linearly with radius, at least out to the disk edge:

$$V_h(r) \propto r \qquad 0.05\, r_{opt} \lesssim r \lesssim r_{opt} \tag{9}$$

The halo density profile has a well defined (core) radius within which the density is approximately constant. This is inconsistent with the singular halo density distribution emerging in the Cold Dark Matter (CDM) halo formation scenario.

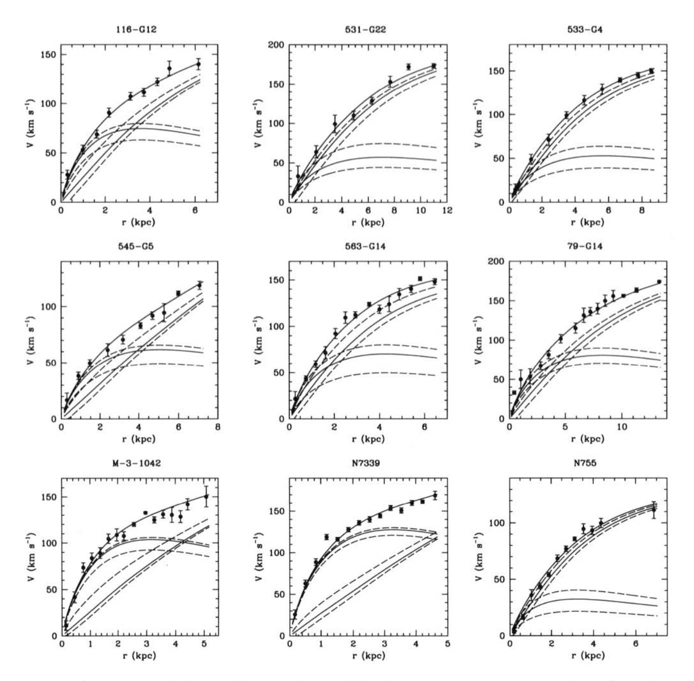

Fig. 5. BBS fits (*thick solid line*) to the RC's (*points with errorbars*). Thin solid lines represent the disk and halo contributions. Notice the steep halo velocity profiles. The maximum and minimum disk solutions (*dashed lines*) provide the theoretical uncertainties.

More precisely, since the CDM halos are, at small radii, likely more cuspy than the NFW profile: $\rho_{CDM} \propto r^{-1.5}$ [e.g. 14], the steepest CDM halo velocity profile $V_h(r) \propto r^{1/4}$ results too shallow with respect to observations. Although the mass models of (3) converge to a distribution with an inner core rather than with a central spike, it is worth, given the importance of such result, checking in a direct way the (in)compatibility of the CDM models with galaxy kinematics.

So, we assume the NFW functional form for the halo density given by (1), leaving c and r_s as free independent parameters, although N–body simulations and semi-analytic investigations indicate that they correlate. This choice to increase the chance of a good fit. We also imposed to the object under study a conservative halo mass upper limit of $2 \times 10^{12} M_\odot$. The fits to the data are shown in Fig. 6, together with the BBS fits: for seven out of nine objects the NFW models are unacceptably worse than the BBS solutions. Moreover, in all objects, the CDM virial mass is too high: $M_{vir} \sim 2 \times 10^{12} M_\odot$ and the resulting disk mass–to–light ratio too low ($\lesssim 10^{-1}$ in the I-band).

4 The Intriguing Evidence from Dark Matter Halos

The dark halos around spirals emerge as an one–parameter family; the order parameter (either the central density or the core radius) correlates with the lu-

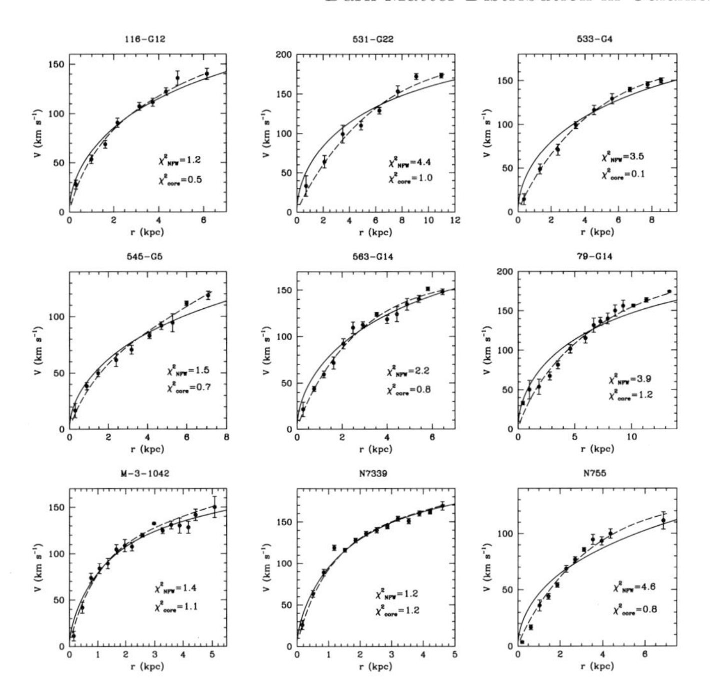

Fig. 6. NFW best–fits *solid lines* of the rotation curves *(filled circles)* compared with the BBS fits *(dashed lines)*. The χ^2 values are also indicated.

minous mass. However, we do not know how it is related to the global structural properties of the dark halo, like the virial radius or the virial mass, unless we extrapolate out over the BBS profile. That is because the halo RC, out to the outermost data, is completely determined by physical parameters, the central core density and the core radius, which have little counterpart in the gravitational instability/hierarchical clustering picture.

Caveat the above extrapolation, the location of spiral galaxies in the virtual space of virial mass/halo "central" density/stellar mass, that, on theorethical basis, should be roughly 3-D random determined by several different and non–linear physical processes, is remarkably found to degenerate and to lie on a curve. Indeed, in Fig.7 we show the dark–to–luminous mass ratio as function of the normalized radius and the total disk mass. The surface has been obtained by adopting the correlations between the halo and the disk parameters we found in our previous works (see Fig.3 [19]):

$$\log r_0 = 9.10 + 0.28\ \log \rho_0 - 3.49 \times 10^{10}\ \rho_0^{0.43} \tag{10}$$

$$\log \rho_0 = -23.0 - 0.077\ \log M_d - 9.98 \times 10^{-6}\ M_d^{0.43} \tag{11}$$

and:

$$\log r_d = 4.96 - 1.17\ \log M_d + 0.070\ (\log M_d)^2 \tag{12}$$

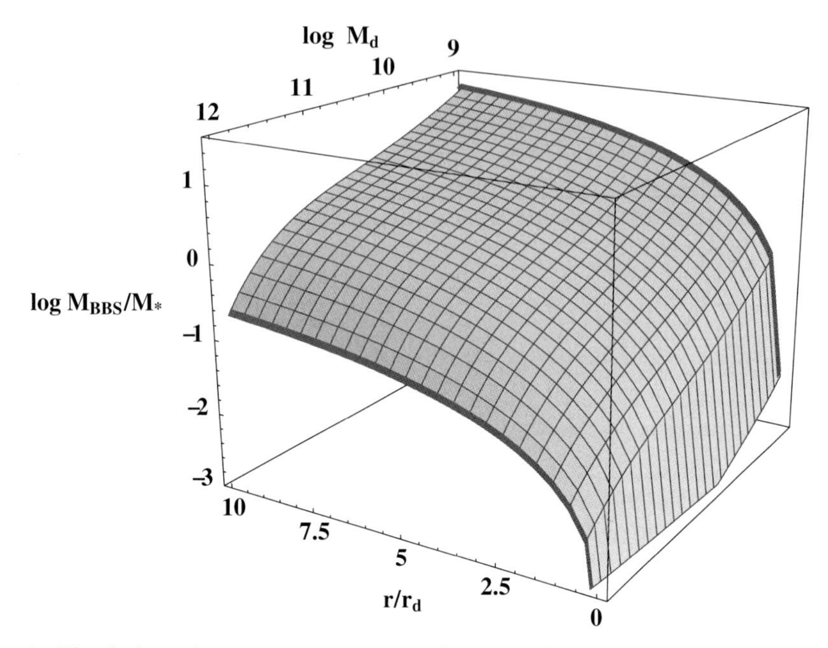

Fig. 7. The dark–to–luminous mass ratio as function of the normalized radius and the total disk mass.

from data in [16]. The dark–to–luminous mass ratio at fixed ratio increases as the total disk mass decreases; for example at $r = 10\ r_d$ it raises from 20% for massive disks ($M_d = 10^{12} M_\odot$) to 220% for smaller disks ($M_d = 10^9 M_\odot$).

Two conclusive statements can be drawn: dark matter halos have an inner constant–density region, whose size exceeds the stellar disk length–scale. Second, there is no evidence that dark halos converge, at large radii, to a $\rho \sim r^{-2}$ (or steeper) profile.

The existence of a region of "constant" density $\rho_0 \simeq \frac{\pi}{24}\ \frac{M_{\rm vir}}{r_0^3}$ and size r_0 is hardly explained within current theories of galaxy formation. Moreover, the evidence of a smooth halo profile is growing more an more in recent literature [e.g. 26, 27, 28, 29] and a number of different solutions have been proposed to solve this problem [e.g. 30, 31, 32, 33]. Let us stress, however, that we should incorporate all the intriguing halo properties described in this review.

References

1. A. Borriello, P. Salucci, MNRAS **323**, 285 (2001).
2. J.S. Bullock, T.S. Kolatt, Y. Sigad, R.S. Somerville, A.V. Kravtsov, A.A. Klypin, J.R. Primack, A. Dekel, MNRAS **321**, 559 (2001).
3. A. Burkert, ApJ **447**, L25 (1995).
4. A. Burkert, J. Silk, ApJ **488**, L55 (1997).
5. S. Cole, C. Lacey, MNRAS **281**, 716 (1997).
6. E. Corbelli, P. Salucci, MNRAS **311**, 411C (2000).

7. R. Flores, J.R. Primack, ApJ **427**, L1 (1994).
8. T. Fukushige, J. Makino, ApJ **477**, L9 (1997).
9. A.V. Kravtsov, A.A. Klypin, J.S. Bullock, J.R. Primack, ApJ **502**, 48 (1998).
10. S.S. McGaugh, W.J.G. de Block, ApJ **499**, 41 (1998).
11. B. Moore, Nature **370**, 629 (1994).
12. B. Moore, F. Governato, T. Quinn, J. Stadel, G. Lake, ApJ **499**, L5 (1998).
13. J.F. Navarro, C.S. Frenk, S.D.M. White, ApJ **462**, 563 (1996).
14. M. Persic, P. Salucci, MNRAS **245**, 577 (1990b).
15. M. Persic, P. Salucci, ApJS **99**, 501 (1995).
16. M. Persic, P. Salucci, F. Stel, MNRAS **281**, 27P (1996).
17. M.-H. Rhee, PhD thesis, Groningen University (1996).
18. P. Salucci, C. Ratnam, P. Monaco, L. Danese, MNRAS **317**, 488S (2000).
19. P. Salucci, A. Burkert, ApJ **537L**, 9S (2000).
20. P. Salucci, M. Persic, MNRAS **309**, 923 (1999).
21. J. Stil, Ph.D. Thesis, Leiden University (1999).
22. T.S. van Albada, J.S. Bahcall, K. Begeman, R. Sancisi, ApJ **295**, 305 (1985).
23. F.C. van den Bosch, B.E. Robertson, J. Dalcanton, W.J.G. de Blok, AJ **119**, 1579V (2000).
24. M. Persic, P. Salucci, in *Dark and Visible Matter in Galaxies*, ASP Conf. Ser., **117**, eds. Persic & Salucci (1997).
25. P. Salucci, A. Burkert, ApJ **537**, L9 (2000).
26. W.J.G. de Blok, S. S. McGaugh, A. Bosma, V. C. Rubin, ApJ **552**, 23 (2001).
27. C. M. Trott, R. L. Webs, astro-ph/0203196, in press (2002).
28. L. D. Matthews, J. S. Gallagher, III, astro-ph/0203188, in press (2002).
29. S. Blais-Ouellette, C. Carignan, P. Amram, astro-ph/0203146 to be published in 'Galaxies: the Third Dimension', ASP Conf. Ser. (2002).
30. M. White , R.A.C.Croft, AJ **539**, 497 (2000).
31. P.J.E. Peebles, ApJ **534**, 127 (2000).
32. C. Firmani, E. D'Onghia, V. Avila–Reese, G. Chincarini, X. Hernandez, MNRAS **315**, 29 (2000).
33. A. El–Zant, I. Shlosman, Y. Hoffman, ApJ **560**, 636 (2001).

Part II

Particle Physics Phenomenology

Some Aspects of B Decays

Nilendra G. Deshpande

Institute of Theoretical Science, University of Oregon, Eugene, Oregon 97403-5203

Abstract. We briefly discuss measurements of angles β and α of the Unitarity triangle. We then review rate asymmetries using SU(3) relationships in the standard model. Some methods to measure angle γ using SU(3) are then discussed. We note that rate for $b \to s\gamma$ can be used to set limits on extra dimensions in which standard model particles propagate.

Measurement of angles of the unitarity triangle is one of the prime goals of the B factories. Both Belle [1] and BaBar [2,3] collaborations have reported values of $\sin 2\beta$. The experimental world average is given by:

$$\sin 2\beta = 0.79 \pm 0.10 \tag{1}$$

This value is in excellent agreement with the Standard Model. A recent theoretical analysis by Ciuchini et. al. [4] yields the value:

$$\sin 2\beta = 0.698 \pm 0.066 \tag{2}$$

It is still important to measure the other angles by different techniques to make sure that CP violation arises only through the CKM matrix. The next measurement is likely to be $\sin 2\alpha$ through the study of time dependent asymmetries in $B_d \to \pi^+\pi^-$ decays. The coefficient of $\sin(\Delta mt)$ in this case yields $\sin(2\alpha_{\text{measured}})$ where α_{measured} is not the same as α because of large penguin contamination. It is possible to estimate the deviation $\delta\alpha = \alpha_{\text{measured}} - \alpha$ with some theory input. For example, using QCD improved factorization of Beneke et. al. [5], it is possible deduce $\sin 2\alpha$ from $\sin(2\alpha_{\text{measured}})$ provided $|V_{ub}/V_{cb}|$ is known. This was done in [6]. The following plot (Fig. 1) summarizes the result of the analysis.

Interesting relationships can be obtained between CP violating rate differences in the standard model if one uses flavor SU(3) symmetry. The quark level effective Hamiltonian up to one loop level in electroweak interaction for hadronic charmless B decays, can be written as

$$H_{eff}^q = \frac{4G_F}{\sqrt{2}}[V_{ub}V_{uq}^*(c_1O_1 + c_2O_2) - \sum_{i=3}^{12}(V_{ub}V_{uq}^*c_i^{uc} + V_{tb}V_{tq}^*c_i^{tc})O_i]. \tag{3}$$

The operators are defined in [7]. The coefficients $c_{1,2}$ and $c_i^{jk} = c_i^j - c_i^k$, with j indicates the internal quark, are the Wilson Coefficients (WC). These WC's have

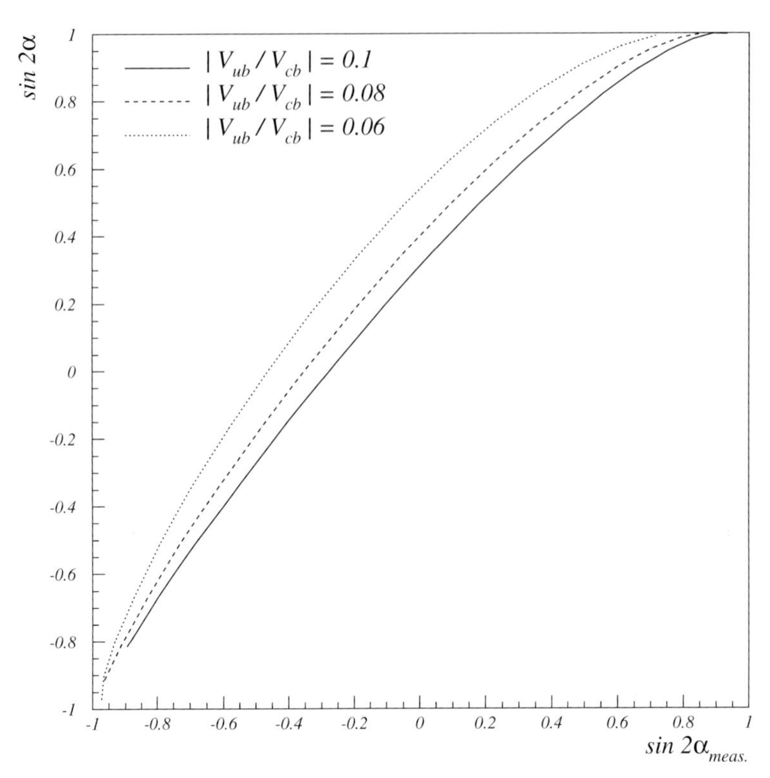

Fig. 1. The "true" value of $\sin 2\alpha$ as a function of the value of $\sin 2\alpha$ "measured" in $B \to \pi^+\pi^-$ decays for $|V_{ub}/V_{cb}| = 0.1$ (solid curve), 0.08 (dashed curve) and 0.06 (dotted curve).

been evaluated by several groups [7], with $|c_{1,2}| \gg |c_i^j|$. In the above the factor $V_{cb}V_{cq}^*$ has been eliminated using the unitarity property of the KM matrix.

The operators $O_{1,2}$, $O_{3-6,11,12}$, and O_{7-10} transform under SU(3) symmetry as $\bar{3}_a + \bar{3}_b + 6 + \overline{15}$, $\bar{3}$, and $\bar{3}_a + \bar{3}_b + 6 + \overline{15}$, respectively. These properties enable us to write the decay amplitudes for $B \to PP$ in only a few SU(3) invariant amplitudes.

For the $T(q)$ amplitude, for example, we have [8]

$$\begin{aligned} T(q) = & A_{\bar{3}}^T B_i H(\bar{3})^i (M_l^k M_k^l) + C_{\bar{3}}^T B_i M_k^i M_j^k H(\bar{3})^j \\ & + A_6^T B_i H(6)_k^{ij} M_j^l M_l^k + C_6^T B_i M_j^i H(6)_l^{jk} M_k^l \\ & + A_{\overline{15}}^T B_i H(\overline{15})_k^{ij} M_j^l M_l^k + C_{\overline{15}}^T B_i M_j^i H(\overline{15})_l^{jk} M_k^l \,, \end{aligned} \tag{4}$$

where $B_i = (B_u, B_d, B_s) = (B^-, \bar{B}^0, \bar{B}_s^0)$ is a SU(3) triplet, M_i^j is the SU(3) pseudoscalar octet, and the matrices $H(i)$ contain information about the transformation properties of the operators O_{1-12}.

For $q = d$, the non-zero entries of the matrices $H(i)$ are given by

$$H(\bar{3})^2 = 1 \,, \quad H(6)_1^{12} = H(6)_3^{23} = 1 \,, \quad H(6)_1^{21} = H(6)_3^{32} = -1 \,,$$

$$H(\overline{15})_1^{12} = H(\overline{15})_1^{21} = 3\,,\ H(\overline{15})_2^{22} = -2\,,\ H(\overline{15})_3^{32} = H(\overline{15})_3^{23} = -1\,. \quad (5)$$

And for $q = s$, the non-zero entries are

$$\begin{aligned} H(\bar{3})^3 \quad &= 1\,,\ H(6)_1^{13} = H(6)_2^{32} = 1\,,\ H(6)_1^{31} = H(6)_2^{23} = -1\,, \\ H(\overline{15})_1^{13} &= H(\overline{15})_1^{31} = 3\,,\ H(\overline{15})_3^{33} = -2\,,\ H(\overline{15})_2^{32} = H(\overline{15})_2^{23} = -1\,. \quad (6) \end{aligned}$$

Due to the anti-symmetric property of $H(6)$ in exchanging the upper two indices, A_6 and C_6 are not independent. For individual decay amplitude, A_6 and C_6 always appear together in the form $C_6 - A_6$. We will absorb A_6 in the definition of C_6. In terms of the SU(3) invariant amplitudes, the decay amplitudes for various B meson decays are given by [9]

$$\begin{aligned}
&\Delta S = 0 \\
&T^{B_u}_{\pi^-\pi^0}(d) = \frac{8}{\sqrt{2}} C^T_{\overline{15}}, \\
&T^{B_u}_{\pi^-\eta_8}(d) = \frac{2}{\sqrt{6}}(C^T_{\bar{3}} - C^T_6 + 3A^T_{\overline{15}} + 3C_{\overline{15}}), \\
&T^{B_u}_{K^-K^0}(d) = C^T_{\bar{3}} - C^T_6 + 3A^T_{\overline{15}} - C^T_{\overline{15}}, \\
&T^{B_d}_{\pi^+\pi^-}(d) = 2A^T_{\bar{3}} + C^T_{\bar{3}} + C^T_6 + A^T_{\overline{15}} + 3C^T_{\overline{15}}, \\
&T^{B_d}_{\pi^0\pi^0}(d) = \frac{1}{\sqrt{2}}(2A^T_{\bar{3}} + C^T_{\bar{3}} + C^T_6 + A^T_{\overline{15}} - 5C^T_{\overline{15}}), \\
&T^{B_d}_{K^-K^+}(d) = 2(A^T_{\bar{3}} + A^T_{\overline{15}}), \\
&T^{B_d}_{\bar{K}^0K^0}(d) = 2A_{\bar{3}} + C^T_{\bar{3}} - C^T_6 - 3A^T_{\overline{15}} - C_{\overline{15}}, \\
&T^{B_d}_{\pi^0\eta_8}(d) = \frac{1}{\sqrt{3}}(-C^T_{\bar{3}} + C^T_6 + 5A^T_{\overline{15}} + C_{\overline{15}}), \\
&T^{B_d}_{\eta_8\eta_8}(d) = \frac{1}{\sqrt{2}}(2A_{\bar{3}} + \frac{1}{3}C^T_{\bar{3}} - C^T_6 - A^T_{\overline{15}} + C_{\overline{15}}), \\
&T^{B_s}_{K^+\pi^-}(d) = C^T_{\bar{3}} + C^T_6 - A^T_{\overline{15}} + 3C_{\overline{15}}, \\
&T^{B_s}_{K^0\pi^0}(d) = -\frac{1}{\sqrt{2}}(C^T_{\bar{3}} + C^T_6 - A^T_{\overline{15}} - 5C_{\overline{15}}), \\
&T^{B_s}_{K^0\eta_8}(d) = -\frac{1}{\sqrt{6}}(C^T_{\bar{3}} + C^T_6 - A^T_{\overline{15}} - 5C_{\overline{15}}).
\end{aligned}$$

$$\begin{aligned}
&\Delta S = -1 \\
&T^{B_u}_{\pi^-\bar{K}^0}(s) = C^T_{\bar{3}} - C^T_6 + 3A^T_{\overline{15}} - C^T_{\overline{15}}, \\
&T^{B_u}_{\pi^0K^-}(s) = \frac{1}{\sqrt{2}}(C^T_{\bar{3}} - C^T_6 + 3A^T_{\overline{15}} + 7C^T_{\overline{15}})\,, \\
&T^{B_u}_{\eta_8K^-}(s) = \frac{1}{\sqrt{6}}(-C^T_{\bar{3}} + C^T_6 - 3A^T_{\overline{15}} + 9C^T_{\overline{15}}), \\
&T^{B_d}_{\pi^+K^-}(s) = C^T_{\bar{3}} + C^T_6 - A^T_{\overline{15}} + 3C^T_{\overline{15}},
\end{aligned}$$

$$T^{B_d}_{\pi^0\bar{K}^0}(s) = -\frac{1}{\sqrt{2}}(C^T_{\bar{3}} + C^T_6 - A^T_{\overline{15}} - 5C^T_{\overline{15}}),$$
$$T^{B_d}_{\eta_8\bar{K}^0}(s) = -\frac{1}{\sqrt{6}}(C^T_{\bar{3}} + C^T_6 - A^T_{\overline{15}} - 5C^T_{\overline{15}}),$$
$$T^{B_s}_{\pi^+\pi^-}(s) = 2(A^T_{\bar{3}} + A^T_{\overline{15}}),$$
$$T^{B_s}_{\pi^0\pi^0}(s) = \sqrt{2}(A^T_{\bar{3}} + A^T_{\overline{15}}),$$
$$T^{B_s}_{K^+K^-}(s) = 2A^T_{\bar{3}} + C^T_{\bar{3}} + C^T_6 + A^T_{\overline{15}} + 3C^T_{\overline{15}},$$
$$T^{B_s}_{K^0\bar{K}^0}(s) = 2A^T_{\bar{3}} + C^T_{\bar{3}} - C^T_6 - 3A^T_{\overline{15}} - C^T_{\overline{15}},$$
$$T^{B_s}_{\pi^0\eta_8}(s) = \frac{2}{\sqrt{3}}(C^T_6 + 2A^T_{\overline{15}} - 2C^T_{\overline{15}}),$$
$$T^{B_s}_{\eta_8\eta_8}(s) = \sqrt{2}(A^T_{\bar{3}} + \frac{2}{3}C^T_{\bar{3}} - A^T_{\overline{15}} - 2C^T_{\overline{15}}).$$

The amplitudes for $P(q)$ in terms of SU(3) invariant amplitudes can be obtained in a similar way. We will indicate the corresponding amplitudes by A^P_i and C^P_i. Rate differences are defined as follows

$$\Delta(B \to PP) = \Gamma(B \to PP) - \Gamma(\bar{B} \to \bar{P}\bar{P}). \tag{7}$$

SU(3) symmetry relates $\Delta S = 0$ and $\Delta S = -1$ decays. One particularly interesting class of relations are the ones with $T(d) = T(s) = T$ and $P(d) = P(s) = P$. For this class of decays, we have [8]

$$A(d) = V_{ub}V^*_{ud}T + V_{tb}V^*_{td}P,$$
$$A(s) = V_{ub}V^*_{us}T + V_{tb}V^*_{ts}P. \tag{8}$$

Due to different KM matrix elements involved in $A(d)$ and $A(s)$, although the amplitudes have some similarities, the branching ratios are not simply related. However, when considering rate difference, $\Delta(B \to PP)$, the situation is dramatically different. Because a simple property of the KM matrix element, $Im(V_{ub}V^*_{ud}V^*_{tb}V_{td}) = -Im(V_{ub}V^*_{us}V^*_{tb}V_{ts})$, we find that in the SU(3) limit,

$$\Delta(d) = -\Delta(s), \tag{9}$$

where $\Delta(i) = (|A(i)|^2 - |\bar{A}(i)|^2)\lambda_{ab}/(8\pi m_B)$ is the CP violating rate difference defined earlier and $\lambda_{ab} = \sqrt{1 - 2(m^2_a + m^2_b)/m^2_B + (m^2_a - m^2_b)^2/m^4_B}$ with $m_{a,b}$ being the masses of the two particles in the final state.

In the SU(3) limit we find the following equalities:

$$\begin{aligned}
&1)\quad \Delta(B^- \to K^-K^0) = -\Delta(B^- \to \pi^-\bar{K}^0)\,,\\
&2)\quad \Delta(\bar{B}^0 \to \pi^-\pi^+) = -\Delta(B_s \to K^-K^+)\,,\\
&3)\quad \Delta(\bar{B}^0 \to K^-K^+) = -\Delta(B_s \to \pi^-\pi^+)\\
&\qquad = -2\Delta(B_s \to \pi^0\pi^0)\,,\\
&4)\quad \Delta(\bar{B}^0 \to \bar{K}^0K^0) = -\Delta(B_s \to K^0\bar{K}^0)\,,
\end{aligned}$$

$$\begin{aligned} 5) \quad & \Delta(\bar{B}^0 \to \pi^+ K^-) = -\Delta(B_s \to K^+ \pi^-), \\ 6) \quad & \Delta(\bar{B}^0 \to \pi^0 \bar{K}^0) = -\Delta(B_s \to K^0 \pi^0) \\ & = 3\Delta(\bar{B}^0 \to \eta_8 \bar{K}^0) = -3\Delta(B_s \to K^0 \eta_8). \end{aligned} \tag{10}$$

If it turns out that the annihilation contributions are all small as can be tested in $B^- \to K^- K^0$, $B_s \to \pi^+\pi^-$ and $B_s \to \pi^0\pi^0$, there are additional relations for rate differences. We find

$$\begin{aligned} & 1) \approx 4), \\ & 2) \approx -5), \\ & 6) \approx \Delta(\bar{B}^0 \to \pi^0 \pi^0) \end{aligned} \tag{11}$$

In the limit that annihilation contributions are small, it is difficult to perform tests related to 1), 3) and 4) because the decay rates involved are all small. The equalities of 2) and 5) provide the best chances to test the SM.

We can use factorization assumption to estimate SU(3) breaking. This should be a good approximation because corrections to factorization are $O(\alpha_s)$, thus neglected terms of of order SU(3) breaking times $O(\alpha_s)$

$$\begin{aligned} \Delta(\bar{B}^0 \to \pi^+\pi^-) &\approx -\frac{f_\pi^2}{f_K^2}\Delta(\bar{B}^0 \to \pi^+ K^-), \\ \Delta(B_s \to K^+ K^-) &\approx -\frac{f_K^2}{f_\pi^2}\Delta(B_s \to \pi^- K^+). \end{aligned} \tag{12}$$

Similar method can be applied to $B \to PV$ decays [10]. We find the following interesting relations

$$\begin{aligned} \Delta(B_d \to \pi^- \rho^+) &\approx -\frac{f_\pi^2}{f_K^2}\Delta(B_d \to K^- \rho^+), \\ \Delta(B_d \to \pi^+ \rho^-) &\approx \Delta(B_d \to \pi^+ K^*). \end{aligned} \tag{13}$$

Buras and Fleischer [11] gave a method to determine γ *without* neglecting rescattering using $B_d \to \pi^- K^+$, $B^+ \to \pi^+\pi^0$ decays and time *dependent* measurements of the $B_d \to \pi^0 K_S$ decay. For this method, they also require time dependent analysis of, for example, $B_d \to J/\psi K_S$ to measure β. Gronau and Pirjol [12] suggested a method using time independent measurements of *all* the $B_d \to \pi K$ and $B_s \to \pi K$ modes. In their method also rescattering effects are included. However, it might be difficult to measure the neutral modes of B_s decays since that will involve tagging at hadron machines.

We have discussed a technique to determine γ *including* rescattering effects (and the EWP operators) using B meson decays to π's and K's [13]. We will illustrate this technique for one of the cases discussed.

We do *not* require any time dependent studies. The strategy is as follows. In cases 1 and 2, using 5 $\Delta S = 0$ decay modes, we determine the strong phases and magnitudes of the tree level and penguin contributions as functions of γ (assuming flavor $SU(2)$ symmetry). Then, using flavor $SU(3)$ symmetry, we

Table 1. The 6 (or 8) B decay modes used by each of the 4 cases to determine γ.

Case	Modes used	
	$\Delta S = 0$	$\Delta S = 1$
1	$B^+ \to \pi^+\pi^0$, $B_d \to \pi^+\pi^-, \pi^0\pi^0$ $\bar{B}_d \to \pi^+\pi^-, \pi^0\pi^0$	$B_d \to \pi^- K^+, \pi^0 K^0$ $B_s \to \pi^+\pi^-$ (or $\pi^0\pi^0$)
2	$B^+ \to \pi^+\pi^0$, $B_d \to \pi^+\pi^-, \pi^0\pi^0$ $\bar{B}_d \to \pi^+\pi^-, \pi^0\pi^0$	$B_s \to K^+K^-$ (CP-averaged)
3	$B^+ \to \pi^+\pi^0$, $B_d \to \pi^+\pi^-, \pi^0\pi^0$ $B_d \to K^+K^-$	$B_d \to \pi^0 K^0, \pi^- K^+$ $\bar{B}_d \to \pi^0 \bar{K}^0, \pi^+ K^-$
4	$B^+ \to \pi^+\pi^0$ $B_s \to \pi^+ K^-$ (or $\pi^0 \bar{K}^0$) (CP-averaged)	$B_d \to \pi^0 K^0, \pi^- K^+$ $\bar{B}_d \to \pi^0 \bar{K}^0, \pi^+ K^-$

predict the rate for *one* $\Delta S = 1$ mode in case 2. In case 1, two $\Delta S = 1$ modes have to be measured to make a prediction for a third $\Delta S = 1$ mode. The measurement of the decay for which we have a prediction (as a function of γ) then determines γ. A similar idea can be applied to predict a $\Delta S = 0$ decay mode as a function of γ using measurements of $\Delta S = 1$ (and some $\Delta S = 0$) modes (cases 3 and 4).

The decay amplitudes for $B_d \to \pi K$ can be written as

$$\sqrt{2}\, \mathcal{A}(B_d \to \pi^0 K^0) = -I_{1/2} + 2I_{3/2}, \tag{14}$$

$$\mathcal{A}(B_d \to \pi^- K^+) = I_{1/2} + I_{3/2}, \tag{15}$$

where $I_{1/2}$ and $I_{3/2}$ are the amplitudes for B_d decay to $\pi\ K$ $(I = 1/2)$ and $(I = 3/2)$ respectively. Then,

$$\begin{aligned} 3I_{3/2} &= \sqrt{2}\mathcal{A}(B_d \to \pi^0 K^0) + \mathcal{A}(B_d \to \pi^- K^+) \\ &= -\lambda_u^{(s)} 8\tilde{C}_{15}^T + 8\lambda_c^{(s)} C_{15}^P \end{aligned}$$

$$= -8C_{15}^T \left(\lambda_u^{(s)} - \frac{3}{2}\kappa\lambda_c^{(s)} \right)$$
$$= -8\,|C_{15}^T||\lambda_u^{(s)}|\left(e^{i\gamma} + \delta_{EW}\right), \tag{16}$$

Here δ_{EW} is given by $-|\lambda_c^{(s)}|/|\lambda_u^{(s)}|\;3/2\;\kappa \sim -0.66$, and the EWP contribution *is* important for $B_d \to \pi K$ decays. $|C_{15}^T|$ can be obtained from the $B^+ \to \pi^+\pi^0$ decay rate.

$I_{1/2}$ is given by

$$I_{1/2} = -\lambda_u^{(s)} \left(\tilde{C}_3^T + \tilde{C}_6^T + \frac{1}{3}\tilde{C}_{15}^T \right) + \lambda_c^{(s)}(C_3^P + C_6^P + \frac{1}{3}C_{15}^P)$$
$$+\lambda_u^{(s)}\tilde{A}_{15}^T - \lambda_c^{(s)}A_{15}^P$$
$$\equiv e^{i\phi'_{\tilde{T}}}|\lambda_u^{(s)}|e^{i\gamma}\tilde{T}' - |\lambda_c^{(s)}|e^{i\phi'_P}P'. \tag{17}$$

The four quantities: $\tilde{T}'$, P', $\phi'_{\tilde{T}}$ and ϕ'_P can thus be determined as functions of γ from the measurements of the four decay rates: $B_d \to \pi^-K^+$, $B_d \to \pi^0K^0$ and their CP-conjugates.

Due to the EWP contribution (see (16)), the triangle construction is a bit different in this case as shown below.

We multiply the CP-conjugate amplitudes by $e^{i2\gamma}$ to get the "barred" amplitudes. In this case there is an angle between $I_{3/2}$ and $\bar{I}_{3/2}$ denoted by $2\tilde{\gamma}$ and their magnitudes are functions of γ (see (16)):

$$|I_{3/2}| = |\bar{I}_{3/2}| = \frac{8}{3}|C_{15}^T||\lambda_u^{(s)}|\sqrt{(1 + \delta_{EW}^2 + 2\delta_{EW}\cos\gamma)}, \tag{18}$$

$$\tan\tilde{\gamma} = \frac{\delta_{EW}\sin\gamma}{1 + \delta_{EW}\cos\gamma}. \tag{19}$$

Given γ, we can thus construct the triangles of (16) and its CP-conjugate (see Fig. 2). Knowing the magnitudes and orientations of $I_{1/2}$ and $\bar{I}_{1/2}$ from Fig. 2, we can determine $\tilde{T}'$, P', $\phi'_{\tilde{T}}$ and ϕ'_P as functions of γ.

The $B_d \to K^+K^-$ amplitude is given by:

$$\mathcal{A}\left(B_d \to K^+K^-\right) = -\lambda_u^{(d)}\left(2A_3^T + 2A_{15}^T\right) - \sum_q \lambda_q^{(d)}\left(2A_{3,q}^P + 2A_{15,q}^P\right)$$
$$\equiv ae^{i\phi_a}. \tag{20}$$

We can see that $\sqrt{2}\mathcal{A}\left(B_d \to \pi^0\pi^0\right) + \mathcal{A}\left(B_d \to K^-K^+\right)$ can be obtained from $\sqrt{2}\mathcal{A}\left(B_d \to K^0\pi^0\right)$ and
$\mathcal{A}\left(B_d \to \pi^+\pi^-\right) - \mathcal{A}\left(B_d \to K^-K^+\right)$ can be obtained from $\mathcal{A}\left(B_d \to \pi^-K^+\right)$ by scaling the $\Delta S = 1$ amplitudes by appropriate CKM factors. Thus, we can determine γ, including *all* rescattering effects, by measuring the 8 decay modes: $B^+ \to \pi^+\pi^0$, B_d and $\bar{B}_d \to \pi K$ (all), $B_d \to K^+K^-$, $B_d \to \pi^0\pi^0$ and $B_d \to \pi^-\pi^+$ (or CP-conjugates of the last three modes).

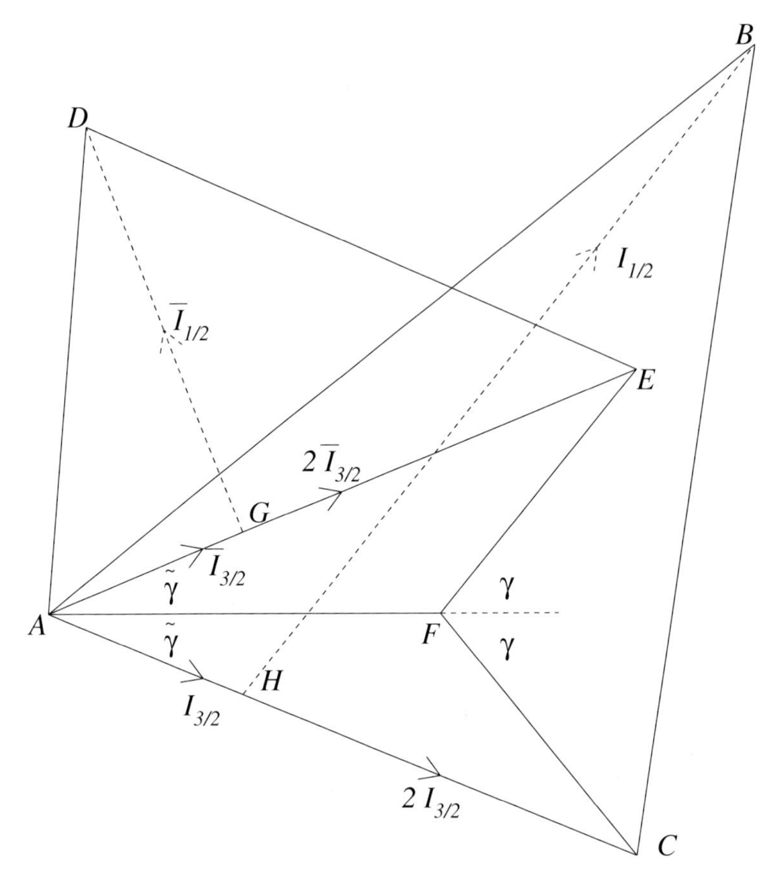

Fig. 2. The triangles formed by the $B_d \to \pi K$ amplitudes: $AB = |\mathcal{A}\left(B_d \to K^+\pi^-\right)|$, $BC = |\sqrt{2}\mathcal{A}\left(B_d \to \pi^0 K^0\right)|$, $AD = |\mathcal{A}\left(\bar{B}_d \to K^-\pi^+\right)|$, $DE = |\sqrt{2}\mathcal{A}\left(\bar{B}_d \to \pi^0 \bar{K}^0\right)|$. $AF = |\sqrt{2}\,\mathcal{A}\left(B^+ \to \pi^+\pi^0\right)|\,|\lambda_u^{(s)}|/|\lambda_u^{(d)}|$ and $FC = FE = AF\ \delta_{EW}$ (see (16)). In the phase convention where the strong phase of C_{15}^T is zero, the angle between AF and the real axis is $\pi + \gamma$.

If the annihilation amplitudes are small, we can determine γ by measuring any *one* $B_d \to \pi\pi$ decay mode, in addition to the $B^+ \to \pi^+\pi^0$, B_d (and $\bar{B}_d$) $\to \pi K$ decay modes. If we measure the CP-conjugate $B_d \to \pi\pi$ rates as well, then a CP-averaged measurement of the decay rate $B_d \to K^+K^-$ suffices.

We now show [14] how $b \to s\gamma$ decay can be used to set limits on the size of extra dimensions. The motivations for studying theories with *flat* extra dimensions of size $(\mathrm{TeV})^{-1}$ accessible to (at least some of) the SM fields are varied.

From the $4D$ point of view, these extra dimensions take the form of Kaluza-Klein (KK) excitations of SM fields with masses $\sim n/R$, where R is a typical size of an extra dimension. In models with *only* SM gauge fields in the bulk, there are contributions to muon decay, atomic parity violation (APV) etc. from tree-level exchange of KK states of gauge bosons [15,16]. Then, precision electroweak measurements result in a strong constraint on the size of extra dimensions and, in turn, imply that the effect on the process $b \to s\gamma$ is small.

To avoid these constraints, we will focus on models with *universal* extra dimensions, i.e., extra dimensions accessible to *all* the SM fields. In this case, due to conservation of extra dimensional momentum, there are *no* vertices with only one KK state, i.e., coupling of KK state of gauge boson to quarks and leptons always involves (at least one) *KK* mode of quark or lepton.

This, in turn, implies that there is no tree-level contribution to weak decays of quarks and leptons, APV $e^+e^- \to \mu^+\mu^-$ etc. from exchange of KK states of gauge bosons [17,18].

However, there is a constraint on R^{-1} from *one-loop* contribution of KK states of (mainly) the top quark to the T parameter. For $m_t \ll R^{-1}$, this constraint is roughly given by $\sum_n m_t^2 \Big/ \left(m_t^2 + (n/R)^2 \right) \lesssim 0.5-0.6$ (depending on the neutral Higgs mass) [18]. For the case of one extra dimension, this gives $R^{-1} \gtrsim 300$ GeV. The KK excitations of quarks appear as heavy stable quarks at hadron colliders and searches by the CDF collaboration also imply $R^{-1} \gtrsim 300$ GeV for one extra dimension [18]. We consider in this talk $b \to s\gamma$ for the case of only minimal SM with one Higgs doublet in extra dimensions.

The effective Hamiltonian for $\Delta S = 1$ B meson decays is

$$\mathcal{H}_{\text{eff}} = \frac{4G_F}{\sqrt{2}} V_{tb} V_{ts}^* \sum_{j=1}^{8} C_j(\mu) \mathcal{O}_j, \tag{21}$$

where the operator relevant for the transition $b \to s\gamma$ is

$$\mathcal{O}_7 = \frac{e}{16\pi^2} m_b \, \bar{s}_{L\alpha} \sigma^{\mu\nu} b_{R\alpha} F_{\mu\nu}. \tag{22}$$

The coefficient of this operator from $W - t$ exchange in the SM is

$$C_7^W(m_W) = -\frac{1}{2} A\left(\frac{m_t^2}{m_W^2}\right), \tag{23}$$

where the loop function A is given by

$$A(x) = x \left[\frac{\frac{2}{3}x^2 + \frac{5}{12}x - \frac{7}{12}}{(x-1)^3} - \frac{\left(\frac{3}{2}x^2 - x\right) \ln x}{(x-1)^4} \right]. \tag{24}$$

Of course, this includes the contribution from the charged would-be-Goldstone boson (WGB) (i.e., longitudinal W). With extra dimensions, there is a one-loop contribution from KK states of W (accompanied by KK states of top quark, $t^{(n)}$), but as we show below, this is smaller than that from KK states of charged WGB. In the limit $m_W \ll R^{-1}$, the KK states of W get a mass $\sim n/R$ by "eating" the field corresponding to extra polarization of W in higher dimensions – this field is a scalar from the $4D$ point of view. Thus, the coupling of *all* components of $W^{(n)}$ to fermions is g, unlike the case of the zero-mode, where the coupling of *longitudinal* W to fermions is given by the Yukawa coupling of Higgs to fermions. Therefore, the contribution of $W^{(n)}$ to the coefficient of the dimension-5 operator $\bar{s}\sigma_{\mu\nu}bF^{\mu\nu}$ is $\sim e\ m_b\ g^2/\left(16\pi^2\right) m_t^2 \sum_n 1/\left(n/R\right)^4$, wherethe factor m_t^2 reflects GIM cancelation. In terms of the operator $\mathcal{O}_7$, the contribution of each KK state of W to C_7 is $\sim m_t^2 m_W^2/\left(n/R\right)^4$.

From the above discussion, it is clear that the *KK* states of charged would-be-Goldstone boson (denoted by $\mathrm{WGB}^{(n)}$) are physical (unlike the *zero*-mode). The loop contribution of $\mathrm{WGB}^{(n)}$ with mass n/R (and $t^{(n)}$ with mass $\sqrt{m_t^2 + (n/R)^2}$) is of the same form as that of physical charged Higgs in 2 Higgs doublet models [19] with the appropriate modification of masses and couplings of virtual particles in the loop integral

$$C_7^{\mathrm{WGB}^{(n)}}\left(R^{-1}\right) \approx \frac{m_t^2}{m_t^2 + (n/R)^2} \times$$
$$\left[B\left(\frac{m_t^2 + (n/R)^2}{(n/R)^2}\right) - \frac{1}{6}A\left(\frac{m_t^2 + (n/R)^2}{(n/R)^2}\right)\right]. \tag{25}$$

Here, the factor $m_t^2/\left(m_t^2 + (n/R)^2\right)$ accounts for (a) the coupling of $\mathrm{WGB}^{(n)}$ to $t^{(n)}$ which is $\lambda_t \sim m_t/v$, i.e., the same as that of $\mathrm{WGB}^{(0)}$ (longitudinal W), and (b) the fact that this contribution decouples in the limit of large KK mass – the functions A and B (see below) in the above expression approach a constant as n/R becomes large.

The loop function B is given by [19]

$$B(y) = \frac{y}{2}\left[\frac{\frac{5}{6}y - \frac{1}{2}}{(y-1)^2} - \frac{\left(y - \frac{2}{3}\right)\ln y}{(y-1)^3}\right]. \tag{26}$$

It is clear that the ratio of the contribution of $W^{(n)}$ and that of $\mathrm{WGB}^{(n)}$ is $\sim \left(m_W R/n\right)^2 \lesssim O(1/10)$ since $R^{-1} \gtrsim 300$ GeV (due to constraints from the T parameter and searches for heavy quarks). In what follows, we will neglect the $W^{(n)}$ contribution.

At NLO, the coefficient of the operator at the scale$\mu \sim m_b$ is given by [20]

$$C_7\left(m_b\right) \approx 0.698\, C_7\left(m_W\right) - 0.156\, C_2\left(m_W\right) + 0.086\, C_8\left(m_W\right). \tag{27}$$

Here, C_2 is the coefficient of the operator $\mathcal{O}_2 = \left(\bar{c}_{L\alpha}\gamma^\mu b_{L\alpha}\right)\left(\bar{s}_{L\beta}\gamma_\mu c_{L\beta}\right)$ and is approximately same as in the SM (i.e., 1) since the KK states of W do not contribute to it at tree-level. C_8 is the coefficient of the chromomagnetic operator $\mathcal{O}_8 = g_s/\left(16\pi^2\right)\ m_b \times$
$\bar{s}_{L\alpha}\sigma^{\mu\nu}T^a_{\alpha\beta}b_{R\beta}G^a_{\mu\nu}$.

In the SM, $C_8\left(m_W\right) \approx -0.097$ [20] due to the contribution of $W - t$ loop (using $m_t \approx 174$ GeV). The coefficient of this operator also gets a loop contribution from KK states which is of the same order as the contribution to C_7. Since the coefficient of C_8 in (27) is small, we neglect the contribution of KK states to C_8.

The coefficient of $\mathcal{O}_7$ at the scale m_W is given by the sum of the contributions of $W^{(0)}$ (23) and that of $\mathrm{WGB}^{(n)}$ (25) summed over n.

Since $C_7^W(m_W) < 0$ and $C_7^{\mathrm{WGB}^{(n)}}\left(R^{-1}\right) > 0$, we see that contribution from $\mathrm{WGB}^{(n)}$ interferes destructively with the W contribution. The SM prediction for $\Gamma(b \to s\gamma)/\Gamma(b \to cl\nu)$ has an uncertainty of about 10% and the experimental error is about 15% (both are 1σ errors) [21]. The central values of theory and experiment agree to within $1/2\ \sigma$. The semileptonic decay is not affected by the KK states (at tree-level).

Combining theory and experiment 2σ errors in quadrature, this means that the 95% CL constraint on the contribution of KK states is that it should not modify the SM prediction for $\Gamma(b \to s\gamma)$ by more than 36%. Since $\Gamma(b \to s\gamma) \propto [C_7(m_b)]^2$, the constraint is $\left| \left[C_7^{\mathrm{total}}(m_b)\right]^2 / \left[C_7^{\mathrm{SM}}(m_b)\right]^2 - 1 \right| \lesssim 36\%$.

Using $m_t \approx 174$ GeV, we get $A \approx 0.39$ in (23) and $C_7^{\mathrm{SM}}(m_b) \approx -0.3$ from (27). Assuming $m_t \ll R^{-1}$, we get $B \approx 0.19$ and $A \approx 0.21$ in (25). Then, using (27) and the above criterion, we get the constraint

$$\sum_n m_t^2 \Big/ \left(m_t^2 + (n/R)^2\right) \lesssim 0.5 \tag{28}$$

which is comparable to that from the T parameter. For one extra dimension, performing the sum over KK states with the exact expressions for A and B in (25), the constraint is $R^{-1} \gtrsim 280$ GeV.

This work was supported, in part, by the U. S. Department of Energy under Grant No. DE-FG03-96ER40969. I wish to thank Professor Josip Trampetic for excellent hospitality at the Adriatic Meeting.

References

1. K. Abe, et. al., Belle Collaboration, hep-ex/0202027.
2. B. Aubertm, et. al., BABAR Collaboration, hep-ex/0201020.
3. B. Aubertm, et. al., BABAR Collaboration, hep-ex/0202007, gives a more current value for $\sin 2\beta$ as $0.75 \pm 0.09 \pm 0.04$.
4. M. Ciuchini, et. al., J. High Energy Phys. **0107**, 013 (2001).
5. M. Beneke, et. al., Nucl. Phys. B **606**, 245 (2001).
6. K. Agashe and N. G. Deshpande, Phys. Rev. D **61**, 071301 (2000).
7. G. Buchalla, A. Buras and M. Lautenbacher, Rev. Mod. Phys. **68**, 1125 (1996); M. Ciuchini, et. al., Nucl. Phys. B **415**, 403 (1994).
8. N. G. Deshpande and X.-G. He, Phys. Rev. Lett. **75**, 1703 (1995).
9. X.-G. He, Eur. Phys. J. C **9**, 443 (1999).
10. N. G. Deshpande, X.-G. He and J.-O. Shi, Phys. Rev. D **62**, 034018 (2000).
11. A. J. Buras and R. Fleischer, Eur. Phys. J. C **11**, 93 (1999).
12. M. Gronau and D. Pirjol, hep-ph/9811335.
13. K. Agashe and N. G. Deshpande, Phys. Lett. B **451**, 215 (1999).
14. N. G. Deshpande, K. Agashe and G.-H. Wu, Phys. Lett. B **514**, 309 (2001).
15. M. Graesser, hep-ph/9902310, Phys. Rev. D **61**, 074019 (2000).
16. P. Nath and M. Yamaguchi, hep-ph/9902323, Phys. Rev. D **60**, 116004 (1999).

17. R. Barbieri, L. J. Hall and Y. Nomura, hep-ph/0011311, Phys. Rev. D **63**, 105007 (2001).
18. T. Appelquist, H.-C. Cheng and B. A. Dobrescu, hep-ph/0012100.
19. B. Grinstein and M. B. Wise, Phys. Lett. B **201**, 274 (1988); W.-S. Hou and R. S. Willey, Phys. Lett. B **202**, 591 (1988).
20. G. Buchalla, A. Buras and M. Lautenbacher hep-ph/9512380, Rev. Mod. Phys. **68**, 1125 (1996).
21. A. L. Kagan and M. Neubert, hep-ph/9805303, Eur. Phys. J. C **7**, 5 (1999); D. E. Groom et. al., (Particle Data Group), Eur. Phys. J. C **15**, 1 (2000).

The Flavour and CP Problems in SUSY

Antonio Masiero[1] and Oscar Vives[2]

[1] Dip. di Fisica "G. Galilei", Univ. di Padova and INFN, Sezione di Padova, Via Marzolo 8, I-35121, Padua, Italy
[2] Theoretical Physics Department, University of Oxford, Oxford, OX1 3NP, United Kingdom

Abstract. Although direct searches of supersymmetry (SUSY) constitute the only way we have to clearly verify the existence of a low-energy SUSY extension of the standard model, yet, in particular in our pre-LHC era, it is of utmost importance to study any possible signal where SUSY manifests itself indirectly in discrepancies with the SM expectations in rare processes. In this talk we'll consider a wide range of flavor changing neutral current and/or CP violating phenomena where, indeed, SUSY contributions are comparable to the SM ones. Such analysis provides stringent constraints on different SUSY model parameter spaces and, at the same time, it individuates possible windows for SUSY signals in spite of all the existing constraints. Our attention will focus in particular on the CP violating processes which are the most sensitive place for SUSY effects in the vast class of rare phenomena of the SM

1 CP Violation in SUSY

CP violation has major potentialities to exhibit manifestations of new physics beyond the standard model. Indeed, it is quite a general feature that new physics possesses new CP violating phases in addition to the Cabibbo–Kobayashi–Maskawa (CKM) phase ($\delta_{\rm CKM}$) or, even in those cases where this does not occur, $\delta_{\rm CKM}$ shows up in interactions of the new particles, hence with potential departures from the SM expectations. Moreover, although the SM is able to account for the observed CP violation in the kaon system, we cannot say that we have tested so far the SM predictions for CP violation. The detection of CP violation in B physics will constitute a crucial test of the standard CKM picture within the SM. Again, on general grounds, we expect new physics to provide departures from the SM CKM scenario for CP violation in B physics. A final remark on reasons that make us optimistic in having new physics playing a major role in CP violation concerns the matter–antimatter asymmetry in the universe. Starting from a baryon–antibaryon symmetric universe, the SM is unable to account for the observed baryon asymmetry. The presence of new CP–violating contributions when one goes beyond the SM looks crucial to produce an efficient mechanism for the generation of a satisfactory ΔB asymmetry.

The above considerations apply well to the new physics represented by low–energy supersymmetric extensions of the SM. Indeed, as we will see below, supersymmetry introduces CP violating phases in addition to $\delta_{\rm CKM}$ and, even if one envisages particular situations where such extra–phases vanish, the phase $\delta_{\rm CKM}$ itself leads to new CP–violating contributions in processes where SUSY particles are exchanged. CP violation in B decays has all potentialities to exhibit departures from the SM CKM picture in low–energy SUSY extensions, although, as we will discuss, the detectability of such deviations strongly depends on the regions of the SUSY parameter space under consideration.

In any MSSM, at least two new "genuine" SUSY CP–violating phases are present. They originate from the SUSY parameters μ, M, A and B. The first of these parameters is the dimensionful coefficient of the $H_u H_d$ term of the superpotential. The remaining three parameters are present in the sector that softly breaks the N=1 global SUSY. M denotes the common value of the gaugino masses, A is the trilinear scalar coupling, while B denotes the bilinear scalar coupling. In our notation, all these three parameters are dimensionful. The simplest way to see which combinations of the phases of these four parameters are physical [1] is to notice that for vanishing values of μ, M, A and B the theory possesses two additional symmetries [2]. Indeed, letting B and μ vanish, a $U(1)$ Peccei–Quinn symmetry originates, which in particular rotates H_u and H_d. If M, A and B are set to zero, the Lagrangian acquires a continuous $U(1)$ R symmetry. Then we can consider μ, M, A and B as spurions which break the $U(1)_{PQ}$ and $U(1)_R$ symmetries. In this way, the question concerning the number and nature of the meaningful phases translates into the problem of finding the independent combinations of the four parameters which are invariant under $U(1)_{PQ}$ and $U(1)_R$ and determining their independent phases. There are three such independent combinations, but only two of their phases are independent. We use here the commonly adopted choice:

$$\varphi_A = \arg\left(A^* M\right), \qquad \varphi_B = \arg\left(B^* M\right). \tag{1}$$

where also $\arg\left(B\mu\right) = 0$, i.e. $\varphi_\mu = -\varphi_B$.

The main constraints on φ_A and φ_B come from their contribution to the electric dipole moments of the neutron and of the electron. For instance, the effect of φ_A and φ_B on the electric and chromoelectric dipole moments of the light quarks (u, d, s) lead to a contribution to d^e_N of order [3]

$$d^e_N \sim 2\left(\frac{100\mathrm{GeV}}{\tilde{m}}\right)^2 \sin\varphi_{A,B} \times 10^{-23}\mathrm{e\,cm}, \tag{2}$$

where $\tilde{m}$ here denotes a common mass for squarks and gluinos. The present experimental bound, $d^e_N < 1.1\times 10^{-25}$ e cm, implies that $\varphi_{A,B}$ should be $< 10^{-2}$, unless one pushes SUSY masses up to $\mathcal{O}(1$ TeV$)$. A possible caveat to such an argument calling for a fine–tuning of $\varphi_{A,B}$ is that uncertainties in the estimate of the hadronic matrix elements could relax the severe bound in (2) [4].

In view of the previous considerations, most authors dealing with the MSSM prefer to simply put φ_A and φ_B equal to zero. Actually, one may argue in favor of this choice by considering the soft breaking sector of the MSSM as resulting from SUSY breaking mechanisms which force φ_A and φ_B to vanish. For instance, it is conceivable that both A and M originate from one same source of $U(1)_R$ breaking. Since φ_A "measures" the relative phase of A and M, in this case it would "naturally" vanish. In some specific models, it has been shown [5] that through an analogous mechanism also φ_B may vanish.

If $\varphi_A = \varphi_B = 0$, then the novelty of SUSY in CP violating contributions merely arises from the presence of the CKM phase in loops where SUSY particles run [6]. The crucial point is that the usual GIM suppression, which plays a major role in evaluating ε_K and ε'/ε in the SM, in the MSSM case (or more exactly in the CMSSM) is replaced by a super–GIM cancellation which has the same "power" of suppression as the original GIM (see previous section). Again, also in the CMSSM, as it is the case in the SM, the smallness of ε_K and ε'/ε is guaranteed not by the smallness of δ_{CKM}, but rather by the small CKM angles and/or small Yukawa couplings. By the same token, we do not expect any significant departure of the CMSSM from the SM predictions also concerning CP violation in B physics. As a matter of fact, given the large lower bounds on squark and gluino masses, one expects relatively tiny contributions of the SUSY loops in ε_K or ε'/ε in comparison with the normal W loops of the SM. Let us be more detailed on this point.

In the CMSSM, the gluino exchange contribution to FCNC is subleading with respect to chargino ($\chi^{\pm}$) and charged Higgs ($H^{\pm}$) exchanges. Hence, when dealing with CP violating FCNC processes in the CMSSM with $\varphi_A = \varphi_B = 0$, one can confine the analysis to $\chi^{\pm}$ and $H^{\pm}$ loops. If one takes all squarks to be degenerate in mass and heavier than ~ 200 GeV, then $\chi^{\pm} - \tilde{q}$ loops are obviously severely penalized with respect to the SM W^{+}–q loops (remember that at the vertices the same CKM angles occur in both cases).

The only chance for the CMSSM to produce some sizeable departure from the SM situation in CP violation is in the particular region of the parameter space where one has light $\tilde{q}$, $\chi^{\pm}$ and/or $H^{\pm}$. The best candidate (indeed the only one unless $\tan\beta \sim m_t/m_b$) for a light squark is the stop. Hence one can ask the following question: can the CMSSM present some novelties in CP–violating phenomena when we consider χ^{+}–$\tilde{t}$ loops with light $\tilde{t}$, χ^{+} and/or H^{+}?

Several analyses in the literature tackle the above question or, to be more precise, the more general problem of the effect of light $\tilde{t}$ and χ^{+} on FCNC processes [7–9]. A first important observation concerns the relative sign of the W^{+}–t loop with respect to the χ^{+}–$\tilde{t}$ and H^{+}–t contributions. As it is well known, the latter contribution always interferes positively with the SM one. Interestingly enough, in the region of the MSSM parameter space that we consider here, also the χ^{+}–$\tilde{t}$ contribution interferes constructively with the SM contribution. The second point regards the composition of the lightest chargino, i.e. whether the gaugino or higgsino component prevails. This is crucial since the light stop is predominantly $\tilde{t}_R$ and, hence, if the lightest chargino is mainly a wino, it couples to $\tilde{t}_R$ mostly through the LR mixing in the stop sector. Consequently, a suppression in the contribution to box diagrams going as $\sin^4\theta_{LR}$ is present (θ_{LR} denotes the mixing angle between the lighter and heavier stops). On the other hand, if the lightest chargino is predominantly a higgsino (i.e. $M_2 \gg \mu$ in the chargino mass matrix), then the χ^{+}–lighter $\tilde{t}$ contribution grows. In this case, contributions $\propto \theta_{LR}$ become negligible and, moreover, it can be shown that they are independent on the sign of μ. A detailed study is provided in reference [8,9]. For instance, for $M_2/\mu = 10$, they find that the inclusion of the SUSY contribution to the box diagrams doubles the usual SM contribution for values of the lighter $\tilde{t}$ mass up to 100–120 GeV, using $\tan\beta = 1.8$, $M_{H^+} = 100$ TeV, $m_\chi = 90$ GeV and the mass of the heavier $\tilde{t}$ of 250 GeV. However, if m_χ is pushed up to 300 GeV, the χ^+–$\tilde{t}$ loop yields a contribution which is roughly 3 times less than in the case $m_\chi = 90$ GeV, hence leading to negligible departures from the SM expectation. In the cases where the SUSY contributions are sizeable, one obtains relevant restrictions on the ρ and η parameters of the CKM matrix by making a fit of the parameters A, ρ and η of the CKM matrix and of the total loop contribution to the experimental values of ε_K and ΔM_{B_d}. For instance, in the above–mentioned case in which the SUSY loop contribution equals the SM W^+–t loop, hence giving a total loop contribution which is twice as large as in the pure SM case, combining the ε_K and ΔM_{B_d} constraints leads to a region in the ρ–η plane with $0.15 < \rho < 0.40$ and $0.18 < \eta < 0.32$, excluding negative values of ρ.

In conclusion, the situation concerning CP violation in the MSSM case with $\varphi_A = \varphi_B = 0$ and exact universality in the soft–breaking sector can be summarized in the following way: the MSSM does not lead to any significant deviation from the SM expectation for CP–violating phenomena as d_N^e, ε_K, ε'/ε and CP violation in B physics; the only exception to this statement concerns a small portion of the MSSM parameter space where a very light $\tilde{t}$ ($m_{\tilde{t}} < 100$ GeV) and χ^+ ($m_\chi \sim 90$ GeV) are present. In this latter particular situation, sizeable SUSY contributions to ε_K are possible and, consequently, major restrictions in the ρ–η plane can be inferred. Obviously, CP violation in B physics becomes a crucial test for this MSSM case with very light $\tilde{t}$ and χ^+. Interestingly enough, such low values of SUSY masses are at the border of the detectability region at LEP II.

In next Section, we will move to the case where, still keeping the minimality of the model, we switch on the new CP violating phases. Later on we will give up also the strict minimality related to the absence of new flavor structure in the SUSY breaking sector and we will see that, in those more general contexts, we can expect SUSY to significantly depart from the SM predictions in CP violating phenomena.

2 Flavour Blind SUSY Breaking and *CP* Violation

We have seen in the previous section that in any MSSM there are additional phases which can cause deviations from the predictions of the SM in CP violation experiments. In fact, in the CMSSM, there are already two new phases present, (1), and for most of the MSSM parameter space, the experimental bounds on the electric dipole moments (EDM) of the electron and neutron constrain these phases to be at most $\mathcal{O}(10^{-2})$. However, in the last few years, the possibility of having non–zero SUSY phases has again attracted a great deal of attention. Several new mechanisms have been proposed to suppress supersymmetric contributions to EDMs below the experimental bounds while allowing SUSY phases $\mathcal{O}(1)$. Methods of suppressing the EDMs consist of cancellation of various SUSY contributions among themselves [10], non universality of the soft breaking parameters at the unification scale [11] and approximately degenerate heavy sfermions for the first two generations [12]. In the presence of one of these mechanisms, large supersymmetric phases are naturally expected and EDMs should be generally close to the experimental bounds. [1]

In this section we will study the effects of these phases in CP violation observables as ε_K, ε'/ε and B^0 CP asymmetries. Following our work of [14] it is clear that the presence of large SUSY phases is not enough to produce sizeable supersymmetric contributions to these observables. In fact, *in the absence of the CKM phase, a general MSSM with all possible phases in the soft–breaking terms, but no new flavor structure beyond the usual Yukawa matrices, can never give a sizeable contribution to ε_K, ε'/ε or hadronic B^0 CP asymmetries.* However, we will see in the next section, that as soon as one introduces some new flavor structure in the soft SUSY–breaking sector, even if the CP violating phases are flavor independent, it is indeed possible to get sizeable CP contribution for large SUSY phases and $\delta_{CKM} = 0$. Then, we can rephrase our sentence above in a different way: *A new result in hadronic B^0 CP asymmetries in the framework of supersymmetry would be a direct proof of the existence of a completely new flavor structure in the soft–breaking terms.* This means that B–factories will probe the flavor structure of the supersymmetry soft–breaking terms even before the direct discovery of the supersymmetric partners [14].

3 CP Violation in the Presence of New Flavour Structures

In Sect. 2, we have shown that CP violation effects are always small in models with flavor blind soft–breaking terms. However, as soon as one introduces some new flavor structure in the soft breaking sector, it is indeed possible to get sizeable CP contribution for large SUSY phases and $\delta_{CKM} = 0$ [11,29,30]. To show this, we will mainly concentrate in new supersymmetric contributions to ε'/ε.

In the CMSSM, the SUSY contribution to ε'/ε is small [31,14]. However in a MSSM with a more general framework of flavor structure it is relatively easy to obtain larger SUSY effects to ε'/ε. In [32] it was shown that such large SUSY contributions arise once one assumes that: i) hierarchical quark Yukawa matrices are protected by flavor symmetry, ii) a generic dependence of Yukawa matrices on Polonyi/moduli fields is present (as expected in many supergravity/superstring theories), iii) the Cabibbo rotation originates from the down–sector and iv) the phases are of order unity. In fact, in [32], it was illustrated how the observed ε'/ε could be mostly or entirely due to the SUSY contribution.

[1] In a more general (and maybe more natural) MSSM there are many other CP violating phases [13] that contribute to CP violating observables.

The universality of the breaking is a strong assumption and is known not to be true in many supergravity and string inspired models [33]. In these models, we expect at least some non–universality in the squark mass matrices or tri–linear terms at the supersymmetry breaking scale. Hence, sizeable flavor–off-diagonal entries will appear in the squark mass matrices. In this regard, gluino contributions to ε'/ε are especially sensitive to $(\delta^d_{12})_{LR}$; even $|\mathrm{Im}(\delta^d_{12})^2_{LR}| \sim 10^{-5}$ gives a significant contribution to ε'/ε while keeping the contributions from this MI to Δm_K and ε_K well bellow the phenomenological bounds. The situation is the opposite for L–L and R–R mass insertions; the stringent bounds on $(\delta^d_{12})_{LL}$ and $(\delta^d_{12})_{RR}$ from Δm_K and ε_K prevent them to contribute significantly to ε'/ε.

The LR squark mass matrix has the same flavor structure as the fermion Yukawa matrix and both, in fact, originate from the superpotential couplings. It may be appealing to invoke the presence of an underlying flavor symmetry restricting the form of the Yukawa matrices to explain their hierarchical forms. Then, the LR mass matrix is expected to have a very similar form as the Yukawa matrix. Indeed, we expect the components of the LR mass matrix to be roughly the SUSY breaking scale (e.g., the gravitino mass) times the corresponding component of the quark mass matrix. However, there is no reason for them to be simultaneously diagonalizable based on this general argument. To make an order of magnitude estimate, we take the down quark mass matrix for the first and second generations to be (following our assumption iii)),

$$Y^d v_1 \simeq \begin{pmatrix} m_d & m_s V_{us} \\ & m_s \end{pmatrix}, \tag{3}$$

where the (2,1) element is unknown due to our lack of knowledge on the mixings among right–handed quarks (if we neglect small terms $m_d V_{cd}$). Based on the general considerations on the LR mass matrix above, we expect

$$m^{2\,(d)}_{LR} \simeq m_{3/2} \begin{pmatrix} a m_d & b m_s V_{us} \\ & c m_s \end{pmatrix}, \tag{4}$$

where a, b, c are constants of order unity. Unless $a = b = c$ exactly, M_d and $m^{2,d}_{LR}$ are not simultaneously diagonalizable and we find

$$(\delta^d_{12})_{LR} \simeq \frac{m_{3/2} m_s V_{us}}{m^2_{\tilde{q}}} = 2 \times 10^{-5} \cdot \\ \left(\frac{m_s(M_{Pl})}{50\ \mathrm{MeV}}\right) \left(\frac{m_{3/2}}{m_{\tilde{q}}}\right) \left(\frac{500\ \mathrm{GeV}}{m_{\tilde{q}}}\right) \tag{5}$$

It turns out that, following the simplest implementation along the lines of the above described idea, the amount of flavor changing LR mass insertion in the s and d–squark propagator results to roughly saturate the bound from ε'/ε if a SUSY phase of order unity is present [32].

This line of work has received a great deal of attention in recent times, after the last experimental measurements of ε'/ε in KTeV and NA31 [34,35]. The effects of non–universal A terms in CP violation experiments were previously analyzed by Abel and Frere [36] and after this new measurement discussed in many different works [11]. In the following we show a complete realization of the above Masiero–Murayama (MM) mechanism from a Type I string–derived model recently presented by one of the authors [37].

3.1 Type I String Model and ε'/ε

In first place we explain our starting model, which is based on type I string models. Our purpose is to study explicitly CP violation effects in models with non–universal gaugino masses and A–terms. Type I models can realize such initial conditions. These models contain nine–branes and three types of five–branes (5_a, $a = 1, 2, 3$). Here we assume that the gauge group $SU(3) \times U(1)_Y$ is on a 9–brane and the gauge group $SU(2)$ on the 5_1–brane like in [29,38], in order to get non–universal gaugino masses between $SU(3)$ and $SU(2)$. We call these branes the $SU(3)$–brane and the $SU(2)$–brane, respectively.

Chiral matter fields correspond to open strings spanning between branes. Thus, they must be assigned accordingly to their quantum numbers. For example, the chiral field corresponding to the open string between the $SU(3)$ and $SU(2)$ branes has non–trivial representations under both $SU(3)$ and $SU(2)$, while the chiral field corresponding to the open string, which starts and ends on the $SU(3)$–brane, should be an $SU(2)$–singlet.

There is only one type of the open string that spans between the 9 and 5–branes, that we denote as the C^{95_1}. However, there are three types of open strings which start and end on the 9–brane, that is, the C^9_i sectors (i=1,2,3), corresponding to the i–th complex compact dimension among the three complex dimensions. If we assign the three families to the different C^9_i sectors we obtain non–universality in the right–handed sector. Notice that, in this model, we can not derive non–universality for the squark doublets, i.e. the left–handed sector. In particular, we assign the C^9_1 sector to the third family and the C^9_3 and C^9_2, to the first and second families, respectively.

Under the above assignment of the gauge multiplets and the matter fields, soft SUSY breaking terms are obtained, following the formulae in [16]. The gaugino masses are obtained

$$M_3 = M_1 = \sqrt{3} m_{3/2} \sin\theta e^{-i\alpha_S}, \tag{6}$$

$$M_2 = \sqrt{3} m_{3/2} \cos\theta \Theta_1 e^{-i\alpha_1}. \tag{7}$$

While the A–terms are obtained as

$$A_{C^9_1} = -\sqrt{3} m_{3/2} \sin\theta e^{-i\alpha_S} = -M_3, \tag{8}$$

for the coupling including C^9_1, i.e. the third family,

$$\begin{aligned} A_{C^9_2} &= -M_3 - \sqrt{3} m_{3/2} \cos\theta (\Theta_1 e^{-i\alpha_1} - \Theta_2 e^{-i\alpha_2}), \\ A_{C^9_3} &= -M_3 - \sqrt{3} m_{3/2} \cos\theta (\Theta_1 e^{-i\alpha_1} - \Theta_3 e^{-i\alpha_3}), \end{aligned} \tag{9}$$

for the second and first families. Here $m_{3/2}$ is the gravitino mass, α_S and α_i are the CP phases of the F–terms of the dilaton field S and the three moduli fields T_i, and θ and Θ_i are goldstino angles, and we have the constraint, $\sum \Theta_i^2 = 1$.

Thus, if quark fields correspond to different C_i^9 sectors, we have non–universal A–terms. We obtain the following trilinear SUSY breaking matrix, $(Y^A)_{ij} = (Y)_{ij}(A)_{ij}$,

$$Y^A = \begin{pmatrix} & & \\ & Y_{ij} & \\ & & \end{pmatrix} \cdot \begin{pmatrix} A_{C_3^9} & 0 & 0 \\ 0 & A_{C_2^9} & 0 \\ 0 & 0 & A_{C_1^9} \end{pmatrix} \tag{10}$$

In addition, soft scalar masses for quark doublets and the Higgs fields are obtained,

$$m^2_{C^{95_1}} = m^2_{3/2}(1 - \frac{3}{2}\cos^2\theta(1 - \Theta_1^2)). \tag{11}$$

The soft scalar masses for quark singlets are obtained as

$$m^2_{C_i^9} = m^2_{3/2}(1 - 3\cos^2\theta\Theta_i^2), \tag{12}$$

if it corresponds to the C_i^9 sector.

Now, below the string or SUSY breaking scale, this model is simply a MSSM with non–trivial soft–breaking terms from the point of view of flavor. Scalar mass matrices and tri–linear terms have completely new flavor structures, as opposed to the super–gravity inspired CMSSM or the SM, where the only connection between different generations is provided by the Yukawa matrices.

This model includes, in the quark sector, 7 different structures of flavor, M_Q^2, M_U^2, M_D^2, Y_d, Y_u, Y_d^A and Y_u^A. From these matrices, M_Q^2, the squark doublet mass matrix, is proportional to the identity matrix, and hence trivial, then we are left with 6 non–trivial flavor matrices. Notice that we have always the freedom to diagonalize the hermitian squark mass matrices (as we have done in the previous section, (11,12)) and fix some general form for the Yukawa and tri–linear matrices. In this case, these four matrices are completely observable, unlike in the SM or CMSSM case.

At this point, to specify completely the model, we need not only the soft–breaking terms but also the complete Yukawa textures. The only available experimental information is the Cabbibo–Kobayashi–Maskawa (CKM) mixing matrix and the quark masses. Here, we choose our Yukawa texture following two simple assumptions : i) the CKM mixing matrix originates from the down Yukawa couplings (as done in the MM case) and ii) our Yukawa matrices are hermitian [39]. With these two assumptions we fix completely the Yukawa matrices as $v_1 Y_d = K^\dagger \cdot M_d \cdot K$ and $v_2 Y_u = M_u$, with M_d and M_u diagonal quark mass matrices, K the Cabibbo–Kobayashi–Maskawa (CKM) mixing matrix and $v = v_1/(\cos\beta) = v_2/(\sin\beta) = \sqrt{2}M_W/g$. We take $\tan\beta = v_2/v_1 = 2$ in the following in all numerical examples. In this basis we can analyze the down tri–linear matrix at the string scale,

$$Y_d^A = K^\dagger \cdot \frac{M_d}{v_1} \cdot K \cdot \begin{pmatrix} A_{C_3^9} & 0 & 0 \\ 0 & A_{C_2^9} & 0 \\ 0 & 0 & A_{C_1^9} \end{pmatrix} \quad (13)$$

Hence, together with the up tri–linear matrix we have our MSSM completely defined. The next step is simply to use the MSSM Renormalization Group Equations [22,21] to obtain the whole spectrum and couplings at the low scale, M_W. The dominant effect in the tri–linear terms renormalization is due to the gluino mass which produces the well–known alignment among A–terms and gaugino phases. However, this renormalization is always proportional to the Yukawa couplings and not to the tri–linear terms. This implies that, in the SCKM basis, the gluino effects will be diagonalized in excellent approximation, while due to the different flavor structure of the tri–linear terms large off–diagonal elements will remain with phases $\mathcal{O}(1)$ [32]. To see this more explicitly, we can roughly approximate the RGE effects at M_W as,

$$Y_d^A = c_{\tilde{g}} m_{\tilde{g}} Y_d + c_A Y_d \cdot \begin{pmatrix} A_{C_3^9} & 0 & 0 \\ 0 & A_{C_2^9} & 0 \\ 0 & 0 & A_{C_1^9} \end{pmatrix} \quad (14)$$

with $m_{\tilde{g}}$ the gluino mass and $c_{\tilde{g}}$, c_A coefficients order 1 (typically $c_{\tilde{g}} \simeq 5$ and $c_A \simeq 1$).

We go to the SCKM basis after diagonalizing all the Yukawa matrices (that is, $K.Y_d.K^\dagger = M_d/v_1$). In this basis, we obtain the tri–linear couplings as,

$$Y_d^A = \Big(c_A \frac{M_d}{v_1} \cdot K \cdot \text{Diag}\,(A_{C_3^9}, A_{C_2^9}, A_{C_1^9}) \cdot K^\dagger \\ + c_{\tilde{g}}\; m_{\tilde{g}} \frac{M_d}{v_1} \Big) \quad (15)$$

From this equation we can get the L–R down squark mass matrix $m_{LR}^{2\;(d)} = v_1\; Y_d^{A^*} - \mu e^{i\varphi_\mu} \tan\beta\, M_d$. And finally using unitarity of K we obtain for the L–R Mass Insertions,

$$(\delta_{LR}^{(d)})_{ij} = \frac{m_i}{m_{\tilde{q}}^2} \Big(\delta_{ij}\; (c_A A_{C_3^9}^* \;+\; c_{\tilde{g}}\; m_{\tilde{g}}^*) - \\ \delta_{ij} \mu e^{i\varphi_\mu} \tan\beta + K_{i2}\; K_{j2}^*\; c_A\; (A_{C_2^9}^* - A_{C_3^9}^*) + \\ K_{i3}\; K_{j3}^*\; c_A\; (A_{C_1^9}^* - A_{C_3^9}^*) \Big) \quad (16)$$

where $m_{\tilde{q}}^2$ is an average squark mass and m_i the quark mass. The same rotation must be applied to the L–L and R–R squark mass matrices,

$$M_{LL}^{(d)\,2}(M_W) = K\;.\;M_Q^2(M_W)\;.\;K^\dagger \\ M_{RR}^{(d)\,2}(M_W) = K\;.\;M_D^2(M_W)\;.\;K^\dagger \quad (17)$$

However, the off–diagonal MI in these matrices are sufficiently small in this case thanks to the universal and dominant contribution from gluino to the squark mass matrices in the RGE.

At this point, with the explicit expressions for $(\delta^{(d)}_{LR})_{ij}$, we can study the gluino mediated contributions to EDMs and ε'/ε. In this non–universal scenario, it is relatively easy to maintain the SUSY contributions to the EDM of the electron and the neutron below the experimental bounds while having large SUSY phases that contribute to ε'/ε. This is due to the fact the EDM are mainly controled by flavor–diagonal MI, while gluino contributions to ε'/ε are controled by $(\delta^{(d)}_{LR})_{12}$ and $(\delta^{(d)}_{LR})_{21}$. Here, we can have a very small phase for $(\delta^{(d)}_{LR})_{11}$ and $(\delta^{(u)}_{LR})_{11}$ and phases $\mathcal{O}(1)$ for the off–diagonal elements without any fine–tuning [37]. It is important to remember that the observable phase is always the relative phase between these mass insertions and the relevant gaugino mass involved. In (16) we can see that the diagonal elements tend to align with the gluino phase, hence to have a small EDM, it is enough to have the phases of the gauginos and the μ term approximately equal, $\alpha_S = \alpha_1 = -\varphi_\mu$. However α_2 and α_3 can still contribute to off–diagonal elements. In Fig. 1 we show the allowed values for α_S, α_2 and α_3 assuming $\alpha_1 = \varphi_\mu = 0$. We impose the EDM, ε_K and $b \to s\gamma$ bounds separately for gluino and chargino contributions together with

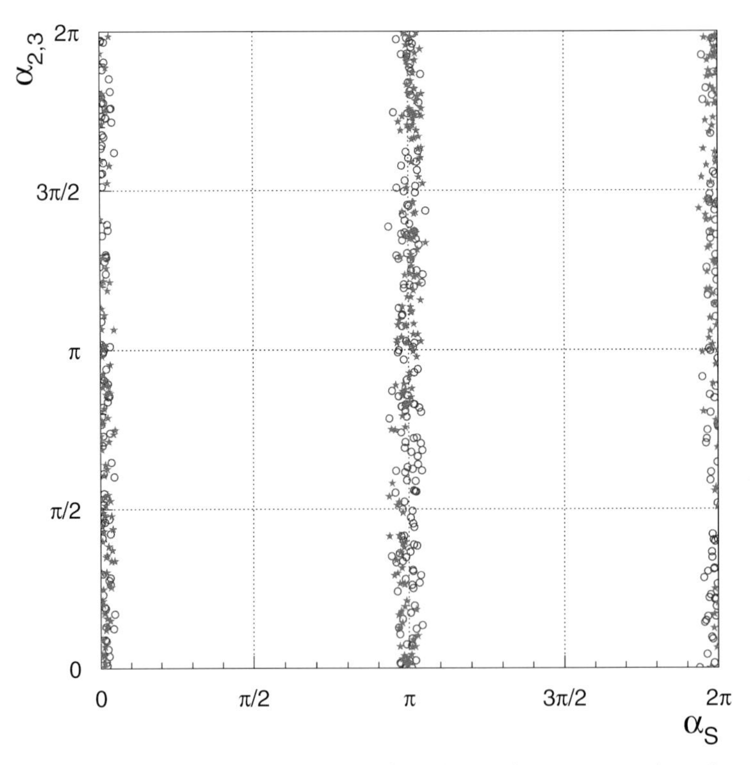

Fig. 1. Allowed values for α_2–α_S (open circles) and α_3–α_S (stars)

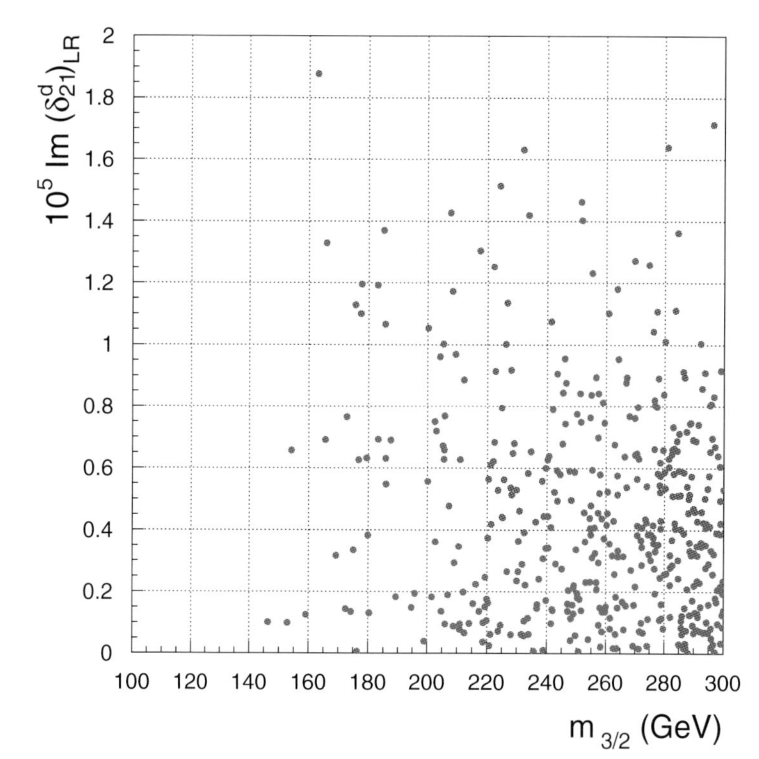

Fig. 2. $(\delta_{LR}^{(d)})_{21}$ versus $m_{3/2}$ for experimentally allowed regions of the SUSY parameter space

the usual bounds on SUSY masses. We can see that, similarly to the CMSSM situation, φ_μ is constrained to be very close to the gluino and chargino phases (in the plot $\alpha_S \simeq 0, \pi$), but α_2 and α_3 are completely unconstrained.

Finally, in Fig. 2, we show the effects of these phases in the $(\delta_{LR}^{(d)})_{21}$ MI as a function of the gravitino mass. All the points in this plot satisfy all CP–conserving constraints besides EDM and ε_K constraints. We must remember that a value of $|\mathrm{Im}(\delta_{12}^d)_{LR}^2| \sim 10^{-5}$ gives a significant contribution to ε'/ε. In this plot, we can see a large percentage of points above or close to 1×10^{-5}. Hence, we can conclude that, in the presence of new flavor structures in the SUSY soft–breaking terms, it is not difficult to obtain sizeable SUSY contributions to CP violation observables and specially to ε'/ε [32,37].[2]

[2] With these L–R mass insertions alone, it is in general difficult to saturate ε_K [20]. However, in some special situations, it is still possible to have large contributions [29,40]. On the other hand, the L–L mass insertions can naturally contribute to ε_K [41]

4 Conclusions and Outlook

Here we summarize the main points of this talk:

- Flavor and CP problems constrain low–energy SUSY, but, at the same time, provide new tools to search for SUSY indirectly.
- In all generality, we expect new CP violating phases in the SUSY sector. However, these new phases are not going to produce sizeable effects as long as the SUSY model we consider does not exhibit a new flavor structure in addition to the SM Yukawa matrices.
- In the presence of a new flavor structure in SUSY, we showed that large contributions to CP violating observables are indeed possible.

In summary, given the fact that LEP searches for SUSY particles are close to their conclusion and that for Tevatron it may be rather challenging to find a SUSY evidence, we consider CP violation a potentially precious ground for SUSY searches before the advent of the "SUSY machine", LHC.

References

1. M. Dugan, B. Grinstein and L. Hall, Nucl. Phys. B **255**, 413 (1985).
2. S. Dimopoulos and S. Thomas, Nucl. Phys. B **465**, 23 (1996).
3. W. Buchmuller and D. Wyler, Phys. Lett. B **121**, 321 (1983). J. Polchinski and M. Wise, Phys. Lett. B **125**, 393 (1983). W. Fischler, S. Paban and S. Thomas, Phys. Lett. B **289**, 373 (1992).
4. J. Ellis and R. Flores, Phys. Lett. B **377**, 83 (1996).
5. M. Dine, A. Nelson and Y. Shirman, Phys. Rev. D **51**, 1362 (1995); M. Dine, Y. Nir and Y. Shirman, Phys. Rev. D **55**, 1501 (1997).
6. M. J. Duncan and J. Trampetic, Phys. Lett. B **134**, 439 (1984). E. Franco and M. Mangano, Phys. Lett. B **135**, 445 (1984). J.M. Gerard, W. Grimus, A. Raychaudhuri and G. Zoupanos, Phys. Lett. B **140**, 349 (1984). J. M. Gerard, W. Grimus, A. Masiero, D. V. Nanopoulos and A. Raychaudhuri, Phys. Lett. B **141**, 79 (1984). Nucl. Phys. B **253**, 93 (1985). P. Langacker and R. Sathiapalan, Phys. Lett. B **144**, 401 (1984). M. Dugan, B. Grinstein and L. Hall, in [1].
7. A. Brignole, F. Feruglio and F. Zwirner, Z. Phys. C **71**, 679 (1996).
8. M. Misiak, S. Pokorski and J. Rosiek, hep-ph/9703442.
9. G.C. Branco, G.C. Cho, Y. Kizukuri and N. Oshimo, Nucl. Phys. B **449**, 483 (1995); G.C. Branco, G.C. Cho, Y. Kizukuri and N. Oshimo, Phys. Lett. B **337**, 316 (1994).
10. T. Ibrahim and P. Nath, Phys. Rev. D **58**, 111301 (1998); M. Brhlik, G.J. Good and G.L. Kane, Phys. Rev. D **59**, 115004 (1999); A. Bartl, T. Gajdosik, W. Porod, P. Stockinger and H. Stremnitzer, Phys. Rev. D **60**, 073003 (1999).
11. S.A. Abel and J.M. Frere, Phys. Rev. D **55**, 1623 (1997); S. Khalil, T. Kobayashi and A. Masiero, Phys. Rev. D **60**, 075003 (1999); S. Khalil and T. Kobayashi, Phys. Lett. B **460**, 341 (1999).
12. S. Dimopoulos and G.F. Giudice, Phys. Lett. B **357**, 573 (1995); A. Cohen, D.B. Kaplan and A.E. Nelson, Phys. Lett. B **388**, 599 (1996); A. Pomarol and D. Tommasini, Nucl. Phys. B **466**, 3 (1996).

13. Y. Grossman, Y. Nir and R. Rattazzi, hep-ph/9701231.
14. D. Demir, A. Masiero and O. Vives, Phys. Lett. B **479**, 230 (2000).
15. D. Demir, A. Masiero and O. Vives, Phys. Rev. D **61**, 075009 (2000).
16. L.E. Ibanez, C. Munoz and S. Rigolin, Nucl. Phys. B **553**, 43 (1999).
17. M. Brhlik, L. Everett, G.L. Kane and J. Lykken, Phys. Rev. Lett. **83**, 2124 (1999); M. Brhlik, L. Everett, G.L. Kane and J. Lykken, Phys. Rev. D **62**, 035005 (2000). T. Ibrahim and P. Nath, Phys. Rev. D **61**, 093004 (2000).
18. D. Demir, A. Masiero and O. Vives, Phys. Rev. Lett. **82**, 2447 (1999), Err. ibid **83**, 2093 (1999).
19. S. Baek and P. Ko, Phys. Rev. Lett. **83**, 488 (1999); S. Baek and P. Ko, Phys. Lett. B **462**, 95 (1999).
20. F. Gabbiani, E. Gabrielli, A. Masiero and L. Silvestrini, Nucl. Phys. B **477**, 321 (1996).
21. S. Bertolini, F. Borzumati, A. Masiero and G. Ridolfi, Nucl. Phys. B **353**, 591 (1991).
22. N.K. Falck, Z. Phys. C **30**, 247 (1986).
23. L.J. Hall, V.A. Kostelecky and S. Raby, Nucl. Phys. B **267**, 415 (1986).
24. H.E. Haber and G.L Kane, Phys. Rep. **117**, 75 (1985).
25. P. Cho, M. Misiak and D. Wyler, Phys. Rev. D **54**, 3329 (1996).
26. F.M. Borzumati, Z. Phys. C **63**, 291 (1994); S. Bertolini and F. Vissani, Z. Phys. C **67**, 513 (1995); T. Goto, Y.Y. Keum, T. Nihei, Y. Okada and Y. Shimizu, Phys. Lett. B **460**, 333 (1999).
27. A.L. Kagan and M. Neubert, Eur. Phys. J. **C7** (1999) 5; A.L. Kagan and M. Neubert, Phys. Rev. D **58**, 094012 (1998).
28. M. Ciuchini *et al.* J. High Energy Phys. **10**, 008 (1998); R. Contino and I. Scimemi, Eur. Phys. J. C **10**, 347 (1999).
29. M. Brhlik, L. Everett, G.L. Kane, S.F. King and O. Lebedev, Phys. Rev. Lett. **84**, 3041 (2000).
30. R. Barbieri, R. Contino and A. Strumia, Nucl. Phys. B **578**, 153 (2000). A. L. Kagan and M. Neubert, Phys. Rev. Lett. **83**, 4429 (1999); K. S. Babu, B. Dutta and R. N. Mohapatra, Phys. Rev. D **61**, 091701 (2000).
31. E. Gabrielli and G. F. Giudice, Nucl. Phys. B **433**, 3 (1995).
32. A. Masiero and H. Murayama, Phys. Rev. Lett. **83**, 907 (1999).
33. See, *e.g.*, Y. Kawamura, H. Murayama, and M. Yamaguchi, Phys. Rev. D **51**, 1337 (1995). A. Brignole, L.E. Ibañez, and C. Muñoz, hep-ph/9707209.
34. KTeV Collaboration, A. Alavi-Harati *et al.*, Phys. Rev. Lett. **83**, 22 (1999).
35. NA31 Collaboration (G.D. Barr *et al.*), Phys. Lett. B **317**, 233 (1993).
36. S.A. Abel and J.M. Frere in [11].
37. S. Khalil, T. Kobayashi and O. Vives, Nucl. Phys. B **580**, 275 (2000).
38. T. Ibrahim and P. Nath in [17].
39. P. Ramond, R. G. Roberts and G. G. Ross, Nucl. Phys. B **406**, 19 (1993).
40. G. D'Ambrosio, G. Isidori and G. Martinelli, hep-ph/9911522.
41. A. Masiero and O. Vives, Phys. Rev. Lett. **86**, 26 (2001). A. Masiero, M. Piai and O. Vives, Phys. Rev. D **64**, 055008 (2001).

Family Replicated Fit of All Quark and Lepton Masses and Mixings

Holger B. Nielsen and Yasutaka Takanishi

[1] Deutsches Elektronen-Synchrotron DESY, Notkestraße 85, D-22603 Hamburg, Germany
[2] The Niels Bohr Institute, Blegdamsvej 17, DK-2100 Copenhagen Ø, Denmark

Abstract. We review our recent development of family replicated gauge group model, which generates the Large Mixing Angle MSW solution. The model is based on each family of quarks and leptons having its own set of gauge fields, each containing a replica of the Standard Model gauge fields plus a $(B-L)$-coupled gauge field. A fit of all the seventeen quark-lepton mass and mixing angle observables, using just six new Higgs field vacuum expectation values, agrees with the experimental data order of magnitudewise. However, this model can not predict the baryogenesis in right order, therefore, we discuss further modification of our model and present a preliminary result of baryon number to entropy ratio.

1 Introduction

We have previously attempted to fit all the fermion masses and their mixing angles [1,2] including baryogenesis [3] in a model without supersymmetry or grand unification. This model has the maximum number of gauge fields consistent with maintaining the irreduciblity of the usual Standard Model fermion representations, added three right-handed neutrinos. The predictions of this previous model are in order of magnitude agreement with all existing experimental data, however, only provided we use the Small Mixing Angle MSW [4] (SMA-MSW) solution. But, for the reasons given below, the SMA-MSW solution is now disfavoured by experiments. So here we review a modified version of the previous model, which manages to accommodate the Large Mixing Angle MSW (LMA-MSW) solution for solar neutrino oscillations using 6 additional Higgs fields (relative to the Standard Model) vacuum expectation values (VEVs) as adjustable parameters.

A neutrino oscillation solution to the solar neutrino problem and a favouring of the LMA-MSW solution is supported by SNO results [5]: The measurement of the ^{8}B and *hep* solar neutrino fluxes shows no significant energy dependence of the electron neutrino survival probability in the Super-Kamiokande and SNO energy ranges.

Moreover, the important result which also supports LMA-MSW solution on the solar neutrino problem, reported by the Super-Kamiokande collaboration [6], that the day-night asymmetry data disfavour the SMA-MSW solution at the 95% C.L..

In fact, global analyses [7–10] of all solar neutrino data have confirmed that the LMA-MSW solution gives the best fit to the data and that the SMA-MSW solution is very strongly disfavoured and only acceptable at the 3σ level. Typical best fit values of the mass squared difference and mixing angle parameters in the two flavour LMA-MSW solution are $\Delta m^2_\odot \approx 4.5 \times 10^{-5}\ \mathrm{eV}^2$ and $\tan^2\theta_\odot \approx 0.35$.

This paper is organised as follows: In the next section, we present our gauge group – the family replicated gauge group – and the quantum numbers of fermion and Higgs fields. Then, in section 3 we discuss our philosophy of all gauge- and Yukawa couplings at Planck scale being of order unity. In section 4 we address how the family replicated gauge group breaks down to Standard Model gauge group, and we add a small review of see-saw mechanism. The mass matrices of all sectors are presented in section 5, the renormalisation group equations – renormalisable and also 5 dimensional non-renormalisable ones – are shown in section 6. The calculation is described in section 7 and the results are presented in section 8. We discuss further modification of our model and present a preliminary results of baryon number to entropy ratio in section 9. Finally, section 10 contains our conclusion.

2 Quantum Numbers of Model

Our model has, as its back-bone, the property that there are generations (or families) not only for fermions but also for the gauge bosons, *i.e.*, we have a generation (family) replicated gauge group namely

$$\times_{i=1,2,3} (SMG_i \times U(1)_{B-L,i}) \ , \tag{1}$$

where SMG denotes the Standard Model gauge group $\equiv SU(3) \times SU(2) \times U(1)$, $\times$ denotes the Cartesian product and i runs through the generations, $i = 1, 2, 3$.

Note that this family replicated gauge group, (1), is the maximal gauge group under the following assumptions:

- It should only contain transformations which change the known 45 (= 3 generations of 15 Weyl particles each) Weyl fermions of the Standard Model and the additional three heavy see-saw (right-handed) neutrinos. That is our gauge group is assumed to be a subgroup of $U(48)$.
- We avoid any new gauge transformation that would transform a Weyl state from one irreducible representation of the Standard Model group into another irreducible representation: there is no gauge coupling unification.
- The gauge group does not contain any anomalies in the gauge symmetry – neither gauge nor mixed anomalies even without using the Green-Schwarz anomaly cancelation mechanism.
- It should be as big as possible under the foregoing assumptions.

The quantum numbers of the particles/fields in our model are found in Table 1 and use of the following procedure: In Table 1 one finds the charges under the six $U(1)$ groups in the gauge group 1. Then for each particle one should take the representation under the $SU(2)_i$ and $SU(3)_i$ groups ($i = 1, 2, 3$) with lowest dimension matching to $y_i/2$ according to the requirement

$$\frac{t_i}{3} + \frac{d_i}{2} + \frac{y_i}{2} = 0 \pmod 1 \ , \tag{2}$$

Table 1. All $U(1)$ quantum charges in the family replicated model. The symbols for the fermions shall be considered to mean "proto"-particles. Non-abelian representations are given by a rule from the abelian ones (see (2)).

	SMG_1	SMG_2	SMG_3	$U_{B-L,1}$	$U_{B-L,2}$	$U_{B-L,3}$
u_L, d_L	$\frac{1}{6}$	0	0	$\frac{1}{3}$	0	0
u_R	$\frac{2}{3}$	0	0	$\frac{1}{3}$	0	0
d_R	$-\frac{1}{3}$	0	0	$\frac{1}{3}$	0	0
e_L, ν_{e_L}	$-\frac{1}{2}$	0	0	-1	0	0
e_R	-1	0	0	-1	0	0
ν_{e_R}	0	0	0	-1	0	0
c_L, s_L	0	$\frac{1}{6}$	0	0	$\frac{1}{3}$	0
c_R	0	$\frac{2}{3}$	0	0	$\frac{1}{3}$	0
s_R	0	$-\frac{1}{3}$	0	0	$\frac{1}{3}$	0
μ_L, ν_{μ_L}	0	$-\frac{1}{2}$	0	0	-1	0
μ_R	0	-1	0	0	-1	0
ν_{μ_R}	0	0	0	0	-1	0
t_L, b_L	0	0	$\frac{1}{6}$	0	0	$\frac{1}{3}$
t_R	0	0	$\frac{2}{3}$	0	0	$\frac{1}{3}$
b_R	0	0	$-\frac{1}{3}$	0	0	$\frac{1}{3}$
τ_L, ν_{τ_L}	0	0	$-\frac{1}{2}$	0	0	-1
τ_R	0	0	-1	0	0	-1
ν_{τ_R}	0	0	0	0	0	-1
ϕ_{WS}	0	$\frac{2}{3}$	$-\frac{1}{6}$	0	$\frac{1}{3}$	$-\frac{1}{3}$
ω	$\frac{1}{6}$	$-\frac{1}{6}$	0	0	0	0
ρ	0	0	0	$-\frac{1}{3}$	$\frac{1}{3}$	0
W	0	$-\frac{1}{2}$	$\frac{1}{2}$	0	$-\frac{1}{3}$	$\frac{1}{3}$
T	0	$-\frac{1}{6}$	$\frac{1}{6}$	0	0	0
χ	0	0	0	0	-1	1
ϕ_{B-L}	0	0	0	0	0	2

where t_i and d_i are the triality and duality for the i'th proto-generation gauge groups $SU(3)_i$ and $SU(2)_i$ respectively.

3 The Philosophy of All Couplings Being Order Unity

Any realistic model and at least certainly our model tends to get far more fundamental couplings than we have parameters in the Standard Model and thus pieces of data to fit. This is especially so for our model based on many $U(1)$ charges [11] because we take it to have practically any not mass protected particles one may propose at the fundamental mass scale, taken to be the Planck mass. Especially we assume the existence of Dirac fermions with order of fundamental scale masses needed to allow the quark and lepton Weyl particles to take up successively gauge charges from the Higgs fields VEVs. So unless we make assumptions about the many coupling constants and fundamental masses we have no chance to predict anything. Almost the only chance of making an assumption about all these couplings, which is not very model dependent, is to assume that they are *all of order unity* in the fundamental unit. This is the same type of assumption that is really behind use of dimensional arguments to estimate sizes of quantities. A procedure very often used successfully. If we really assumed every coupling and mass of order unity we would get the effective Yukawa couplings of the quarks and leptons to the Weinberg-Salam Higgs field to be also of order unity what is phenomenologically not true. To avoid this prediction we then blame the smallness of all but the top-Yukawa coupling on smallness in fundamental Higgs VEVs. That is to say we assume that the VEVs of the Higgs fields in Table 1, ρ, ω, T, W, χ, ϕ_{B-L} and ϕ_{WS} are (possibly) very small compared to the fundamental/Planck unit, and these are the quantities we have to fit.

Technically we implement these unknown – but of order unity according to our assumption – couplings and masses by a Monte Carlo technique: we put them equal to random numbers with a distribution dominated by numbers of order unity and then perform the calculation of the observable quantities such as quark or lepton masses and mixing angles again and again. At the end we average the logarithmic of these quantities and exponentiate them. In this way we expect to get the typical order of magnitude predicted for the observable quantities. In praxis we do not have to put random numbers in for all the many couplings in the fundamental model, but can instead just provide each mass matrix element with a single random number factors.

After all a product of several of order unity factors is just an order unity factor again. To resume our model philosophy: *Only Higgs field VEVs are not of order unity. We must be satisfied with order of magnitude results.*

4 Breaking of the Family Replicated Gauge Group to the Standard Model

The family replicated gauge group broken down to its diagonal subgroup at scales about one or half order of magnitude under the Planck scale by Higgs fields – W, T, ω, ρ and χ (in Table 1):

$$\times_{i=1,2,3}(SMG_i \times U(1)_{B-L,i}) \to SMG \times U(1)_{B-L} \,. \tag{3}$$

This diagonal subgroup is further broken down by yet two more Higgs fields — the Weinberg-Salam Higgs field ϕ_{WS} and another Higgs field ϕ_{B-L} — to $SU(3) \times U(1)_{em}$.

4.1 See-Saw Mechanism

See-saw mechanics is build into our model to fit the scale of the neutrino oscillations, *i.e.*, we use the right-handed neutrinos with heavy Majorana masses (10^{11} GeV).

In order to mass-protect the right-handed neutrino from getting Planck scale masses, we have to introduce ϕ_{B-L} which breaks the $B-L$ quantum charge spontaneously, and using this new Higgs filed we are able to deal the neutrino oscillations, *i.e.*, to fit the scale of the see-saw particle masses. However, due to mass-protection by the Standard Model gauge symmetry, the left-handed Majorana mass terms should be negligible in our model. Then, naturally, the light neutrino mass matrix – effective left-left transition Majorana mass matrix – can be obtained via the see-saw mechanism [12]:

$$M_{\rm eff} \approx M_\nu^D \, M_R^{-1} \, (M_\nu^D)^T \, . \tag{4}$$

5 Mass Matrices

Using the $U(1)$ fermion quantum charges and Higgs field (presented in Table 1) we can calculate the degrees of suppressions of the left-right transition – Dirac mass – matrices and also Majorana mass matrix (right-right transition).

Note that the random complex order of unity numbers which are supposed to multiply all the mass matrix elements are not represented in following matrices: the up-type quarks:

$$M_{\scriptscriptstyle U} \simeq \frac{\langle(\phi_{\rm WS})^\dagger\rangle}{\sqrt{2}} \begin{pmatrix} (\omega^\dagger)^3 W^\dagger T^2 & \omega\rho^\dagger W^\dagger T^2 & \omega\rho^\dagger (W^\dagger)^2 T \\ (\omega^\dagger)^4 \rho W^\dagger T^2 & W^\dagger T^2 & (W^\dagger)^2 T \\ (\omega^\dagger)^4 \rho & 1 & W^\dagger T^\dagger \end{pmatrix} \tag{5}$$

the down-type quarks:

$$M_{\scriptscriptstyle D} \simeq \frac{\langle\phi_{\rm WS}\rangle}{\sqrt{2}} \begin{pmatrix} \omega^3 W (T^\dagger)^2 & \omega\rho^\dagger W (T^\dagger)^2 & \omega\rho^\dagger T^3 \\ \omega^2 \rho W (T^\dagger)^2 & W (T^\dagger)^2 & T^3 \\ \omega^2 \rho W^2 (T^\dagger)^4 & W^2 (T^\dagger)^4 & W T \end{pmatrix} \tag{6}$$

the charged leptons:

$$M_{\scriptscriptstyle E} \simeq \frac{\langle\phi_{\rm WS}\rangle}{\sqrt{2}} \begin{pmatrix} \omega^3 W (T^\dagger)^2 & (\omega^\dagger)^3 \rho^3 W (T^\dagger)^2 & (\omega^\dagger)^3 \rho^3 W T^4 \chi \\ \omega^6 (\rho^\dagger)^3 W (T^\dagger)^2 & W (T^\dagger)^2 & W T^4 \chi \\ \omega^6 (\rho^\dagger)^3 (W^\dagger)^2 T^4 & (W^\dagger)^2 T^4 & W T \end{pmatrix} \tag{7}$$

the Dirac neutrinos:

$$M_\nu^D \simeq \frac{\langle(\phi_{\rm WS})^\dagger\rangle}{\sqrt{2}} \begin{pmatrix} (\omega^\dagger)^3 W^\dagger T^2 & (\omega^\dagger)^3 \rho^3 W^\dagger T^2 & (\omega^\dagger)^3 \rho^3 W^\dagger T^2 \chi \\ (\rho^\dagger)^3 W^\dagger T^2 & W^\dagger T^2 & W^\dagger T^2 \chi \\ (\rho^\dagger)^3 W^\dagger T^\dagger \chi^\dagger & W^\dagger T^\dagger \chi^\dagger & W^\dagger T^\dagger \end{pmatrix} \tag{8}$$

and the Majorana (right-handed) neutrinos:

$$M_R \simeq \langle\phi_{\rm B-L}\rangle \begin{pmatrix} (\rho^\dagger)^6 (\chi^\dagger)^2 & (\rho^\dagger)^3 (\chi^\dagger)^2 & (\rho^\dagger)^3 \chi^\dagger \\ (\rho^\dagger)^3 (\chi^\dagger)^2 & (\chi^\dagger)^2 & \chi^\dagger \\ (\rho^\dagger)^3 \chi^\dagger & \chi^\dagger & 1 \end{pmatrix} \tag{9}$$

6 Renormalisation Group Equations from Planck Scale to Week Scale via See-Saw Scale

It should be kept in mind that the effective Yukawa couplings for the Weinberg-Salam Higgs field, which are given by the Higgs field factors in the above mass matrices multiplied by order unity factors, are the running Yukawa couplings at a scale *near the Planck scale.* In this way, we had to use the renormalisation group (one-loop) β-functions to run these couplings down to the experimentally observable scale which we took for the charged fermion masses to be compared to "measurements" at the scale of 1 GeV, except for the top quark mass prediction. We define the top quark pole mass:

$$M_t = m_t(M)\left(1 + \frac{4}{3}\frac{\alpha_s(M)}{\pi}\right) \ , \tag{10}$$

where we put $M = 180$ GeV for simplicity.

We use the one-loop β functions for the gauge couplings and the charged fermion Yukawa matrices [13] as follows:

$$\begin{aligned}
16\pi^2 \frac{dg_1}{dt} &= \frac{41}{10}\, g_1^3 \\
16\pi^2 \frac{dg_2}{dt} &= -\frac{19}{16}\, g_2^3 \\
16\pi^2 \frac{dg_3}{dt} &= -7\, g_3^3 \\
16\pi^2 \frac{dY_U}{dt} &= \frac{3}{2}\,\left(Y_U (Y_U)^\dagger - Y_D (Y_D)^\dagger\right)\, Y_U \\
&\quad + \left\{ Y_S - \left(\frac{17}{20} g_1^2 + \frac{9}{4} g_2^2 + 8 g_3^2\right)\right\}\, Y_U \\
16\pi^2 \frac{dY_D}{dt} &= \frac{3}{2}\,\left(Y_D (Y_D)^\dagger - Y_U (Y_U)^\dagger\right)\, Y_D \\
&\quad + \left\{ Y_S - \left(\frac{1}{4} g_1^2 + \frac{9}{4} g_2^2 + 8 g_3^2\right)\right\}\, Y_D \\
16\pi^2 \frac{dY_E}{dt} &= \frac{3}{2}\,\left(Y_E (Y_E)^\dagger\right)\, Y_E \\
&\quad + \left\{ Y_S - \left(\frac{9}{4} g_1^2 + \frac{9}{4} g_2^2\right)\right\}\, Y_E \\
Y_S &= \mathrm{Tr}(\, 3\, Y_U^\dagger\, Y_U + 3\, Y_D^\dagger\, Y_D + Y_E^\dagger\, Y_E\,) \ ,
\end{aligned} \tag{11}$$

where $t = \ln\mu$.

By calculation we use the following initial values of gauge coupling constants:

$$U(1): \quad g_1(M_Z) = 0.462 \ , \quad g_1(M_{\mathrm{Planck}}) = 0.614 \tag{12}$$

$$SU(2): \quad g_2(M_Z) = 0.651 \ , \quad g_2(M_{\mathrm{Planck}}) = 0.504 \tag{13}$$

$$SU(3): \quad g_3(M_Z) = 1.22 \ , \quad g_3(M_{\mathrm{Planck}}) = 0.491 \tag{14}$$

6.1 The Renormalisation Group Equations for the Effective Neutrino Mass Matrix

The effective light neutrino masses are given by an irrelevant, nonrenormalisable (5 dimensional term) – effective mass matrix $M_{\rm eff}$ – for which the running formula is [14]:

$$16\pi^2 \frac{dM_{\rm eff}}{dt} = (-3g_2^2+2\lambda+2Y_S)\, M_{\rm eff} - \frac{3}{2}\left(M_{\rm eff}\,(Y_E Y_E^\dagger)^T + (Y_E Y_E^\dagger)\, M_{\rm eff}\right) \quad , \quad (15)$$

where λ is the Weinberg-Salam Higgs self-coupling constant and the mass of the Standard Model Higgs boson is given by $M_H^2 = \lambda \left\langle \phi_{WS} \right\rangle^2$. We just for simplicity take $M_H = 115$ GeV thereby we ignore the running of the Higgs self-coupling and fixed as $\lambda = 0.2185$.

Note that the renormalisation group equations are used to evolve the effective neutrino mass matrix from the see-saw sale, set by $\langle \phi_{B-L} \rangle$ in our model, to 1 GeV.

7 Method of Numerical Computation

In the philosophy of order unity numbers spelled out in Sect. 3 we evaluate the product of mass-protecting Higgs VEVs required for each mass matrix element and provide it with a random complex number, λ_{ij}, of order one as a factor taken to have Gaussian distribution with mean value zero. But we hope the exact form of distribution does not matter much provided we have $\langle \ln |\lambda_{\rm ij}| \rangle = 0$. In this way, we simulate a long chain of fundamental Yukawa couplings and propagators making the transition corresponding to an effective Yukawa coupling in the Standard Model and the parameters in neutrino sector. In the numerical computation we then calculate the masses and mixing angles time after time, using different sets of random numbers and, in the end, we take the logarithmic average of the calculated quantities according to the following formula:

$$\langle m \rangle = \exp\left(\sum_{i=1}^{N} \frac{\ln \rm m_i}{N} \right) . \quad (16)$$

Here $\langle m \rangle$ is what we take to be the prediction for one of the masses or mixing angles, m_i is the result of the calculation done with one set of random number combinations and N is the total number of random number combinations used.

Since we only expect to make order of magnitude fits, we should of course not use ordinary χ^2 defined form the experimental uncertainties by rather the χ^2 that would correspond to a relative uncertainly – an uncertain factor of order unity. Since the normalisation of such a χ^2 is not so easy to choose exactly we define instead a quantity which we call the goodness of fit (g.o.f.). Since our model can only make predictions order of magnitudewise, this quantity g.o.f. should only depend on the ratios of the fitted masses and mixing angles to the

Table 2. Best fit to conventional experimental data. All masses are running masses at 1 GeV except the top quark mass which is the pole mass. Note that we use the square roots of the neutrino data in this table, as the fitted neutrino mass and mixing parameters $\langle m \rangle$, in our goodness of fit (g.o.f.) definition, (17).

	Fitted	Experimental
m_u	4.4 MeV	4 MeV
m_d	4.3 MeV	9 MeV
m_e	1.0 MeV	0.5 MeV
m_c	0.63 GeV	1.4 GeV
m_s	340 MeV	200 MeV
m_μ	80 MeV	105 MeV
M_t	208 GeV	180 GeV
m_b	7.2 GeV	6.3 GeV
m_τ	1.1 GeV	1.78 GeV
V_{us}	0.093	0.22
V_{cb}	0.027	0.041
V_{ub}	0.0025	0.0035
$\Delta m^2_\odot$	9.5×10^{-5} eV^2	4.5×10^{-5} eV^2
$\Delta m^2_{\rm atm}$	2.6×10^{-3} eV^2	3.0×10^{-3} eV^2
$\tan^2 \theta_\odot$	0.23	0.35
$\tan^2 \theta_{\rm atm}$	0.65	1.0
$\tan^2 \theta_{13}$	4.8×10^{-2}	$\lesssim 2.6 \times 10^{-2}$
g.o.f.	3.63	–

experimentally determined masses and mixing angles:

$$\text{g.o.f.} \equiv \sum \left[\ln \left(\frac{\langle \mathrm{m} \rangle}{\mathrm{m}_{\exp}} \right) \right]^2 , \tag{17}$$

where $m_{\exp}$ are the corresponding experimental values presented in Table 2.

We should emphasise that we do not adjust the order of one numbers by selection, *i.e.*, the complex random numbers are needed for only calculational purposes. That means that we have only six adjustable parameters – VEVs of Higgs fields – and, on the other hand, that the averages of the predicted quantities, $\langle m \rangle$, are just results of integration over the "dummy" variables – random numbers – therefore, the random numbers are not at all parameters!

Strictly speaking, however, one could consider the choice of the distribution of the random order unity numbers as parameters. But we hope that provided we impose on the distribution the conditions that the average be zero and the average of the logarithm of the numerical value be zero, too, any reasonably smooth distribution would give similar results for the $\langle m \rangle$ values at the end. In our early work [2] we did see that a couple of different proposals did not make too much difference.

8 Results

We averaged over $N = 10,000$ complex order unity random number combinations. These complex numbers are chosen to be a number picked from a Gaussian distribution, with mean value zero and standard deviation one, multiplied by a random phase factor. We put them as factors into the mass matrices (5-9). Then we computed averages according to (16) and used (17) as a χ^2 to fit the 6 free parameters and found:

$$\begin{aligned} &\langle\phi_{WS}\rangle = 246\ \text{GeV}\,,\ \langle\phi_{B-L}\rangle = 1.64\times 10^{11}\ \text{GeV}\,,\ \langle\omega\rangle = 0.233\,, \\ &\langle\rho\rangle = 0.246\,,\ \langle W\rangle = 0.134\,,\ \langle T\rangle = 0.0758\,,\ \langle\chi\rangle = 0.0737\,, \end{aligned} \qquad (18)$$

where, except for the Weinberg-Salam Higgs field and $\langle\phi_{B-L}\rangle$, the VEVs are expressed in Planck units. Hereby we have considered that the Weinberg-Salam Higgs field VEV is already fixed by the Fermi constant. The results of the best fit, with the VEVs in (18), are shown in Table 2 and the fit has g.o.f. $= 3.63$.

We have $11 = 17 - 6$ degrees of freedom – predictions – leaving each of them with a logarithmic error of $\sqrt{3.63/11} \simeq 0.57$, which is very close to the theoretically expected value 64% [15]. This means that we can fit all quantities within a factor $\exp\left(\sqrt{3.63/11}\right) \simeq 1.78$ of the experimental values.

From the Table 2 the experimental mass values are a factor two higher than predicted for down, charm and for the Cabibbo angle V_{us} while they are smaller by a factor for strange and electron. Thinking only on the angles and masses (not squared) the agreement is in other cases better than a factor two.

Experimental results for the values of neutrino mixing angles are often presented in terms of the function $\sin^2 2\theta$ rather than $\tan^2\theta$ (which, contrary to $\sin^2 2\theta$, does not have a maximum at $\theta = \pi/4$ and thus still varies in this region). Transforming from $\tan^2\theta$ variables to $\sin^2 2\theta$ variables, our predictions for the neutrino mixing angles become:

$$\sin^2 2\theta_{\odot} = 0.61\,, \qquad (19)$$

$$\sin^2 2\theta_{\text{atm}} = 0.96\,, \qquad (20)$$

$$\sin^2 2\theta_{13} = 0.17\,. \qquad (21)$$

We also give here our predicted hierarchical neutrino mass spectrum:

$$m_1 = 4.9\times 10^{-4}\ \text{eV}\,, \qquad (22)$$

$$m_2 = 9.7\times 10^{-3}\ \text{eV}\,, \qquad (23)$$

$$m_3 = 5.2\times 10^{-2}\ \text{eV}\,. \qquad (24)$$

Our agreement with experiment is excellent: all of our order of magnitude neutrino predictions lie inside the 99% C.L. border determined from phenomenological fits to the neutrino data, even including the CHOOZ upper bound. Our prediction of the solar mass squared difference is about a factor of 2 larger than the global data fit even though the prediction is inside of the LMA-MSW region, giving a contribution to our goodness of fit of g.o.f. ≈ 0.14. Our CHOOZ angle also turns out to be about a factor of 2 larger than the experimental limit at 90% C.L., delivering another contribution of g.o.f. ≈ 0.14. In summary our predictions for the neutrino sector agree extremely well with the data, giving a contribution of only 0.34 to g.o.f. while the charged fermion sector contributes 3.29 to g.o.f..

8.1 *CP* Violation

Since we have taken our random couplings to be – whenever allowed – complex we have order of unity or essentially maximal CP-violation so a unitary triangle with angles of order one is a success of our model. After our fitting of masses and of mixings we can simply predict order of magnitudewise of CP-violation in *e.g.* $K^0 - \bar{K}^0$ decay or in CKM and MNS mixing matrices in general.

The Jarlskog area J_{CP} provides a measure of the amount of CP violation in the quark sector [16] and, in the approximation of setting cosines of mixing angles to unity, is just twice the area of the unitarity triangle:

$$J_{CP} = V_{us}\, V_{cb}\, V_{ub}\, \sin\delta\,, \tag{25}$$

where δ is the CP violation phase in the CKM matrix. In our model the quark mass matrix elements have random phases, so we expect δ (and also the three angles α, β and γ of the unitarity triangle) to be of order unity and, taking an average value of $|\sin\delta| \approx 1/2$, the area of the unitarity triangle becomes

$$J_{CP} \approx \frac{1}{2}\, V_{us}\, V_{cb}\, V_{ub}\,. \tag{26}$$

Using the best fit values for the CKM elements from Table 2, we predict $J_{CP} \approx 3.1 \times 10^{-6}$ to be compared with the experimental value $(2 - 3.5) \times 10^{-5}$. Since our result for the Jarlskog area is the product of four quantities, we do not expect the usual $\pm 64\%$ logarithmic uncertainty but rather $\pm\sqrt{4}\cdot 64\% = 128\%$ logarithmic uncertainty. This means our result deviates from the experimental value by $\ln(\frac{2.7\times 10^{-5}}{3.1\times 10^{-6}})/1.28 = 1.7$ "standard deviations".

The Jarlskog area has been calculated from the best fit parameters in Table 2, it is also possible to calculate them directly while making the fit. So we have calculated J_{CP} for $N = 10,000$ complex order unity random number combinations. Then we took the logarithmic average of these $10,000$ samples of J_{CP} and obtained the following result:

$$J_{CP} = 3.1 \times 10^{-6}\,, \tag{27}$$

in good agreement with the values given above.

8.2 Neutrinoless Double Beta Decay

Another prediction, which can also be made from this model, is the electron "effective Majorana mass" – the parameter in neutrinoless beta decay – defined by:

$$|\langle m \rangle| \equiv \left| \sum_{i=1}^{3} U_{ei}^2 \, m_i \right| \,, \tag{28}$$

where m_i are the masses of the neutrinos ν_i and U_{ei} are the MNS mixing matrix elements for the electron flavour to the mass eigenstates i. We can substitute values for the neutrino masses m_i from (22-24) and for the fitted neutrino mixing angles from Table 2 into the left hand side of (28). As already mentioned, the CP violating phases in the MNS mixing matrix are essentially random in our model. So we combine the three terms in (28) by taking the square root of the sum of the modulus squared of each term, which gives our prediction:

$$|\langle m \rangle| \approx 3.1 \times 10^{-3} \ \ \mathrm{eV} \,. \tag{29}$$

In the same way as being calculated the Jarlskog area we can compute using $N = 10,000$ complex order unity random number combinations to get the $|\langle m \rangle|$. Then we took the logarithmic average of these 10,000 samples of $|\langle m \rangle|$ as usual:

$$|\langle m \rangle| = 4.4 \times 10^{-3} \ \ \mathrm{eV} \,. \tag{30}$$

This result does not agree with the central value of recent result – "evidence" – from the Heidelberg-Moscow collaboration [17].

9 Baryogenesis via Lepton Number Violation

Having now a well fitted model giving orders of magnitude for all the Yukawa couplings and having the see-saw mechanism, it is obvious that we ought to calculate the amount of baryons Y_B relative to entropy being produced via the Fukugita-Yanagida mechanism [18]. According to this mechanism the decay of the right-handed neutrinos by CP-violating couplings lead to an excess of the $B - L$ charge (meaning baryon number minus lepton number), the relative excess in the decay from Majorana neutrino generation number i being called ϵ_i. This excess is then immediately – and continuously back and forth – being converted partially to a baryon number excess, although it starts out as being a lepton number L asymmetry, since the right-handed neutrinos decay to leptons and Weinberg-Salam Higgs particles. It is a complicated discussion to estimate to what extend the $B - L$ asymmetry is washed out later in the cosmological development, but our estimates goes that there is not enough baryon number excess left to fit the Big Bang development at the stage of formation of the light elements primordially (nuclearsynthesis).

Recently we have, however, developed a modified version [19] of our model – only deviating in the right-handed sector – characterized by changing the quantum numbers assumed for the see-saw scale producing Higgs field ϕ_{B-L} in such a way that the biggest matrix elements in the right-handed mass matrix (9) becomes the pair of – because of the symmetry – identical off diagonal elements (row, column)=(2,3) and (3,2). Thereby we obtain two almost degenerate right-handed neutrinos and that helps for making the $B - L$ asymmetry in the decay bigger. In this modified model that turns out to fit the rest of our predictions approximately equally well or even better we then get a very satisfactory baryon number relative to entropy prediction

$$Y_B \approx 2.5 \times 10^{-11} \,. \tag{31}$$

In the same time as making this modification of the ϕ_{B-L} quantum numbers we also made some improvements in the calculation by taking into account the running of the Dirac neutrino Yukawa couplings from the Planck scale to the corresponding right-handed neutrino scales. Also, we calculated more accurate dilution factors than previous our work [3]. However, foregoing work was based on the mass matrices which predicted the SMA-MSW, so we must investigate the baryogenesis using the present mass matrices, of course, with the modified right-handed Majorana mass matrix.

10 Conclusion

We have reviewed our model which is able to predict the experimentally favored LMA-MSW solution rather than the SMA-MSW solution for solar neutrino oscillations after careful choice of the $U(1)$ charges for the Higgs fields causing transitions between 1st and 2nd generations. However, the fits of charged lepton quantities become worse compare to our "old" model that can predict SMA-MSW solar neutrino solution. On the other hand, we now can fit the neutrino quantities very well: the price paid for the greatly improved neutrino mass matrix fit – the neutrino parameters now contribute only very little to the g.o.f. – is a slight deterioration in the fit to the charged fermion mass matrices. In particular the predicted values of the quark masses m_d and m_c and the Cabibbo angle V_{us} are reduced compared to our previous fits. However the overall fit agrees with the seventeen measured quark-lepton mass and mixing angle parameters in Table 2 within the theoretically expected uncertainty [15] of about 64%; it is a perfect fit order of magnitudewise. It should be remarked that our model provides an order of magnitude fit/understanding of all the effective Yukawa couplings of the Standard Model and the neutrino oscillation parameters in terms of only 6 parameters – the Higgs field vacuum expectation values!

Acknowledgments

We wish to thank the organisers of the 8th Adriatic Meeting for the wonderful organisation and for the hospitality extended to us during the symposia. We thank to the Volkswagenstiftung for financial support. Y.T. wishes to thank S. Koch for friendship and afternoon swimming lessons at Adriatic beach.

References

1. C. D. Froggatt, H. B. Nielsen and Y. Takanishi, Nucl. Phys. B **631**, 285 (2002).
2. H. B. Nielsen and Y. Takanishi, Nucl. Phys. B **588**, 281 (2000); Nucl. Phys. B **604**, 405 (2001).
3. H. B. Nielsen and Y. Takanishi, Phys. Lett. B **507**, 241 (2001).
4. L. Wolfenstein, Phys. Rev. D **17**, 2369 (1978); Phys. Rev. D **20**, 2634 (1979); S. P. Mikheev and A. Yu. Smirnov, Sov. J. Nucl. Phys. **42**, 913 (1985); Nuovo Cim. C **9**, 17 (1986).
5. Q. R. Ahmad *et al.*, SNO Collaboration, Phys. Rev. Lett. **87**, 071301 (2001).
6. S. Fukuda *et al.*, Super-Kamiokande Collaboration, Phys. Rev. Lett. **86**, 5656 (2001).
7. G. L. Fogli, E. Lisi, D. Montanino and A. Palazzo, Phys. Rev. D **64**, 093007 (2001).
8. J. N. Bahcall, M. C. Gonzalez-Garcia and C. Peña-Garay, J. High Energy Phys. **0108**, 014 (2001).
9. A. Bandyopadhyay, S. Choubey, S. Goswami and K. Kar, Phys. Lett. B **519**, 83 (2001).
10. P. I. Krastev and A. Yu. Smirnov, hep-ph/0108177.
11. C. D. Froggatt and H. B. Nielsen, Nucl. Phys. B **147**, 277 (1979).
12. T. Yanagida, in Proceedings of the Workshop on Unified Theories and Baryon Number in the Universe, Tsukuba, Japan (1979), eds. O. Sawada and A. Sugamoto, KEK Report No. 79-18; M. Gell-Mann, P. Ramond and R. Slansky in Supergravity, Proceedings of the Workshop at Stony Brook, NY (1979), eds. P. van Nieuwenhuizen and D. Freedman (North-Holland, Amsterdam, 1979).
13. H. Arason, D. J. Castaño, B. Keszthelyi, S. Mikaelian, E. J. Piard, P. Ramond and B. D. Wright, Phys. Rev. D **46**, 3945 (1992).
14. S. Antusch, M. Drees, J. Kersten, M. Lindner and M. Ratz, Phys. Lett. B **519**, 238 (2001); P. H. Chankowski and P. Wasowicz, hep-ph/0110237.
15. C. D. Froggatt, H. B. Nielsen and D. J. Smith, hep-ph/0108262.
16. C. Jarlskog, Phys. Rev. Lett. **55**, 1039 (1985).
17. H. V. Klapdor-Kleingrothaus, A. Dietz, H. L. Harney and I. V. Krivosheina, Mod. Phys. Lett. A **16**, 2409 (2002).
18. M. Fukugita and T. Yanagida, Phys. Lett. B **174**, 45 (1986).
19. H. B. Nielsen and Y. Takanishi, "Baryogenesis via Lepton Number Violation in Family Replicated Gauge Groups Model", in progress.

Nonleptonic Two Body B Decays and CP Violation

Anthony I. Sanda

Nagoya University, Japan

Abstract. We discuss perturbative QCD method to compute branching ratios of two body decay modes. We emphasize that penguin annihilation diagrams give non-negligible contributions. They also give final state interaction phases which in turn leads to large **CP** asymmetries in many two body decays.

1 Introduction

There are two curious questions:

(1) Why does the factorization approximation work so well? For this question we refer the reader to [1] which describes the approximation and shows that it must be a first approximation to some systematic expansion.

(2) Why penguin amplitudes are so large? The fact that $Br(B \to K\pi) \sim Br(B \to \pi\pi)$ implies[2] that $\frac{Penguin}{Tree} \sim \lambda$(where $\lambda = \sin\theta_C$, the Cabibbo angle), while we would have guessed that this ratio is $\mathcal{O}(\lambda^2)$.

In this lecture I will introduce perturbative QCD(PQCD) method which attempts to understand these questions in a systematic manner. To introduce PQCD, it is most convenient to start with a discussion of the pion form factor.

2 Feynman's Approach

In the seventies, Feynman visited Fermilab where I was a postdoc. He told me one day that he figured out how to look at the form factor with in the context of the parton model. His picture for pion form factor is given in Fig. 1. Consider a photon colliding with a pion head on and a final state pion flies off in the opposite direction. The only way for the final state to remain a pion is to hit a parton with $x \sim 1$, where the partons that were not hit are all "wee" partons and they don't know which way they are going. These wee prtons could just as well be moving in the opposite direction. So, the final state is an opositely moving parton with $x \sim 1$ and wee partons moving in the same direction. This state remains a pion.

The pion form factor is related to the probability that all of its momentum is carried by a single parton. If Feynman is correct, form factors can not be computed with perturbative QCD, because the probability that the pion momentum is carried by a single parton requires a detailed knowledge of wee dynamics. Soft hadronic physics is highly non-perturbative.

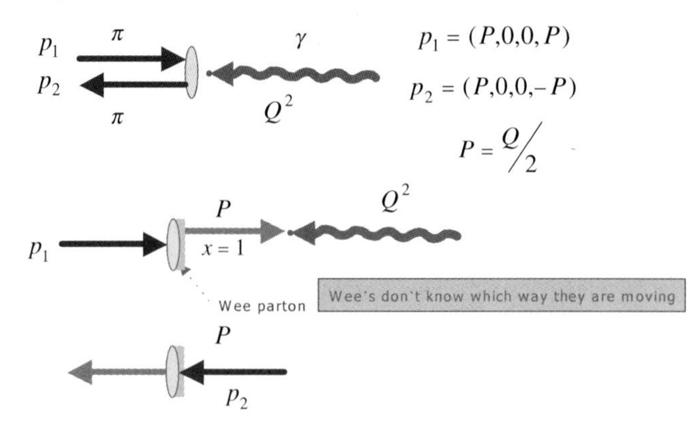

Fig. 1. Feynman's view of the pion form factor. Photon collides with the pion head on and the final state pion flies off in the opposite direction. The only way the final state remains a pion is to hit a parton with $x \sim 1$ where the partons that were not hit are all "wee" partons and they don't know which way they are going. So, the final state also remains a pion.

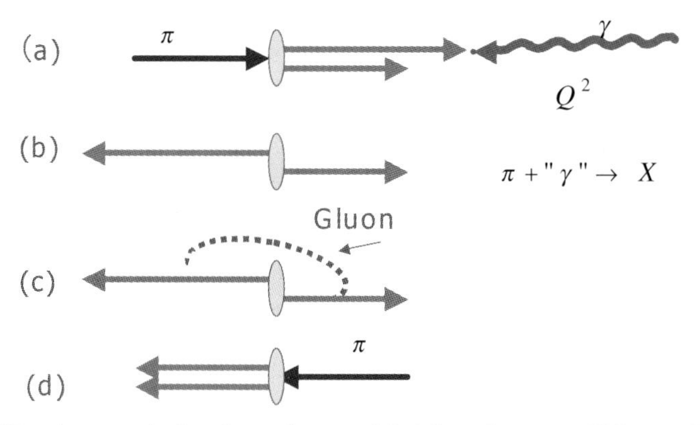

Fig. 2. QCD picture of the form factor. (a) The photon collides with a parton with momentum fraction x. (b) In general, it leads to a deep inelastic scattering $\pi +$ "γ" $\to$ *anything*. (c) In order for the final state to remain a pion, a hard gluon must be exchanged with the spectator parton and the parton which collided and got its momentum reversed. (d) To have a pion in the final state, momenta of these two partons must be aligned.

To understand Feynman's picture, let us write the analytic formula for the reaction shown in Fig. 2.

$$\langle \pi(P')|J_\mu(0)|\pi(P)\rangle = g^2 \int \frac{d^4k_1}{(2\pi)^4}\frac{d^4k_2}{(2\pi)^4} d^4x d^4y e^{-ik_2y} \langle \pi|\bar{u}_\gamma(0)d_\beta(y)|0\rangle$$
$$e^{ik_1x}\langle 0|\bar{d}_\alpha(x)u_\delta(0)|\pi\rangle \, T_H^{\gamma\beta;\alpha\delta} \, , \tag{1}$$

with

$$T_H^{\gamma\beta;\alpha\delta} = [\gamma_\sigma]^{\alpha\beta} \frac{1}{(k_2 - k_1)^2} \left[\gamma_\mu \frac{\not{P} - \not{k}_2}{(P - k_2)^2} \gamma^\sigma \right]^{\gamma\delta} . \tag{2}$$

This integral is infrared divrgent. Taking $k_1 = x_1 P$ and $k_2 = x_2 P'$, which leads to $(k_2 - k_1)^2 \sim x_1 x_2$, the above integral is dominated by the end-point regions $x_1,\ x_2 \to 0$. In the region $x_1 \to 0$, the momentum carried by the other quark is $(1 - x_1)P$ or it carries all the momentum of the pion. Similarly, $x_2 \sim 0$ is the region where the other quark carries all the momentum. Probably Feynman wrote this formula and realized that this integral contains an infrared divergence[3] near $x \sim 1$. Feynman's picture in which the major contribution to the integral comes from $x \sim 1$ is obtained if the we can regulate his integral some how, and the infrared divergent region contributed most to the integral.

This is the Feynman picture. We can not compute the form factor but it is related to the probability that a pion is in a state where one of the parton carries all the momentum of the pion.

This picture is challenged by [4,5]. In [5], it was shown that when the end-point contributions are important, the transverse momenta $k_\perp$ of the partons must be taken into account, since $(k_2 - k_1)^2 \sim -2x_1x_2P \cdot P' + 2k_{2\perp} \cdot k_{1\perp}$. The inclusion of the parton transverse degrees of freedom then leads to the existence of double logarithms $\alpha_s \ln^2(\mathrm{Q}/\mathrm{k}_\perp)$. The resummation of these double logarithms [5–7], which is required by QCD and gives a Sudakov form factor, strongly suppresses the end-point contributions where $k_{2\perp} \cdot k_{1\perp}$ is small. Thus the kinematic region favored by Feynman does not contribute very much. The major contribution comes from the region where the exchanged gluon in Fig. 2c is hard and $T_H^{\gamma\beta;\alpha\delta}$ which contains the dynamic of hard scattering can be computed by PQCD. All soft non-perturbative dynamics can be absorbed in wave functions.

The suggestion that a form factor can be calculated in PQCD as an expansion in twists and in powers of the coupling constant α_s was first made in [4]. The form factor $F(Q)$ is then expressed as

$$F(Q) = \psi^\Lambda(q\bar{q}) \Gamma(q\bar{q} : q\bar{q}) \psi^\Lambda(q\bar{q}) + \cdots , \tag{3}$$

with Q a large energy scale. The first term contains leading-twsit (twist-2) contributions, and $\cdots$ represents those from higher Fock states, which are down by powers of Q in the light-cone gauge and by powers of α_s. Hence, a Fock state with more partons is of higher twist.

We have also found that some higher twist terms give important contributions and thus can not be neglected. For example, chiral symmetry demands that the pion wave function must contain

$$\phi_\pi(x) = \not{P} \phi_A(x) + m_0 \phi_P(x) \tag{4}$$

where $m_0 = \frac{m_\pi^2}{m_u + m_d}$. Since $m_0 \sim 1.4 GeV$, while it is strictly a higher twist term, we need to keep it since $\frac{m_0}{M_B} \sim .3$.

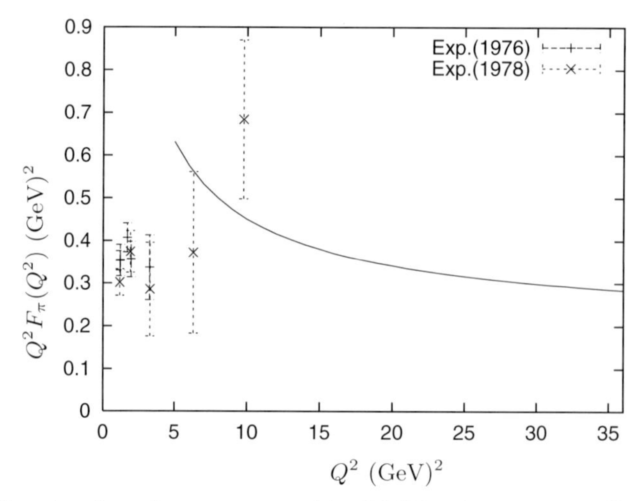

Fig. 3. The pion form factor computed in PQCD using $m_0 = 1.4GeV$. Our message here is that the pion wave function can be chosen so thet the pion form factor is in the ball park of the experimental data. The figure was produced by K. Ukai.

In Fig. 3 we show the pion form factor computed using PQCD where the pion wave function obtained by QCD sum rule has been used. Previous computations using similar method but without the component of the wave function which is proportional to m_0 have yielded result which is too small.

3 Sudakov Form Factor

Since the infrared singularity which was in the back of Feynman's mind was avoided by the Sudakov factor, and it is rather crucial in the validity of the PQCD method, we shall discuss its physical meaning here. First let me point out that QCD requires us to put the Sudakov factor in. It is not something we put it in by hand.

We all know that when an electron scatters off a photon with large Q^2, the probability that it remains an electron vanishes with $Q^2 \to \infty$. This is because it is bound to emit a collinear photon. So, the final state is an electron plus arbitrary many photons. The probability for the final state to remain an electron without any photon, is suppressed by the Sudakov form factor in QED.

If a single quark has a hard interaction with momentum transfer Q^2, the probability that it remains a single quark also vanishes as $Q^2 \to \infty$. This is because the quark will emit collinear gluons. Just like in QED electron scattering. If quark and anti-quark separated by a transverse distance b undergo hard scattering, then the probability for emitting a soft gluon depends on b. If b is small, it will not emit a collinear gluon, as they shield each other's color quantum numbers - they are almost colorless. If b is large, it can emit a collinear gluon, because the quark and anti-quark can be seen as separate particles. So, we have a b dependent Sudakov factor. If b is large, the Sudakov factor suppresses the amplitude for exclusive reaction because, by definition, exclusive reaction can not have gluons being emitted. If amplitude for large b is suppressed, in momentum space, it implies that small $k_\perp$ region is suppressed. Thus Sudakov form factor saves the amplitude from having infrared singularity.

4 $B \to \pi$ Transition Form Factor

The same line of argument used in computing the pion form factor can be used for computing $B \to \pi$ transition form factor $F^{B\pi}(0)$. The relevant diagram is shown in Fig. 4. The result is depicted in Fig. 5.

[8] has obtained a value, which is very small compared to the results obtained from light-cone QCD sum rules or from lattice QCD, which is $F^{B\pi}(0) \sim 0.3$. This has led some authors to conclude that $F^{B\pi}(0)$ is dominated by non-perturbative dynamics and PQCD prediction is incorrect.

We point out that terms involving m_0 has been ignored and the asymptotic form for the pion wave function has been used in [8], in contradiction with the finding of the light cone sum rule.

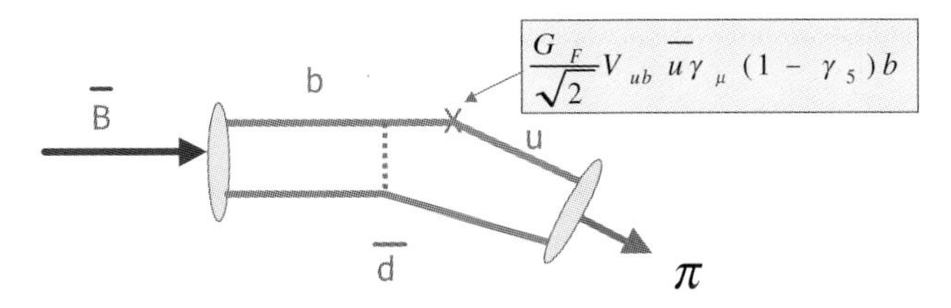

Fig. 4. The b quark is struck by the current and u quark flies out. The spectator quark exchanges a hard gluon and it lines itself with the fast u quark so that it forms a pion.

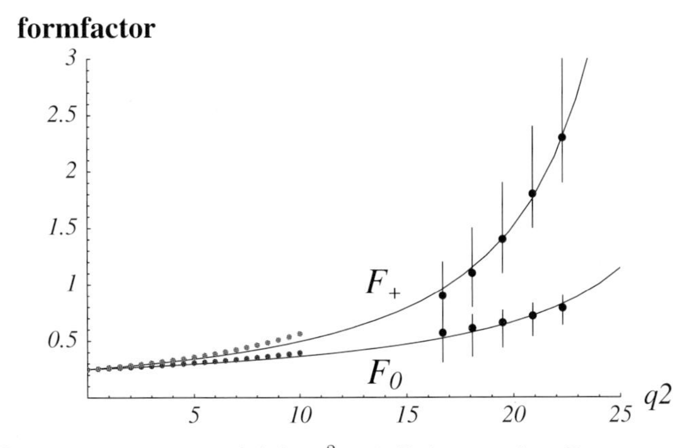

Fig. 5. PQCD computation is valid for $q^2 \sim 0$. It is seen that B meson wave function may be chosen so that PQCD computation is consistent with the extrapolation of the lattice result. This figure was produced by T. Kurimoto.

5 $B \to K\pi$

We have presented some theoretical anticipations for the $B \to K\pi$ [9] and $B \to \pi\pi$ decays [10] in the PQCD framework. Table 1 gives relative importance of each amplitude which contribute to $B \to K\pi$. F_e: tree; F_e^P: penguin; F_a^P: penguin annihilation. These are factorizable contributions and M's are non-factorizable amplitudes. In factorization approximation, we ignore non-factorizable amplitudes. We see that this is indeed a good approximation.

In factorization approximation, annihilation diagrams were assumed to be small. The reason for this assumption is that the form factor in the time like region ($Q^2 \sim M_B^2$) is not known and the factorizable annihilation diagram can not be computed. We see that the penguin annihilation diagrams give a relatively large contribution which is imaginary and factorization approximation should fail here.

Table 1. The numerical value of amplitudes contributing to $B^0 \to K^+\pi^-$ decay. They will be multiplied by appropriate KM matrix elements.

F_e	5.577×10^{-1}
F_e^P	-5.537×10^{-2}
F_a^P	$3.333 \times 10^{-3} + i\, 3.181 \times 10^{-2}$
M_e	$-0.942 \times 10^{-3} + i\, 3.385 \times 10^{-3}$
M_e^P	$2.931 \times 10^{-5} - i\, 1.304 \times 10^{-4}$
M_a^P	$-9.397 \times 10^{-5} - i\, 1.918 \times 10^{-4}$

It can be seen that F_e^P, the penguin annihilation diagram gives relatively large imaginary part.

This leads to our prediction that **CP**violation:

$$\frac{Br(B^0 \to K^+\pi^-) - Br(\overline{B}^0 \to K^-\pi^+)}{Br(B^0 \to K^+\pi^-) + Br(\overline{B}^0 \to K^-\pi^+)} \approx (15 \sim 30)\%$$
$$\frac{Br(B^0 \to K^+\pi^-) - Br(\overline{B}^0 \to K^-\pi^+)}{Br(B^+ \to K^+\pi^0) + Br(B^- \to K^-\pi^0)} \approx (10 \sim 20)\% \tag{5}$$

The range corresponds to possible variation in the KM phase ϕ_3.

6 Summary

We have described a PQCD method for computing two body B decays. We have just began to explore this method and still much work remains. Results depend crucially on hadron wave function we use. This is not surprising. Two body decays must depend on non-perturbative dynamics. Surprising thing is that the

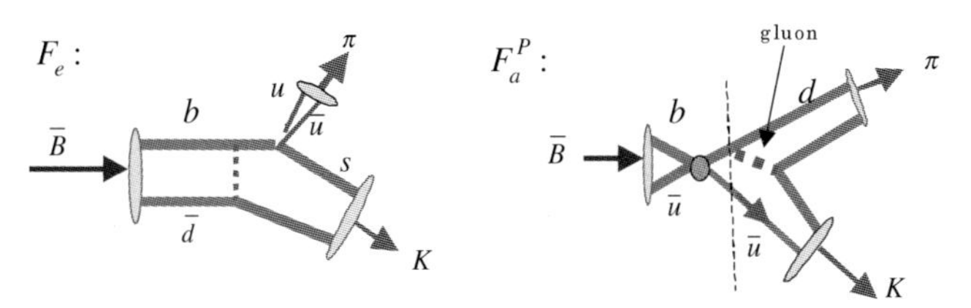

Fig. 6. F_e: tree and F_a^P penguin annihilation diagram. F_e^P produces large imaginary part to $B \to K\pi$ amplitude. It produces large **CP** asymmetry.

non-perturbative effects can be introduced through hadronic wave functions. Once these wave functions are know then we understand these decays. In particular, we now understand why vacuum saturation approximation works so well. The nonfactorizable diagrams M's in Table 1 are small. Penguins are dynamically enhanced. We thus understand why penguins are larger than what we have originally guessed. Finally we predict large **CP** asymmetry, which is less sensitive to parameters in wave functions.

Acknowledgements

The work was supported in part by Grant-in Aid for Special Project Research (Physics of CP violation); Grant-in Aid for Scientific from the Ministry of Education, Science and Culture of Japan. WI thank my collaborators: Y. Keum, E. Kou, T. Kurimoto, H-n. Li, C. D. Lu, T. Morozumi, N. Shinha, R. Shinha, K. Ukai, M. Yang, T. Yoshikawa.

References

1. M. Neubert and B. Stech in *Heavy flavours II*, eds. A. J. Buras and M. Lindner (World Scientific, Singapore).
2. I. I. Bigi and A. I. Sanda *CP Violation*,p.205. (Cambridge University Press, Cambridge, 2000).
3. N. Isgur, C.H. Llewellyn Smith Nucl. Phys. B **317**, 526 (1989).
4. G.P. Lepage and S.J. Brodsky, Phys. Lett. B **87**, 359 (1979); G.P. Lepage and S.J. Brodsky, Phys. Rev. D **22**, 2157 (1980);Phys. Lett. B **243**, 287 (1990) ; A. Szczepaniak, E.M. Henley, and S.J. Brodsky, (1987); Z. Phys. C **29**, 637 (1985).
5. H-n. Li and G.Sterman, Nucl. Phys. B **381**, 129 (1992).
6. J.C. Collins and D.E. Soper, Nucl. Phys. B **193**, 381 (1981).
7. J. Botts and G.Sterman, Nucl. Phys. B **225**, 62 (1989).
8. T. Feldmann, P. Kroll Eur. Phys.J. **C12**, 99 (2000).
9. Y.Y. Keum, H-n. Li, and A.I. Sanda, hep-ph/0004004; hep-ph/0004173.
10. C. D. Lü, K. Ukai, and M. Z. Yang, hep-ph/0004213.

States of Strongly Interacting Matter

Helmut Satz

Fakultät für Physik, Universität Bielefeld, D-33501 Bielefeld, Germany

Abstract. I discuss the phase structure of strongly interacting matter at high temperatures and densities, as predicted by statistical QCD, and consider in particular the nature of the transition of hot hadronic matter to a plasma of deconfined quarks and gluons.

1 Hadronic Matter and Beyond

Hadrons have an intrinsic size, with a radius of about 1 fm. Hence a hadron needs a volume $V_h = (4\pi/3)r_h^3 \simeq 4 \text{ fm}^3$ to exist. This implies an upper limit n_c to the density of hadronic matter, $n_h < n_c$, with $n_c = V_h^{-1} \simeq 0.25 \text{ fm}^{-3} \simeq 1.5\, n_0$, where $n_0 \simeq 0.17 \text{ fm}^{-3}$ denotes standard nuclear density. Fifty years ago, Pomeranchuk pointed out that this also leads to an upper limit for the temperature of hadronic matter [1]. An overall volume $V = NV_h$ causes the grand canonical partition function to diverge when $T \geq T_c \simeq 1/r_h \simeq 0.2$ GeV.

This conclusion was subsequently confirmed by more detailed dynamical accounts of hadron dynamics. Hagedorn proposed a self-similar composition pattern for hadronic resonances, the statistical bootstrap model, in which the degeneracy of a given resonant state is determined by the number of ways of partitioning it into more elementary constituents [2]. The solution of this classical partitioning problem [3] is a level density increasing exponentially with mass, $\rho(m) \sim \exp\{am\}$, which leads to a diverging partition function for an ideal resonance gas once its temperature exceeds the value $T_H = 1/a$, which turns out to be close to the pion mass. A yet more complete and detailed description of hadron dynamics, the dual resonance model, confirmed this exponential increase of the resonance level density [4,5]. While Hagedorn had speculated that T_H might be an upper limit of the temperature of all matter, Cabbibo and Parisi pointed out that T_H could be a critical temperature signalling the onset of a new quark phase of strongly interacting matter [6]. In any case, it seems clear today that hadron thermodynamics, based on what we know about hadron dynamics, contains its own intrinsic limit [7].

On one hand, the quark infrastructure of hadrons provides a natural explanation of such a limit; on the other hand, it does so in a new way, different from all previous reductionist approaches: quarks do not have an independendent existence, and so reductionalism is at the end of the line, in just the way proposed by Lucretius.

The limit of hadron thermodynamics can be approached in two ways. One is by compressing cold nuclear matter, thus increasing the baryon density beyond values of one baryon per baryon volume. The other is by heating a meson gas to temperatures at which collisions produce further hadrons and thus increase the hadron density beyond values allowing each hadron its own volume. In either case, the medium will undergo a transition from a state in which its constituents were colorless, i.e., color-singlet bound states of colored quarks and gluons, to a state in which the constituents are colored. This end of hadronic matter is generally referred to as deconfinement.

The colored constituents of deconfined matter

- could be massive *constituent quarks*, obtained if the liberated quarks dress themselves with gluon clouds;
- or the liberated quarks could couple pairwise to form bosonic colored *di-quarks*;
- or the system could consist of unbound quarks and gluons, the *quark-gluon plasma* (QGP).

One of the tasks of statistical QCD is to determine if and when these different possible states can exist.

In an idealized world, the potential binding a heavy quark-antiquark pair into a color-neutral hadron has the form of a string,

$$V(r) \sim \sigma r, \tag{1}$$

where σ specifies the string tension. For $r \to \infty$, $V(r)$ also diverges, indicating that a hadron cannot be dissociated into its quark constituents: quarks are confined. In a hot medium, however, thermal effects are expected to soften and eventually melt the string at some deconfinement temperature T_c. This would provide the string tension with the temperature behavior

$$\sigma(T) = \begin{cases} \sigma(0)\ [T_c - T]^a & T < T_c, \\ 0 & T > T_c, \end{cases} \tag{2}$$

with a as critical exponent for the order parameter $\sigma(T)$. For $T < T_c$, we then have a medium consisting of color-neutral hadrons, for $T > T_c$ a plasma of colored quarks and gluons. The confinement/deconfinement transition is thus the QCD version of the insulator/conductor transition in atomic matter.

In the real world, the string breaks when $V(r)$ becomes larger than the energy of two separate color singlet bound states, i.e., when the 'stretched' hadron becomes energetically more expensive than two hadrons of normal size. It is thus possible to study the behavior of (2) only in quenched QCD, without dynamical quarks and hence without the possibility of creating new $q\bar{q}$ pairs. The result [8] is shown in Fig. 1, indicating that $a \simeq 0.5$. We shall return to the case of full QCD and string breaking in Sect. 4.

The insulator-conductor transition in atomic matter is accompanied by a shift in the effective constituent mass: collective effects due to lattice oscillations, mean electron fields etc. give the conduction electron a mass different from the electron mass in vacuum. In QCD, a similar phenomenon is expected. At $T = 0$,

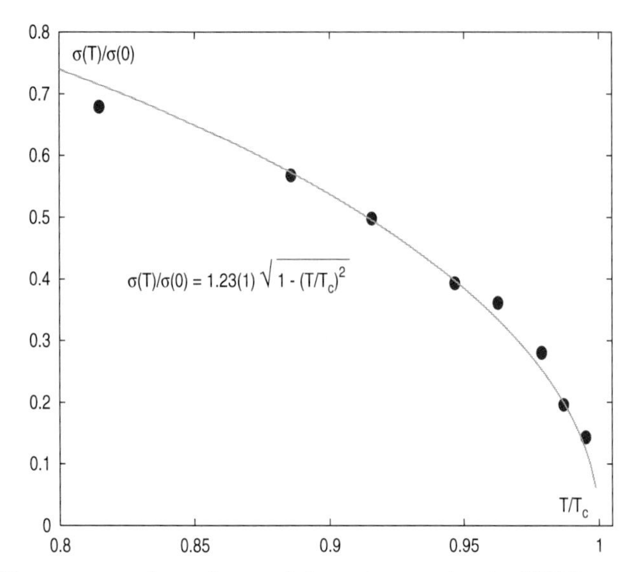

Fig. 1. Temperature dependence of the string tension in $SU(3)$ gauge theory

the bare quarks which make up the hadrons 'dress' themselves with gluons to form constituent quarks of mass $M_q \simeq 300 - 350$ MeV. The mass of a nucleon then is basically that of three constituent quarks, that of the ρ meson twice M_q. With increasing temperature, as the medium gets hotter, the quarks tend to shed their dressing. In the idealized case of massless bare quarks, the QCD Lagrangian $\mathcal{L}_{\rm QCD}$ possesses chiral symmetry: four-spinors effectively reduce to two independent two-spinors. The dynamically created constituent quark mass at low T thus corresponds to a spontaneous breaking of this chiral symmetry, and if at some high $T = T_\chi$ the dressing and hence the constituent quark mass disappears, the chiral symmetry of $\mathcal{L}_{\rm QCD}$ is restored. Similar to the string tension behavior of (2) we thus expect

$$M_q(T) = \begin{cases} M_q(0)\ [T_\chi - T]^b & T < T_\chi, \\ 0 & T > T_\chi. \end{cases} \tag{3}$$

for the constituent quark mass: T_χ separates the low temperature phase of broken chiral symmetry and the high temperature phase in which this is restored, with b as the critical exponent for the chiral order parameter $M_q(T)$.

An obvious basic problem for statistical QCD is thus the clarification of the relation between T_c and T_χ. In atomic physics the electron mass shift occurs at the insulator-conductor transition; is that also the case in QCD?

The deconfined QGP is a color conductor; what about a color superconductor? In QED, collective effects of the medium bind electrons into Cooper pairs, overcoming the repulsive Coulomb force between like charges. These Cooper pairs, as bosons, condense at low temperatures and form a superconductor. In contrast to the collective binding effective in QED, in QCD there is already a microscopic qq-binding, coupling two color triplet quarks to an antitriplet di-

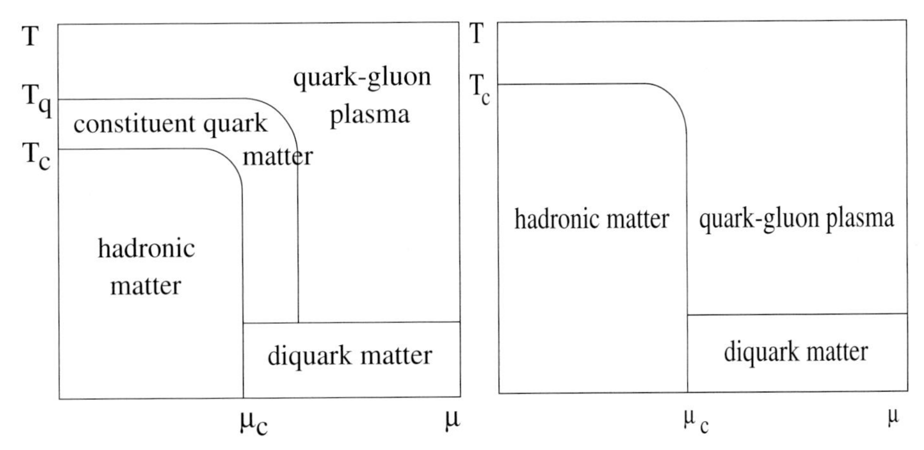

Fig. 2. Four-phase and three-phase structure for strongly interacting matter

quark. A nucleon can thus be considered as a bound state of this diquark with the remaining third quark,

$$[3 \oplus 3 \oplus 3]_1 \sim [(3 \oplus 3 \rightarrow \overline{3}) \oplus 3]_1, \tag{4}$$

leading to a color singlet state. Hence QCD provides a specific dynamical mechanism for the formation of colored diquark bosons and thus for color superconductivity. This possibility [9] has created much interest and activity over the past few years [10].

We thus have color deconfinement, chiral symmetry restoration and diquark condensation as possible transitions of strongly interacting matter for increasing temperature and/or density. This could suggest a phase diagram of the form shown on the left in Fig. 2, with four different phases. The results of finite temperature lattice QCD show that at least at vanishing baryochemical potential ($\mu = 0$) this is wrong, since there deconfinement and chiral symmetry restoration coincide, $T_c = T_\chi$, as the corresponding transitions in atomic physics do. In Sect. 4 we shall elucidate the underlying reason for this.

A second guess could thus be a three-phase diagram as shown on the right in Fig. 2, and this is in fact not in contradiction to anything so far. In passing, we should note, however, that what we have here called the diquark state is most likely more complex and may well consist of more than one phase [10].

After this conceptual introduction to the states of strongly interacting matter, we now turn to the quantitative study of QCD at finite temperature and vanishing baryochemical potential. In this case, along the $\mu = 0$ axis of the phase diagram 2, the computer simulation of lattice QCD has provided a solid quantitative basis.

2 Statistical QCD

The fundamental dynamics of strong interactions is defined by the QCD Lagrangian

$$\mathcal{L}_{\rm QCD} = -\frac{1}{4}\left(\partial_\mu A_\nu^a - \partial_\nu A_\mu^a - g f_{bc}^a A_\mu^b A_\nu^c\right)^2 - \sum_f \overline{\psi}_\alpha^f \left(i\gamma_\mu \partial^\mu + m_f - g\gamma_\mu A^\mu\right)\psi_\beta^f, \tag{5}$$

in terms of the gluon vector fields A and the quark spinors ψ. The corresponding thermodynamics is obtained from the partition function

$$\mathcal{Z}(T,V) = \int \mathcal{D}A\ \mathcal{D}\psi\ \mathcal{D}\overline{\psi}\ \exp\{-S(A,\psi,\overline{\psi};T,V)\}, \tag{6}$$

here defined as functional field integral, in which

$$S(A,\psi,\overline{\psi};T,V) = \int_0^{1/T} d\tau \int_V d^3x\ \mathcal{L}_{\rm QCD}(\tau = ix_0, \mathbf{x}) \tag{7}$$

specifies the QCD action. As usual, derivatives of $\log \mathcal{Z}$ lead to thermodynamic observables; e.g., the temperature derivative provides the energy density, the volume derivative the pressure of the thermal system.

Since this system consists of interacting relativistic quantum fields, the evaluation of the resulting expressions is highly non-trivial. Strong interactions (no small coupling constant) and criticality (correlations of all length scales) rule out a perturbative treatment in the transition regions, which are of course of particular interest. So far, the only *ab initio* results are obtained through the lattice formulation of the theory, which leads to something like a generalized spin problem and hence can be evaluated by computer simulation. A discussion of this approach is beyond the scope of this survey; for an overview, see e.g. [11]. We shall here just summarize the main results; it is to be noted that for computational reasons, the lattice approach is so far viable only for vanishing baryochemical potential, so that all results given in this section are valid only for $\mu = 0$.

As reference, it is useful to recall the energy density of an ideal gas of massless pions of three charge states,

$$\epsilon_\pi(T) = \frac{\pi^2}{30}\ 3\ T^4 \simeq T^4, \tag{8}$$

to be compared to that of an ideal QGP, which for three massless quark flavors becomes

$$\epsilon_{\rm QCD}(T) \simeq 16\ T^4. \tag{9}$$

The corresponding pressures are obtained through the ideal gas form $3P(T) = \epsilon(T)$. The main point to note is that the much larger number of degrees of freedom of the QGP as compared to a pion gas leads at fixed temperatures to much higher energy densities and pressures.

The energy density and pressure have been studied in detail in finite temperature lattice QCD with two and three light dynamical quark species, as well as for the more realistic case of two light and one heavier species. The results are shown in Fig. 3, where it is seen that in all cases there is a sudden increase from a state of low to one of high values, as expected at the confinement-deconfinement transition. To confirm the connection between the transition and the increase of energy density or pressure, we make use of the order parameters for deconfinement and chiral symmetry restoration; these first have to be specified somewhat more precisely than was done in the more conceptual discussion of Sect. 2.

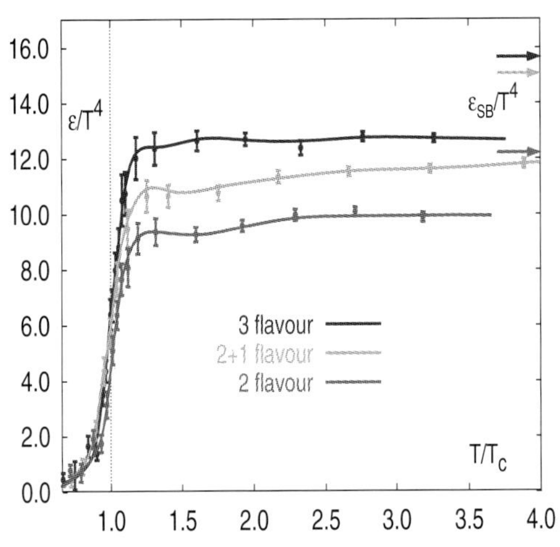

Fig. 3. Energy density and pressure in full QCD with light dynamical quarks

In the absence of light dynamical quarks, for $m_q \to \infty$, QCD reduces to pure SU(3) gauge theory; the potential between two static test quarks then has the form shown in (1) when $T < T_c$ and vanishes for $T \geq T_c$. The Polyakov loop expectation value defined by

$$\langle |L(T)| \rangle \equiv \lim_{r \to \infty} \exp\{-V(r,T)/T\} = \begin{cases} 0, & \text{confinement} \\ L(T) > 0, & \text{deconfinement} \end{cases} \tag{10}$$

thus also constitutes an order parameter for the confinement state of the medium, and it is easier to determine than the string tension $\sigma(T)$. In lattice QCD, $L(T)$ becomes very similar to the magnetization in spin systems; it essentially determines whether a global $Z_3 \in SU(3)$ symmetry of the Lagrangian is present or is spontaneously broken for a given state of the medium.

In the other extreme, for $m_q \to 0$, $\mathcal{L}_{\rm QCD}$ has intrinsic chiral symmetry, and the chiral condensate $\langle \psi\bar{\psi} \rangle$ provides a measure of the effective mass term in $\mathcal{L}_{\rm QCD}$. Through

$$\langle \psi\bar{\psi} \rangle = \begin{cases} K(T) > 0, & \text{broken chiral symmetry,} \\ 0, & \text{restored chiral symmetry.} \end{cases} \tag{11}$$

we can determine the temperature range in which the state of the medium shares and in which it spontaneously breaks the chiral symmetry of the Lagrangian with $m_q = 0$.

There are thus two *bona fide* phase transitions in finite temperature QCD at vanishing baryochemical potential.

For $m_q = \infty$, $L(T)$ provides a true order parameter which specifies the temperature range $0 \leq T \leq T_c$ in which the Z_3 symmetry of the Lagrangian is present, implying confinement, and the range $T > T_c$, with spontaneously broken Z_3 symmetry and hence deconfinement.

For $m_q = 0$, the chiral condensate defines a range $0 \leq T \leq T_\chi$ in which the chiral symmetry of the Lagrangian is spontaneously broken (quarks acquire an effective dynamical mass), and one for $T > T_\chi$ in which $\langle\psi\bar{\psi}\rangle(T) = 0$, so that the chiral symmetry is restored. Hence here $\langle\psi\bar{\psi}\rangle(T)$ is a true order parameter.

In the real world, the (light) quark mass is small but finite: $0 < m_q < \infty$. This means that the string breaks for all temperatures, even for $T = 0$, so that $L(T)$ never vanishes. On the other hand, with $m_q \neq 0$, the chiral symmetry of $\mathcal{L}_{\mathrm{QCD}}$ is explicitly broken, so that $\langle\psi\bar{\psi}\rangle$ never vanishes. It is thus not clear if some form of critical behavior remains, and we are therefore confronted by two basic questions:

- how do $L(T)$ and $\langle\psi\bar{\psi}\rangle(T)$ behave for small but finite m_q? Is it still possible to identify transition points, and if so,
- what if any relation exists between T_c and T_χ?

In Fig. 4 we show the lattice results for two light quark species; it is seen that $L(T)$ as well as $\langle\psi\bar{\psi}\rangle(T)$ still experience very strong variations, so that clear transition temperatures can be identified through the peaks in the corresponding susceptibilities, also shown in the figure. Moreover, the two peaks occur at the same temperature; one thus finds here (and in fact for all small values of m_q) that

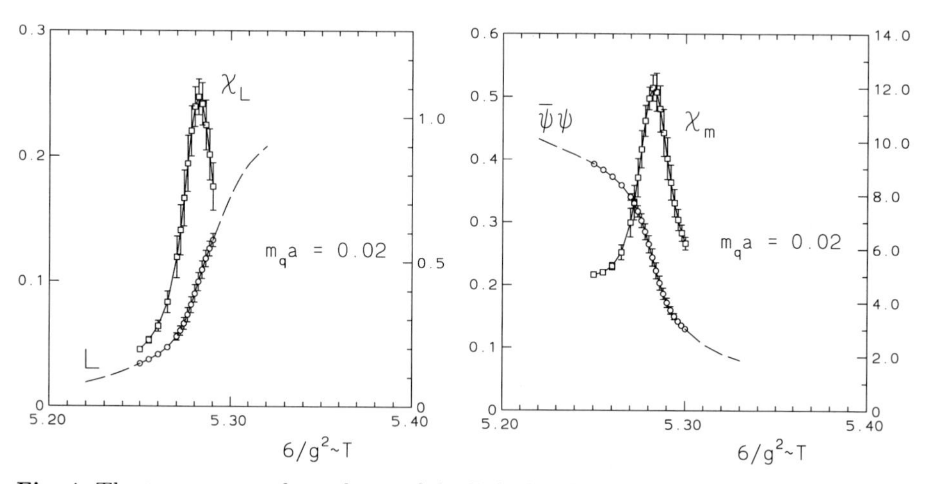

Fig. 4. The temperature dependence of the Polyakov loop L and the chiral condensate $\psi\bar{\psi}$, as well as of the corresponding susceptibilities

$T_c = T_\chi$, so that the two 'quasi-critical' transitions of deconfinement and chiral symmetry restoration coincide. Although all lattice calculations are performed for non-vanishing bare quark mass in the Lagrangian, results obtained with different m_q values can be extrapolated to the chiral limit $m_q = 0$. The resulting transition temperatures are found to be $T_c(N_f = 2) \simeq 175$ MeV and $T_c(N_f = 3) \simeq 155$ MeV for two and three light quark flavors, respectively. The order of the transition is still not fully determined. For $N_f = 3$ light quark species, one obtains a first order transition. For two light flavors, a second order transition is predicted [12], but not yet unambiguously established.

3 The Nature of Deconfinement

In this last section I want to consider in some more detail two basic aspects which came up in the previous discussion of deconfinement:

- Why do deconfinement and chiral symmetry restoration coincide for all (small) values of the input quark mass?
- Is there still some form of critical behavior when $m_q \neq 0$?

Both features have recently been addressed, leading to some first and still somewhat speculative conclusions which could, however, be more firmly established by further lattice studies.

In the confined phase of pure gauge theory, we have $L(T) = 0$, the Polyakov loop as generalized spin is disordered, so that the state of the system shares the Z_3 symmetry of the Lagrangian. Deconfinement then corresponds to ordering through spontaneous breaking of this Z_3 symmetry, making $L \neq 0$. In going to full QCD, the introduction of dynamical quarks effectively brings in an external field $H(m_q)$, which in principle could order L in a temperature range where it was previously disordered.

Since $H \to 0$ for $m_q \to \infty$, H must for large quark masses be inversely proportional to m_q. On the other hand, since $L(T)$ shows a rapid variation signalling an onset of deconfinement even in the chiral limit, the relation between H and m_q must be different for $m_q \to 0$. We therefore conjecture [13,14] that H is determined by the effective constituent quark mass M_q, setting

$$H \sim \frac{1}{m_q + c\langle\psi\bar{\psi}\rangle}, \tag{12}$$

since the value of M_q is determined by the amount of chiral symmetry breaking and hence by the chiral condensate. From (12) we obtain

- for $m_q \to \infty$, $H \to 0$, so that we recover the pure gauge theory limit;
- for $m_q \to 0$, we have

$$\langle\psi\bar{\psi}\rangle = \begin{cases} \text{large}, \ H \text{ small}, L \text{ disordered}, & \text{for } T \leq T_\chi; \\ \text{small}, \ H \text{ large}, L \text{ ordered}, & \text{for } T > T_\chi. \end{cases} \tag{13}$$

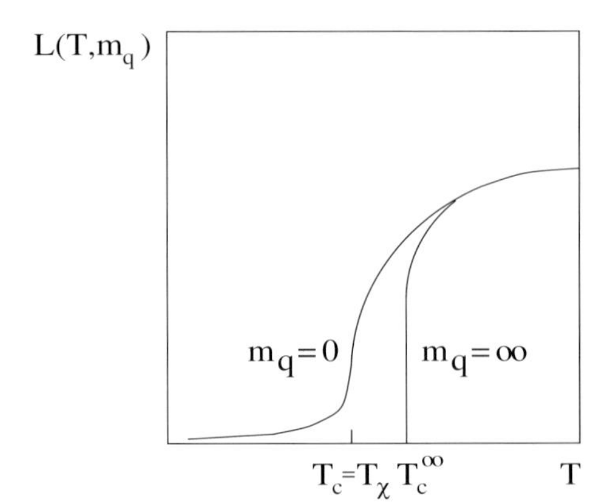

Fig. 5. Temperature dependence of the Polyakov loop in the chiral and the pure gauge theory limits

In full QCD, it is thus the onset of chiral symmetry restoration that drives the onset of deconfinement, by ordering the Polyakov loop at a temperature value below the point of spontaneous symmetry breaking [14]. In Fig. 5 we compare the behavior of $L(T)$ in pure gauge theory to that in the chiral limit of QCD. In both cases, we have a rapid variation at some temperature T_c. This variation is for $m_Q \to \infty$ due to the spontaneous breaking of the Z_3 symmetry of the Lagrangian at $T = T_c^{\infty}$; for $m_q \to 0$, the Lagrangian retains at low temperatures an approximate Z_3 symmetry which is explicitly broken at T_χ by an external field which becomes strong when the chiral condensate vanishes. For this reason, the peaks in the Polyakov loop and the chiral susceptibility coincide and we have $T_\chi = T_c < T_c^{\infty}$.

A quantitative test of this picture can be obtained from finite temperature lattice QCD. It is clear that in the chiral limit $m_q \to 0$, the chiral susceptibilities (derivatives of the chiral condensate $\langle\psi\bar{\psi}\rangle$) will diverge at $T = T_\chi$. If deconfinement is indeed driven by chiral symmetry restoration, i.e., if $L(T, m_q) = L(H(T), m_q)$ with $H(T) = H(\langle\psi\bar{\psi}\rangle(T))$ as given in (12), than also the Polyakov loop susceptibilities (derivatives of L) must diverge in the chiral limit. Moreover, these divergences must be governed by the critical exponents of the chiral transition.

Preliminary lattice studies support our picture [14]. In Fig. 6 we see that the peaks in the Polyakov loop susceptibilities as function of the effective temperature increases as m_q decreases, suggesting divergences in the chiral limit. Further lattice calculations for smaller m_q (which requires larger lattices) would certainly be helpful. The question of critical exponents remains so far completely open, even for the chiral condensate and its susceptibilities.

Next we want to consider the nature of the transition for $0 < m_q < \infty$. For finite quark mass neither the Polyakov loop nor the chiral condensate constitute genuine order parameters, since both are non-zero at all finite temperatures. Is there then any critical behavior? For pure $SU(3)$ gauge theory, the deconfinement transition is of first order, and the associated discontinuity in $L(T)$ at T_c cannot

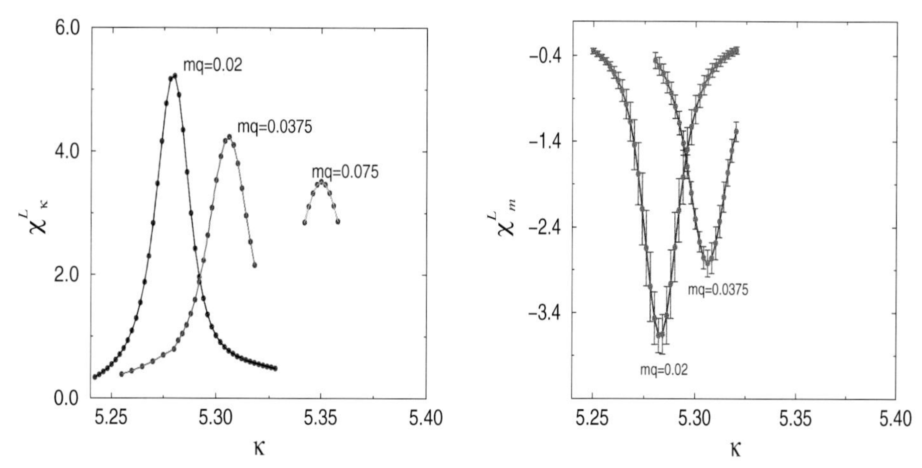

Fig. 6. The Polyakov loop susceptibilities with respect to temperature χ_κ^L (left) and to quark mass χ_m^L (right) as function of the temperature variable $\kappa = 6/g^2$ for different quark massses.

disappear immediately for $m_q < \infty$. Hence in a certain mass range $m_q^0 < m_q \leq \infty$, a discontinuity in $L(T)$ remains; it vanishes for m_q^0 at the endpoint $T_c(m_q^0)$ in the $T - m_q$ plane; see Fig. 7. For $m_q = 0$, we have the genuine chiral transition (perhaps of second order [12]) at T_χ, which, as we just saw, leads to critical behavior also for the Polyakov loop, so that here $T_c(m_q = 0) = T_\chi$ is a true critical temperature. What happens between $T_c(m_q^0)$ and $T_c(m_q = 0) = T_\chi$? The dashed line in Fig. 7 separating the hadronic phase from the quark-gluon plasma is not easy to define unambiguously: it could be obtained from the peak position of chiral and/or Polyakov loop susceptibilities [15], or from maximizing

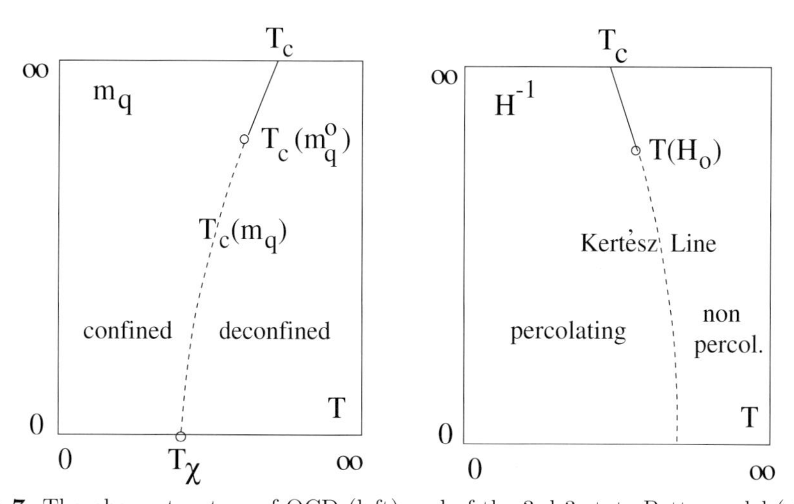

Fig. 7. The phase structure of QCD (left) and of the 3-d 3-state Potts model (right)

the correlation length in the medium [16]. In any case, it does not appear to be related to thermal critical behavior in a strict mathematical sense.

An interesting new approach to the behavior along this line could be provided by cluster percolation [17]. For spin systems without external field, the thermal magnetization transition can be equivalently described as a percolation transition of suitably defined clusters [18,19]. We recall that a system is said to percolate once the size of clusters reach the size of the system (in the infinite volume limit). One can thus characterize the Curie point of a spin system either as the point where with decreasing temperature spontaneous symmetry breaking sets in, or as the point where the size of suitably bonded like-spin clusters diverges: the critical indices of the percolation transition are identical to those of the magnetization transition.

For non-vanishing external field H, there is no more thermal critical behavior; for the 2d Ising model, as illustration, the partition function now is analytic. In a purely geometric description, however, the percolation transition persists for all H, but the critical indices now are those of random percolation and hence differ from the thermal (magnetization) indices. For the 3d three state Potts' model (which also has a first order magnetization transition), the resulting phase diagram is shown on the right of Fig. 7; here the dashed line, the so-called Kertész line [20], is defined as the line of the geometric critical behavior obtained from cluster percolation. The phase on the low temperature side of the Kertész line contains percolating clusters, the high temperature phase does not [21]. Comparing this result to the $T - m_q$ diagram of QCD, one is tempted to speculate that deconfinement for $0 < m_q < \infty$ corresponds to the Kertész line of QCD [22]. First studies have shown that in pure gauge theory, one can in fact describe deconfinement through Polyakov loop percolation [23,24]. It will indeed be interesting to see if this can be extended to full QCD.

4 Summary

We have seen that at high temperatures and vanishing baryon density, hadronic matter becomes a plasma of deconfined colored quarks and gluons. In contrast, at high baryon densities and low temperatures, one expects a condensate of colored diquarks. The quark-gluon plasma constitutes the conducting, the diquark condensate the superconducting phase of QCD.

For vanishing baryon density, the deconfinement transition has been studied extensively in finite temperature lattice QCD. In pure $SU(N)$ gauge theory (QCD for $m_q \to \infty$), deconfinement is due to the spontaneous breaking of a globel Z_N symmetry of the Lagrangian and structurally of the same nature as the magnetization transition in Z_N spin systems. In full QCD, deconfinement is triggered by a strong explicit breaking of the Z_N symmetry through an external field induced by the chiral condensate $\langle\psi\bar{\psi}\rangle$. Hence for $m_q = 0$ deconfinement coincides with chiral symmetry restoration.

For finite quark mass, $0, <, m_q, <, \infty$, it does not seem possible to define thermal critical behavior in QCD. On the other hand, spin systems under similar conditions retain geometric cluster percolation as a form of critical behavior even when there is no more thermal criticality. It is thus tempting to speculate that cluster percolation will allow a definition of color deconfinement in full QCD as genuine but geometric critical behavior.

Acknowledgements

It is a pleasure to thank S. Digal, J. Engels, S. Fortunato, F. Karsch and E. Laermann for many helpful discussions. The financial support of the German Ministry of Science (contract 06BI902) and of the GSI Darmstadt (contract BI-SAT) is gratefully acknowledged.

References

1. I. Ya. Pomeranchuk, Dokl. Akad. Nauk SSSR **78**, 889 (1951).
2. R. Hagedorn, Nuovo Cim. Suppl. **3**, 147 (1965).
3. L. Euler, Novi Commentarii Academiae Scientiarum Petropolitanae **3**, 125 (1753); E. Schröder, Z. Math. Phys. **15**, 361 (1870); G. H. Hardy and S. Ramanujan, Proc. London Math. Soc. **17**, 75 (1918).
4. S. Fubini and G. Veneziano, Nuovo Cim. **64A**, 811 (1969).
5. K. Bardakci and S. Mandelstam, Phys. Rev. **184**, 1640 (1969).
6. N. Cabibbo and G. Parisi, Phys. Lett. **59B**, 67 (1975).
7. H. Satz, Fortschr. Phys. **33**, 259 (1985).
8. O. Kaczmarek et al., Phys. Rev. D **62**, 034021 (2000).
9. D. Bailin and A. Love, Phys. Rep. **107**, 325 (1984).
10. K. Rajagopal, Nucl. Phys. A **642**, 26c (1998); Nucl. Phys. A **661**, 150c (1999); E. Shuryak, Nucl. Phys. A **642**, 14c (1998); Th. Schäfer, Nucl. Phys. A **642**, 45c (1998).
11. F. Karsch, hep-lat/0106019.
12. R. D. Pisarski and F. Wilczek, Phys. Rev. D **29**, 338 (1984).
13. R. V. Gavai, A. Gocksch and M. Ogilvie, Phys. Rev. Lett. **56**, 815 (1986).
14. S. Digal, E. Laermann and H. Satz, Eur. Phys. J. C **18**, 583 (2001).
15. S. Fortunato and H. Satz, Phys. Lett. B **509**, 189 (2001).
16. M. Caselle et al., hep-lat/0110160, to appear in the Proceedings of *Lattice 2001*, Oct. 2001, Berlin, Germany.
17. For a recent survey, see D. Stauffer and A. Aharony, *Introduction to Percolation Theory*, (Taylor & Francis, London 1994).
18. C. M. Fortuin and P. W. Kasteleyn, J. Phys. Soc. Japan **26** (Suppl.), 11 (1969); Physica **57**, 536 (1972).
19. A. Coniglio and W. Klein, J. Phys. A **13**, 2775 (1980).
20. J. Kertész, Physica A **161**, 58 (1989).
21. S. Fortunato and H. Satz, hep-ph/0108058; Nucl. Phys. B, in press.
22. H. Satz, Nucl. Phys. A **642**, 130c (1998).
23. S. Fortunato and H. Satz, Phys. Lett. B **475**, 311 (2000); Nucl. Phys. A **681**, 466c (2001).
24. H. Satz, hep-lat/0110013, to appear in the *Proceedings of the Europhysics Conference on Computational Physics*, Aachen, Germany, September 2001.

Ghost-Free APT Analysis of Perturbative QCD Observables

Dmitry V. Shirkov

Bogoliubov Laboratory, JINR, 141980 Dubna, Russia, e-address: shirkovd@thsun1.jinr.ru

Abstract. The review of the essence and of application of recently devised ghost–free Analytic Perturbation Theory (APT) is presented. First, we discuss the main intrinsic problem of perturbative QCD – ghost singularities and with the resume of its resolving within the APT. By examples for diverse energy and momentum transfer values we show the property of better convergence for the APT modified QCD expansion.

It is shown that in the APT analysis the three-loop contribution ($\sim \alpha_s^3$) is numerically inessential. This gives raise a hope for practical solution of the well–known problem of non–satisfactory convergence of QFT perturbation series due to its asymptotic nature. Our next result is that a usual perturbative analysis of time-like events is not adequate at $s \leq 2\,\mathrm{GeV}^2$. In particular, this relates to τ decay.

Then, for the "high" ($f = 5$) region it is shown that the common NLO, NLLA perturbation approximation widely used there (at 10 GeV$\lesssim\sqrt{s}\lesssim$170 GeV) yields a systematic theoretic negative error of a couple per cent level for the $\bar{\alpha}_s^{(2)}$ values. This results in a conclusion that the $\bar{\alpha}_s(M_Z^2)$ value averaged over the $f = 5$ data appreciably differs $\langle\bar{\alpha}_s(M_Z^2)\rangle_{f=5} \simeq 0.124$ from the currently popular "world average" ($= 0.118$).

1 Introduction

In this talk, we first discuss this main problem of the perturbative QCD (=pQCD) – the unphysical (so called "ghost" or "Landau") singularities lying in the physically accessible domain and associated with the QCD scale parameter $\Lambda_{\overline{\mathrm{MS}}}^{f=3} \simeq 300 - 400\,\mathrm{MeV}$.

In practice, they complicate interpretation of data in the "small energy" and "small momentum transfer" regions ($\sqrt{s},\, Q \equiv \sqrt{Q^2} \lesssim 1 \div 1.5$ GeV). Meanwhile, as it is well known, their existence contradicts some basic statements of the local QFT.

Then, we give a resume of this problem solution without any additional adjustable parameters within the Analytic Perturbation Theory (APT). To illustrate, we present some impressive results of the APT application for analysis of QCD observables.

This contribution is based on our recent papers [1] and [2].

1.1 Common QCD Coupling and Observables

The perturbative QCD contribution to an observable $\mathcal{O}$ in both the space– and time–like cases is usually taken in the form of two– or three–term expansion

$$\frac{\mathcal{O}(x)}{\mathcal{O}_0} = 1 + r(x)\,; \quad r(x) = c_1\,\bar{\alpha}_s(x) + c_2\,\bar{\alpha}_s^2 + c_3\,\bar{\alpha}_s^3 + \ldots\,; \quad x = Q^2 \ \text{or} = s \quad (1)$$

over powers of an effective QCD coupling $\bar{\alpha}_s$ which is of the same form in both the channels, e.g., in the three–loop case (see, [15]) at the massless $\overline{\rm MS}$ scheme

$$\bar{\alpha}_s^{(3)}(x) = \frac{1}{\beta_0 L} - \frac{b_1}{\beta_0^2}\frac{\ln L}{L^2} + \frac{1}{\beta_0^3 L^3}\left[b_1^2(\ln^2 L - \ln L - 1) + b_2\right] +$$

$$+\frac{1}{\beta_0^4 L^4}\left[b_1^3\left(-\ln^3 L + \frac{5}{2}\ln^2 L + 2\ln L - \frac{1}{2}\right) - 3b_1 b_2 \ln L + \frac{b_3}{2}\right]; \quad L = \ln(\frac{x}{\Lambda^2}). \qquad (2)$$

Here, for the beta-function coefficients we use the normalization

$$\beta(\alpha) = -\beta_0\,\alpha^2 - \beta_1\,\alpha^3 - \beta_2\,\alpha^4 + \ldots = -\beta_0\,\alpha^2\left(1 + b_1\,\alpha + b_2\,\alpha^2 + \ldots\right),$$

that is free of the π powers. Numerically, they are of an order of unity

$$\beta_0(f) = \frac{33 - 2\,f}{12\pi}; \quad b_1(f) = \frac{153 - 19f}{2\pi(33 - 2f)};$$

$$\beta_0(4 \pm 1) = 0.875 \pm 0.005\,; \; b_1(4 \pm 1) = 0.490^{-0.089}_{+0.076}\,.$$

Meanwhile, in the RG formalism[1] the notion of invariant coupling $\bar{g}(q)$ is *defined only in the space–like domain.* In particular, this means that if some observable $\mathcal{O}(Q^2)$ is a function of one invariant space–like argument $Q^2 > 0$ and a coupling g_μ, then, due to its renormalization invariance, it should be a function of RG invariants only. In the one–coupling massless case it is a function of only one argument, the *invariant coupling function* $\bar{g}$

$$\mathcal{O}(Q^2/\mu^2, g_\mu) = F\left(\bar{g}(Q^2/\mu^2, g_\mu)\right) \quad \text{with} \quad F(g) = \mathcal{O}(1, g)\,.$$

Just due to this important property, in the weak coupling case we deal with the *functional expansion* of an observable $\mathcal{O}(Q^2)$ in powers of $\bar{g}$. This is a real foundation of QCD power expansion (1) in the Euclidean case with $x = Q^2$. At the same time, inside the RG formalism, there is no natural means for defining an invariant coupling $\tilde{g}(s)$; $s = -Q^2$ and perturbative expansion for an observable $\tilde{O}(s)$ in the time–like region.

Nevertheless, in current practice, people commonly use the same singular expression for the QCD effective coupling $\bar{\alpha}_s$, like (2), in both the space– and time–like domains. The only price paid for the transferring from the Euclidean to Minkowskian region is the change of numerical expansion coefficients. The time–like ones $c_{k\geq 3} = d_k - \delta_k$ include negative terms proportional to π^2 and lower expansion coefficients c_k

$$\delta_3 = (\pi\beta_0(f))^2\,(c_1/3)\,, \quad \delta_4 = (\pi\beta_0)^2\,(c_2 + (5/6)\,b_1\,c_1)\;\ldots\,. \qquad (3)$$

These essential (as far as $\pi^2\beta_0^2(f = 4 \pm 1) = 4.340^{-.666}_{+.723}$) structures δ_k arise [7] – [10] in the course of analytic continuation from the Euclidean to Minkowskian

[1] For details, see, e.g., the chapter "Renormalization group" in the monograph [4], Appendix IX in the textbook [5] or Sect. 1 in [6].

region. The coefficients d_k are genuine ones as far as they are calculated via the relevant Feynman diagrams.

To demonstrate the importance of the "π^2 terms", we give their values for the τ–decay, the e^+e^- hadron annihilation and the Z_0 decay – see Table 1. In our normalization (1), all coefficients c_k, d_k and δ_k are of an order of unity. In the $f = 4, 5$ region the contribution δ_3 prevails in c_3 and $|d_3| \ll |c_3|$.

Table 1. Minkowskian c_k and Euclidean d_i expansion coefficients and their differences δ_i.

Process	f	$c_1 = d_1$	$c_2 = d_2$	c_3	d_3	δ_3	δ_4
τ decay	3	$1/\pi$	.526	0.852	1.389	0.537	5.01
e^+e^-	4	.318	.155	-0.351	0.111	0.462	2.451
e^+e^-	5	.318	.143	-0.413	-0.023	0.390	1.752
Z_0 decay	5	.318	.095	-0.483	-0.094	0.390	1.576

1.2 Unphysical Singularities Problem Resolving

In QCD, the unphysical singularity lies in the quite physical, the infrared (IR) region. This means that, if one believes in QCD as in a consistent physically meaningful theory, one has no other choice as to treat these singularities as an artefact of some approximations used in current pQCD. This point of view is supported by lattice simulations and solution of Schwinger–Dyson equations – see, e.g., Sect. 5.3. in a recent review [11].

To illustrate of the severe inconsistency of common pQCD practice of treating singularities, take the relation between the Adler function D and the cross–section ratio R

$$D(Q^2) = Q^2 \int_0^\infty \frac{R(s)\, ds}{(s+Q^2)^2} \,. \tag{4}$$

In the case of inclusive e^+e^- hadron annihilation, $R(s)$ is the ratio of cross–sections usually presented in the form $R(s) = 1 + r(s)$ with a function r expandable in $\bar{\alpha}_s(s)$ powers like in (1). In parallel, the Adler function is also used to be presented in the form $D = 1 + d$ with d expanded in powers of $\bar{\alpha}_s(Q^2)$. Here, we face two paradoxes.

First, $\bar{\alpha}_s(Q^2)$, as expressed by (2), and, hence, the perturbative $D(Q^2)$ obeys non-physical singularity at $Q^2 = \Lambda^2$ in evident contradiction with representation (4). Second, the integrand $R(s)$, being expressed via powers of $\bar{\alpha}_s(s)$, obeys non-integrable singularities at $s = \Lambda^2$, which makes the r.h.s. of (4) senseless.

This second problem is typical of inclusive cross–sections, e.g., for the τ hadronic decay. Generally, in the current literature it is treated in a very peculiar way – by shifting the integration contour from the real axis with strong singularities on it into a complex plane. However, such a "trick" cannot be justified within the theory of complex variable.

Meanwhile, as it is known from the early 80s, the perturbation expansion (1) for the Minkowskian observable with the coefficients modified by the π^2–terms is valid only at a small parameter $\pi^2/\ln^2(s/\Lambda^2)$ values, that is in the region of sufficiently high energies $W \equiv \sqrt{s} \gg \Lambda e^{\pi/2} \simeq 2\,\mathrm{GeV}$. Here, it is appropriate to remind "the RKP construction" devised[7,8] in early 80s. There, the integral transformation

$$R(s) = \frac{i}{2\pi}\int_{s-i\varepsilon}^{s+i\varepsilon}\frac{dz}{z}\,D_{\mathrm{pt}}(-z) \equiv \mathbf{R}\left[D(q^2)\right] \tag{5}$$

reverse to the Adler relation (4) – that is treated now as an integral transformation –

$$R(s) \to D(Q^2) = Q^2\int_0^\infty \frac{R(s)\,ds}{(s+Q^2)^2} \equiv \mathbf{D}\left\{R(s)\right\} \tag{6}$$

was used in defining modified expansion functions $\mathfrak{A}_k(s) = \mathbf{R}[\alpha_s^k(q^2)]$ for the pQCD contribution

$$r(s) = d_1\mathfrak{A}_1(s) + d_2\mathfrak{A}_2(s) + d_3\mathfrak{A}_3(s) \tag{7}$$

to an observable in the time–like region.

At the one-loop level, with the common effective coupling $\bar{\alpha}_s^{(1)} = \left[\beta_0 \ln(\mathrm{q}^2/\Lambda^2)\right]^{-1}$ one has

$$\mathfrak{A}_1^{(1)}(s) = \mathbf{R}\left[\bar{\alpha}_s^{(1)}\right] = \frac{1}{\pi\beta_0}\arccos\frac{L}{\sqrt{L^2+\pi^2}} = \frac{1}{\beta_0}\left[\frac{1}{2} - \frac{1}{\pi}\arctan\frac{L}{\pi}\right];\; L = \ln\frac{\mathrm{s}}{\Lambda^2} \tag{8}$$

and for higher functions

$$\mathfrak{A}_2^{(1)}(s) = \frac{1}{\beta_0^2\left[L^2+\pi^2\right]};\quad \mathfrak{A}_3^{(1)}(s) = \frac{L}{\beta_0^3\left[L^2+\pi^2\right]^2};\quad \mathfrak{A}_4^{(1)}(s) = \frac{L^2-\pi^2/3}{\beta_0^4\left[L^2+\pi^2\right]^3}, \tag{9}$$

which *are not powers* of $\mathfrak{A}_1^{(1)}(s)$. The r.h.s of (8) at $L \geq 0$ can also be presented in the form

$$\mathfrak{A}_1^{(1)}(s) = \frac{1}{\pi\beta_0}\arctan\frac{\pi}{L}$$

convenient for the UV analysis. Just this form was discovered in the early 80s in [12] and [7], while (9) in [7] and [8].

On the other hand, expression (8) was first discussed only 15 years later by Milton and Solovtsov [9]. Thee authors made an important observation that this expression represents a continuous monotone function without unphysical singularity and proposed to use it as an effective "Minkowskian QCD coupling" $\tilde{\alpha}(s) \equiv \mathfrak{A}_1(s)$ in the time–like region.

For the two–loop case, to the popular approximation

$$\beta_0\bar{\alpha}_{s,pop}^{(2)}(Q^2) = 1/l - b_1(f)\ln l/l^2\,;\quad l = \ln\left(\mathrm{Q}^2/\Lambda^2\right)$$

there corresponds [7,13] suitably modified expression $\tilde{\alpha}^{(2)}_{pop}(s)$ with structures $(L^2+\pi^2)$ and $\ln(\mathrm{L}^2+\pi^2)$ – see (11) in [2]. It is also possible to construct regular expression $\tilde{\alpha}^{(2)}(s)$ in the closed analytical form[14] corresponding to the exact two–loop solution expressed[15] via Lambert function.

At $L \gg \pi$, by expanding $\tilde{\alpha}^{(2)}$ and $\mathfrak{A}_2$ of (9) in powers of π^2/L^2 we arrive at the π^2–terms (3). All these functions $\tilde{\alpha}^{(1,2)}$ are monotonically decreasing with a finite IR value $\tilde{\alpha}(0)=1/\beta_0(f=3)\simeq 1.4$. Higher functions go to zero in the IR limit $\mathfrak{A}_k(0)=0\,;\,k\geq 2$. They have no singularity at $L=0$.

As it has been noticed in [1], by applying $\mathbf{D}$ (6) to functions $\mathfrak{A}_k(s)$, instead of $\bar{\alpha}_s(Q^2)$ powers, we obtain expressions $\mathbf{D}[\mathfrak{A}_k(s)] = \mathcal{A}_k(Q^2)$ that are free of singularities. These functions have first been discussed at 90s [16] – [20] in the context of the "Analytic approach" to perturbative QCD.

Therefore, this Analytic approach in the Euclidean region and the RKP formulation for Minkowskian observables can be united in the single scheme, the "Analytic Perturbation Theory" – APT, that has been formulated quite recently in [1].

2 The Analytic Perturbation Theory

The APT scheme closely relates two ghost–free formulations of modified perturbation expansion for observables.

The first one changes the usual power expansion (1) in the time-like region into the nonpower one (7). It uses operation (5), that is reverse $\mathbf{R}=[\mathbf{D}]^{-1}$ to the one defined by the "Adler relation" (6) and transforms a real function $R(s)$ of a positive (time–like) argument into a real function $D(Q^2)$ of a positive (space–like) argument.

By operation $\mathbf{R}$, one can define [9] the RG–invariant Minkowskian coupling $\tilde{\alpha}(s)=\mathbf{R}\,[\bar{\alpha}_s]$, and its "effective powers" $\mathfrak{A}_k$ that are free of ghost singularities. They are *are not powers* of $\mathfrak{A}_1^{(1)}$. Some examples are given by (8) and (9).

By applying $\mathbf{D}$ to $\mathfrak{A}_k(s)$, one can "try to return" to the Euclidean domain. However, instead of α_s powers, we arrive at some other functions

$$\mathcal{A}_k(Q^2)=\mathbf{D}\,[\mathfrak{A}_k]\;, \tag{10}$$

analytic in the cut Q^2-plane and free of ghost singularities. At the one–loop case

$$\beta_0\mathcal{A}_1^{(1)}(Q^2)=\frac{1}{\ln(\mathrm{Q}^2/\Lambda^2)}-\frac{\Lambda^2}{Q^2-\Lambda^2}\,,\;\beta_0^2\mathcal{A}_2^{(1)}(Q^2)=\frac{1}{\ln^2(\mathrm{Q}^2/\Lambda^2)}+\frac{q^2\Lambda^2}{(Q^2-\Lambda^2)^2},\dots \tag{11}$$

These expressions have been discovered by other means [16,17] in the mid–90s. The first function $\mathcal{A}_1=\alpha_{\rm an}(Q^2)$, an analytic invariant Euclidean coupling, is a *counterpart* of the Minkowskian coupling $\tilde{\alpha}(s)=\mathfrak{A}_1(s)$. Both $\alpha_{\rm an}$ and $\tilde{\alpha}$ are real monotonically decreasing functions with the same maximum value

$$\alpha_{\rm an}(0)=\tilde{\alpha}(0)=1/\beta_0(f=3)\simeq 1.4\,.$$

All higher functions vanish $\mathcal{A}_k(0) = \mathfrak{A}_k(0) = 0$ in this limit. For $k \geq 2$, they oscillate in the IR region and form [19,21] an asymptotic sequence à lá Erdélyi.

The same properties remain valid for a higher–loop case. Explicit expressions for $\mathcal{A}_k$ and $\mathfrak{A}_k$ at the two–loop case can be written down (see, [15] and [14]) in terms of a special Lambert function. They are illustrated below in Figs. 1 and 2. Note here that to relate Euclidean and Minkowskian functions, instead of integral expressions (5) and (6) one can use simpler relations, in terms of spectral functions $\rho(\sigma) = \mathcal{I}m\,\mathcal{A}(-\sigma)$,

$$\mathcal{A}_k(Q^2;f) = \frac{1}{\pi}\int_0^\infty \frac{d\sigma}{\sigma+Q^2}\,\rho_k(\sigma;f)\,; \quad \mathfrak{A}_k(s;f) = \frac{1}{\pi}\int_s^\infty \frac{d\sigma}{\sigma}\rho_k(\sigma;f)\,, \tag{12}$$

equivalent to expressions $\mathcal{A}_k(q^2) = \mathbf{D}\left[\mathfrak{A}_k\right]$, and $\mathfrak{A}_k(s) = \mathbf{R}[\mathcal{A}_k]$.

The mechanism of liberation of singularities is quite different. While in the space-like domain it involves nonperturbative, power in Q^2, structures, in the time-like region it is based only upon resummation of the "π^2 terms". Figuratively, (non-perturbative !) *analyticization* [16,17,19] in the Q^2–channel can be treated as a quantitatively distorted reflection (under $Q^2 \to s = -Q^2$) of π^2–resummation in the s–channel. This effect of "distorting mirror", first discussed in [9] and [22], is clearly seen in the Figs. 1, 2 mentioned above.

This means also that origin of nonperturbative $1/Q^2$ structures now has got another motivation, (10), independent of the analyticization prescription.

In reality, a physical domain includes regions with various "numbers of active quarks", i.e., with diverse flavor numbers $f = 3,4,5$ and 6. In each of these regions, we deal with a different amount of quark quantum fields, that is with distinct QFT models with corresponding Lagrangians. To combine them into a joint picture, one uses the procedure of the threshold matching which establishes relations between renormalization procedures with different f values.

For example, in the $\overline{\text{MS}}$ scheme the matching relation has a simple form

$$\bar{\alpha}_s(Q^2 = \xi M_f^2; f-1) = \bar{\alpha}_s(Q^2 = \xi M_f^2; f)\,; \quad 1 \leq \xi \leq 2\,. \tag{13}$$

It defines a "global effective coupling"

$$\bar{\alpha}_s(Q^2) = \bar{\alpha}_s(Q^2;f) \quad \text{at} \quad \xi M_{f-1}^2 \leq Q^2 \leq \xi M_f^2\,,$$

continuous in the space-like region of positive Q^2 values with discontinuity of derivatives at matching points ξM_f^2. To this global $\bar{\alpha}_s$, there corresponds a discontinuous spectral density

$$\rho_k(\sigma) = \rho_k(\sigma;3) + \sum_{f\geq 4}\theta(\sigma - \xi M_f^2)\left\{\rho_k(\sigma;f) - \rho_k(\sigma;f-1)\right\} \tag{14}$$

with $\rho_k(\sigma;f) = \mathcal{I}m\;\bar{\alpha}_s^k(-\sigma,f)$ which yields [1] via relations analogous to (46)

$$\mathcal{A}_k(Q^2) = \frac{1}{\pi}\int_0^\infty \frac{d\sigma}{\sigma+Q^2}\,\rho_k(\sigma)\,; \quad \mathfrak{A}_k(s) = \frac{1}{\pi}\int_s^\infty \frac{d\sigma}{\sigma}\rho_k(\sigma)\,, \tag{15}$$

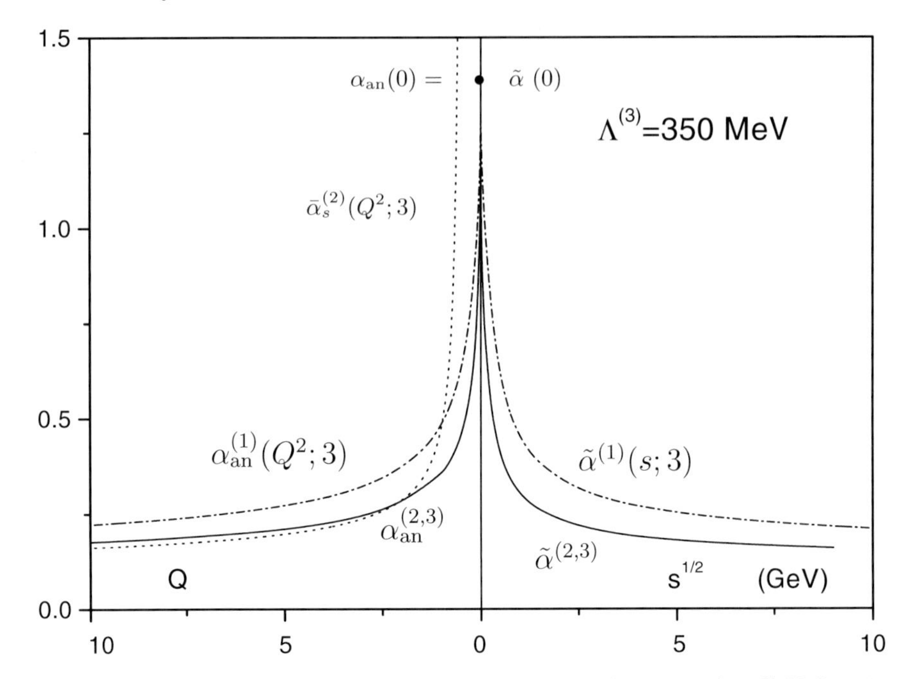

Fig. 1. Space-like and time-like invariant analytic couplings in a few GeV domain

the smooth global Euclidean and spline–continuous global Minkowskian expansion functions.

Here, in Fig. 1, by the dotted line we give a usual two-loop QCD coupling $\bar{\alpha}_s(Q^2)$ with a singularity at $Q^2 = \Lambda^2$. The dash–dotted curves represent the one-loop APT expressions (8) and (35). The solid APT curves are based on the exact two-loop (Lambert) solutions and approximate three–loop solutions in the $\overline{\text{MS}}$ scheme. Their practical coincidence (within the 2–4 per cent limit) demonstrates reduced sensitivity of the APT approach (see [17,18,23]) with respect to higher–loop effects in the whole Euclidean and Minkowskian regions. Figure 2 shows higher two–loop functions in comparison with α_{an} and $\tilde{\alpha}$ powers.

Generally, functions $\mathfrak{A}_k$ and $\mathcal{A}_k$ differ from the local ones with a fixed f value. Minkowskian global functions $\mathfrak{A}_k$ can be presented via $\mathfrak{A}_k(s,f)$ by relations

$$\tilde{\alpha}(s) = \tilde{\alpha}(s;f) + c(f)\,;\ \mathfrak{A}_2(s) = \mathfrak{A}_2(s;f) + \ \mathfrak{c}_2(f) \text{ at } M_f^2 \le s \le M_{f+1}^2 \qquad (16)$$

with *shift constants* $c(f)$, $\mathfrak{c}_2(f)$. Numerical estimate performed in [15,1] (see also [2]) for $\Lambda_3 \sim 300 - 400\,\text{MeV}$ reveals that these constants are essential in the $f = 3,4$ region at a few per cent level for $\tilde{\alpha}$ and at ca 10% level for $\mathfrak{A}_2$.

Meanwhile, global Eucledean functions $\mathcal{A}_k(Q^2)$ cannot be related to the local ones $\mathcal{A}_k(Q^2,f)$ by simple relations. Nevertheless, numerical calculation shows [15,14] that in the $f = 3$ region one has approximate relations similar to (16).

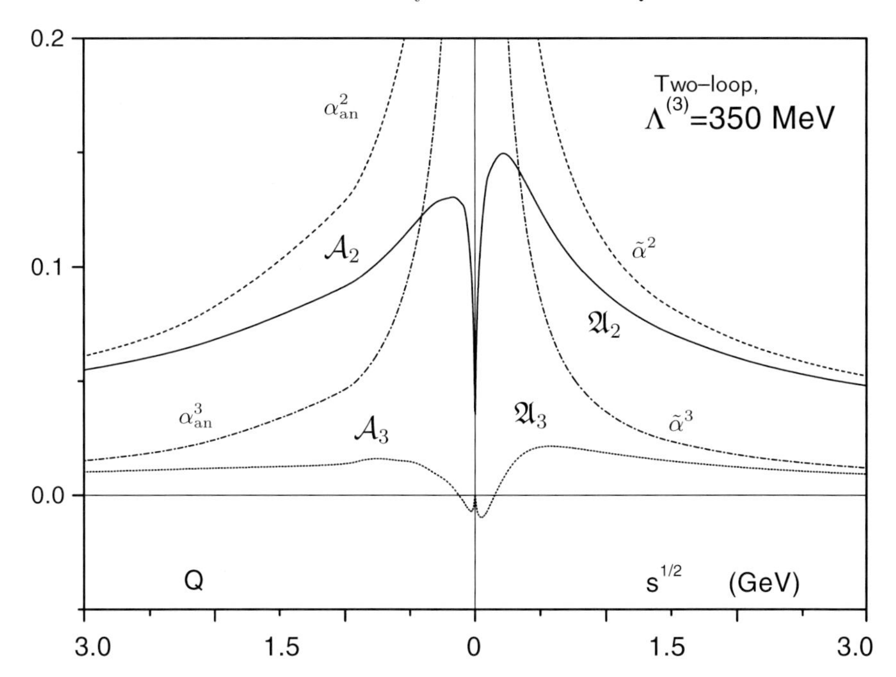

Fig. 2. "Distorted mirror symmetry" for global expansion functions

3 The APT Applications

In the usual treatment – see, e.g., [15] – (the QCD perturbative part of) a Minkowskian observable, like e^+e^- annihilation or Z_0 decay cross–section ratio, is presented as

$$R(s) = R_0\,(1 + r(s))\;; \quad r_{PT}(s) = c_1\,\bar{\alpha}_s(s) + c_2\,\bar{\alpha}^2_s(s) + c_3\,\bar{\alpha}^3_s(s) + \dots\;. \tag{17}$$

Here, the coefficients c_1 , c_2 and c_3 are not diminishing numerically – see Table 1. A rather big negative c_3 value comes mainly from the $-c_1\pi^2\beta_0^2/3$ term.In the APT, we have

$$r_{APT}(s) = d_1\tilde{\alpha}(s) + d_2\,\mathfrak{A}_2(s) + d_3\,\mathfrak{A}_3(s)\; + \dots \tag{18}$$

with reasonably decreasing coefficients $d_{1,2} = c_{1,2}$ and $d_3 = c_3 + c_1\pi^2\beta_0^2/3$, the mentioned π^2 term of c_3 being "swallowed" by $\tilde{\alpha}(s)$.

In the Euclidean channel, instead of power expansion similar to (17), we typically have

$$d_{APT}(Q^2) = d_1\alpha_{\rm an}(Q^2) + d_2\,\mathcal{A}_2(Q^2) + d_3\,\mathcal{A}_3(Q^2)\; + \dots \tag{19}$$

with the same coefficients d_k . Here, the modification is related to nonperturbative, power in Q^2, structures $\mathcal{A}_k$ like in (35).

In Table 2, we give relative values of the first, second, and third terms of the r.h.s. in (17),(18) and (19) for the "Euclidean" [24,25] sum rules, τ – decay in

Table 2. Relative contributions (in %) of 1– , 2– and 3–loop terms to observables

Process	Q or $\sqrt{s}$	f	*PT*			*APT*		
GLS sum rule	1.73 GeV	4	65	24	11	75	21	**4**
Bjorken. s.r.	1.73 GeV	3	55	26	19	80	19	**1**
Incl. τ-decay	0 - 2 GeV	3	55	29	16	88	11	**1**
$e^+e^- \to$ hadr.	10 GeV	4	96	8	-4	92	7	**.5**
$Z_o \to$ hadr.	89 GeV	5	98.6	3.7	-2.3	96.9	3.5	**-.4**

the vector channel [26], as well as for e^+e^- and Z_0 inclusive cross-sections. As it follows from this Table, in the APT case, *the three–loop term is very small, and numerically unessential.*

This conclusion is valuable when the three–loop term d_3 is unknown. Here, mainly in the $f = 5$ region, people use the NLLA approximation. For the Minkowskian observable, e.g., in the event–shape (see, [28]) analysis there corresponds the two-term expression

$$r(s) = c_1 \alpha_s(s) + c_2 \alpha_s^2(s) \,. \tag{20}$$

On the basis of the numerical estimates of Table 1, in such a case, we recommend instead *to use the two-term APT representation*

$$r^{(2)}_{APT}(s) = d_1 \tilde{\alpha}(s) + d_2 \,\mathfrak{A}_2(s) \tag{21}$$

which, at $L^2 \gg \pi^2$, is equivalent to the three-term expression

$$r_3^{\Delta}(s) = d_1 \left\{ \bar{\alpha}_s - \frac{\pi^2 \beta_0^2}{3} \bar{\alpha}_s^3 \right\} + d_2 \bar{\alpha}_s^2 = c_1 \, \bar{\alpha}_s + c_2 \bar{\alpha}_s^2 - \underline{\boldsymbol{\delta_3 \, \bar{\alpha}_s^3}} \tag{22}$$

i.e., to take into account the known predominant π^2 part of the next coefficient c_3 . As it follows from the comparison of the last expression with the previous, two–term one (20), the $\bar{\alpha}_s$ numerical value extracted from (22), for the same measured value r_{obs}, will differ mainly by a positive quantity (e.g., at $\bar{\alpha}_s \simeq 0.12 \div 0.15$)

$$(\Delta \bar{\alpha}_s)_3 = \left. \frac{\pi \delta_3 \, \bar{\alpha}_s^3}{1 + 2\pi d_2 \bar{\alpha}_s} \right|_{20 \div 100\mathrm{GeV}}^{f=5} = \frac{1.225 \, \bar{\alpha}_s^3}{1 + 0.90 \, \bar{\alpha}_s} \simeq 0.002 \div 0.003 \,. \tag{23}$$

Moreover, in the $f = 4$ region, where the three-loop (NNLLA) approximation is commonly used in the data analysis, the π^2 term δ_4 of the next order turns out also to be essential. Hence, we propose there, instead of (17), *to use the APT three–term expression*

$$r^{(3)}_{APT}(s) = d_1 \tilde{\alpha}(s) + d_2 \,\mathfrak{A}_2(s) + d_3 \,\mathfrak{A}_3(s) \tag{24}$$

approximately equivalent to the four-term one

$$r_4^{\Delta}(s) = d_1 \, \bar{\alpha}_s + d_2 \, \bar{\alpha}_s^2 + c_3 \, \bar{\alpha}_s^3 - \underline{\boldsymbol{\delta_4 \, \bar{\alpha}_s^4}} \,; \quad c_3 = d_3 - \delta_3 \tag{25}$$

with δ_3 and δ_4 defined [7,10] in (3), or to

$$r_4^{\Delta}(s) = d_1 \left\{ \bar{\alpha}_s - \frac{\pi^2 \beta_0^2}{3} \bar{\alpha}_s^3 - b_1 \frac{5}{6} \pi^2 \beta_0^2 \bar{\alpha}_s^4 \right\} + d_2 \left\{ \bar{\alpha}_s^2 - \pi^2 \beta_0^2 \bar{\alpha}_s^4 \right\} + d_3 \bar{\alpha}_s^3 .$$

The three– and two–term structures in braces are related to specific expansion functions $\tilde{\alpha} = \mathfrak{A}_1$ and $\mathfrak{A}_2$ defined above (15) and entering into the non-power expansion (24).

To estimate roughly the numerical effect of using this last modified expression (25), we take the e^+e^- inclusive annihilation. For $\sqrt{s} \simeq 3 \div 5$ GeV with $\bar{\alpha}_s \simeq 0.28 \div 0.22$ one has $(\Delta\bar{\alpha}_s)_4|_{3\div5\mathrm{GeV}}^{f=4} \simeq 0.85\,\bar{\alpha}_s^4 \simeq 0.005 \div 0.002$ – an important effect on the level of ca $1 \div 2\%$.

Moreover, the $(\Delta\bar{\alpha}_s)_4$ correction is noticeable even in the lower part of the $f = 5$ region! Indeed, to $\sqrt{s} \simeq 10 \div 40$ GeV with $\bar{\alpha}_s \simeq 0.20 \div 0.15$ there corresponds

$$(\Delta\bar{\alpha}_s)_4|_{10\div40\ \mathrm{GeV}}^{f=5} \simeq 0.71\,\bar{\alpha}_s^4 \simeq (1.1 \div 0.3) \cdot 10^{-3} \quad (\lesssim 0.5\%) .$$

It is essential to note that approximate expressions (22) and (25) are equivalent to the exact ones (21) and (24) only in the region $L = \ln\left(\mathrm{s}/\Lambda^2\right) \gg \pi$.

In particular, at $\sqrt{s} \leq 1.5\,\mathrm{GeV}$ it is rather desorienting to refer to $\bar{\alpha}_s(s)$ and it is erroneous to use $\tilde{\alpha}_{appr}(s)$ and common expansion (17). This means that *below* $s = 2\,\mathrm{GeV}^2$ *it is nonadequate to use common* $\bar{\alpha}_s(s)$ *and power expansion (17).*

In other words, we claim that below $s = 2\,\mathrm{GeV}^2$ it is an intricate business to analyze data in terms of the "old good" (but singular) α_s . Here, approximate relation $\tilde{\alpha}(s) \simeq \bar{\alpha}_s - \pi^2\beta^2\bar{\alpha}_s^3/3$ – compare with (22) – does not work as it is illustrated in Fig. 2 of paper [2].

This refers to analysis of τ decay. Here, we would like to attract attention to the important paper [26] that treats the τ decay within the APT approach (with effective mass of light quarks ($\simeq m_\pi$) and the threshold resummation factor) and results in $\Lambda^{(3)} = 420$ MeV that corresponds to $\alpha_{\mathrm{an}}(M_\tau^2) = 0.32$ or $\tilde{\alpha}(M_\tau^2) = 0.30$. At the same time, attempts to interpret results of APT for τ decay in terms of α_s, needs some special precaution[2] – see a more detailed comment on the τ decay theoretical analysis in [27].

In this, low–energy Minkowskian/Euclidean region data have to be analyzed in terms of nonpower expansion (18)/(19) and extracted parameter should be $\alpha_{\mathrm{an}}(s)\,/\tilde{\alpha}(Q^2)$ or $\Lambda^{(3)}$. In Table 3 of paper [2] we give few numerical examples for the chain

$$\alpha_{\mathrm{an}}(M_\tau) \leftrightarrow \tilde{\alpha}(M_\tau) \leftrightarrow \Lambda^{(3)} \rightarrow \Lambda^{(5)} \leftrightarrow \bar{\alpha}_s(M_Z)$$

[2] Generally, it is possible to use correspondence between α_{an} , $\tilde{\alpha}$ and α_s as expressed by relations that follow from (8) and (35) – see, e.g., (14) in [2]. However, the use of $\alpha_s^{\overline{\mathrm{MS}}}(\mu^2)$ at $\mu \lesssim 1$ *GeV* as a QCD parameter could be misleading due to vicinity to singularity. For example, at $\Lambda^{(3)} = 400$ MeV one has $\alpha_s(M_\tau^2) \simeq 0.35$ and $\alpha_s(1\ \mathrm{GeV}^2) \simeq 0.55$ to be compared with $\alpha_{\mathrm{an}}(M_\tau^2) \simeq 0.31$ and $\alpha_{\mathrm{an}}(1\ \mathrm{GeV}^2) \simeq 0.40$.

that gives the means to study the QCD theoretical compatibility of LE data with the HE ones in the APT analysis. Here, the main element of correlation is the chain $\Lambda^{(3)} \leftrightarrow \Lambda^{(3)} \leftrightarrow \Lambda^{(5)}$ that follows from the matching condition (13).

4 Quantitative Illustration

Consider now a few cases in the $f = 5$ region.

Υ Decay. According to the Particle Data Group (PDG) overview (see Fig. 9.1 on page 88 of [15]), this is (with $\alpha_s(M_\Upsilon^2) \simeq 0.170$ and $\bar{\alpha}_s(M_Z^2) = 0.114$) one of the most "annoying" points of their summary of $\bar{\alpha}_s(M_Z^2)$ values. It is also singled out theoretically. The expression for the ratio of decay widths starts with the cubic term

$$R(\Upsilon) = R_0\, \alpha_s^3(M_\Upsilon^2)(1 + e_1\, \alpha_s) \quad \text{with} \quad e_1 \simeq 1\,.$$

Due to this, the π^2 correction[3] is rather big here $\mathfrak{A}_3 \simeq \alpha_s^3\left(1 - 2(\pi\beta_0)^2\alpha_s^2\right)$. Accordingly, $\Delta\alpha_s(M_\Upsilon^2) = (2/3)\,(\pi\beta_0)^2\,\alpha_s^3(M_\Upsilon^2) \simeq 0.0123$, which corresponds to $\Delta\bar{\alpha}_s(M_Z^2) = 0.006$ with resulting $\bar{\alpha}_s(M_Z^2) = 0.120$.

The NNLO Case. Now, let us turn to a few cases usually analyzed by the three-term expansion formula (1). For the first example, take e^+e^- hadron annihilation at $\sqrt{s} = 42\,\text{GeV}$ and $11\,\text{GeV}$.

A common form (see, e.g., (21) in [29]) of theoretical presentation of the QCD correction in our normalization looks like

$$r_{e^+e^-}(\sqrt{s}) = 0.318\bar{\alpha}_s(s) + 0.143\,\bar{\alpha}_s^2 - 0.413\,\bar{\alpha}_s^3\,.$$

In the standard PT analysis, one has (see, e.g., Table 3) $\bar{\alpha}_s(42^2) = 0.144$ that corresponds to $r_{e^+e^-}(42) \simeq 0.0476$. Along with the APT prescription, one should use

$$r_{e^+e^-}(\sqrt{s}) = 0.318\,\tilde{\alpha}(s) + 0.143\,\mathfrak{A}_2(s) - 0.023\,\mathfrak{A}_3(s)\,, \tag{26}$$

which yields $\tilde{\alpha}(42^2) = 0.142 \;\rightarrow\; \alpha_s(42^2) = 0.145$ and $\bar{\alpha}_s(M_Z^2) = 0.127$ to be compared with $\bar{\alpha}_s(M_Z^2) = 0.126$ under a usual analysis.

Quite analogously, with $\bar{\alpha}_s(11^2) = 0.200$ and $r_{e^+e^-}(11) \simeq 0.0661$ we obtain via (26) $\tilde{\alpha}(11^2) = 0.190$ that corresponds to $\bar{\alpha}_s(M_Z^2) = 0.129$ instead of 0.130.

For the next example, take the Z_0 inclusive decay. The observed ratio

$$R_Z = \Gamma(Z_0 \to hadrons)/\Gamma(Z_0 \to leptons) = 20.783 \pm .029$$

can be written down as follows: $R_Z = R_0\left(1 + r_Z(M_Z^2)\right)$ with $R_0 = 19.93$. A common form (see, e.g., (15) in [29]) of presenting the QCD correction r_Z looks like

$$r_Z(M_Z) = 0.3326\bar{\alpha}_s + 0.0952\,\bar{\alpha}_s^2 - 0.483\,\bar{\alpha}_s^3\,.$$

To $[r_Z]_{obs} = 0.04184$ there corresponds $\bar{\alpha}_s(M_Z^2) = 0.124$. In the APT case, from

$$r_Z(M_Z) = 0.3326\,\tilde{\alpha}(M_Z^2) + 0.0952\,\mathfrak{A}_2(M_Z^2) - 0.094\,\mathfrak{A}_3(M_Z^2) \tag{27}$$

[3] First proposal of taking into account this effect in the Υ decay was made[8] twenty years ago. Meanwhile, in current practice it is completely forgotten.

we obtain $\tilde{\alpha}(M_Z^2) = 0.122$ and $\bar{\alpha}_s(M_Z^2) = 0.124$. Note that here the three-term approximation (7) gives the same relation between the $\bar{\alpha}_s(M_Z^2)$ and $\tilde{\alpha}(M_Z^2)$ values.

Nevertheless, in accordance with our preliminary estimate for the $(\Delta\bar{\alpha}_s)_4$ role, even the so-called NNLO theory needs some π^2 correction in the $W = \sqrt{s} \lesssim 50$ GeV region.

The NLO Case. Some experiments in the HE ($f = 5$) Minkowskian region (mainly with a shape analysis) usually are confronted with the two-term expression (20). As it has been shown above (23), the main theoretical error here can be expressed in the form

$$(\Delta\bar{\alpha}_s(s)|_{20\div 100\mathrm{GeV}}^{f=5} \simeq 1.225\,\bar{\alpha}_s^3(s) \simeq 0.002 \div 0.003\,. \tag{28}$$

An adequate expression for the equivalent shift of the $\bar{\alpha}_s(M_Z^2)$ value is

$$[\Delta\bar{\alpha}_s(M_Z^2)]_3 = 1.225\bar{\alpha}_s(s)\bar{\alpha}_s(M_Z^2)^2\,. \tag{29}$$

We give the results of our approximate APT calculations, mainly by (28) and (29), in the form of Table 3. In the last column of Table 3 in brackets, we indicate the difference between the APT and usual analysis. The three–loop cases are marked by bold figures. Dots in the lower part of the Table correspond to shape–events data for energies $W = 133, 161, 172$ and 183 GeV with the same positive shift 0.002 for the the extracted $\bar{\alpha}_s$ values.

Our average 0.121 over $f = 5$ events from Bethke's Table 6 [29] correlates with data of the same author (see Summary of [30]). The best χ^2 fit yields

Table 3. The APT revised[4]part ($f = 5$) of Bethke's [29] Table 6

Process	$\sqrt{s}$ GeV	loops No	$\bar{\alpha}_s$ (s) [2]	$\bar{\alpha}_s(M_Z^2)$ [2]	$\bar{\alpha}_s$ (s) APT	$\bar{\alpha}_s(M_Z^2)$ APT
Υ-decay [5]	9.5	2	.170	.114	.182	.120 (+6)
$e^+e^-[\sigma_{had}]$	10.5	**3**	.200	.130	.198	.129(**-1**)
$e^+e^-[j\,\&\,sh]$	22.0	2	.161	.124	.166	.127(+3)
$e^+e^-[j\,\&\,sh]$	35.0	2	.145	.123	.149	.126(+3)
$e^+e^-[\sigma_{had}]$	42.4	**3**	.144	.126	.145	.127(**+1**)
$e^+e^-[j\,\&sh]$	44.0	2	.139	.123	.142	.126(+3)
$e^+e^-[j\,\&sh]$	58	2	.132	.123	.135	.125(+2)
$Z_0 \to$ **had.**	91.2	**3**	.124	.124	.124	.124(**0**)
$e^+e^-[j\,\&\,sh]$	91.2	2	.121	.121	.123	.123(+2)
-"-	....	2	...	...	...	...(+2)
$e^+e^-[j\,\&\,sh]$	189	2	.110	.123	.112	.125(+2)

Averaged $\langle\bar{\alpha}_s(M_Z^2)\rangle_{f=5}$ values 0.121; 0.124

[5] "j & sh" = jets and shapes; Figures in brackets in the last column give the difference $\Delta\bar{\alpha}_s(M_Z^2)$ between common and the APT values.

$\bar{\alpha}_s(M_Z^2)_{[2]} = 0.1214$ and[6]

$$\bar{\alpha}_s(M_Z^2)_{APT} = 0.1235\,.$$

Our new $\chi^2_{\rm APT}$ is smaller $\chi^2_{APT}/\chi^2_{PT} \simeq 0.73$ than the usual one. This illustrates the effectiveness of the APT procedure in the region far enough from the ghost singularity.

5 Conclusion

It is a common standpoint that in QCD it is legitimate to use the power in α_s expansion for observables in the low energy (low momentum transfer) region. At the same time, there exist rather general (and old [31]) arguments in favor of nonanalyticity of the S matrix elements at the origin of the complex plane of the α variable, with α being an expansion parameter[32]. This implies that common power expansion, mathematically, has no domain of convergence. It corresponds to the factorial growth ($\sim n!$) of expansion coefficients at large n [33,34]. Nevertheless, "practically" one can use such divergent series for obtaining numerical information in the case when few first subsequent terms of series do diminish.

It is a popular belief that in QCD, with its "not small enough" α_s values, one does face an asymptotic nature of perturbation expansion by observing approximate equality of relative contributions of the second (α_s^2) and the third (α_s^3) terms into observable, like in all PT columns of Table 2.

Our first qualitative result consists in observation that dimishing properties of the APT series drastically differ from the usual PT ones.

The better "practical convergence" of the APT series for the Euclidean observable, as it has been demonstrated in the right part of Table 2, probably means that essential singularity at $\alpha_s = 0$ is adequately taken into account by new expansion functions $\mathcal{A}_k(Q^2)$. On the other hand, in the time–like region the improved approximation property of the APT expansion over $\mathfrak{A}_k(s)$ has a bit different nature, being related, in our opinion, to the nonuniform convergence of the usual PT series for Minkowskian observables. In any case, from a practical point of view:

1. In the APT approach one can use the nonpower expansions (18) and (19) without the last term.

The next point, discussed in Sect. 3, refers to a more specific issue connected with current practice of the Minkowskian observable analysis in the low–energy ($s \lesssim 3\,\mathrm{GeV}^2$) region (like, e.g., inclusive τ decay). As it has been shown

2. Below $s = 2\,\mathrm{GeV}^2$ it is impossible to use the power expansion (1) for a time–like observable.

[6] This value, corresponding to $\Lambda^{(5)} = 290\,\mathrm{MeV}$, correlates with fresh APT analysis [26] of τ decay that gives $\Lambda^{(3)} = 420$ MeV.

Second group of our results is of a quantitative nature:

3. Effective positive shift $\Delta\bar{\alpha}_s = +0.002$ in the upper half ($\geq 50\,\mathrm{GeV}$) of the $f = 5$ region for all time-like events that have been analyzed up to now in the NLO mode.

4. Effective shift $\Delta\bar{\alpha}_s \simeq +0.003$ in the lower half ($10 \div 50\,\mathrm{GeV}$) of the $f = 5$ region for time-like events that have been analyzed in the NLO mode.

5. The new value

$$\bar{\alpha}_s(M_Z^2) = 0.124\,, \tag{30}$$

obtained by averaging the APT results over the $f = 5$ region.

The quantitative results are based on the new APT nonpower expansion (7) and plausible hypothesis on the π^2–term prevalence in common expansion coefficients for observables in the Minkowskian domain. The hypothesis has some preliminary support – see Table 1.

Nevertheless, our result (30) being taken as granted raises two physical questions:

– The issue of self-consistency of QCD invariant coupling behavior between the "medium ($f = 3, 4$)" and "high ($f = 5, 6$)" regions.

– The new "enlarged value" (30) can influence various physical speculations in the several hundred GeV region.

Acknowledgements

The author is indebted to A.P. Bakulev, R. Kögerler, S.V. Mikhailov, I. L. Solovtsov, O.P. Solovtsova and N. Stefanis for useful discussions and comments. This work was partially supported by Russian Foundation for Basic Research (RFBR grants Nos 99-01-00091 and 00-15-96691), by INTAS grant No 96-0842 and by CERN–INTAS grant No 99-0377.

References

1. D.V. Shirkov, Theor. Math. Phys.**127**, 409 (2001); JINR preprint E2-2000-298; hep-ph/0012283; see also JINR preprint E2-2000-46; hep-ph/0003242.
2. D.V. Shirkov, Eur. Phys. J. C **22**. 331 (2001).
3. D.E.Groom *et al.*, Eur. Phys. J. C **15**, 1 (2000).
4. N.N. Bogoliubov and D.V. Shirkov, Introduction to the theory of Quantized Fields, (Wiley & Intersc. New York, 1959,1980).
5. N.N. Bogoliubov and D.V. Shirkov, Quantum Fields, (Benjamin/Cummings, Reading, 1983).
6. Yu.L. Dokshitzer and D.V. Shirkov, Z. Phys. C **67**, 449 (1995).
7. A. Radyushkin, Dubna JINR preprint E2-82-159 (1982); see also JINR Rapid Comm. No.4[78]-96, 9 (1996); hep–ph/9907228.
8. N.V. Krasnikov, A.A. Pivovarov, Phys. Lett. B **116**, 168 (1982).
9. K.A. Milton and I.L. Solovtsov, Phys. Rev. D **55**, 5295 (1997).

10. A.L. Kataev and V.V. Starshenko, Mod. Phys. Lett. A **10**, 235 (1995).
11. L. Alkofer and L. von Smekal, Phys. Repts. **353**, 281 (2001).
12. B. Schrempp and F. Schrempp, Z. Physik C **6**, 7 (1980).
13. A.P. Bakulev, A.V. Radyushkin and N.G. Stefanis, Phys. Rev. D **62**, 113001 (2000).
14. D. S. Kourashev and B. A. Magradze, Report RMI–2001–18; hep-ph/0104142.
15. B. Magradze, Preprint JINR, E2-2000-222; hep-ph/0010070.
16. D.V. Shirkov and I.L. Solovtsov, JINR Rapid Comm. No.2[76]-96, pp 5-10; hep-ph/9604363.
17. D.V. Shirkov and I.L. Solovtsov, Phys.Rev.Lett. **79**, 1209 (1997).
18. D.V. Shirkov and I.L. Solovtsov, Phys.Lett. B **442**, 344 (1998).
19. D.V. Shirkov, *TMP* **119** (1999) 438–447, hep-th/9810246.
20. I.L. Solovtsov and D.V. Shirkov, Theor. Math. Phys. **120**, 1220 (1999).
21. D.V. Shirkov, Lett. Math. Phys. **48**, 135 (1999).
22. K.A. Milton and O.P. Solovtsova, Phys. Rev. D **57**, 5402 (1998).
23. A.I. Alekseev, hep-ph/0105338.
24. K.A. Milton, I.L. Solovtsov and O.P. Solovtsova, Phys. Lett. B **439**, 421 (1998).
25. K.A. Milton, I.L. Solovtsov and O.P. Solovtsova, Phys. Rev. D **60**, 016001 (1999).
26. K.A. Milton, I.L. Solovtsov and O.P. Solovtsova, Phys. Rev. D **64**, 016005 (2001).
27. K.A. Milton, I.L. Solovtsov and O.P. Solovtsova, hep-ph/0111197.
28. Delphi Collaboration, " Consistent measurements of α_s from ...", CERN-EP/99-133.
29. Z. Bethke, J. Phys. G **26**, R27 (2000); Preprint MPI-PhE/2000-07; hep-ex/0004021.
30. Z. Bethke, "QCD at LEP", (Oct 11, 2000) in http:/cern.wed.cern.ch/CERN/Announcement/2000/LEPFest/.
31. F.J. Dyson, Phys. Rev. **85**, 631 (1952).
32. D.V. Shirkov, Lett. Math. Phys. **1**, 179 (1976); Lett. Nuovo Cim. **18**, 452 (1977).
33. L.N. Lipatov, Sov. Phys. JETP **45**, 216 (1977).
34. D.I. Kazakov and D.V. Shirkov, Fortsch. der Physik **28**, 456 (1980).

Perturbative Logarithms and Power Corrections in QCD Hadronic Functions. A Unifying Approach

Nikolaos G. Stefanis

Institut für Theoretische Physik II, Ruhr-Universität Bochum, D-44780 Bochum, Germany

Abstract. I present a unifying scheme for hadronic functions that comprises logarithmic corrections due to gluon emission in perturbative QCD, as well as power-behaved corrections of nonperturbative origin. The latter are derived by demanding that perturbatively resummed partonic observables should be analytic in the whole Q^2-plane if they are to be related to physical observables measured in experiments. I also show phenomenological consequences of this approach. The focus is on the electromagnetic pion form factor to illustrate both effects, Sudakov logarithms and power corrections in leading order of $\Lambda^2_{\rm QCD}/Q^2$. The same approach applied to the inclusive Drell-Yan cross section enables us to perform an absolutely normalized calculation of the leading power correction in $b^2\Lambda^2_{\rm QCD}$ (b being the impact parameter), which after exponentiation, gives rise to a nonperturbative Sudakov-type contribution that provides enhancement rather than suppression, hence partly counteracting the perturbative Sudakov suppression.

1 Introduction

In recent years, effort in QCD has turned increasingly toward the problem of including resummation effects due to multiple soft gluon emission, both in perturbation theory, as well as in the nonperturbative regime. The first effect is related to Sudakov suppression [1], well-known from QED, whereas those in the nonperturbative regime manifest themselves as power-behaved corrections [2], which, after exponentiation, amount to a Sudakov-like form factor [3]. However, as it turns out [4] this contribution provides enhancement rather than suppression. The hope is that improving the perturbative and nonperturbative structure of the theory this way, it will be possible to get better agreement with the existing hadronic data in terms of both correct overall shape and also normalization. In these investigations the crucial organizing principle is QCD factorization, which provides a handle to separate the short-distance (hard) component of a reaction (controlled by the large mass scale in the process, Q) - that will be treated perturbatively - from its long-distance (soft) nonperturbative part, related to the nontrivial QCD vacuum structure (and field condensates).

In processes which involve the emission of virtual gluon quanta of low momentum, one must resum their contributions to all orders of the strong coupling constant. This gives rise to exponentially suppressing factors in b-space (where b is the impact parameter conjugate to the transverse momentum $Q_\perp$) of the reaction amplitude (or cross-section) of the Sudakov type with exponents containing double and single logarithms of the large mass scale of the process [1]. However, because of the Landau singularity of the running coupling at transverse distances $b \propto 1/\Lambda_{\rm QCD}$, an essential singularity appears in the Sudakov factor. Thus, one has to consider power corrections of $\mathcal{O}\left(b^2 \Lambda_{\rm QCD}^2\right)$, which, though negligible for small b relative to logarithmic corrections $\propto \ln\left(\mathrm{b}^2\Lambda_{\rm QCD}^2\right)$, may become important for larger values of the impact parameter.

In this talk, I will discuss a general methodology to treat (power) series in the running strong coupling in connection with gluon emission. To be more precise, I will address this issue in terms of two processes: one to which the OPE applies, viz. the pion electromagnetic form factor at leading perturbative order, and another, the Drell-Yan process, to which the OPE is not applicable. The first is a typical example of an exclusive process with registered intact hadrons in the initial and final states (for a recent review and references, see, e.g., [5]). Such processes provide a "window" to view the detailed structure of hadrons in terms of quarks and gluons at Fermi level (*Hadron Femptoscopy*). The Drell-Yan mechanism, on the other hand, has two identified hadrons in the initial state and a lepton pair (plus unspecified particles) in the final state, whose transverse momentum distribution is proportional to the large invariant mass of the materialized photon.

The goal in the second case will be to obtain not only the usual resummed (Sudakov) expression (which comprises logarithmic corrections due to soft-gluon radiation), but also to include the leading power correction as well, specifying, in particular, its concomitant coefficient. This becomes possible within a theoretical scheme, which models the IR behavior of the running coupling by demanding analyticity of physical observables (in the complex Q^2 plane) as a *whole* – as opposed to imposing analyticity of individual powers, i.e., order by order in perturbation theory –, while preserving renormalization-group invariance (references and additional information can be found in the recent surveys [6,7] and D.V. Shirkov, these proceedings). The underlying idea behind our method [4], is to demand that if hadronic observables, calculated at the partonic level, are to be compared with experimental data, they have to be analytic in the entire Q^2 plane. This "analytization" procedure encompasses Renormalization Group (RG) invariance (i.e., resummation of UV logarithms and correct UV asymptotics) and causality (which imposes a spectral representation). As we shall see below, *analytization* removes all unphysical singularities in the the IR region, rendering perturbatively calculated hadronic observables IR-renormalon free.

2 Analytic Factorization Scheme (AFS)

2.1 Perturbative Pion Form Factor with Sudakov Corrections

Let us conduct our investigation by considering the space-like electromagnetic pion's form factor in the transverse (impact) configuration space:

$$F_\pi\left(Q^2\right) = \int_0^1 dx dy \int_{-\infty}^{\infty} \frac{d^2\mathbf{b}}{(4\pi)^2}\, \mathcal{P}_\pi^{\rm out}\left(y, b, P'; C_1, C_2, C_4\right) T_{\rm H}\left(x, y, b, Q; C_3, C_4\right)$$

$$\times \mathcal{P}_{\pi}^{\text{in}}(x, b, P; C_1, C_2, C_4) + \ldots, \tag{1}$$

where the modified pion wave function is defined in terms of matrix elements, viz.,

$$\begin{aligned} \mathcal{P}_{\pi}(x, b, P, \mu) &= \int^{|\mathbf{k}_\perp|<\mu} d^2\mathbf{k}_\perp \mathrm{e}^{-i\mathbf{k}_\perp \cdot \mathbf{b}_\perp} \tilde{\mathcal{P}}_{\pi}(x, \mathbf{k}_\perp, P) \\ &= \int \frac{dz^-}{2\pi} \mathrm{e}^{-ixP^+z^-} \left\langle 0 \left| \mathrm{T}\left(\bar{q}(0)\gamma^+\gamma_5 q\left(0, z^-, \mathbf{b}_\perp\right)\right) \right| \pi(P) \right\rangle_{A^+=0} \end{aligned} \tag{2}$$

with $P^+ = Q/\sqrt{2} = P^{-\prime}$, $Q^2 = -(P' - P)^2$, whereas the dependence on the renormalization scale μ on the RHS of (2) enters through the normalization scale of the current operator, evaluated on the light cone, and the dependence on the effective quark mass has not been displayed explicitly. In (2), T_{H} is the amplitude for a quark and an anti-quark to scatter via a series of hard-gluon exchanges with gluonic transverse momenta (alias inter-quark transverse distances) not neglected from the outset. In the above, the ellipsis indicates the non-factorizing soft part, as well as disregarded higher-order corrections. The scheme constants C_i emerge from the truncation of the perturbative series and would be absent if one was able to derive all-order expressions in the coupling constant. The scale C_1/b ($C_1 = C_3$) serves to separate perturbative from non-perturbative transverse distances (lower factorization scale of the Sudakov regime and *transverse* cutoff). The re-summation range in the Sudakov form factor is limited from above by the scale $C_2\xi Q$ (upper factorization scale of the Sudakov regime and *collinear* cutoff).[1] The arbitrary constant C_4 serves to define the renormalization scale $C_4 f(x, y)Q = \mu_{\text{R}}$, which appears in the argument of the analytic running coupling $\alpha_{\text{s}}^{\text{an}}$ [8] (choice of renormalization prescription):

$$\begin{aligned} \bar{\alpha}_{\text{s}}^{\text{an}(1)}(Q^2) &\equiv \bar{\alpha}_{\text{s}}^{\text{pert}(1)}(Q^2) + \bar{\alpha}_{\text{s}}^{\text{npert}(1)}(Q^2) \\ &= \frac{4\pi}{\beta_0}\left[\frac{1}{\ln(\mathrm{Q}^2/\Lambda^2)} + \frac{\Lambda^2}{\Lambda^2 - Q^2}\right], \end{aligned} \tag{3}$$

where here and below $\Lambda \equiv \Lambda_{\text{QCD}}$ is the QCD scale parameter.

To leading order in analytic perturbation theory (APT), one has

$$T_{\text{H}}(x, y, b, Q; \mu_{\text{R}}) = 8C_{\text{F}}\alpha_{\text{s}}^{\text{an}}(\mu_{\text{R}}^2) K_0\left(\sqrt{xy}\, bQ\right), \tag{4}$$

where $C_{\text{F}} = (N_{\text{c}}^2 - 1)/2N_{\text{c}} = 4/3$ for $SU(3)_{\text{c}}$. The amplitude

$$\begin{aligned} \mathcal{P}_{\pi}(x, b, P \simeq Q, C_1, C_2, \mu) = \exp\Bigg[&-s(x, b, Q, C_1, C_2) - s(\bar{x}, b, Q, C_1, C_2) \\ &-2\int_{C_1/b}^{\mu} \frac{d\bar{\mu}}{\bar{\mu}} \gamma_{\text{q}}\left(\alpha_{\text{s}}^{\text{an}}(\bar{\mu})\right)\Bigg] \mathcal{P}_{\pi}(x, b, C_1/b) \end{aligned} \tag{5}$$

[1] Note that $\sqrt{2}C_2 = C_2^{\text{CSS}}$ [1].

describes the distribution of longitudinal momentum fractions of the $q\bar{q}$ pair, taking into account the intrinsic transverse size of the pion state and comprising corrections due to soft real and virtual gluons, including also evolution from the initial amplitude $\mathcal{P}_\pi(x, b, C_1/b)$ at scale C_1/b to the renormalization scale $\mu \propto Q$ (more details and references are relegated to [9]). The main effect of the absence of a Landau pole in the running coupling α_s^{an} is to make the functions $s(x, b, Q, C_1, C_2)$, $s(\bar{x}, b, Q, C_1, C_2)$ well-defined (analytic) in the IR region and to slow down evolution by extending soft-gluon cancellation down to the scale $C_1/b \simeq \Lambda_{QCD}$, where the full Sudakov form factor acquires a finite value, modulo its Q^2 dependence (see LHS of Fig. 1). In addition, as we shall see below, the Sudakov exponent contains power-behaved corrections in $(C_1/b\Lambda)^{2p}$ and $(C_2/\xi Q\Lambda)^{2p}$, starting with $p = 1$. Such contributions are the footprints of soft gluon emission at the kinematic boundaries to the non-perturbative QCD regime, characterized by the transversal (or IR) and the longitudinal (or collinear) cutoffs.

The pion distribution amplitude evaluated at the (low) factorization scale C_1/b is approximately given by

$$\mathcal{P}_\pi(x, b, C_1/b, m_q) \simeq \frac{f_\pi/\sqrt{2}}{2\sqrt{N_c}}\,\phi_\pi(x, C_1/b)\,\Sigma(x, b, m_q)\,. \tag{6}$$

To model the intrinsic transverse momenta of the pion bound state, we have to make an ansatz for their distribution. (For a recent derivation from

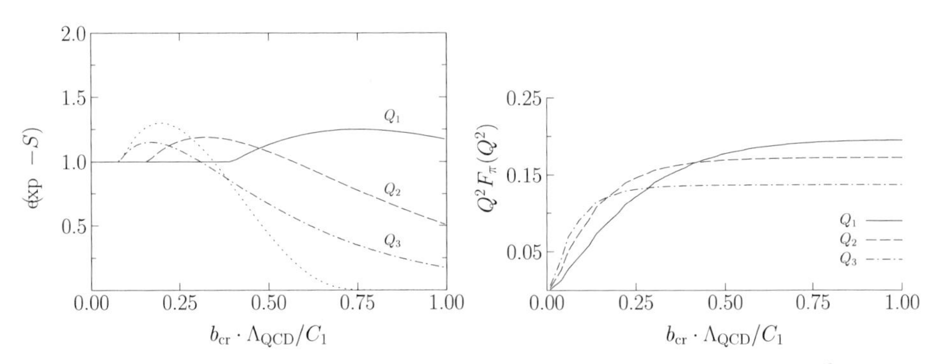

Fig. 1. (**a**) Sudakov form factor versus transverse separation b for three Q^2 values: $Q_1 = 2$ GeV, $Q_2 = 5$ GeV, and $Q_3 = 10$ GeV, with all $\xi_i = 1/2$, and where we have set $C_1 = 2e^{-\gamma_E}$, $C_2 = e^{-1/2}$ and $\Lambda_{QCD} = 0.242$ GeV. The dotted curve shows the result obtained with $\alpha_s^{\overline{MS}}$, and $\Lambda_{QCD} = 0.2$ GeV for $Q_2 = 5$ GeV, using the same set of C_i. In that case, evolution is limited by the (renormalization) scale $\mu_R = t = \{\max\sqrt{xy}\,Q, C_1/b\}$, as proposed in [10], albeit the enhancement at small b-values due to the quark anomalous dimension is not neglected here. (**b**) Saturation behavior of pion's electromagnetic form factor, calculated in the AFS at NLO with commensurate scale setting (see text) and including a mass term (with $m_q = 0.33$ GeV) in the BHL ansatz [11] for the soft pion wave function. The scheme parameters are defined in (17). Here b_{cr} denotes the integration cutoff over transverse distances in (23). The momentum transfer values are as in part (a)

an instanton-based model, see [12]). Here, I employ the Brodsky-Huang-Lepage (BHL) ansatz [11] and parameterize the distribution $\Sigma(x, \mathbf{k}_\perp, m_q)$ in the intrinsic transverse momentum $k_\perp$ (or equivalently the intrinsic inter-quark transverse distance b) in the form of a *non-factorizing* in the variables x and $k_\perp$ (or x and b) Gaussian function:

$$\Psi_\pi\left(x, \mathbf{k}_\perp, C_1/b, m_q\right) = \frac{f_\pi/\sqrt{2}}{2N_c} \Phi_\pi(x, C_1/b) \Sigma\left(x, \mathbf{k}_\perp, m_q\right), \tag{7}$$

where

$$\Phi_\pi\left(x, C_1/b\right) = A\, \Phi_{\mathrm{as}}(x) = A\, 6x(1-x) \tag{8}$$

is the asymptotic distribution amplitude, with A being an appropriate normalization factor, and where

$$\Sigma\left(x, \mathbf{k}_\perp, m_q\right) = 16\pi^2 \beta_\pi^2 g(x) \hat{\Sigma}\left(x, \mathbf{k}_\perp\right) \hat{\Sigma}\left(x, m_q\right) \tag{9}$$

with

$$\hat{\Sigma}\left(x, \mathbf{k}_\perp\right) = \exp\left[-\beta_\pi^2 \mathbf{k}_\perp^2 g(x)\right], \tag{10}$$

and

$$\hat{\Sigma}\left(x, m_q\right) = \exp\left[-\beta_\pi^2 m_q^2 g(x)\right]. \tag{11}$$

By inputting f_π and the value of the quark mass m_q and using $g(x) = 1/\left(x\bar{x}\right)$, with $\bar{x} \equiv (1-x)$, we determine the parameters (we refer for more details to [9]) A, β_π^2, $P_{q\bar{q}}$, and $\langle \mathbf{k}_\perp^2 \rangle^{1/2}$, tabulated in Table 1.

We have now to calculate the Sudakov contribution within the AFS. Generically, the Sudakov form factor $F_S\left(\xi, b, Q, C_1, C_2\right)$, i.e., the exponential factor in front of the wave function, will be expressed as the expectation value of an open Wilson (world) line along a contour of finite extent, C, which follows the bent quark line in the hard-scattering process from the segment with direction (four-momentum) P to that with direction P' after being abruptly derailed by the hard interaction which creates a "cusp" in C. It is to be evaluated within the range of momenta from C_1/b (IR cutoff) to $C_2\xi Q$ (longitudinal cutoff) (where $\xi = x, \bar{x}, y, \bar{y}$) and the region of hard interaction of the Wilson line with the off-shell photon is factorized out. Then the Sudakov functions, entering (5), can

Table 1. Values of parameters entering the pion wave function [9]. The values in parentheses refer to the case $m_q = 0$ and the subscript "as" on β_π^2 to the asymptotic distribution amplitude

Input parameters	Determined parameters
$m_q = 0.33$ GeV	$A = \frac{1}{6} \cdot 10.01$ $(\frac{1}{6} \cdot 6)$
$f_\pi = 0.1307$ GeV	$\beta_{\mathrm{as}}^2 = 0.871$ GeV^{-2} (0.743 GeV^{-2})
	$\langle \mathbf{k}^2 \rangle^{1/2} = 0.352$ GeV (0.367 GeV)
	$P_{q\bar{q}} = 0.306$ (0.250)

be expressed in terms of the momentum-dependent cusp anomalous dimension of the bent contour [1,13–15] to read

$$s(\xi, b, Q, C_1, C_2) = \frac{1}{2} \int_{C_1/b}^{C_2 \xi Q} \frac{d\mu}{\mu} \, \Gamma_{\text{cusp}}\left(\gamma, \alpha_s^{\text{an}}(\mu)\right) \tag{12}$$

with the anomalous dimension of the cusp given by

$$\begin{aligned} \Gamma_{\text{cusp}}\left(\gamma, \alpha_s^{\text{an}}(\mu)\right) &= 2\ln\left(\frac{\mathrm{C}_2 \xi \mathrm{Q}}{\mu}\right) \mathrm{A}\left(\alpha_s^{\text{an}}(\mu)\right) + \mathrm{B}\left(\alpha_s^{\text{an}}(\mu)\right) , \\ &\equiv \Gamma_{\text{cusp}}^{\text{pert}} + \Gamma_{\text{cusp}}^{\text{npert}} , \end{aligned} \tag{13}$$

$\gamma = \ln(\mathrm{C}_2\xi\mathrm{Q}/\mu)$ being the cusp angle, i.e., the emission angle of a soft gluon and the bent eikonalized quark line after the external (large) momentum Q has been injected at the cusp point by the off-mass-shell photon, and where in the second line of (13) the superscripts relate to the origin of the corresponding terms in the running coupling. The functions A and B are known at two-loop order:

$$\begin{aligned} A\left(\alpha_s^{\text{an}}(\mu)\right) &= \frac{1}{2}\left[\gamma_{\mathcal{K}}\left(\alpha_s^{\text{an}}(\mu)\right) + \beta(g)\frac{\partial}{\partial g}\mathcal{K}(C_1, \alpha_s^{\text{an}}(\mu))\right] \\ &= C_{\mathrm{F}} \frac{\alpha_s^{\text{an}}(g(\mu))}{\pi} + \frac{1}{2} K(C_1)\, C_{\mathrm{F}} \left(\frac{\alpha_s^{\text{an}}(g(\mu))}{\pi}\right)^2 , \end{aligned} \tag{14}$$

and

$$\begin{aligned} B\left(\alpha_s^{\text{an}}(\mu)\right) &= -\frac{1}{2}\left[\mathcal{K}\left(C_1, \alpha_s^{\text{an}}(\mu)\right) + \mathcal{G}\left(\xi, C_2, \alpha_s^{\text{an}}(\mu)\right)\right] \\ &= \frac{2}{3}\frac{\alpha_s^{\text{an}}(g(\mu))}{\pi} \ln\left(\frac{\mathrm{C}_1^2}{\mathrm{C}_2^2}\frac{\mathrm{e}^{2\gamma_{\mathrm{E}}-1}}{4}\right) . \end{aligned} \tag{15}$$

The first term in (14) is universal,[2] while the second one as well as the contribution termed B are scheme dependent. The K-factor in the $\overline{\text{MS}}$ scheme to two-loop order is given by [1,13,16]

$$K(C_1) = \left(\frac{67}{18} - \frac{\pi^2}{6}\right) C_{\mathrm{A}} - \frac{10}{9} n_{\mathrm{f}} T_{\mathrm{F}} + \beta_0 \ln\left(\mathrm{C}_1 \mathrm{e}^{\gamma_{\mathrm{E}}}/2\right) \tag{16}$$

with $C_{\mathrm{A}} = N_{\mathrm{C}} = 3$, $n_{\mathrm{f}} = 3$, $T_{\mathrm{F}} = 1/2$, and γ_{E} being the Euler-Mascheroni constant. A set of constants C_i, $(i = 1, 2, 3)$, which eliminate artifacts of dimensional regularization while practically preserving the matching between the re-summed and the fixed-order calculation, are [9]

$$\begin{aligned} &C_1 = 2\exp(-\gamma_{\mathrm{E}}), \quad C_2 = \exp(-1/2), \quad C_3 = 2\exp(-\gamma_{\mathrm{E}}), \quad C_4 = \exp(-4/3), \\ &K = 4.565, \quad \kappa = 0. \end{aligned} \tag{17}$$

[2] In works quoted above, the cusp anomalous dimension is identified with the universal term, whereas the other (scheme and/or process dependent) terms are considered as additional anomalous dimensions. Here this distinction is irrelevant.

The quantities $\mathcal{K}$, $\mathcal{G}$ in (15) are calculable using the non-Abelian extension to QCD [1] of the Grammer-Yennie method for QED or employing the Wilson (world) lines approach [13–15]. The soft (Sudakov-type) form factor depends only on the cusp angle which varies with the inter-quark transverse distance b ranging between C_1/b and $C_2\xi Q$. The corresponding anomalous dimensions are inter-linked through the relation $2\Gamma_{\rm cusp}\left(\alpha_{\rm s}^{\rm an}(\mu)\right) = \gamma_{\mathcal{K}}\left(\alpha_{\rm s}^{\rm an}(\mu)\right)$ with $\Gamma_{\rm cusp}(\alpha_{\rm s}^{\rm an}(\mu)) = C_{\rm F}\,\alpha_{\rm s}^{\rm an}(\mu^2)/\pi$, which shows that $\frac{1}{2}\gamma_{\mathcal{K}} = A\left(\alpha_{\rm s}^{\rm an}(\mu)\right)$. (Note that $\gamma_{\mathcal{G}} = -\gamma_{\mathcal{K}}$ and $\gamma_{\rm q}\left(\alpha_{\rm s}^{\rm an}(\mu)\right) = -\alpha_{\rm s}^{\rm an}(\mu^2)/\pi$.)

The leading contribution to the Sudakov functions $s\,(\xi, b, Q, C_1, C_2)$ (where $\xi = x, \bar{x}, y, \bar{y}$) within our framework, is obtained by expanding the functions A and B in a power series in $\alpha_{\rm s}^{\rm an}$ and collecting together all large logarithms $\left(\frac{\alpha_{\rm s}^{\rm an}}{\pi}\right)^n \ln\left(\frac{\rm C_2}{\rm C_1}\xi{\rm bQ}\right)^{\rm m}$, which correspond to large logarithms $\ln\left(\frac{\rm Q^2}{\rm k_\perp^2}\right)$ in transverse momentum space. The leading contribution results from the expression

$$s\,(\xi, b, Q, C_1, C_2) = \frac{1}{2}\int_{C_1/b}^{C_2\xi Q}\frac{d\mu}{\mu}\left\{2\ln\left(\frac{\rm C_2\xi Q}{\mu}\right)\left[\frac{\alpha_{\rm s}^{\rm an(2)}(\mu)}{\pi}{\rm A}^{(1)} + \left(\frac{\alpha_{\rm s}^{\rm an(1)}(\mu)}{\pi}\right)^2 A^{(2)}\,(C_1)\right] + \frac{\alpha_{\rm s}^{\rm an(1)}(\mu)}{\pi}B^{(1)}\,(C_1, C_2) + \dots\right\}, \quad (18)$$

where the two-loop expression [8] for the strong coupling is to be used in front of $A^{(1)}$, whereas the other two terms are to be evaluated with the one-loop result. Let me remark at this point that in the following we ignore the difference between the analytic strong coupling squared and its "analytized" second power. These issues will be considered elsewhere. The specific values of the coefficients $A^{(i)}$, $B^{(i)}$ are

$$\begin{aligned} A^{(1)} &= C_{\rm F}\,, \\ A^{(2)}\,(C_1) &= \frac{1}{2}\,C_{\rm F}\,K\,(C_1)\,, \\ B^{(1)}\,(C_1, C_2) &= \frac{2}{3}\ln\left(\frac{\rm C_1^2}{\rm C_2^2}\frac{{\rm e}^{2\gamma_{\rm E}-1}}{4}\right), \end{aligned} \quad (19)$$

in which the term proportional to $A^{(1)}$ represents the universal part. The *universal* part of the Sudakov factor in LLA and including power corrections, reads

$$F_{\rm S}^{\rm univ}\,(\mu_{\rm F}, Q) = \exp\left\{-\frac{C_{\rm F}}{\beta_0}\left[\ln\left(\frac{\tilde{\rm Q}^2}{\Lambda^2}\right)\ln\frac{\ln\tilde{\rm Q}^2/\Lambda^2}{\ln\mu_{\rm F}^2/\Lambda^2} - \ln\frac{\tilde{\rm Q}^2}{\mu_{\rm F}^2} + \ln\left(\frac{\tilde{\rm Q}^2}{\mu_{\rm F}^2}\right) \times\ln\frac{\Lambda^2-\mu_{\rm F}^2}{\Lambda^2} + \frac{1}{2}\ln^2\frac{\tilde{\rm Q}^2}{\mu_{\rm F}^2} + {\rm Li}_2\left(\frac{\tilde{\rm Q}^2}{\Lambda^2}\right) - {\rm Li}_2\left(\frac{\mu_{\rm F}^2}{\Lambda^2}\right)\right]\right\}, \quad (20)$$

where $\tilde{Q}$ represents the scale $C_2\xi Q$ and the IR matching (factorization) scale $\mu_{\rm F}$ varies with the inverse transverse distance b, i.e., $\mu_{\rm F} = C_1/b$. Note that the four last terms in this equation originate from the non-perturbative power correction (cf. (13)), and that Li_2 is the dilogarithm (Spence) function which comprises power-behaved corrections of the IR-cutoff ($b\Lambda$) and the longitudinal cutoff (Q/Λ). To complete the discussion about the Sudakov factor, I display the result obtained by neglecting power corrections:

$$
\begin{aligned}
s(\xi, b, Q, C_1, C_2) = \; & \frac{1}{\beta_0}\left[\left(2A^{(1)}\hat{Q} + B^{(1)}\right)\ln\frac{\hat{\mathrm{Q}}}{\hat{\mathrm{b}}} - 2\mathrm{A}^{(1)}\left(\hat{\mathrm{Q}} - \hat{\mathrm{b}}\right)\right] - \frac{4}{\beta_0^2}A^{(2)} \\
& \times\left(\ln\frac{\hat{\mathrm{Q}}}{\hat{\mathrm{b}}} - \frac{\hat{\mathrm{Q}} - \hat{\mathrm{b}}}{\hat{\mathrm{b}}}\right) + \frac{\beta_1}{\beta_0^3}A^{(1)}\left\{\ln\frac{\hat{\mathrm{Q}}}{\hat{\mathrm{b}}} - \frac{\hat{\mathrm{Q}} - \hat{\mathrm{b}}}{\hat{\mathrm{b}}}\left[1 + \ln\left(2\hat{\mathrm{b}}\right)\right]\right. \\
& \left. + \frac{1}{2}\left[\ln^2\left(2\hat{\mathrm{Q}}\right) - \ln^2\left(2\hat{\mathrm{b}}\right)\right]\right\},
\end{aligned} \tag{21}
$$

where the convenient abbreviations [10] $\hat{Q} \equiv \ln\frac{\mathrm{C_2\xi Q}}{\Lambda}$ and $\hat{b} \equiv \ln\frac{\mathrm{C_1}}{\mathrm{b}\Lambda}$ have been used. Note that expressions given in the literature by other authors are erroneous.

In the following, (18) is evaluated numerically to NLLA with appropriate kinematic bounds [9] to ensure proper factorization at the numerical level. The electromagnetic pion form factor in next-to-leading logarithmic order has the following form in LO of T_{H}:

$$
\begin{aligned}
F_\pi(Q^2) = \; & \frac{2}{3}A^2\pi\, C_{\mathrm{F}}\, f_\pi^2 \int_0^1 dx \int_0^1 dy \int_0^\infty b\, db\, \alpha_{\mathrm{s}}^{\mathrm{an}(1)}(\mu_{\mathrm{R}})\, \Phi_\pi(x)\Phi_\pi(y) \\
& \times \exp\left[-\frac{b^2\left(x\bar{x} + y\bar{y}\right)}{4\beta_\pi^2}\right] \exp\left[-\beta_\pi^2 m_{\mathrm{q}}^2\left(\frac{1}{x\bar{x}} + \frac{1}{y\bar{y}}\right)\right] K_0\left(\sqrt{xy}Qb\right) \\
& \times \exp\left[-S\left(x, y, b, Q, C_1, C_2, C_4\right)\right],
\end{aligned} \tag{22}
$$

whereas in NLO it reads

$$
\begin{aligned}
F_\pi\left(Q^2\right) = \; & 16A^2\pi C_{\mathrm{F}}\left(\frac{f_\pi/\sqrt{2}}{2\sqrt{N_{\mathrm{c}}}}\right)^2 \int_0^1 dx \int_0^1 dy \int_0^\infty b\, db\, \alpha_{\mathrm{s}}^{\mathrm{an}}\left(\mu_{\mathrm{R}}^2\right)\Phi_\pi(x)\Phi_\pi(y) \\
& \times \exp\left[-\frac{b^2\left(x\bar{x} + y\bar{y}\right)}{4\beta_\pi^2}\right] \exp\left[-\beta_\pi^2 m_{\mathrm{q}}^2\left(\frac{1}{x\bar{x}} + \frac{1}{y\bar{y}}\right)\right] K\left(\sqrt{xy}Qb\right) \\
& \times \exp\left(-S\left(x, y, b, Q, C_1, C_2, C_4\right)\right)\left\{1 + \frac{\alpha_{\mathrm{s}}^{\mathrm{an}}}{\pi}\left[f_{\mathrm{UV}}\left(x, y, Q^2/\mu_{\mathrm{R}}^2\right)\right.\right. \\
& \left.\left. + f_{\mathrm{IR}}\left(x, y, Q^2/\mu_{\mathrm{F}}^2\right) + f_{\mathrm{C}}(x, y)\right]\right\}.
\end{aligned} \tag{23}
$$

In these equations the Sudakov form factor, including evolution, is given by

$$
\begin{aligned}
S\left(x, y, b, Q, C_1, C_2, C_4\right) \equiv \; & s\left(x, b, Q, C_1, C_2\right) + s\left(\bar{x}, b, Q, C_1, C_2\right) + \left(x \leftrightarrow y\right) \\
& - 8\,\tau\left(C_1/b, \mu_{\mathrm{R}}\right)
\end{aligned} \tag{24}
$$

with the "evolution time" [9]

$$\tau\left(\frac{C_1}{b},\mu\right) = \int_{C_1^2/b^2}^{\mu^2} \frac{dk^2}{k^2} \frac{\alpha_s^{\mathrm{an}(1)}(k^2)}{4\pi}$$
$$= \frac{1}{\beta_0}\ln\frac{\ln\left(\mu^2/\Lambda^2\right)}{\ln\left(\mathrm{C}_1^2/\left(\mathrm{b}\Lambda\right)^2\right)} + \frac{1}{\beta_0}\left[\ln\frac{\mu^2}{\left(\mathrm{C}_1/\mathrm{b}\right)^2} - \ln\frac{\left|\mu^2-\Lambda^2\right|}{\left|\frac{\mathrm{C}_1^2}{\mathrm{b}^2}-\Lambda^2\right|}\right] \quad (25)$$

and the functions f_i taken from [17]. I present predictions for F_π in Fig. 2, adopting the BLM commensurate-scale method [18], and setting $\mu_{\mathrm{F}} = C_1/b$ and $\mu_{\mathrm{BLM}} = \mu_{\mathrm{R}}\exp(-5/6)$, where $\mu_{\mathrm{R}} = C_4 f(x,y)Q = C_4\sqrt{xy}Q$.

As one sees, the hard contribution to $F_\pi(Q^2)$ within the AFS and with a BLM-optimized choice of scales provides a sizeable fraction of the magnitude of the form factor – especially at NLO. No artificial rising at low Q^2 of the hard form factor appears, as in conventional approaches, so that this region is dominated by the Feynman-type contribution [21]. Moreover, the self-consistency of perturbation theory has been improved, as one infers from the saturation behavior of the scaled form factor, presented on the RHS of Fig. 1. Indeed, $Q^2F_\pi\left(Q^2\right)$ accumulates the bulk of its magnitude below $b_{\mathrm{cr}}\Lambda_{\mathrm{QCD}}/C_1 \leq 0.5$, i.e., for short transverse distances, where the application of perturbative QCD is sound. Even better predictions can be obtained, using a more accurate pion distribution amplitude, recently derived in [22] with QCD sum rules and non-local condensates.

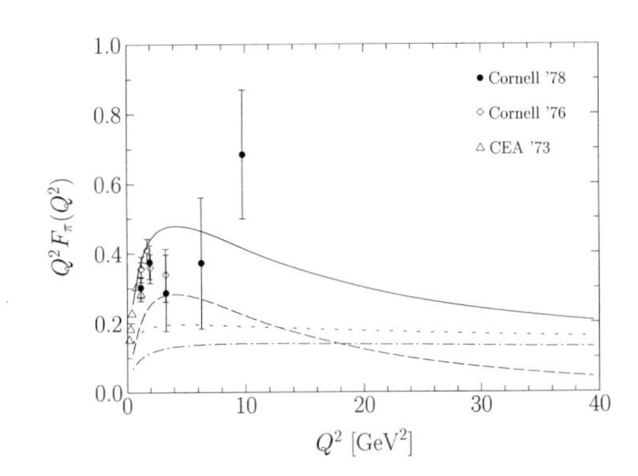

Fig. 2. Space-like pion form factor calculated within the AFS. Further details are provided in the text. LO calculation (dashed-dotted line); NLO calculation (dotted line). The dashed line gives the result for the soft, Feynman-type contribution, computed with $m_{\mathrm{q}} = 0.33$ GeV in the pion wave function, and the solid line represents the sum of the NLO hard contribution and the soft one [9]. The data are taken from [19,20].

2.2 Power Corrections to Pion Form Factor

The rationale of *global analyticity*, i.e., analyticity as a whole, implies

$$
\begin{aligned}
\left[Q^2 F_\pi\left(Q^2\right)\right]_{\text{an}} &= \int_0^1 dx \int_0^1 dy \left[\Phi_\pi^{\text{out}}\left(y, Q^2\right) T_{\text{H}}\left(x, y, Q^2, \alpha_s(\hat{Q}^2)\right) \Phi_\pi^{\text{in}}\left(x, Q^2\right)\right]_{\text{an}} \\
&= A \int_0^1 dx \int_0^1 dy\, x\, y [T_{\text{H}}(x, y, Q^2, \alpha_{\text{s}}(\hat{Q}^2))]_{\text{an}}
\end{aligned}
\tag{26}
$$

wherre A is a normalization constant for the pion distribution amplitude, taken again to be the asymptotic one. Without the analytization requirement, the pion form factor is not Borel-summable (see, e.g., [23]), but only an asymptotic series in the coupling constant. Analytization entails

$$
\left[\alpha_{\text{s}}^n\left(Q^2\right)\right]_{\text{an}} \equiv \frac{1}{\pi} \int_0^\infty \frac{d\xi}{\xi + Q^2 - i\epsilon} \rho^{(n)}(\xi),
\tag{27}
$$

where the spectral density $\rho^{(n)}(\xi)$ is the *dispersive conjugate of all powers n* of α_{s}. For the leading-order expression under consideration the spectral density becomes [8,24,25]

$$
\rho\left(Q^2\right) = \text{Im}\alpha_{\text{s}}\left(-Q^2\right) = \frac{\pi}{\beta_1} \frac{1}{\ln^2(\text{Q}^2/\Lambda^2) + \pi^2}
\tag{28}
$$

Then (27) reduces to

$$
\left[\alpha_{\text{s}}\left(Q^2\right)\right]_{\text{an}} = \frac{1}{2\pi i} \int_C \frac{dz}{z - Q^2 + i\epsilon} \alpha_{\text{s}}(z),
\tag{29}
$$

where C is a closed contour in the complex z-plane with a branch cut along the negative real axis, assuming exactly the form of (3), as proposed by Shirkov and Solovtsov [8]. Recasting the strong coupling in the form

$$
\alpha_{\text{s}}(z) = \frac{1}{\beta_1} \frac{1}{\ln \frac{\text{z}}{\Lambda^2}} = \pm \int_0^\infty d\sigma \exp\left(\mp \sigma \beta_1 \ln\Lambda^2/\text{z}\right)
\tag{30}
$$

with the plus sign corresponding to the case $|z|/\Lambda^2 > 1$ and the minus one to $|z|/\Lambda^2 < 1$, and inserting it into (26), we find after some standard manipulations the Borel transform of the scaled pion form factor at leading perturbative order [4]:

$$
\left[Q^2 F_\pi\left(Q^2\right)\right]_{\text{an}}^{(1)} = \int_0^\infty d\sigma \exp\left(-\sigma\beta_1 \ln\text{Q}^2/\Lambda^2\right) \tilde{\pi}(\sigma)_{\text{an}}^{(1)}.
\tag{31}
$$

The Borel image of the form factor reads

$$
\begin{aligned}
\tilde{\pi}(\sigma)_{\text{an}}^{(1)} &= 16\pi C_{\text{F}} A \frac{\sin\left(\pi\beta_1\sigma\right)}{\pi} \int_0^1 dx \int_0^1 dy \bar{x}\bar{y} \int_0^\infty \frac{d\xi}{\xi + xy} \\
&\times \left[\xi^{-\sigma\beta_1} \Theta\left(\xi - \frac{\Lambda^2}{Q^2}\right) + \left(\frac{Q^2}{\Lambda^2}\right)^{2\sigma\beta_1} \xi^{\sigma\beta_1} \Theta\left(\frac{\Lambda^2}{Q^2} - \xi\right)\right].
\end{aligned}
\tag{32}
$$

This expression has no IR renormalons in contrast to approaches that use the conventional one-loop α_s parameterization [23].

Hence, the integration over the Borel parameter σ can be performed without any ambiguity to arrive at the following result for the pion form factor

$$\left[Q^2 F_\pi\left(Q^2\right)\right]_{\mathrm{an}}^{(1)} = 16\pi C_{\mathrm{F}} A \frac{1}{\beta_1} \int_0^1 dw\, \phi(w) \left[\frac{1}{\ln\left(\frac{\mathrm{w}Q^2}{\Lambda^2}\right)} + \frac{1}{1-\frac{wQ^2}{\Lambda^2}}\right] . \tag{33}$$

The remaining integration can be carried out analytically to arrive at an expression derived in [4]. Here I only display the expression for the physically relevant case $Q^2 \gg \Lambda^2$:

$$\begin{aligned}\left[Q^2 F_\pi\left(Q^2\right)\right]_{\mathrm{an}}^{(1)} &= 16\pi C_{\mathrm{F}} A \left[\frac{1}{4}\alpha_{\mathrm{s}}\left(Q^2\right) + \mathcal{O}\left(\alpha_{\mathrm{s}}^2\right)\right] - \frac{1}{\beta_1} 16\pi C_{\mathrm{F}} A \frac{\Lambda^2}{Q^2} \\ &\times \left[\frac{1}{2}\ln^2\left(\frac{Q^2}{\Lambda^2}\right) - 2\ln\left(\frac{Q^2}{\Lambda^2}\right) - \frac{\pi^2}{6} + 2\right] + \mathcal{O}\left(\frac{\Lambda^4}{Q^4}\right)\end{aligned} \tag{34}$$

referring for further details to [4].

3 Power Corrections to Drell-Yan Process

As a second example of the AFS, I discuss the derivation of power corrections to the inclusive Drell-Yan cross-section with the large scale Q^2 being here the invariant lepton pair mass. Citations to previous works and full details of the derivation are given in [4]. Consider the logarithmic derivative of the unrenormalized expression of the eikonalized Drell-Yan cross section, with the notations of [2]:

$$\begin{aligned}\frac{d\ln \mathrm{W}_{\mathrm{DY}}}{d\ln \mathrm{Q}^2} &\equiv \Pi^{(1)}\left(Q^2\right) \\ &= 4C_{\mathrm{F}}\mu^{2\epsilon} \int \frac{d^{2-2\epsilon}k_\perp}{(2\pi)^{2-2\epsilon}} \frac{1}{k_\perp^2}\, \alpha_{\mathrm{s}}\left(k_\perp^2\right)\left(\mathrm{e}^{-i\mathbf{k}_\perp\cdot\mathbf{b}} - 1\right) .\end{aligned} \tag{35}$$

The following important remarks are now in place: (i) The argument of α_s is taken to depend on $k_\perp$ to account for higher-order quantum corrections, originating from momentum scales larger than this [16]. (ii) The integral over the transverse momentum is *not well-defined* at very small mass scales owing to the Landau singularity of the QCD running coupling in that region. (iii) The evaluation at the edge of phase space is sensitive to the regularization applied to account for power corrections due to soft gluon emission transient to nonperturbative QCD.

Imposing analytization as a whole and integrating over transverse momenta, we obtain

$$\left[\Pi^{(1)}\left(Q^2\right)\right]_{\mathrm{an}} = \int_0^\infty d\sigma\, \mathrm{e}^{-\sigma\beta_1 \ln\left(4/\mathrm{b}^2\Lambda^2\right)}\, \tilde{\Pi}_{\mathrm{an}}^{(1)}(\sigma) \tag{36}$$

with a Borel transform given by

$$\tilde{\Pi}^{(1)}_{\text{an}}(\sigma) = \frac{4C_{\text{F}}}{\pi}\left(\frac{\mu^2 b^2}{4}\right)^{\epsilon} \sin\left(\pi\sigma\beta_1\right) \int_0^\infty d\xi\, g(\xi)\left[\xi^{-\sigma\beta_1}\Theta\left(\xi - \frac{b^2\Lambda^2}{4}\right)\right.$$
$$\left. + \left(\frac{b^2\Lambda^2}{4}\right)^{-2\sigma\beta_1} \xi^{\sigma\beta_1}\Theta\left(\frac{b^2\Lambda^2}{4} - \xi\right)\right], \tag{37}$$

where

$$g(\xi) = \int \frac{d^{2-2\epsilon}q}{(2\pi)^{2-2\epsilon}} \frac{1}{q^2}\frac{1}{q^2+\xi}\left(\mathrm{e}^{-2i\mathbf{q}\cdot\hat{\mathbf{b}}} - 1\right). \tag{38}$$

Combining denominators in (38) and carrying out the integrations over ξ, we then find

$$\left[\Pi^{(1)}\left(Q^2\right)\right]_{\text{an}} = \frac{C_{\text{F}}}{\pi}\left(\mu^2 b^2 \pi\right)^{\epsilon} \int_0^\infty d\sigma\, \mathrm{e}^{-\sigma\beta_1 \ln\left(4/\mathrm{b}^2\Lambda^2\right)} \frac{1}{\Gamma\left(1+\sigma\beta_1\right)}$$
$$\times\left[-\frac{1}{\sigma\beta_1+\epsilon}\Gamma\left(-1-\sigma\beta_1-\epsilon\right) + \sum_{n=0}^{\infty}\frac{(-1)^n}{(n+1)!}\right.$$
$$\left.\times\left(\frac{b^2\Lambda^2}{4}\right)^{n+1-\sigma\beta_1-\epsilon}\frac{1}{n+1-\sigma\beta_1-\epsilon}\right] - \frac{C_{\text{F}}}{\pi\beta_1} f\left(\frac{b^2\Lambda^2}{4}\right) \tag{39}$$

with $f\left(b^2\Lambda^2/4\right)$ being a complicated expression, provided in [4] and $\Gamma(x,y)$, denoting the incomplete Gamma function. The first term in (39), viz., the integral over σ, diverges for $\sigma\beta_1 = 0$, i.e., for small values of $\alpha_{\text{s}}\left(k_\perp\right)$ (or equivalently for large transverse momenta $k_\perp$). This UV divergence is regulated dimensionally within the $\overline{MS}$ renormalization scheme adopted here. Were it not for the terms containing powers of $b\Lambda$, expression (39) and that found by Korchemsky and Sterman [2] (namely, their equation (18)) would be the same. In our case, however, the imposition of analytization cures all divergences related to IR renormalons that are generated by the Γ-functions whenever $\sigma\beta_1$ is an integer different from zero.

Let us concentrate on the second term in $\left[\Pi^{(1)}\left(Q^2\right)\right]_{\text{an}}$ that gives rise to *power corrections*. Retaining only the leading contribution in $b^2\Lambda^2$, we find

$$f\left(b^2\Lambda^2\right) = -a_0 - a_1\frac{b^2\Lambda^2}{4}\ln\frac{\mathrm{b}^2\Lambda^2}{4} + \mathrm{a}_2\frac{\mathrm{b}^2\Lambda^2}{4} + \mathcal{O}\left(\mathrm{b}^4\Lambda^4\right) \tag{40}$$

with the constant coefficients [4]: $a_0 \simeq 1.795$, $a_1 \simeq 2.179$, $a_2 \simeq 1.394$. Now one can expand the integral in the first term of $\left[\Pi^{(1)}\left(Q^2\right)\right]_{\text{an}}$ in powers of $b^2\Lambda^2$ and regulate the UV pole at $\sigma\beta_1 = 0$ dimensionally. For $\sigma\beta_1$ an integer, both terms inside the bracket have poles, but they *mutually cancel* so that their sum is singularity-free and the integral finite. Retaining terms of order $b^2\Lambda^2$, the main contribution stems from the leading renormalon at $\sigma\beta_1 = 1$:

$$\left[\Pi^{(1)}\left(Q^2\right)\right]_{\text{an}} = \left[\Pi^{(1)}\left(Q^2\right)\right]_{\text{PT}} + \left[\Pi^{(1)}\left(Q^2\right)\right]_{\text{pow}} \tag{41}$$

with the perturbative part being defined by

$$\left[\Pi^{(1)}\left(Q^2\right)\right]_{\rm PT} = \frac{C_{\rm F}}{\pi\beta_1}\ln\frac{\ln\left(\mathrm{C}/\mathrm{b}^2\Lambda^2\right)}{\ln\left(\mathrm{Q}^2/\Lambda^2\right)}, \tag{42}$$

a result coinciding with that obtained in [2]. Power corrections in the impact parameter b are encoded in the second contribution ($b^2\Lambda^2 \ll 1$):

$$\left[\Pi^{(1)}\left(Q^2\right)\right]_{\rm pow} = S_0 + b^2 S_2\left(b^2\Lambda^2\right) + \mathcal{O}\left(b^4\Lambda^4\right), \tag{43}$$

where

$$S_0 = \frac{C_{\rm F}}{\pi\beta_1} a_0 \tag{44}$$

and

$$S_2\left(b^2\Lambda^2\right) = \frac{C_{\rm F}}{4\pi\beta_1}\Lambda^2\left[-1+\gamma_{\rm E}+(1+a_1)\ln\frac{\mathrm{b}^2\Lambda^2}{4} - \mathrm{a}_2\right]. \tag{45}$$

The DY cross-section $W_{\rm DY}$, comprising the leading logarithmic perturbative contribution (Sudakov exponent $S_{\rm PT}$) and the first power correction (in $b^2\Lambda^2$) reads (with the Q-dependence arising due to collinear interactions)

$$W_{\rm DY}(b,Q) = \exp\left[-S_{\rm PT}(b,Q) - S_0(Q) - b^2 S_2(b,Q) + \ldots\right], \tag{46}$$

where

$$S_0(Q) \sim S_0 \ln\mathrm{Q} + \mathrm{const} \tag{47}$$

and

$$S_2(b,Q) \sim S_2\left(b^2\Lambda^2\right)\ln\mathrm{Q} + \mathrm{const}. \tag{48}$$

The Sudakov factor, representing the perturbative tail of the hadronic wave function, *suppresses* constituent configurations which involve large impact space separations, while the exponentiated power corrections in b^2, being of nonperturbative origin, *provide enhancement* of such configurations (since $S_2\left(b^2\Lambda^2\right)$ is always negative). Hence, the net result is less suppression of the DY cross-section and also enhancement of the pion wave function in b space with the endpoint region $b\Lambda \sim 1$ being less enhanced relative to small b transverse distances.

4 Conclusions

I have presented a theoretical framework, based on *analytization* that enables the calculation of perturbative gluonic corrections (Sudakov form factor), as well as power-behaved ones that are linked to nonperturbative effects in QCD. Moreover, one can calculate the absolute normalization of the power corrections to hadronic observables *systematically* without any renormalon ambiguity from the outset.

Acknowledgement

I wish to thank the organizers for the warm hospitality extended to me during this stimulating meeting and the Deutsche Forschungsgemeinschaft for a travel grant. It is a pleasure to thank A.P. Bakulev, A.I. Karanikas, and W. Schroers for collaboration and K. Passek, M. Praszałowicz, D.V. Shirkov, and I.L. Solovtsov for discussions.

References

1. J.C. Collins, D.E. Soper, Nucl. Phys. B **193**, 381 (1981); ibid. B **197**, 446 (1982).
2. G.P. Korchemsky, G. Sterman, Nucl. Phys. B **437**, 415 (1995).
3. J.C. Collins, D.E. Soper, G. Sterman, Nucl. Phys. B **250**, 199 (1985).
4. A.I. Karanikas, N.G. Stefanis, Phys. Lett. B **504**, 225 (2001).
5. N.G. Stefanis, Eur. Phys. J.direct C **7**, 1 (1999).
6. D.V. Shirkov, Eur. Phys. J. C **22**, 331 (2001); Theor. Math. Phys. **127**, 409 (2001).
7. D.V. Shirkov and I.L. Solovtsov, Theor. Math. Phys. 120 (1999) 1210.
8. D.V. Shirkov, I. L. Solovtsov, Phys. Rev. Lett. **79**, 1209 (1997).
9. N.G. Stefanis, W. Schroers, H.-Ch. Kim, Eur. Phys. J. C **18**, 137 (2000).
10. H.-n. Li, G. Sterman, Nucl. Phys. B **381**, 129 (1992).
11. S.J. Brodsky, T. Huang, G.P. Lepage, In: *Banff Summer Institute on Particles and Fields (1981)*, ed. by A.Z. Capri and A.N. Kamal (Plenum Press, New York 1983) pp. 143-199.
12. M. Praszałowicz, A. Rostworowski, Phys. Rev. D **64**, 074003 (2001); hep-ph/0111196.
13. G.P. Korchemsky, A.V. Radyushkin, Nucl. Phys. B **283**, 342 (1987).
14. G.P. Korchemsky, Phys. Lett. B **217**, 330 (1989); ibid. B **220**, 629 (1989).
15. G. Gellas, A.I. Karanikas, C.N. Ktorides, N.G. Stefanis, Phys. Lett. B **412**, 95 (1997); A.I. Karanikas, C.N. Ktorides, N.G. Stefanis, S.M.H. Wong, Phys. Lett. B **455**, 291 (1999).
16. J. Kodaira, L. Trentadue, Phys. Lett. B **112**, 66 (1982).
17. B. Melić, B. Nižić, K. Passek, Phys. Rev. D **60**, 074004 (1999).
18. S.J. Brodsky, G.P. Lepage, P.B. Mackenzie, Phys. Rev. D **28**, 228 (1982).
19. C.N. Brown et al., Phys. Rev. D **8**, 92 (1973).
20. C.J. Bebek et al., Phys. Rev. D **13**, 25 (1976); ibid. D **17**, 1693 (1978).
21. A.V. Radyushkin, Few Body Syst. Suppl. **11**, 57 (1999); A.P. Bakulev, A.V. Radyushkin, N.G. Stefanis, Phys. Rev. D **62**, 113001 (2000).
22. A.P. Bakulev, S.V. Mikhailov, N.G. Stefanis, Phys. Lett. **B508**, 279 (2001); hep-ph/0104290.
23. S. Agaev, hep-ph/9611215; Mod. Phys. Lett. A **13**, 2637 (1998); Eur. Phys. J. C **1**, 321 (1998).
24. A.V. Radyushkin, JINR preprint E2-82-159, JINR Rapid Communications 4[78]-96, 9 (1982).
25. N.V. Krasnikov, A.A. Pivovarov, Phys. Lett. B **116**, 168 (1982).

Bounds on $\tan\beta$ in the MSSM from Top Quark Production at TeV Energies

Claudio Verzegnassi[1,2]

[1] Dipartimento di Fisica Teorica, Università di Trieste, Strada Costiera 14, Miramare (Trieste)
[2] INFN, Sezione di Trieste

Abstract. I consider the process of top-antitop quark production from electron-positron annihilation, for c.m. energies in the few TeV regime, in the MSSM theoretical framework. I show that, at the one loop level, in a moderately "light" SUSY scenario, the slopes of a number of observable quantities in an energy region around $3\,TeV$ would only depend on $\tan\beta$. Under optimal experimental conditions, a combined measurement of slopes might identify with acceptable precision $\tan\beta$ values larger that 20.

1 Introduction

The existence in the SM of large electroweak Sudakov logarithms [1] in four-fermion processes at the one-loop level,for c.m. energies in the TeV range [2,3], and the subsequent efforts for providing a full resummation of the relevant terms [4,5] have been the subject of theoretical discussions in the last couple of years. Quite recently, the extension of this kind of analysis at one loop within the MSSM ,for final massless fermion pairs [6] and also for final top quark pairs production [7] has been accomplished.

One of the conclusions of [7] is that for the process of top pairs from electron-positron annihilation at c.m. energies in the few TeV range,that would be explored by the future CERN CLIC collider [8], the SM component of the logarithmic effects seems to be fully under control. Technically speaking, this is due to the almost total lack of linear and θ dependent (θ is the c.m. scattering angle) logarithmic terms, whose presence might give rise to tedious resummation complications. This is a consequence of the weak isospin characterization of the final state, and it would not apply for instance to final bottom pairs.

The fact that virtual SM effects are fully under control raises the interest of considering the role of possible virtual SUSY effects of Sudakov origin in top production at CLIC energies. Our starting goal will be that of examining the computable one-loop level, given the fact that a resummation of SUSY effects is not yet available, although work along that direction is already in progress [9]. It was shown in [6] that the leading Sudakov SUSY effect at one loop is only of linear logarithmic kind (in the SM, also quadratic logarithms appear), and θ-independent. It is only produced by final vertices, and contains a component of "massless quark" kind and one of "massive top" Yukawa origin that strongly depends on $\tan\beta$. Its numerical effect can vary from a few percent to ten percent, strongly depending on $\tan\beta$, being apparently still sensitive to relatively large $\tan\beta > 10$ values. In [7] it was suggested that this fact might be used to try to fix the value of $\tan\beta$ from a combined analysis of the value of several observables at fixed energy, e.g. around the proposed CLIC "optimal" values $\sqrt{q^2} = 3\ TeV$. The conclusion of that Reference was that a deeper investigation of this possibility would have followed. The aim of this short seminar is precisely that of summarizing the aforementioned investigation. The new proposal will be that of considering, rather than measurements at fixed energies, variations of observables (slopes) with energy around a chosen interesting (e.g. $3\ TeV$) energy value. I will show that the only unknown quantities in the coefficients of the various slopes are functions of $\tan\beta$ alone. All the other MSSM parameters can be incorporated asymptotically into terms that either vanish or remain constant, thus disappearing in the slope. I will then show that, simulating a reasonable experimental situation, it might be possible to identify $\tan\beta$ with "decent" precision (i.e. to better than a relative fifty percent for relatively large ($20 < \tan\beta < 40$) values. This should be compared and combined with other interesting recently proposed $\tan\beta$ detection techniques [10].

2 SUSY Sudakov Logarithms in the *TeV* Region

Following the previous discussion, attitude, I will now rewrite all the relevant formulae of [7] indicating with the "SM" symbol the (supposedly perfectly known) SM component. For my purposes, I will need the asymptotic Sudakov expansion of the various quantities,that contains as the leading term a θ-independent linear logarithm. This is, as I said in the Introduction, the sum of a "massless" and a "massive" component, whose origin is due to the vertex diagrams shown in Fig 1. In the limit when the c.m. energy $\sqrt{q^2}$ becomes very large, they produce the overall massive leading logarithmic SUSY Sudakov term (relevant for our analysis) listed in the following equations for photon or Z exchanges:

$$\Gamma^{\gamma}_{\mu} \to \frac{e\alpha}{12\pi M^2_W s^2_W} \ln\frac{\mathrm{q}^2}{\mathrm{M}^2}\{\mathrm{m}^2_\mathrm{t}(1+\cot^2\beta)[(\gamma_\mu \mathrm{P_L}) + 2(\gamma_\mu \mathrm{P_R})] + m^2_b(1+\tan^2\beta)(\gamma_\mu P_L)\} \tag{1}$$

$$\Gamma^{Z}_{\mu} \to \frac{e\alpha}{48\pi M^2_W s^3_W c_W} \ln\frac{\mathrm{q}^2}{\mathrm{M}^2}\{(3-4\mathrm{s}^2_\mathrm{W})\mathrm{m}^2_\mathrm{t}(1+\cot^2\beta)(\gamma_\mu \mathrm{P_L}) -8s^2_W m^2_t(1+\cot^2\beta)(\gamma_\mu P_R) + (3-4s^2_W)m^2_b(1+\tan^2\beta)(\gamma_\mu P_L)\} \tag{2}$$

One sees that, in the leading term, the only SUSY parameters that appear are $\tan\beta$ and an overall common (unknown) "SUSY scale" M which I only assumed to be "reasonably" smaller than the energy value $\sqrt{q^2} = 3\ TeV$ in which I am interested in this talk. Of course, this assumption might be wrong and heavier SUSY masses might turn out to be produced. In that case, my "asymptotic" expansions would still be valid, obviously in a (suitably) higher energy range.

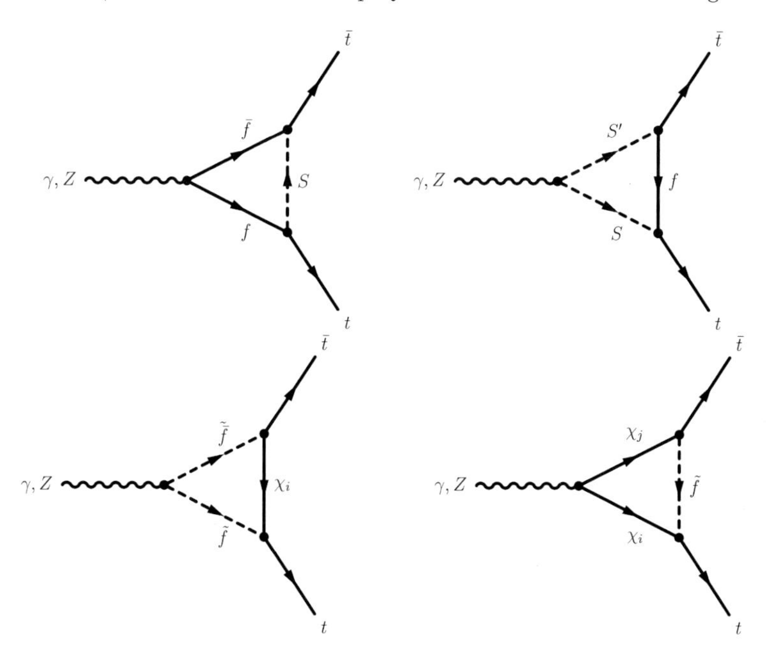

Fig. 1. Triangle diagrams with SUSY Higgs and with SUSY partners contributing to the asymptotic logarithmic behaviour in the energy; f represent t or b quarks; S represent charged or neutral Higgs bosons $H^{\pm}$, A^0, H^0, h^0 or Goldstone G^0; $\tilde{f}$ represent stop or sbottom states; χ represent charginos or neutralinos. The arrow corresponds to the momentum flow of the indicated particle.

Starting from (1), (2) it is a straightforward task to derive the leading SUSY Sudakov contributions to the various observables [7]. I write here, in the previously discussed spirit, the expressions that will be relevant for my purposes, considering, for simplicity, only a set of four observables where the final top quark helicity is not measured. The chosen quantities are σ_t (the cross section for top pair production), $A_{FB,t}$ and $A_{LR,t}$ (the forward-backward and longitudinal polarization asymmetries) and A_t (the polarized forward-backward asymmetry). For the chosen observables I obtain the following asymptotic expansions:

$$\begin{aligned}
\sigma_t &= \sigma_t^{SM}\{1 + \frac{\alpha}{4\pi}\{(4.44\,N + 11.09)\ln\frac{\mathrm{q}^2}{\mu^2} - 10.09\ln\frac{\mathrm{q}^2}{\mathrm{M}^2}\} + \quad (3)\\
&+ F_{\sigma_t}(\tan\beta)\ln\frac{\mathrm{q}^2}{\mathrm{M}'^2}\\
F_{\sigma_t}(t) &= \frac{\alpha}{4\pi}(-29\,t^{-2} - 0.0084\,t^2 - 14),\\
A_{FB,t} &= A_{FB,t}^{SM} + \frac{\alpha}{4\pi}\{(0.22\,N + 1.29)\ln\frac{\mathrm{q}^2}{\mu^2} - 0.23\ln\frac{\mathrm{q}^2}{\mathrm{M}^2}\} + \quad (4)\\
&+ F_{A_{FB,t}}(\tan\beta)\ln\frac{\mathrm{q}^2}{\mathrm{M}'^2}\ .
\end{aligned}$$

$$
\begin{aligned}
F_{A_{FB,t}}(t) &= \frac{\alpha}{4\pi}(1.2\,t^{-2} - 0.00082\,t^2 + 0.62), \\
A_{LR,t} &= A^{SM}_{LR,t} + \frac{\alpha}{4\pi}\{(1.03\,N + 5.95)\ln\frac{\mathrm{q}^2}{\mu^2} - 4.03\ln\frac{\mathrm{q}^2}{\mathrm{M}^2}\} + \qquad (5) \\
&+ F_{A_{LR,t}}(\tan\beta)\ln\frac{\mathrm{q}^2}{\mathrm{M}'^2} \\
F_{A_{LR,t}}(t) &= \frac{\alpha}{4\pi}(7.7\,t^{-2} - 0.0051\,t^2 + 3.8), \\
A_t &= A^{SM}_t + \frac{\alpha}{4\pi}\{(0.91\,N + 5.25)\ln\frac{\mathrm{q}^2}{\mu^2} - 3.20\ln\frac{\mathrm{q}^2}{\mathrm{M}^2}\} + \qquad (6) \\
&+ F_{A_t}(\tan\beta)\ln\frac{\mathrm{q}^2}{\mathrm{M}'^2}, \\
F_{A_t}(t) &= \frac{\alpha}{4\pi}(7.5\,t^{-2} - 0.0049\,t^2 + 3.7).
\end{aligned}
$$

In the previous equations, I have also listed in the first bracket, for sake of completeness, the SUSY asymptotic linear logarithm of RG origin. This contributes a universal term of self-energy origin, where no SUSY parameters (except a SUSY scale) appear. In my procedure I will add this RG logarithm to that of SM origin, and consider it as a part of the ("uninteresting") "Non SUSY Sudakov" structure, that I shall try to eliminate. The second bracket contains the massless Sudakov term (i.e. the one that would have been obtained $\simeq$ for u, c pair production) and the third one contains the massive Sudakov term.

One notices, as stressed in [5], a strong $\tan\beta$ dependence in some observable (particularly σ_t) that might possibly be exploited to perform a determination of this parameter. With this purpose, I will now review this possibility with some caution, in a way that I shall now illustrate.

Clearly, if the logarithmic term were the only relevant one in the SUSY component of the observables, a determination of $\tan\beta$ might proceed in principle via a fit of the various observables at a fixed chosen energy. Quite generally, in an asymptotic expansion like the one that we are assuming, there will be extra non leading contributions, in particular constant terms and terms that vanish (at least as $1/q^2$) asymptotically. I assume (consistently with our philosophy) that the latter ones can be safely neglected and concentrate the attention on possible constant quantities. The approach followed in [7] was that of computing exactly the contributions to the various observables from the considered SUSY vertices, and to try to fit the numerical results with an expansion of the form

$$
(F_1\cot^2\beta + F_2\tan^2\beta + F_3)\ln\frac{\mathrm{q}^2}{\mathrm{M}^2} + \mathrm{B} + \cdots
$$

This has been fully discussed in [7], and in Fig. 2 I show the results that were obtained.

As one can see in the figure, values in the range

$$
\tan\beta < 2, \qquad \tan\beta > 20
$$

can be detected with $N = 10$ c.m. energy values with a relative error smaller than 50%, that we consider qualitatively as a "decent" accuracy. Obviously, if

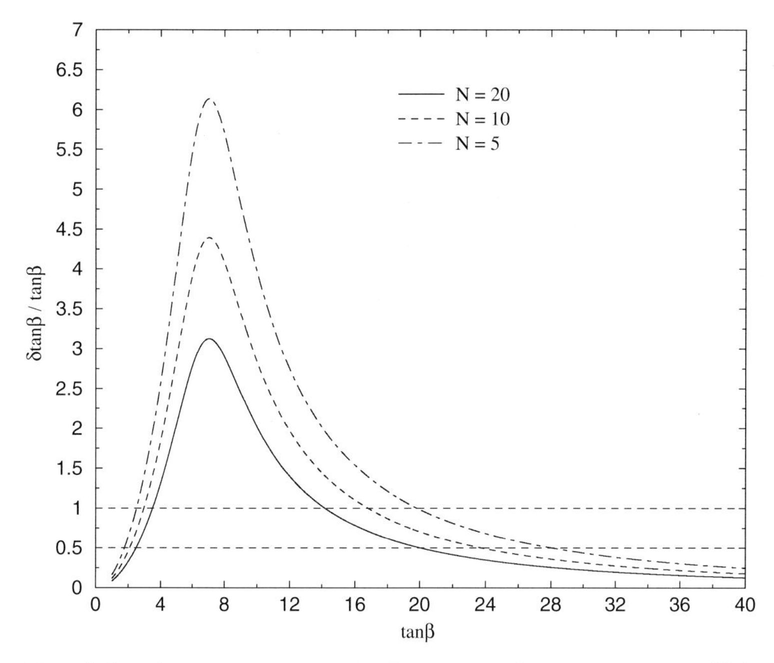

Fig. 2. Plot of the relative error on $\tan\beta$. The statistical accuracy is $\sigma = 1\%$ for all the observables. N is the number of c.m. energy values at which independent measurements are taken. The curves actually depend on the combination $\sigma/\sqrt{N}$. One of the dashed lines corresponds to a 100 % relative error. With $N = 10$ measurements, it determines a region $3 < \tan\beta < 17$ where the determination of $\tan\beta$ is completely unsatisfactory due to the flatness of the coefficient of the SUSY Sudakov logarithms with respect to $\tan\beta$. The second line marks the 50% accuracy level and identify the region $\tan\beta < 2$ or $\tan\beta > 20$.

a higher experimental precision (e.g. a few permille in σ_t) were achievable, the same result could be obtained with a smaller number ($N \simeq 3$) of independent energy measurements.

This seems to us an interesting possibility, particularly for what concerns the second large $\tan\beta$ range. In this case, to our knowledge, the realistic possibilities of measuring $\tan\beta$ are rather restricted and not simple, as exhaustively discussed in a recent paper [10].

References

1. V. V. Sudakov, Sov. Phys. JETP 3, 65 (1956); Landau-Lifshits: *Relativistic Quantum Field theory* IV tome, ed. MIR.
2. M. Kuroda, G. Moultaka and D. Schildknecht, Nucl. Phys. B **350**, 25 (1991); G.Degrassi and A Sirlin, Phys. Rev. D **46**, 3104 (1992); A. Denner, S. Dittmaier and R. Schuster, Nucl. Phys. B **452**, 80 (1995); A. Denner, S. Dittmaier and T. Hahn, Phys. Rev. D **56**, 117 (1997); A. Denner and S. Pozzorini, hep-ph/0101213.

3. P. Ciafaloni and D. Comelli, Phys. Lett. B **446**, 278 (1999); M. Beccaria, P. Ciafaloni, D. Comelli, F. Renard, C. Verzegnassi, Phys. Rev. D **61**, 073005 (2000); hep-ph/9906319; Phys. Rev. D **61**, 011301 (2000).
4. M. Ciafaloni, P. Ciafaloni and D. Comelli, Phys. Lett. B **501**, 216 (2001); M. Melles, Phys. Lett. B **495**, 81 (2000); W. Beenakker, A. Werthenbach, Phys. Lett. B **489**, 148 (2000); M. Hori, H. Kawamura and J. Kodaira, HUPD-003, hep-ph/0007329; M. Melles,hep-ph/001196;hep-ph/000456; J.H. Kühn, A.A. Penin and V.A. Smirnov, Eur. Phys. J C **17**, 97 (2000).
5. V.S. Fadin, L.N. Lipatov, A.D. Martin and M. Melles, Phys. Rev. D **61**,094002(2000); M. Melles, PSI PR-00-18, hep-ph/0012157.
6. M. Beccaria, F.M. Renard and C. Verzegnassi, Phys. Rev. D **63**, 095010 (2001).
7. M. Beccaria, F.M. Renard and C. Verzegnassi, Phys. Rev. D **63**, 053013 (2001).
8. "The CLIC study of a multi-TeV e^+e^- linear collider", CERN-PS-99-005-LP (1999).
9. M. Beccaria, M. Melles, F.M. Renard and C. Verzegnassi, in preparation.
10. A. Datta, A. Djouadi and J.-L. Kneur, Phys. Lett. B **509**, 299, (2001).

Part III

Experimental Particle Physics

Diffractive Physics in the Near Future

Gilvan A. Alves

Centro Brasileiro de Pesquisas Físicas, Rua Xavier Sigaud 150, 22290-180 Rio de Janeiro - RJ, Brazil

Abstract. This paper summarize the present status and near term perspectives for Diffractive Physics at Tevatron. We describe the new detectors that are being installed around the CDF and DØ interaction regions, and discuss the physics topics accessible in view of these upgrades.

1 Introduction

Diffraction is certainly one of the oldest subject in physics. In particle physics we have seen a big development in the second half of the last century [1]. More recently, with the discovery of Hard Diffraction by the UA8-Collaboration [2], we have had a significant development of Diffractive Physics both in theoretical and experimental points of view.

On the experimental side we have now many results coming from HERA (H1 and ZEUS)[3] and the Tevatron (CDF and DØ)[4] in a large number of topics including diffractive structure functions, diffractively produced jets, diffractive production of Heavy Flavors, etc.

On the theoretical side[5], a great deal of progress have been made in several issues, like factorization, diffractive structure functions, Large Rapidity Gap Survival Probability, etc.

Although the experimental results from HERA are showing a consistent improvement in the quality of data, the Tevatron results are still plagued by poor statistics. Not to mention that some processes, like Double Pomeron Exchange, have not been directly observed, and for most of the data sample, the kinematical information (t and ξ) is still missing.

Despite of the theoretical progress for the unification of the soft and hard aspects of the strong interaction, we still need more experimental data to guide some of the theoretical development. It is important to stress that currently about 40% of the $p\bar{p}$ total cross section is due to the Pomeron exchange, and at higher energies this value can be even larger. In this sense, it is extremely important to have precise data from the physical region of the proton anti-proton interaction at the Tevatron, to answer questions like: Is the Pomeron picture universal?, what are its hadronic characteristics?, Is it a glueball?, a dual object?, what is the diffractive contribution for the heavy flavor cross sections?, can the Higgs and Centauros be produced diffractively at Tevatron energies? and so on.

All of these questions lead to the proposal of new devices inserted at Tevatron. CDF reinstalled and improved its Roman Pots Spectrometer, and added a new

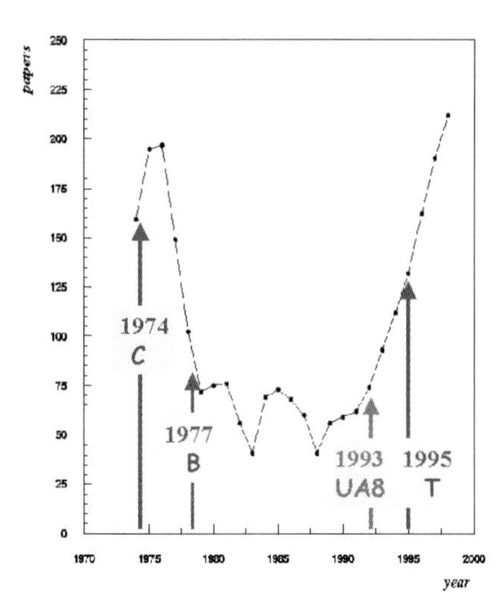

Fig. 1. Number of paper per year in diffractive physics. We show four important dates which have some correlation with diffractive physics papers. 1974 was the year of charm discovery, three years later, in 1977 the bottom, 1993 was the year of the hard diffraction discovery and 1995 the top discovery.

Miniplug Calorimeter plus Beam Shower Counters. DØ added a new Forward Proton Detector (FPD).

Figure 1 shows the growing interest in diffractive physics. There is a clear growth in the number of papers on this subject in particle physics[6].

We are expecting a new period of diffractive results from several experiments in the near future. From the Tevatron, the upcoming Run II with CDF and DØ; From *Brookhaven National Laboratory (BNL)/Relativistic Heavy Ion Collider (RHIC)* with the pp2pp experiment, polarized and unpolarized proton - proton interactions at energies of $\sqrt{s} = 60$ - 500 GeV; From the *Deutsches Elecktronen Synchroton (DESY)*, with H1 and ZEUS, the two detectors at HERA, in electron (positron) (27.5 GeV) proton (820-920 GeV) interactions. For the far future we will have a new era of experiments, with TOTEM [7] being integrated into *Compact Muon Solenoid (CMS)* as one of the detectors of the *Large Hadron Collider (LHC)/CERN*, for proton - proton at 14 TeV center of mass energy.

2 The Tevatron Upgrades for Diffractive Physics CDF and DØ

After the end of Run I, the Tevatron accelerator started its upgrade in order to achieve higher energy and luminosity. The main goals were to achieve energies up to $\sqrt{s} = 2.0$ TeV, and the luminosities of $\approx 3 \times 10^{32} cm^{-2} s^{-1}$. Besides, in order to implement the diffractive program at DØ, some modifications had to be done to the accelerator itself. The two collider detectors CDF and DØ were also submitted to several upgrades, but we will restrict our discussion here only to those that have direct relation with diffractive physics.

The Tevatron beam line had to be modified to accommodate the new DØ leading proton spectrometers. This included modifying the electrostatic separator girder, extending the cryogenic bypass, removal of a quadrupole magnet (Q_1), and other small modifications, like drilling a hole on the floor to allow the insertion and removal of detectors.

2.1 CDF

We will summarize the diffractive physics upgrades for CDF. For more details, please refer to the paper of Goulianos[8]. The CDF upgrades for diffractive physics can be divided in 3 parts:

1. Improve dipole spectrometer (Roman Pots) as is shown in Fig. 2
2. Insertion of a Miniplug Calorimeter for triggering on diffractive events. This device, shown in Fig. 3, is composed by 50 1/4" thick lead plates corresponding to a total of 2 interaction lengths and ≈ 60 radiation lengths. 288 signal towers are viewed by 18 Multi Channel PMTs (16 channels each), covering a pseudorapidity region between 3.5 and 5.5.
3. Install new Beam Shower Counters at large rapidity region (5.5 - 7.5). These counters consist of regular plastic scintillators, read out by PMTs appropriate for high magnetic fields.

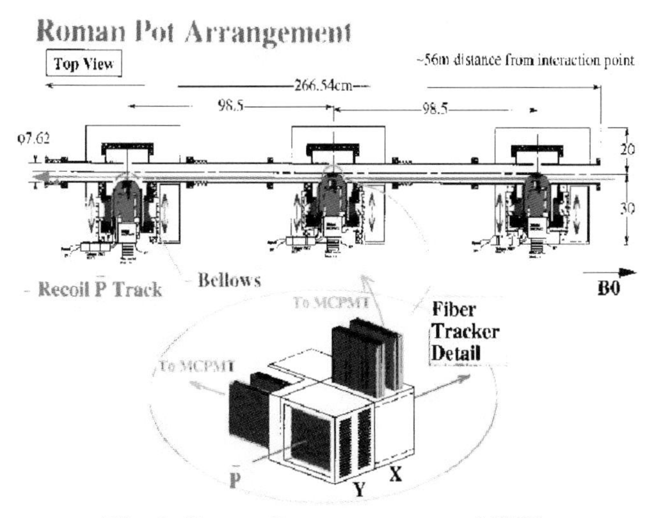

Fig. 2. Roman Pots arrangement of CDF.

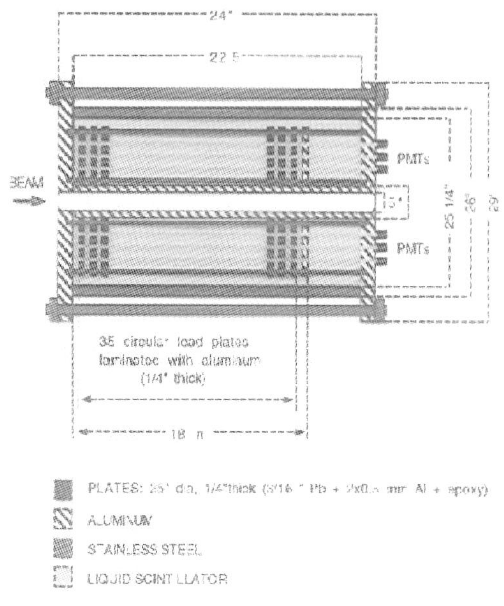

Fig. 3. Side view of the CDF Miniplug Calorimeter.

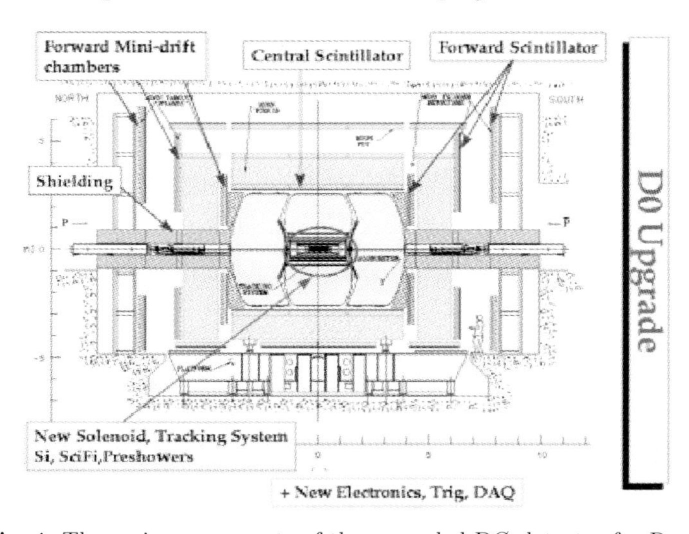

Fig. 4. The main components of the upgraded DØ detector for RunII.

2.2 DØ

The whole DØ detector has been submitted to many modifications for the present run (RunII). Figure 4 summarize the main features.

Now we will describe the Forward Proton Detector, the main component of the DØ Diffractive Upgrade.

3 Forward Proton Detector

The idea of the Forward Proton Detector (FPD) is to cover experimentally a large number of topics which will be very important for the progress of diffractive physics. More details on the FPD can be read in [9].

The Forward Proton Detector consists of 18 Roman Pots arranged on both sides of the DØ detector as shown in Fig. 5, which shows the Roman Pot locations in the Tevatron beam line. There are two castles[1] on the proton side indicated by $P1$ and $P2$ as shown in Fig. 5. The orientation is indicated by the additional letter U for up position, D for down position, I for inside position and O for outside position of the pots ($P1U$, $P1D$, $P1I$, $P1O$, same notation for $P2$). On the antiproton side we have two similar castles, labeled $A1$ and $A2$ followed by the indication of the orientation similar for the proton side. In addition, two others half castles on the antiproton side hold the dipole spectrometer[2] labeled $D1$ and $D2$. The approximated distances of the pots with respect to the interaction point (indicated by 0 on the scale) are also shown.

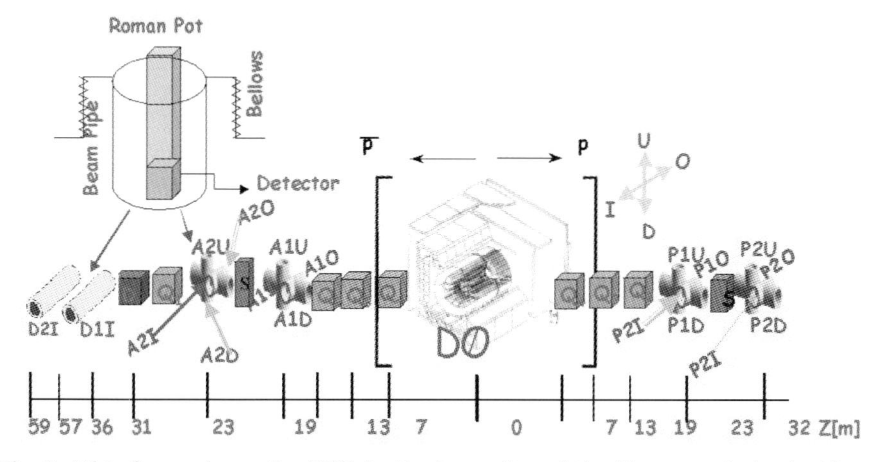

Fig. 5. This figure shows the FPD in the beam line of the Tevatron, in both sides of the DØ detector.

3.1 Roman Pots

The design of the Roman Pots for the DØ FPD is shown in Fig. 6. They were built by LNLS/Brazil (Laboratório Nacional de Luz Synchroton) as part of a regional

[1] Structure that holds the Roman Pots, motors and vacuum equipment.
[2] Named in this way for their position after the dipole magnets.

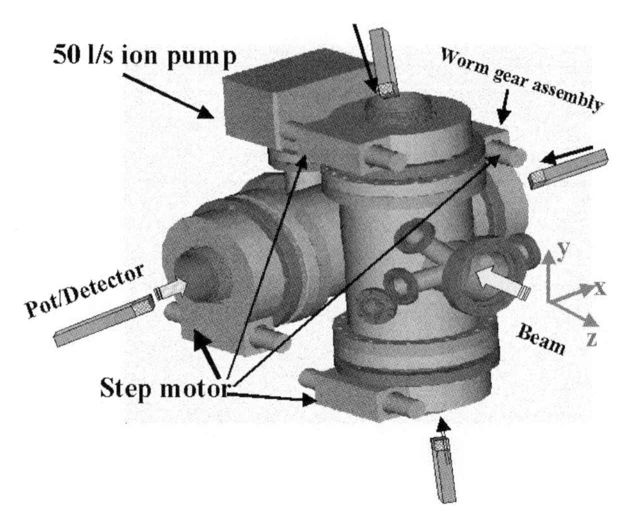

Fig. 6. This figure shows all main parts of the castle. The four "arms" can be seen, each one allowing for a pot to be inserted. The pot movement is done using step motors.

collaboration (LAFEX/CBPF - Centro Brasileiro de Pesquisas Físicas; UFBA - Universidade Federal da Bahia; UFRJ - Universidade Federal do Rio de Janeiro; UERJ - Universidade Estadual do Rio de Janeiro; IFT/UNESP - Universidade Estadual Paulista; and LNLS) During two years we studied many options for the castle and detectors. The castle was made using 316L steel following the technical specifications to achieve a Ultra-High vacuum at the Tevatron. The combination of four view quadrupole stations, as shown in Fig. 6, and the dipole stations, give the possibility to cover a large portion of the available phase space, allowing for a better acceptance. In Fig. 6 we show the castle, indicating its main components. In order to have the best performance, the design of the pot window was submitted to a finite element analysis. The best results were obtained using a 150 microns foil, with elliptical cutout.

3.2 The Detectors

The FPD position detectors are based on square scintillating fibers, 800 micron thick, as shown in Fig. 7. In this figure we can see the frame of the scintillating fibers and 6 planes X X', U U' and V V' which compose one detector. The scintillating fibers are spliced (fused) to clear fibers which guide the signal up to the multi-anode photomultipliers (MAPMT H6568 from Hamamatsu). There are 16 channels per X X' plane and 20 channels for the U U' and V V' planes, giving a total of 112 channels per detector and 2016 channels in total. Studies about the signal, efficiency and resolution have been made. Scintillating fibers are the best option for our detectors among many other possible technologies. The frame is made of polyurethane plastic. The theoretical resolution is 80 microns.

The geometrical acceptance and the pot position acceptance are given in Fig. 8.

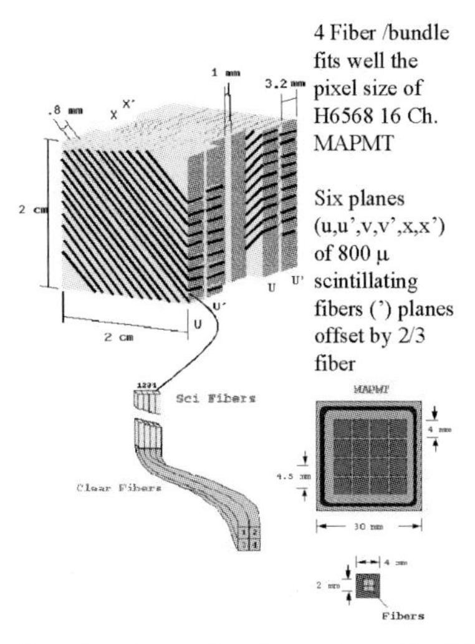

Fig. 7. This figure shows the six planes (u u', v v', x x') of the detector for the FPD spectrometer. Four scintillating fibers are connected together in a single channel of the multi-anode photomultiplier (MAPMT).

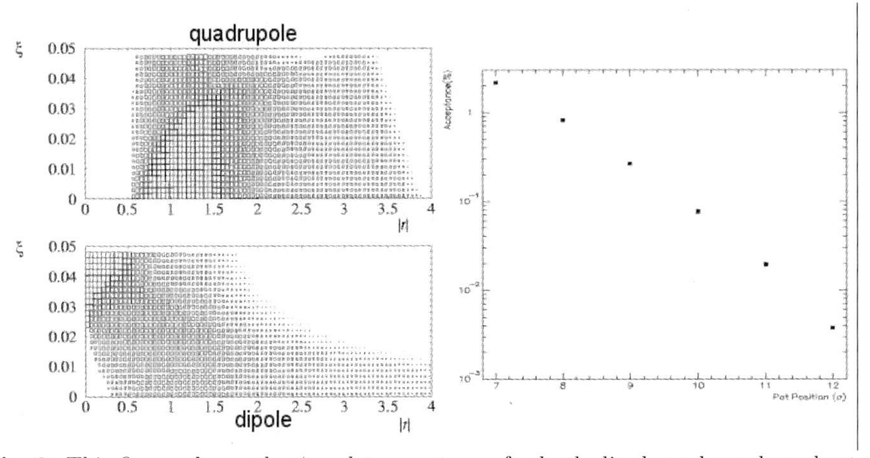

Fig. 8. This figure shows the ξ and t acceptance for both dipole and quadrupole stations. The figure shows also the acceptance versus the pot position. We can see the acceptance variation by moving the pot as close as possible to the beam.

4 Some Physics Topics for the Run II at Tevatron

The new Tevatron detectors were designed to cover a large region of phase space, allowing us to revisit some old and interesting results, and create conditions to observe new physics topics in diffraction. The combination of the Dipole and Quadrupole stations in the DØ/FPD will allow us to get a good sample of data. Early estimates on the size of this sample can be seen on Table 1, where we also make a comparison with the available data to date.

Table 1. This table shows a comparison between the present data on diffractive dijets and our expectation using the FPD at the Tevatron Run II.

Experiment	Dijet Events	E_T [GeV]
UA8	100	8
HERA	Hundreds	5
CDF	Thousands	10
DØ /FPD	500,000	15
	150,000	20
	15,000	30

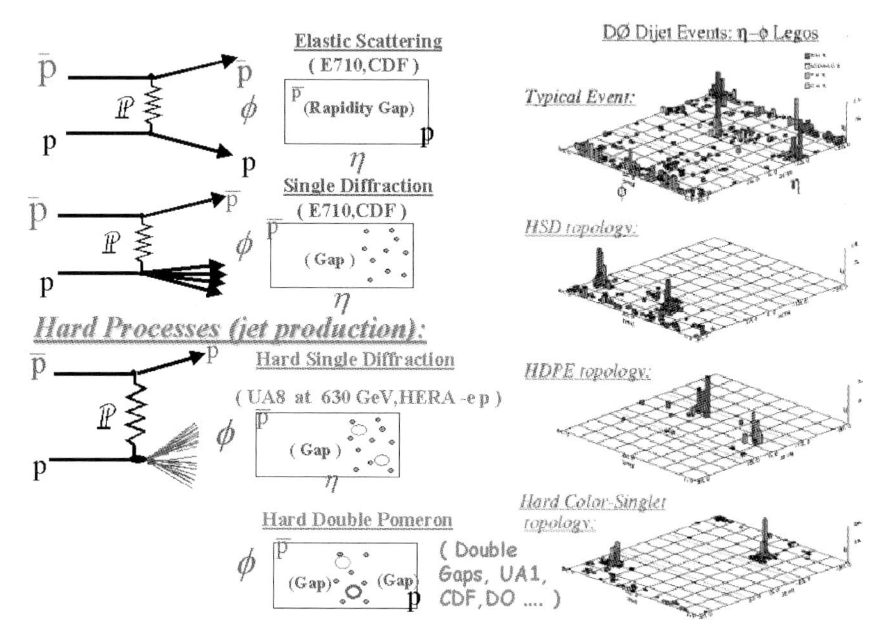

Fig. 9. This figure shows the possible topologies to be studied with FPD in the DØ detector. For each topology we have a corresponding lego plot in pseudo-rapidity versus azimuthal angle. We also show the lego plots corresponding to 3 topologies: the Hard Single Diffraction (HSD), the Hard Double Pomeron Exchange (HDPE) and the Hard Color-Singlet extracted from DØ dijet events.

In Fig. 9 we show several topologies available for study using this detector, some of them, like the Hard Double Pomeron Exchange, have only been studied indirectly using Rapidity Gap techniques.

We will now give a short description of some of the physics topics that can be studied with the new Tevatron Detectors:

4.1 Low and High $|t|$ Elastic Scattering

The acceptance of the FPD spectrometers for low and high t will allow us to extract elastic scattering events in both of these physical regions. The measurements of elastic cross sections can be used to extract the total cross section via Optical Theorem. It is important to know the elastic slope of the differential cross section. The value of the slope for general differential cross sections characterizes a specific process which can be associated to a particular production (e.g. resonance production).

4.2 Total Cross Section

The results from Tevatron [10] experiments are not compatible between them, if we extrapolate to higher energies. It would be very important to have another measurement at these energies. It is important to know the behavior of the cross sections with energy since that is related to the Froissart-Martin bound. After the Tevatron only the LHC will offer a new opportunity to make these measurements.

4.3 Inclusive Single Diffraction

The inclusive single diffraction has many subtopics associated with it. Particularly for the Tevatron detectors, the Diffractive Mass available, for single diffraction events, $M_x = 450\,GeV$, makes the extraction of heavy flavor physics very comfortable. Inclusive single diffraction has been a good laboratory for several problems in diffractive physics. In particular one would like to study the cross section dependence on the diffractive mass (M_x) and four momentum transfered (t).

4.4 Diffractive Jet Production

Jets have been largely studied by QCD. The discovery of diffractively produced jets [2] by UA8 collaboration was very important for diffractive physics. This was the main starting point for hard diffraction. It is expected that single and double diffractive jet production be exhaustively studied in the near future with the Tevatron and HERA detectors, making it possible to verify the universality of Diffractive Parton Distributions. As we see in table 1 it will be possible to get dijet events with higher transverse energies at the Tevatron Run II.

4.5 Hard Double Pomeron Exchange

Due to the interesting topology and exciting physics topics, double Pomeron exchange has been largely discussed as the process for producing many different types of states. [11]. An advantage of the large Diffractive Mass available at the Tevatron, in this case $M_x \simeq 100\,GeV$, is the possibility to study, by direct observation, the Pomeron $\times$ Pomeron interactions and the associated physics. The instrumentation available at DØ is appropriated to face the challenges of the double Pomeron mechanism, and produce several objects not yet observed. Figure 9 shows the double Pomeron exchange graph and the gaps on its corresponding lego plot. Other topics like Glueballs, Centauros and Higgs can be exploited also in this topology.

4.6 Diffractive Heavy Flavor Production

Heavy Flavor Production, has been extensively studied in high p_t physics. Because experimental results are scarce for diffractive heavy flavor production, there was not enough attention to this physics. For the lack of adequate instrumentation we have cross sections without a clear separation of diffractive and non-diffractive events. In principle we can split the studies of diffractive heavy flavor production based on the products: (i) The Charm, (ii) the Bottom and the (iii) Top Physics. Each one having some particularities to be taken into account in the diffractive production [12].

4.7 Diffractive W/Z Boson Production

Diffractively produced W and Z bosons have been observed at the Tevatron. However the present results are not satisfactory, we need more statistics for these kind of events, to help us understand and set constraints on the quark and gluon contents of the Pomeron. Both CDF [13] and DØ have made progress, and the current results are motivating the collaborations to proceed with this measurements.

4.8 Diffractive Structure Functions

The study of Diffractive Structure Functions at the Tevatron allows a comparison with the existing HERA results. To understand the structure of the Pomeron one must know its structure function. This type of study has to be pursued exhaustively to get better and accurate measurements, building a clear interpretation of the Pomeron. One should be able to answer how important are the gluon and the quark components of the Pomeron. The universality of the Pomeron components is also important for this interpretation. Practically all diffractive topics mentioned here depend on this knowledge.

4.9 Glueballs, Centauros and the Higgs

Since the origin of QCD, Glueballs has been studied by theoreticians and experimentalists. However, we do not have a significant progress in this field. Glueballs are perfectly valid states in QCD, so if we do not find them, there must be some unknown suppression rule for this bound states. That means we need dedicated experiments to search for glueballs without ambiguity with quark anti-quark bound states. The family of glueballs is large. Table 2 shows this family (oddballs are also shown). Oddballs should have the priority to be examined, due to the fact that they can not be mistaken by $q\bar{q}$ (meson) or qqq (baryon) bound states with the same quantum numbers. It is a common belief that Glueballs should be largely produced in the double Pomeron Exchange topology.

Another topic is the production of Centauros, which were never observed in accelerator experiments. These objects were discovered in Cosmic Ray Physics as events with several unusual characteristics, like the production of a large

Table 2. This table shows possible glueball state configurations with the mass and the quantum numbers for each one.

Glueballs and Oddballs							
J^{PC}	$(q\bar{q})$	2g	3g	ODD	MASS (GeV)		
					[15]	[16]	[17]
0^{++}	YES	YES	YES	NO	1.58	1.73 ± 0.13	1.74 ± 0.05
0^{+-}	NO	NO	YES	YES			
0^{-+}	YES	YES	YES	NO			
0^{--}	NO	NO	YES	YES	2.56	2.59 ± 0.17	2.37 ± 0.27
1^{++}	YES	YES	YES	NO			
1^{+-}	YES	NO	YES	NO			
1^{-+}	NO	YES	YES	YES			
1^{--}	YES	NO	YES	NO	3.49	3.85 ± 0.24	
2^{++}	YES	YES	YES	NO	2.59	2.40 ± 0.15	2.47 ± 0.08
2^{+-}	NO	NO	YES	YES			
2^{-+}	YES	YES	YES	NO	3.03	3.1 ± 0.18	3.37 ± 0.31
2^{--}	YES	NO	YES	NO	3.71	3.93 ± 0.23	
3^{++}	YES	YES	YES	NO	3.58	3.69 ± 0.22	4.3 ± 0.34
3^{+-}	YES	NO	YES	NO			
3^{-+}	NO	YES	YES	YES			
3^{--}	YES	NO	YES	NO	4.03	4.13 ± 0.29	

multiplicity of charged particles, accompanied by very few photons. For example, as many as 100 charged particles and no more than 3 π^0. [14] There is enough center of mass energy at the Tevatron to produce Centauros, and since the diffractive mass is high enough, it is possible to produce them diffractively. The good calorimetry of the DØ detector can be very useful in observing this kind of events.

Higgs is one of the most exciting subjects for the Tevatron Run II. It has not been excluded the possibility that it can be produced also diffractively. We have two recent studies given by reference [11], showing the possibility of Higgs production by the double Pomeron mechanism.

5 Conclusions

There is a lot of exciting new results coming from the investigation of diffractive phenomena, in particular the ongoing Run II at Tevatron should give us a better picture of diffraction with the new detectors at CDF and DØ. Practically all possible subjects listed here can be studied with the data obtained from these detectors.

The main goal of the experiments in diffractive physics is hard diffraction. Many theoretical and phenomenological progress have been made with the precise data from HERA, but a corresponding set of data from the Tevatron is still missing to allow comparisons between the two regimes of production. This set of data will give us the opportunity to test assumptions like the universality of the Pomeron picture and the Diffractive Parton Distributions.

Another important topic concerns the measurement of the Total Cross Section at the highest energies of the Tevatron, for which there are conflicting results that can be settled by a new measurement from the FPD at DØ. This new measurement will also help to reduce uncertainties on the luminosity for all DØ physics processes.

The double Pomeron exchange and the physics that can be done with this topology is one of the very important subjects of this run at the Tevatron. It has not been excluded, as we called attention, that Centauros and Higgs [11] can be produced diffractively in this topology.

Finally the diffractive physics results from CDF and DØ will be very important for future projects at the LHC, since it is expected that the diffractive contribution becomes more important at increasing energies.

I would like to thank the organizing committee for the superb job in the organization a very enjoyable meeting. In particular I would like to thank Profs. J. Trampetic and J. Wess for the kind invitation and financial support.

References

1. V. Barone and E. Predazzi, *High Energy Particle Diffraction*, (Springer, 2001).
2. A. Brandt et al., UA8 Collaboration, Nucl. Inst. Meth. A **327**, 412 (1993); Phys. Lett. B **297**, 93 (1993); hep-ex/9709015, submitted to Phys. Lett. B.
3. ZEUS collaboration, M. Derrick et.al., Z. Phys. C **65**, 379 (1995); H1 collaboration, T.Ahmed et.al., Nucl. Phys. B **439**, 471 (1995); ZEUS collaboration, M. Derrick et.al. Z. Phys. C **68**, 569, (1995); H1 collaboration, T.Ahmed et.al. Phys. Lett. B **348**, 681 (1995).
4. G.A. Alves, *Diffractive Results from the Tevatron*, Proceedings of DPF99, Los Angeles (1999) hep-ex/9905009; CDF Collab., T. Affolder et al., Phys. Rev. Lett. **84**, 5043 (2000).
5. For a nice review see G.Iacobucci, Invited talk at 20th International Symposium on Lepton and Photon Interactions at High Energies (Lepton Photon 01), hep-ex/0111079.
6. E. M. Gregores, T.L. Lungov and S. F. Novaes, A.Santoro, *Proceedings of Hadron 2000* (S. Paulo, Brazil 2000).
7. G. Matthiae, *The Future of Diffraction at the LHC*, Proceedings of DIFFRACTION 2000, Nucl. Phys. Proc. Suppl. **99** A, 281 (2001).
8. K. Goulianos, *Diffraction in CDF: Run I Results and Plans for Run II*, Proceedings of DIS 2001, hep-ex/0107069 (2001).
9. A. Brandt et al. *A Forward Proton Detector* Fermilab - Pub-97/377, 1997
10. C. Avila, Proceedings of LISHEP98; Thesis: *Measurement the proton - antiproton total cross section at center of mass Energy of 1800 GeV*, Cornell University 1997 and references therein; C. Avila et al., Phys. Lett. B **234**, 158 (1990); Phys. Rev. Lett. **68**, 2433 (1992); Phys. Rev. **50**, 5550 (1994); Phys. Lett. B **234**, 158 (1990); Nucl. Inst. Meth. A **360**, 80 (1995).
11. See for example for Higgs production the paper of D.Kharzeev and E. Levin, hep-ph/0005311; V.A.Khoze, A.D.Martin and M.G.Ryskini, hep-ex/0002072.

12. E.L.Berger, J.C.Collins, D.E. Soper, G.Sterman, Nucl. Phys. B **286**, 704 (1987); A. Kerman and G.Van Dalen, Phys. Rep. **106**, 297 (1984).
13. F.Abe et al. CDF Collaboration, Phys. Rev. Lett. **78**, 2698 (1997).
14. F. Halzen, *FELIX: The Cosmic Ray Connection*, Felix Home Page, http://felix.web.cern.ch/FELIX/Physics/paperi.html, Brazil-Japan Collaboration - Proceedings of the 21st. Int.Conf.-Adelaide-Australia vol.8,259 (1990); C. M. G. Lattes, Y. Fujimoto and S. Hasegawa, Phys. Rep. **65**, 151 (1980); C. E. Navia et al., Phys. Rev. D **40**, 2898 (1989).
15. A. B. Kaidalov and Yu. A. Simonov, hep-ph/9912434.
16. C. Morningstar, M. Peardor, Nucl. Phys. B **63A-C**, 1022 (1998); Phys. Rev. D **60**, 034509 (1999).
17. M. Teper, hep-th/981287.

Observation of Direct CP Violation in Kaon Decays

Konrad Kleinknecht

Institut für Physik, Johannes Gutenberg-Universität, Mainz

Abstract. A small matter-antimatter asymmetry of the weak force was experimentally established. This CP violation may be related to the small excess of matter from the big bang. The nature of CP violation in the K^0 system has been clarified after 37 years of experimentation: it is due to a small part of the weak interaction ("milliweak interaction"). A non-trivial phase in the weak quark mixing matrix generates "direct CP violation" in the weak Hamiltonian. The experiments demonstrating direct CP violation are discussed.

1 The Big Bang and the Baryon Asymmetry in the Expanding Universe

In the big bang, matter and antimatter were made at extremely high temperatures and in equal quantities because the forces which are responsible for their production are completely symmetric with respect to matter and antimatter. In a similar way, the other elementary constituents of matter were produced in pairs during the initial phases of the big bang. In this hot fireball, creation and annihilation of particles and antiparticles led to an equilibrium of approximately equal numbers of particles, antiparticles and photons. The expanding fireball cooled down, and below a certain temperature the creation of particle-antiparticle pairs stopped while annihilation went on. If all forces were symmetric with respect to matter and antimatter, there would be no matter left over in the cold phase but only the photons shifted to the infared regime. These photons are the cosmic microwave background radiation. Today the density of this radiation is 500000 photons per liter.

The big problem however is the complete absence of primordial antimatter in our universe and the small density of matter as compared to photons: about 6×10^{-5} nucleons per liter. The ratio of nucleons over photons is about 10^{-10}, while it was of order one in the early phases of the universe.

A possible explanation for this phenomenon was given by Sacharov and Kuzmin. They stated that this small surplus of matter is possible only if

- one force violates matter-antimatter symmetry
- baryon number is violated as well, and
- the expansion goes through phases when there is no thermodynamic equilibrium.

2 Symmetries

The idea of Sacharov postulated that one of the known forces or a new force violates the symmetry between matter and antimatter and thus produces a small surplus of matter. All the remaining matter annihilates with the corresponding amount of antimatter, and at the end we are left with the surplus (of 10^{-10}) of matter and the red-shifted photons. Such a symmetry violation goes against principles which were cherished for centuries.

Symmetries and conservation laws have long played an important role in physics. The simplest examples of macroscopic relevance are the conservation of energy and momentum, which are due to the invariance of forces under translation in time and space, respectively. In the domain of quantum phenomena, there are also conservation laws corresponding to discrete transformations. One of these is reflection in space ("parity operation") P.

Similarly, the particle-antiparticle conjugation C transforms each particle into its antiparticle, whereby all additive quantum numbers change their sign.

A third transformation of this kind is time reversal T, which reverses momenta and angular momenta. This corresponds formally to an inversion of the direction of time. According to the CPT theorem of Lüders & Pauli [1] [2] there is a connection between these three transformations such that under rather weak assumptions in a local field theory all processes are invariant under the combined operation C P T.

For a long time it was assumed that all elementary processes are also invariant under the application of each of the three operations C, P, and T separately. However, the work of Lee & Yang [3] questioned the assumption, and the subsequent experiments demonstrated the violation of P and C invariance in weak decays of nuclei and of pions and muons. This violation can be visualized by the longitudinal polarization of neutrinos emerging from a weak vertex: they are left-handed when they are particles and right-handed when antiparticles. Application of P or C to a neutrino leads to an unphysical state (Fig. 1).

The combined operation CP, however, transforms a left-handed neutrino into a right-handed antineutrino, thus connecting two physical states. CP invariance therefore was considered [4] to be replacing the separate P and C invariance of weak interactions.

A unique testing ground for CP invariance in the microworld are neutral K mesons. They have the unique property that the particle K^0 and its antiparticle $\overline{K^0}$ differ in one quantum number, strangeness, but still these two particles can mix through a transition mediated by the weak interaction. One consequence of this postulated CP invariance for the neutral K mesons was predicted by Gell-Mann & Pais[5] : there should be a long-lived partner to the known $V^0(K_1^0)$ particle of short lifetime (10^{-10} sec). According to this proposal these two physical particles are mixtures of two strangeness eigenstates, K^0 (S = + 1) and $\overline{K^0}$(S = - 1) produced in strong interactions. Weak interactions do not conserve strangeness and the physical particles should be eigenstates of CP if the weak

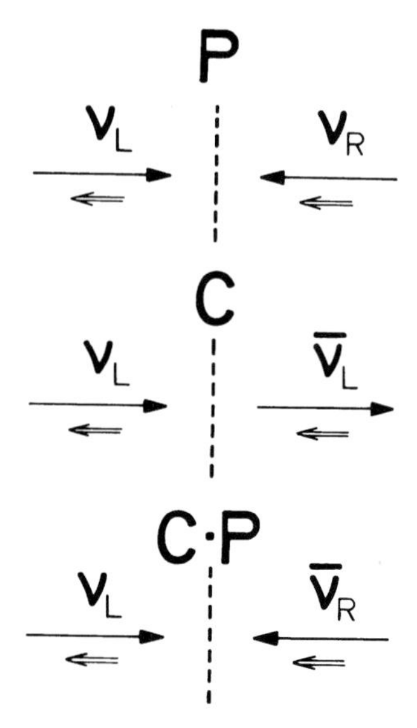

Fig. 1. The mirror image of a left-handed neutrino under P, C, and CP mirror operations

interactions are CP invariant. These eigenstates are (with $\overline{K^0}$ = CP K^0)

$$CPK_1 = CP[(K^0 + \overline{K^0})/\sqrt{2}] = (\overline{K^0} + K^0)/\sqrt{2} = K_1$$
$$CPK_2 = CP[(K^0 - \overline{K^0})/\sqrt{2}] = (\overline{K^0} - K^0)/\sqrt{2} = -K_2.$$

Because of CP $(\pi^+\pi^-) = (\pi^+\pi^-)$ for π mesons in a state with angular momentum zero, the decay into $\pi^+\pi^-$is allowed for the K_1 but forbidden for the K_2; hence the longer lifetime of K_2, which was indeed confirmed when the K_2 was discovered.

In 1964, however, Christenson, Cronin, Fitch & Turlay [6] discovered that the long-lived neutral K meson also decays to $\pi^+\pi^-$ with a branching ratio of $\sim 2 \text{ x } 10^{-3}$. From then on the long-lived state was called K_L because it was no longer identical to the CP eigenstate K_2; similary, the short-lived state was called K_S.

3 Phenomenology and Models of CP Violation

The phenomenon of CP violation in decays of neutral K^0 mesons is now with us for 36 years [6]. The first ten years of intense experimentation after the discovery of the decay $K_L \to \pi^+\pi^-$ were devoted to the observation of other manifestations of the phenomenon, like the decay [7] $K_L \to \pi^0\pi^0$ and the charge asymmetry in the decays [8] $K_L \to \pi^\pm e^\mp \nu$ and $K_L \to \pi^\pm \mu^\mp \nu$, and to precision experiments on the moduli and phases of the CP violating amplitudes [9]. These experimental results excluded a large number of theoretical models proposed to explain CP violation, such that at the time of the London 1974 conference

[10]essentially two classes of models survived: The superweak model postulating a new, very weak, CP violating interaction [11] with $\Delta S = 2$, and milliweak models invoking a small (10^{-3}) part of the normal $\Delta S = 1$ weak interaction as the source of CP violation. In this case, there is a direct decay of a Kaon state with CP quantum number -1 into a two-pion state with CP +1 through a milliweak Hamiltonian. This is called "direct CP violation", as opposed to CP violation by $K^0/\bar{K^0}$ mixing. The key question then became: which of these models is describing the phenomenon? Can one devise experiments distinguishing between those models?

In this context it was very important that a specific milliweak model within the standard model was proposed by Kobayashi and Maskawa [12] in 1973. At the time of the dicovery of CP violation, only 3 quarks were known, and there was no possibility of explaining CP violation as a genuine phenomenon of weak interactions. This situation remained unchanged with the fourth quark because the 2x2 weak quark mixing matrix has only one free parameter, the Cabibbo angle, and no non-trivial complex phase. However, as remarked by Kobayashi and Maskawa, the picture changes if six quarks are present. Then the 3x3 mixing matrix naturally contains a phase, apart from three mixing angles. It is then possible to construct CP violating weak amplitudes from "box-diagrams" of the form shown in Fig. 2.

A necessary consequence of this model of CP violation is the non-equality of the relative decay rates for $K_L \to \pi^+\pi^-$ and $K_L \to \pi^0\pi^0$. This "direct CP violation" is due to "Penguin diagrams" of the form given in Fig. 3.

For a quantitative discussion, we use the conventional notations. Let $\bar{K}^0 =$ CP K^0, then the eigenstates of CP are:

$$K_1 = (K^0 + \bar{K}^0)/\sqrt{2} = +CPK_1$$
$$K_2 = (K^0 - \bar{K}^0)/\sqrt{2} = -CPK_2$$

The physical long-lived (K_L) and short-lived (K_S) states are then

$$K_S = (K_1 + \epsilon_S K_2)/(1 + |\epsilon_S|^2)^{1/2}$$
$$K_L = (K_2 + \epsilon_L K_1)/(1 + |\epsilon_L|^2)^{1/2}$$

With CPT invariance, $\epsilon_S = \epsilon_L = \epsilon$. The admixture parameter

$$\epsilon = \frac{\mathrm{Im}\Gamma_{12}/2 + i\mathrm{Im}M_{12}}{i(\Gamma_S - \Gamma_L)/2 - (m_S - m_L)}$$

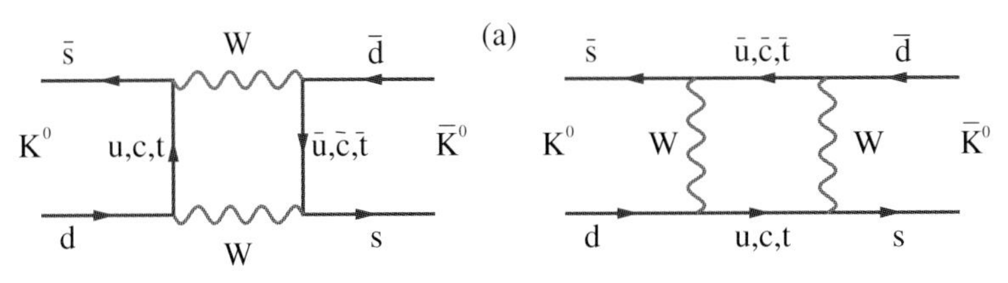

Fig. 2. Box diagram for $K^0 - \bar{K}^0$ mixing connected to CP violating parameter ϵ.

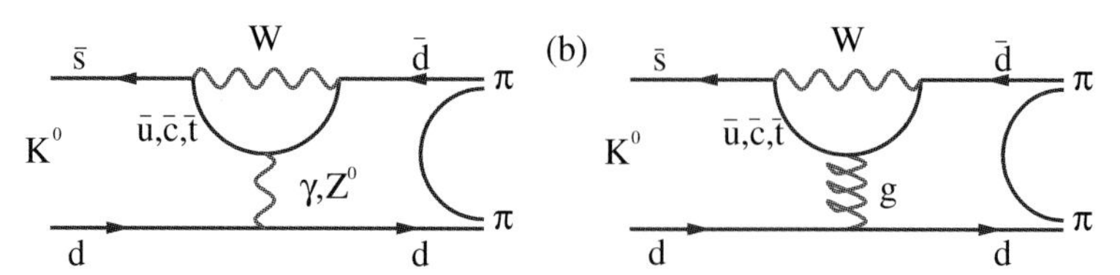

Fig. 3. Penguin diagrams for $K^0 \to 2\pi$ decay with direct CP violation (amplitude ϵ').

is given in terms of the $K^0 - \bar{K}^0$ mass matrix M and decay matrix Γ. The phase of ϵ, assuming CPT, is

$$Arg(\epsilon) = \arctan(2\Delta m/\Gamma_S) = 43.7^\circ \pm 0.2^\circ .$$

The experimentally observable quantities are

$$\begin{aligned} |\eta_{+-}|e^{i\phi_{+-}} &= \eta_{+-} &&= \langle\pi^+\pi^-|T|K_L\rangle/\langle\pi^+\pi^-|T|K_S\rangle , \\ |\eta_{00}|e^{i\phi_{00}} &= \eta_{00} &&= \langle\pi^0\pi^0|T|K_L\rangle/\langle\pi^0\pi^0|T|K_S\rangle . \end{aligned}$$

It can be shown that these amplitude ratios consist of the contribution from CP violation in the $K^0 - \bar{K}^0$ mixing (box diagrams in the KM model), called ϵ above, and another one from CP violation in the weak $K \to 2\pi$ amplitudes (penguin diagrams in the KM model), called ϵ':

$$\begin{aligned} \eta_{+-} &= \epsilon + \epsilon' \\ \eta_{00} &= \epsilon - 2\epsilon' \end{aligned}$$

In this way η_{+-}, η_{00} and $3\epsilon'$ form a triangle in the complex plane, the Wu-Yang triangle. The CP violating decay amplitude ϵ' is due to interference of $\Delta I = 1/2(A_0)$ and $\Delta I = 3/2(A_2)$ amplitudes:

$$\epsilon' = \frac{i\mathrm{Im}A_2}{2A_0}\exp[i(\delta_2 - \delta_0)]$$

and its phase is given by the $\pi\pi$ phase shifts in the $I = 0$ and $I = 2$ states, δ_0 and δ_2 (CPT assumed):

$$Arg(\epsilon') = (\delta_2 - \delta_0) + \pi/2$$

which experimentally is $(61 \pm 9)^0$ [13] or $(45 \pm 9)^0$ [14]. The two models discussed above differ significantly: the superweak model predicts vanishing direct CP violation in weak decays, $\epsilon' = 0$, and therefore $\eta_{+-} = \eta_{00} = \epsilon$, while in milliweak models one expects $\epsilon' \neq 0$.

4 Theoretical Estimates for the Parameter ϵ' of Direct CP Violation

The prediction for ϵ' within the weak quark-mixing model of Kobayashi and Maskawa [12] (KM model) can be estimated if one infers the magnitude of the mixing angles from other experiments and if the hadronic matrix elements for box graphs and penguin graphs are calculated. Typical values for $|\epsilon'/\epsilon|$ are in the range $+(0.2-3)\times 10^{-3}$ for three generations of quarks. A measurement of this quantity to this level of precision therefore becomes the "experimentum crucis" for our understanding of CP violation. Since the phase of ϵ' is close to the one of ϵ, and since $|\epsilon'/\epsilon| \ll 1$ in good approximation, we get:

$$\epsilon'/\epsilon = Re(\epsilon'/\epsilon) = \left(1 - |\eta_{00}/\eta_{+-}|^2\right)/6$$

A measurement of the double ratio

$$R = \frac{|\eta_{00}|^2}{|\eta_{+-}|^2} = \frac{\Gamma(K_L \to 2\pi^0)/\Gamma(K_L \to \pi^+\pi^-)}{\Gamma(K_S \to 2\pi^0)/\Gamma(K_S \to \pi^+\pi^-)}$$

to a precision of better than 0,3% is therefore required to distinguish between the two remaining models.

Various methods are used to calculate the value of $\mathrm{Re}\left(\frac{\epsilon'}{\epsilon}\right)$. Due to the difficulties in calculating hadronic matrix elements, which involve long distance effects, the task turns out to be very difficult. The following results have been obtained recently:

1. The Dortmund group uses the $1/N_C$ expansion and Chiral Pertubation Theory (χPT). They quote a range of $1.5\times 10^{-4} < \epsilon'/\epsilon < 31.6\times 10^{-4}$ [15] from scanning the complete range of input parameters.
2. The Munich group uses a phenomenological approach in which as many parameters as possible are taken from experiment. Their result [16] is $1.5\times 10^{-4} < \epsilon'/\epsilon < 28.8\times 10^{-4}$ from a scanning of the input parameters and $\epsilon'/\epsilon = (7.7^{+6.0}_{-3.5})\times 10^{-4}$ using a Monte Carlo method to determine the error.
3. The Rome group uses lattice calculation results for the input parameters. Their result is $\epsilon'/\epsilon = (4.7^{+6.7}_{-5.9}\times 10^{-4}$ [17].
4. The Trieste group uses a chiral quark model to calculate ϵ'/ϵ. Their result is $7\times 10^{-4} < \epsilon'/\epsilon < 31\times 10^{-4}$ [18] from scanning.

It is hoped that reliable hadronic matrix elements will be obtained in the near future by lattice gauge theory calculations.

5 Experiments on Direct CP Violation

5.1 Early Experiments: The Observation of Direct CP Violation

The first observation of direct CP violation was made by a collaboration of physicists at the european laboratory for particle physics CERN at Geneva, Switzerland in 1988. Their experiment, called NA31, was based on the concurrent detection of $2\pi^0$ and $\pi^+\pi^-$ decays. Collinear beams of K_S and K_L were employed alternately. Kaons with energies around 100 GeV were produced by a 450 GeV proton beam from the proton accelerator SPS at CERN. The energies of the decay products were measured in a combination of a high-resolution Liquid-Argon electromagnetic calorimeter and an iron-scintillator hadronic calorimeter. In the first exposures, about 100000 decays of the type $K_L \to \pi^0\pi^0$ and 295000 decays of $K_L \to \pi^*\pi^-$ were observed, and the result for the CP parameter was $\mathrm{Re}(\epsilon'/\epsilon) = (33 \pm 11) \times 10^{-4}$ [19]. In further improved experimentation, the number of observed $K_L \to \pi^0\pi^0$ decays was increased to 4 x 10^5.

A similar sensitivity was achieved by an experiment at Fermilab near Chicago, called E 731. In 1992/93 the experiments NA31 at CERN and E731 at FNAL presented final results. The CERN result [20] $(23.0\pm6.5)\times10^{-4}$ shows with more than 3 Standard deviations a clear evidence for direct CP violation whereas the Fermilab result [21] with $(7.4 \pm 5.9) \times 10^{-4}$ is consistent with zero. While the CERN experiment had observed direct CP violation, the Fermilab experiment did not concur. As a consequence of this disagreement, two new experiments were constructed in order to verify the earlier result. They were called NA48 at CERN and kTeV at Fermilab.

5.2 The NA48 Detector

The new CERN experiment NA48 (Fig. 4) was designed

- to measure all four decay modes concurrently,
- to register data at a rate 10 times higher than NA31,
- to achieve an improved energy resolution for photons (liquid Krypton calorimeter) and for pions (magnetic spectrometer).

In the design of the NA48 detector the cancellation of systematic uncertainties in the double ratio is exploited as much as possible. Important properties of the experiment are 1. two almost collinear beams which lead to an almost

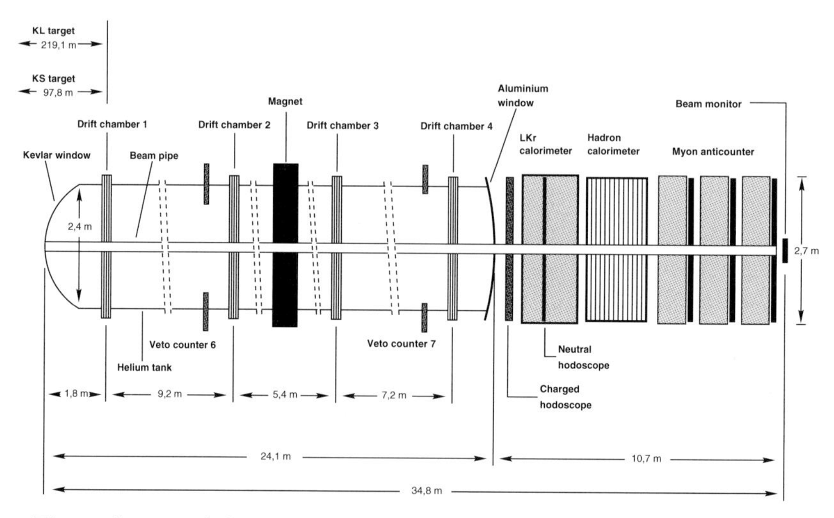

Fig. 4. Layout of the main detector components of the NA48 experiment.

identical illumination of the detector and 2. the lifetime weighting of the events defined as $\mathrm{K_L}$ events.

The $\mathrm{K_L}$ target is located 126 m upstream of the beginning of the decay region. As the decay lengths at the average kaon momentum of 110 GeV/c are $\lambda_\mathrm{S} = 5.9$ m and $\lambda_\mathrm{L} = 3\,400$ m respectively, the neutral beam derived from this target is dominated by $\mathrm{K_L}$. The $\mathrm{K_S}$target is located 6 m upstream of the decay region so that the decays in this beam are dominated by $\mathrm{K_S}$. The two beams are almost collinear: The $\mathrm{K_S}$ target is situated 7.2 cm above the center of the $\mathrm{K_L}$ beam. The relative angle of the beams is 0.6 mrad so that they converge at the position of the electromagnetic calorimeter.

The beginning of the $\mathrm{K_S}$ decay region is defined by an anti-counter (AKS). This detector is used to veto Kaon decays occuring upstream of the counter. The position of the AKS also defines the global Kaon energy scale as the energy is directly correlated to the distance scale. The decay region itself is contained in a 90 m long evacuated tank.

The identification of K_S decays is done by a detector (tagger) consisting of an array of scintillators situated in the proton beam directed to the $\mathrm{K_S}$ target. If a proton signal is detected within a time window of ± 2 ns with respect to a decay, the event is defined as $\mathrm{K_S}$ event. The absence of a proton defines a $\mathrm{K_L}$ event.

A magnetic spectrometer is used to reconstruct $\mathrm{K_{S,L}} \to \pi^+\pi^-$ decays. The spectrometer consists of a dipole magnet with "momentum kick" of 265 MeV/c and four drift chambers which have a spatial resolution of ~ 90 μm. This leads to a mass resolution of 2.5 MeV/c^2. A hodoscope consisting of two planes of plastic scintillator provides the time of a charged event with a resolution of about 200 ps.

A quasi-homogeneous liquid krypton electro-magnetic calorimeter (LKR) is used to identify the four photons from a $\pi^0\pi^0$ event. Liquid krypton has a radiation length of $X_0 = 4.7$ cm which allows one to build a compact calorimeter with high energy resolution ($\Delta E/E = 1.35\,\%$ measured for electrons coming from a $\mathrm{K_L} \to \pi e \nu$ ($\mathrm{K_{e3}}$) decay) and very good time resolution (< 300 ps) and very good linearity. It consists of 13212 2×2 cm^2 cells pointing to the average $\mathrm{K_S}$ decay position. The transverse spatial resolution is better than 1.3 mm.

The electromagnetic calorimeter is complemented by an iron-scintillator sandwich calorimeter with a depth of 6.8 nuclear interaction lenghts which measures the remaining energy of hadrons for use in the trigger for charged events.

A muon veto detector, consisting of three planes of scintillator shielded by 80 cm of iron, is used to identify muons to veto $\mathrm{K_L} \to \pi\mu\nu$ ($\mathrm{K_{\mu 3}}$) events.

5.3 The KTeV Detector

The main elements of the KTeV detector (Fig. 5) are similar to those in NA48 since both experiments work in a similar environment. The main difference is the way in which $\mathrm{K_S}$ mesons are produced.

KTeV uses two parallel well separated kaon beams derived from a single target. In one beam, the K_L mesons from the target pass through a collimator

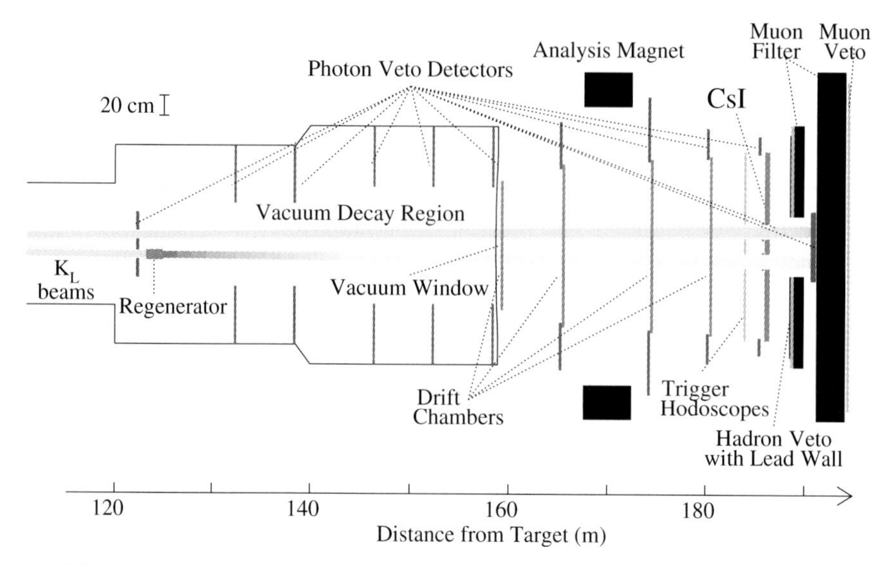

Fig. 5. Layout of the main detector components of the KTeV experiment

and then decay in an evacuated decay region of 30 m length. In the other beam the K_L mesons traverse a slab of matter ("regenerator"). During this passage, the K^0 and $\overline{K^0}$ components of the K_L are affected differently by interactions with matter. Therefore the wave emerging behind the regenerator is a slightly different superposition of these two components. In the forward direction, the energing wave contains a K_S wave coherent with the outgoing K_L wave. This is the regenerated K_S beam used in the experiment. The regenerator alternates between both beams every minute in order to keep the detector illumination identical for the $\mathrm{K_S}$ and the $\mathrm{K_L}$ components.

The decay region of the regenerator beam is defined by a lead-scintillator counter at the downstream end of the regenerator. The decay region of the vacuum beam starts at a mask anti-counter.

Similar to NA48 the KTeV spectrometer consists of four drift chambers; the magnet provides a momentum kick of $p_t = 412$ MeV/c, leading to a mass resolution of $\sigma_{m_{\pi^+\pi^-}} = 1.6$ MeV/c^2. For triggering of charged events a scintillator hodoscope is used.

The electromagnetic calorimeter (CsI) consists of 3100 pure Cesium-Iodide crystals with a radiation length of $X_0 = 1.85$ cm. In the inner region the size is 2.5×2.5 cm^2 and in the outer region 5.0×5.0 cm^2. Two beam holes of 15×15cm^2 allow the two kaon beams to pass to the beam dump. The energy resolution at large photon energies above 10 GeV is 0.75% as measured with $\mathrm{K_{e3}}$ decays.

In addition 10 lead-scintillator "photon veto" counters are used to detect particles escaping the decay volume. The background is further reduced by a muon veto counter consisting of 4 m of steel and a hodoscope.

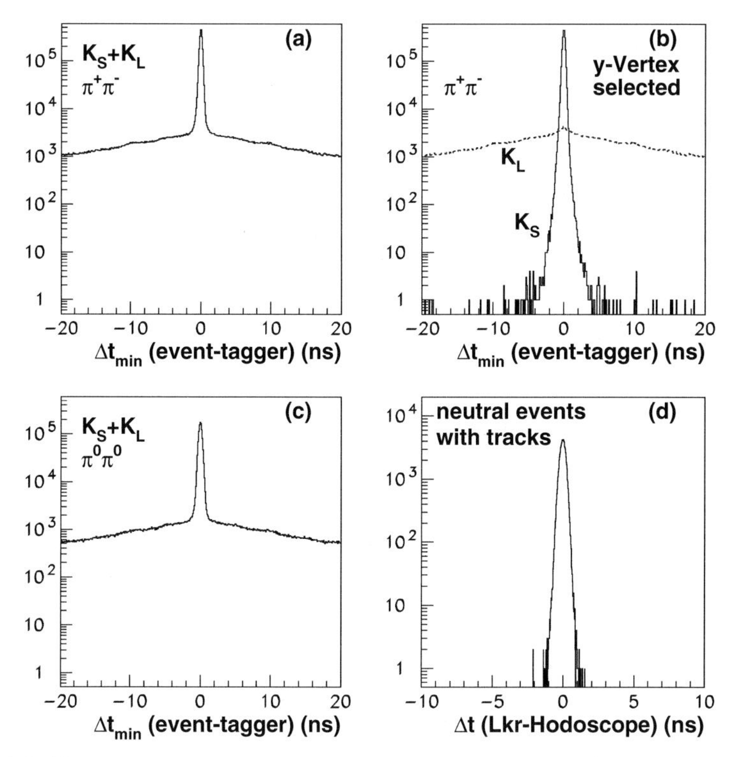

Fig. 6. (a),(c): Minimal difference between tagger time and event time (Δt_{min}). (b) Δt_{min} for charged K_L and K_S events. (d) Comparison between charged and neutral event time. For this measurement decays, selected by the neutral trigger, with tracks are used (γ conversion and Dalitz decays $K_S \to \pi^0\pi^0_D \to \gamma\gamma\gamma e^+e^-$).

5.4 Analysis of NA48 Data: Confirmation of Direct CP Violation

To identify events coming from the K_S target a coincidence window of ± 2 ns between the proton signal in the tagger and the event time is chosen (see Fig. 6 a,c). Due to inefficiencies in the tagger and in the proton reconstruction a fraction α_{SL} of true K_S events are misidentified as K_L events. On the other hand there is a constant background of protons in the tagger which have not led to a good K_S event. If those protons accidentally coincide with a true K_L event, this event is misidentified as a K_S decay. This fraction α_{LS} depends only on the proton rate in the tagger and the width of the coincidence window.

Both effects, α_{SL}^{+-} and α_{LS}^{+-}, can be measured (see Fig. 6 b) in the charged mode as K_S and K_L can be distinguished by the vertical position of the decay vertex. The results are $\alpha_{SL}^{+-} = (1.63 \pm 0.03) \times 10^{-4}$ and $\alpha_{LS}^{+-} = (10.649 \pm 0.008)\%$. This means that about 11% of true K_L events are misidentified as K_S events, however, this quantity is precisely measured to the 10^{-4} level. What is important for the measurement of R is the difference between the charged and the neutral decay modes $\Delta\alpha_{LS} = \alpha_{LS}^{00} - \alpha_{LS}^{+-}$. Proton rates in the sidebands of the tagging window are measured in both modes to measure $\Delta\alpha_{LS}$. The result is $\Delta\alpha_{LS} =$

$(4.3 \pm 1.8) \times 10^{-4}$. Several methods have been used to measure $\Delta\alpha_{SL}$, leading to the conclusion that there is no measurable difference between the mistaggings within an uncertainty of $\pm 0.5 \times 10^{-4}$.

Another important correction is the background subtraction. Decays $\mathrm{K_L} \to \pi e \nu$ and $\mathrm{K_L} \to \pi \mu \nu$ can be misidentified as $\mathrm{K} \to \pi^+\pi^-$ decays, as the ν is undetectable. However, since the ν carries away momentum and energy, these events can be identified by their high transverse momentum p'_t and their reconstructed invariant mass. The remaining background can be measured by extrapolating the shape of the background in the $m - p'^2_\mathrm{t}$-plane into the signal region. In this way the charged background fraction leads to an overall correction on R of $(16.9 \pm 3.0) \times 10^{-4}$.

A similar extrapolation can be done in the neutral decay mode. Here the background comes from $\mathrm{K_L} \to 3\pi^0$ decays, where two γ are not detected. This leads to a misreconstruction of the invariant π^0 masses. In this case the background leads to a correction of R by $(-\ 5.9 \pm 2.0) \times 10^{-4}$.

The number of signal events after these corrections are summarised in table 1.

Table 1. Event numbers after tagging correction (only NA48) and background subtraction.

Event statistics ($\times\ 10^6$)					
	NA48	KTeV		NA48	KTeV
$\mathrm{K_S} \to \pi^+\pi^-$	22.221	4.52	$\mathrm{K_L} \to \pi^+\pi^-$	14.453	2.61
$\mathrm{K_S} \to \pi^0\pi^0$	5.209	1.42	$\mathrm{K_L} \to \pi^0\pi^0$	3.290	0.86

The efficiency of the triggers used to record neutral and charged events have been determined. Independent triggers are used which accept a downscaled fraction of events. In the neutral decay mode the efficiency is measured to be 0.99920 ± 0.00009 without measurable difference between $\mathrm{K_S}$ and $\mathrm{K_L}$ decays. The $\pi^+\pi^-$ trigger efficiency is measured to be $(98.319 \pm 0.038)\%$ for $\mathrm{K_L}$ and $(98.353 \pm 0.022\%)$ for $\mathrm{K_S}$ decays. Here a small difference between the trigger efficiency in $\mathrm{K_S}$ and $\mathrm{K_L}$ decays is found. This leads to a correction to the double ratio of $(-4.5 \pm 4.7) \times 10^{-4}$. The error on this measurement is dominated by the total number of events registered with the independent trigger. This error is one of the main contributions to the systematic error of the measurement of R.

The distance D from the LKR to the decay vertex is reconstructed using the position of the four γ clusters. From the kinematics of the decay one obtains

$$D = \frac{1}{M_\mathrm{K}} \sqrt{\sum_{i,j} E_i\, E_j\, r_{ij}^2},$$

where E_i is the energy of cluster i and r_{ij} the distance between cluster i and cluster j. This formula directly relates the distance scale to the energy scale. It is therefore possible to fix the global energy scale with the measurement of the known AKS position in the neutral decay mode. In addition more checks on the energy scale and the linearity of the energy measurement can be performed, such as the measurement of the invariant π^0 mass and the use of the known position of an added thin CH_2 target (a π^- beam produces $\pi^0 \to 2\gamma$). The comparison of all methods gives an uncertainty of $\pm 3 \times 10^{-4}$ in the global energy scale.

Another systematic problem is the minimization of acceptance corrections by weighting of the K_L events. The difference in the lifetime between K_S and K_L events produce a different illumination of the detector: There are more K_L events decaying closer to the detector and they are therefore also measured at smaller radii closer to the beampipe. NA48 weights the K_L events according to the measured lifetime such that the distribution of the z-position of the decay vertex of K_S and K_L events and the detector acceptances become equal. Using this method the influence of detector inhomogenities is minimised and the analysis becomes nearly independent of acceptance calculations by Monte Carlo methods. In fact the acceptance correction due to small detector differences is quite small. The price to pay for the gain in systematics is the loss in statistics.

Although the acceptance is almost equal there are nevertheless small differences in the beam geometry and detector illumination between decays coming from the K_S and the K_L target. These remaining differences are corrected for with Monte Carlo methods. Using the Monte Carlo to calculate the double ratio R the deviation from the input value 1 is $(26.7 \pm 5.7) \times 10^{-4}$.

Summing up the systematic uncertainties of all different sources in quadrature, they amount to a total 12.4 x 10^{-4} in R, and to 2.1×10^{-4} in ϵ'/ϵ.

The result of NA48 using the data sample from 1997, 1998 and 1999 is [23], [24]

$$\epsilon'/\epsilon = (\ 15.3\ \pm\ 2.6\ ) \times 10^{-4} \tag{1}$$

5.5 Analysis of KTeV Data: Confirmation of Direct CP Violation

KTeV has similar physical backgrounds as NA48, but in addition two-pion decays from K_S mesons produced by incoherent regeneration in the regenerator have to be subtracted. Typical numbers are: a fraction of 6.9×10^{-4} charged background from K_{e3} and $K_{\mu 3}$ decays and a fraction of 27×10^{-4} neutral background from $3\pi^0$ decays. The background levels in the regenerator beam are 107×10^{-4} in the neutral mode (this gives rise to a large systematic error) and 7.2×10^{-4} in the charged mode. The event numbers after background subtraction.

The main difference in the analysis techniques of the two experiments is the treatment of the acceptance correction. KTeV is not using event weighting but uses Monte Carlo studies to correct for detector differences in the K_S and K_L decays. K_{e3} and $3\pi^0$ decays are used to model the detector and the agreement between data and Monte Carlo is good (see Fig. 7). The acceptance correction to R calculated by Monte Carlo simulation is then $(231 \pm 13) \times 10^{-4}$. This can be compared to the size of the total effect of ϵ'/ϵ on R, 168×10^{-4}, as measured by KTeV. The main source of systematic uncertainty is a slight disagreement between data and Monte Carlo comparison in the $\pi^+\pi^-$ decay mode in the vacuum beam. A slope of $(-1.60 \pm 0.63) \times 10^{-4}$ m^{-1} has been found which is applied as a systematic error.

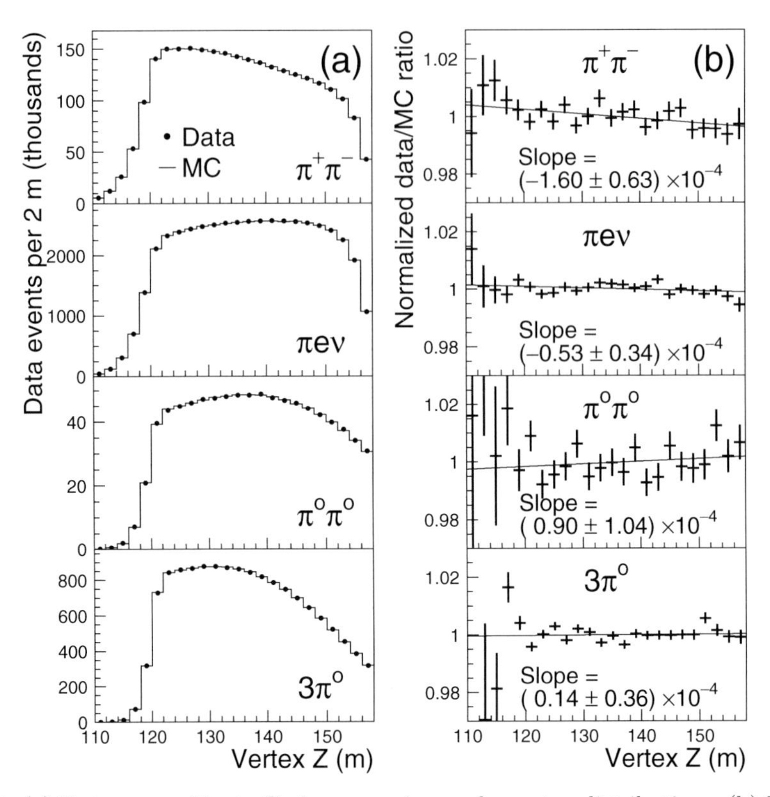

Fig. 7. (a) Data versus Monte Carlo comparisons of z-vertex distributions. (b) Linear fits to the data/MC ratio of (a).

The energy scale is determined using the known position of the regenerator edge. The comparison of the measured position of the vacuum window with the real position gives an uncertainty of the global energy scale of 4.2×10^{-4} on R.

The result is obtained by a fit of $\mathcal{R}e(\epsilon'/\epsilon)$, the regeneration amplitude and phase to the event numbers per energy bin. The result is published in 1999 was [22]

$$\epsilon'/\epsilon = (\ 28.0\ \pm\ 3.0\ (\mathrm{stat})\ \pm\ 2.8\ (\mathrm{sys})\ )\ \times10^{-4} \tag{2}$$

but this was subsequently corrected due to an error in background subtraction:

$$\epsilon'/\epsilon = (\ 23.2\ \pm\ 3.0\ (\mathrm{stat})\ \pm\ 3.2\ (\mathrm{sys})\ \pm\ 0.7\ (\mathrm{MC})\ )\times10^{-4} \tag{3}$$

6 Conclusion

The two experiments kTeV and NA48 have definitively confirmed the original observation of the NA31 team that direct CP violation exists. The results of all published results on ϵ'/ϵ are shown in Fig. 8. Therefore, CP violation as observed in the K meson system is a part of the weak interaction, due to weak

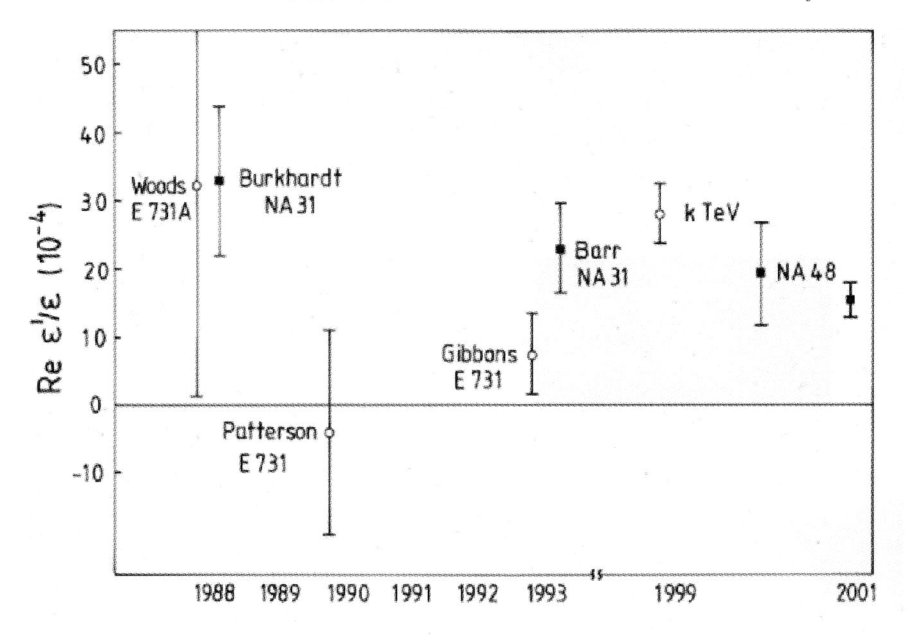

Fig. 8. Measurements of the parameter $\epsilon'/\epsilon$of direct CP violation.

quark mixing. Exotic new interactions like the superweak interaction are not needed. With more data, both experiments will reach a precision of $\mathcal{O}(2 \times 10^{-4})$. We therefore have a very precise experimental result on ϵ'/ϵ. The theoretical calculations of ϵ'/ϵ within the Standard Model however, are still not very precise. This does not change the main conclusion of the experiments, that ϵ' is different from zero and positive, i.e. direct CP violation exists.

However, this milliweak CP violation is probably not large enough to explain the observed matter-antimatter asymmetry from the big bang. An additional, stronger CP violation is needed for this. There are speculations that this might be due to CP violation in the lepton sector in the early universe. Very heavy Majorana neutrinos could play a role in the formation of an antimatter-matter asymmetry at this time.[25]

References

1. G. Lüders, Kgl. Danske Videnskab. Selskab, Matfys. Medd. **28**, 1 (1954).
2. W. Pauli, in *Niels Bohr and the Development of Physics*, ed. W. Pauli, p. 30. (Oxford: Pergamon, 1955).
3. T. D. Lee, C. N. Yang, Phys. Rev. **104**, 254 (1956).
4. L. D. Landau, Nucl. Phys. **3**, 127 (1957).
5. M. Gell-Mann, A. Pais, Phys. Rev. **97**, 1387 (1955).
6. J. H. Christenson, J. W. Cronin, V. L. Fitch, R. Turlay, Phys. Rev. Lett. **13**, 138 (1964).

7. J. M. Gaillard et al., Phys. Rev. Lett. **18**, 20 (1967)
J. W. Cronin et al., Phys. Rev. Lett. **18**, 25 (1967).
8. S. Bennett et al., Phys. Rev. Lett. **19**, 993 (1967)
D. Dorfan et al., Phys. Rev. Lett. **19**, 987 (1967).
9. Reviewed in K. Kleinknecht, Ann. Rev. Nucl. Sci. **26**, 1 (1976).
10. K. Kleinknecht, *Proc. 17th Int. Conf. High Energy Phys.*, ed. J.R. Smith (Chilton, Didcot, UK 1974), p.III-23.
11. L. Wolfenstein, Phys. Rev. Lett. **13**, 562 (1964).
12. M. Kobayashi, T. Maskawa, Prog. Theor. Phys. **49**, 652 (1973);
N. Cabibbo, Phys. Rev. Lett. **10**, 531 (1963).
13. N. N. Biswas et al., Phys. Rev. Lett. **47**, 1378 (1981) and priv. comm. to W. Ochs.
14. W. Ochs, priv. comm. from analysis of CERN-Munich data.
15. T. Hambye, G. O. Köhler, E. A. Paschos, P. H. Soldan, hep-ph/9906434.
16. A. J. Buras, M. Gorbahn, S. Jäger, M. Jamin, M. E. Lautenbacher, L. Silvestrini, hep-ph/9904408, hep-ph/9908395.
17. G. Martinelli, Presentation at Kaon99, Chicago,
http://hep.uchicago.edu/kaon99/talks/martinelli/
18. S. Bertolini, M. Fabbrichesi, J. O. Eeg, hep-ph/9802405.
19. H. Burkhardt et al., Phys. Lett. B **206**, 169 (1988).
20. G.D. Barr et al., Phys. Lett. B **317**, 233 (1993).
21. L.K. Gibbons et al., Phys. Rev. Lett. **70**, 1203 (1993).
22. A. Alavi-Harati et al., Phys. Rev. Lett. **83**, 22 (1999).
23. V. Fanti et al., Phys. Lett. B **465**, 335 (1999).
24. A. Lai et al., A precise measurement of the direct CP violation parameter $Re(\epsilon'/\epsilon)$, Eur. Phys. J. C **22**, 231 (2001).
25. W. Buchmüller, Ann. Phys. (Leipzig) **10**, 95 (2001)

The CMS Experiment and Physics at the LHC

Daniel Denegri

CE Saclay/DAPNIA and CERN/EP, for the CMS Collaboration

Abstract. We discuss the status of the LHC machine, review the progress on the construction of the CMS experiment and the expected performance of CMS in Higgs and SUSY searches, with the expected mass reach for the various MSSM Higgses and sparticles, and some aspects of sparticle mass reconstruction.

1 Introduction

The Large Hadron Collider (LHC) and the experiments now under construction at CERN will surely be the key instruments of particle physics research at the high energy frontier in the coming decades. They will contribute significantly to the elucidation of some of outstanding issues in today's particle physics. The LHC will be a proton-proton collider operating at a centre of mass energy $\sqrt{s} =$ 14 TeX with a luminosity up to $10^{34}\,\mathrm{cm}^{-2}\mathrm{s}^{-1}$. It will also operate as a heavy ions collider with, in case of Pb-Pb collisions, a nucleon-nucleon cm energy of $\sim 5\,\mathrm{TeV}$ and a luminosity of $\sim 10^{27}\,\mathrm{cm}^{-2}\mathrm{s}^{-1}$.

The main motivation to build the LHC is undoubtedly to investigate the mechanism responsible for electroweak symmetry breaking and generating particle masses. The usual way to introduce masses is through the Higgs mechanism [1]. In the Standard Model (SM) one Higgs boson H is expected. In Supersymmetry (SUSY) the particle spectrum is essentially doubled through the introduction of supersymmetric partners to known SM particles, and the Higgs sector is extended to at least five Higgs bosons in the Minimal Supersymmetric Standard Model (MSSM) [2]. SUSY provides a natural explanation why the Higgs mass can remain low, as indicated by present fits to electroweak data. Assuming the existence of superpartners at the TeV scale, the strong, weak and electromagnetic couplings unify at a $\sim 10^{16}$GeV scale; this may be taken as a hint for SUSY at a TeV scale. SUSY also allows for a possible particle-physics origin of cosmological dark matter [3]. Thus the search for the possible supersymmetric partners of SM particles will be a high priority at the LHC. New ideas allowing existence of extra (possibly large) dimensions have emerged recently and which could provide an alternative solution to the energy (mass) scale hierarchy problem [4]. Investigations of possible experimental manifestations of these extra dimensions will be actively pursued at the LHC. The LHC, being an extremely prolific source of B hadrons, will also allow to investigate the matter-antimatter asymmetry in the Universe. Through its heavy ions physics programme the possible quark-gluon phase of matter should be studied in detail.

Two general purpose detectors, CMS [5a] and ATLAS [5b], will be installed at the LHC, Fig. 1. The construction of both these detectors is now proceeding at full speed. The primary goal of these two experiments are the Higgs and SUSY searches. Although the physics goals of these two detectors are very much the same, the experimental techniques are different and complementary, in particular in the B-field configuration and calorimeter choices. These detectors should be completed by mid-2006. More details on the construction of CMS will be given in the following.

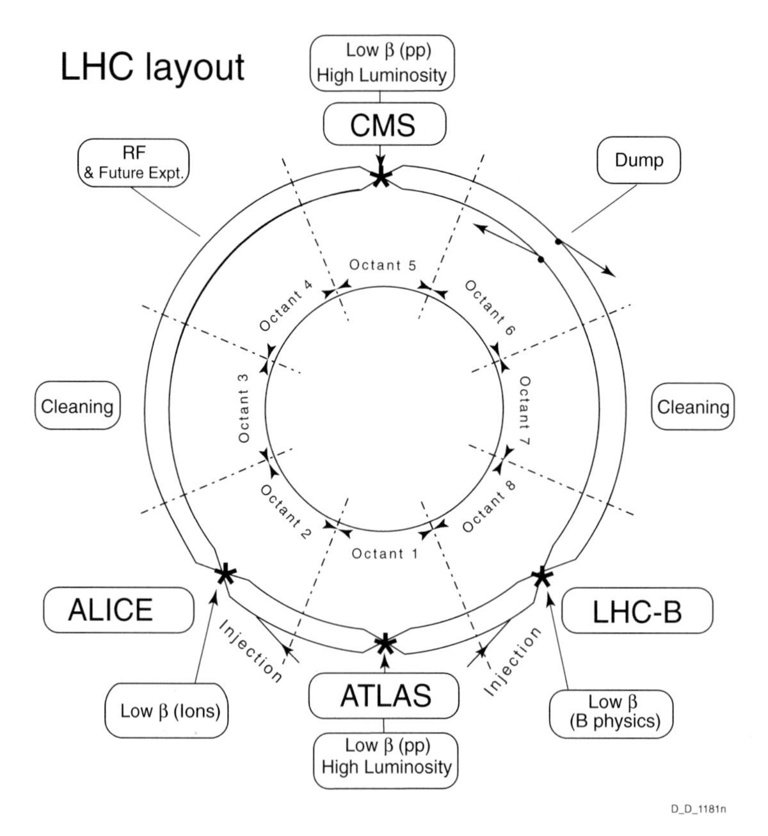

Fig. 1. LHC layout, with the four major experiments.

Two somewhat smaller scale detectors will also be operating at the LHC, the ALICE detector devoted to studies of the expected quark-gluon plasma QCD phase in heavy ion collisions, and the LHC-B detector specialised for B-physics and CP/B violation studies, Fig. 1. These two detectors are now also in the the construction phase. Both ATLAS and CMS have also significant B-physics programs [5,6], and CMS has also an interesting and in part unique heavy ions physics program [7], complementary to the one of ALICE.

2 Status of the LHC

Civil Engineering

The LEP electron- positron collider stopped operating at the end of 2000, it is now being removed and the 27 km long tunnel being prepared to accommodate the LHC machine magnets. The main civil engineering effort at present is on the excavation of the large underground experimental halls which will house the CMS and ATLAS detectors, and of the adjoining service caverns for electronics and computers. The ATLAS caverns are now excavated and will be ready in the course of 2002. At present, i.e. beginning of 2002, the two large shafts giving access to the underground halls of CMS at a depth of 70 m are terminated and excavation of underground caverns is proceeding at full speed. As the CMS detector can be test-assembled in the surface experimental hall, this allows for the underground caverns of CMS to be delivered by mid 2004.

Status of Machine Dipoles

The bending dipoles are the key elements of the LHC. About 65 % of the 27 km circumference of the tunnel will be covered twice with B = 8.3 T magnetic field with 1232 2-in-1 superconducting dipoles of 14.3 m length cooled at 1.9 °K. A number of prototype 10 m long dipoles have been tested in 1999/2000. Five 15 m long 6-block coil (final design) dipoles have been tested and satisfactorily operated above 8.3 T, as well as three second generation final design prototype dipoles in 2001. Figure 2 shows quench curves of some of the 15 m prototypes: the Al-collar Noell (N1) dipole, the steel-collar N2 dipole reaching the nominal

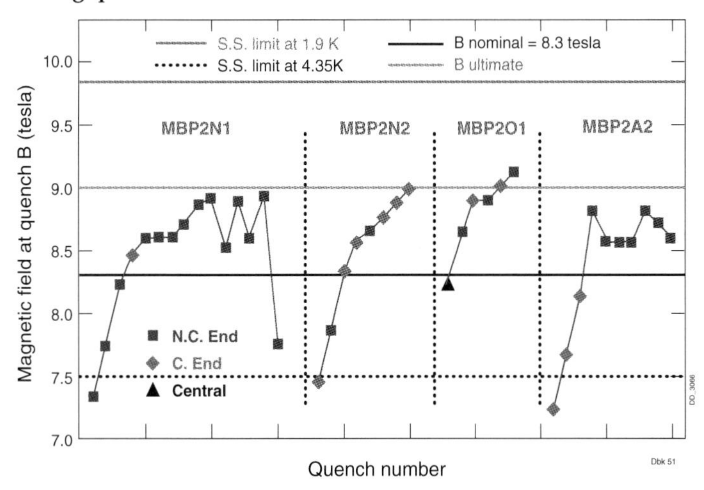

Fig. 2. Quench performance on Main Dipole Prototypes.

8.3 T field in 3 quenches and a 9 T field after additional 5 quenches, the Alsthom/Jeumont (O1) and Ansaldo (A2) dipoles. Thirty final-design industrial pre-production dipoles have been ordered from the three manufacturers, the first ones assembled at CERN and tested in 2001 gave excellent results. Quench performances of some of the pre-series dipoles are shown in Fig. 3. Two prototype final-design quadrupoles have been built and gave excellent results. The limiting factor in dipole production is now superconducting cable procurement. The mass production of LHC dipoles should reach a rate of $\sim$ 20 dipoles/month by mid-2002. The installation of the LHC magnets will go on through the years 2002 till 2006, with machine commissioning end of 2006. A first pilot run is expected at the end of 2006, followed by a real 10 fb^{-1} physics run starting in the first half of 2007.

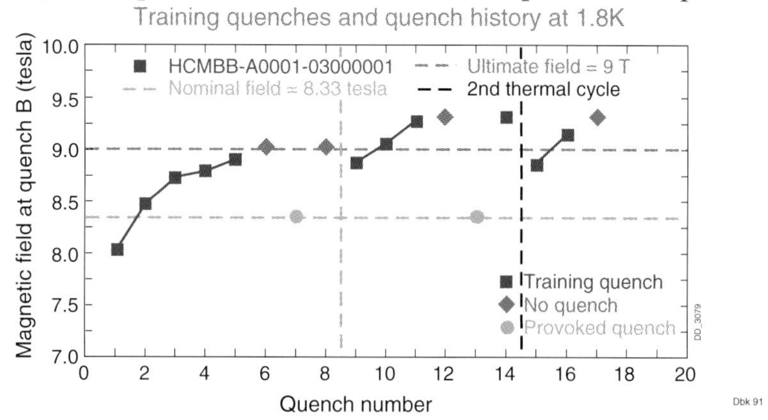

Fig. 3. Quench performance of last tested pre-series dipole (MBPSN01).

3 CMS Detector Construction and Expected Performances

Basic Design Features

A longitudinal cut through the CMS detector is shown in Fig. 4. Basic to the design of CMS is a 13.5 m long, 6.2 m inner diameter solenoid generating a uniform magnetic field of 4 T [8]. The magnetic flux is returned through iron discs in the endcaps and a 1.8 m thick saturated iron barrel yoke instrumented with muon chambers.

A second characteristic feature of the CMS design is that the entire hadronic and electromagnetic calorimeters are located inside the coil to minimise the effects of the "dead volume" of the coil on hadron shower developments i.e. on jet and missing E_t measurements. The innermost part of the detector is occupied by

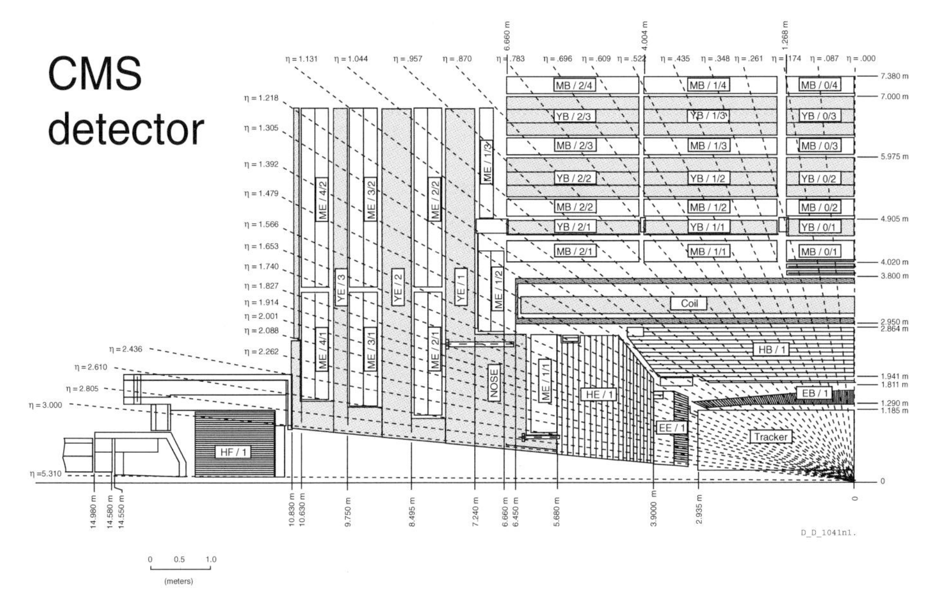

Fig. 4. Longitudinal cut through one quadrant of CMS. The barrel and endcap muon stations are denoted by MB and ME, the tower structure of the hadron calorimeter (HB, HE) and the arrangement of the electromagnetic calorimeter crystals (EB, EE) is indicated.

a 6 m long, 1.2 m radius central tracker for efficient track reconstruction, allowing precise momentum and impact parameter measurements.

The design goal of CMS is to measure muons, electrons and photons with a resolution of $\sim 1\,\%$ over a large momentum range ($\leq 100\,\mathrm{GeV}$), to measure jets with a resolution of $\approx 10\,\%$ at $\mathrm{E_t} = 100\,\mathrm{GeV}$ and to be highly hermetic for missing E_t performance as required for SUSY searches.

Let us now briefly review the main features and status of construction of the major subsystems.

Status of Solenoid Construction, Yoke and Coil

The CMS magnetic return barrel yoke is the main structural element supporting the whole experiment. It is subdivided into 5 "wheels". The mechanical structure of the barrel yoke is shown in Fig. 5. The central wheel also supports the cryostat of the solenoid to which are attached the barrel hadronic and electromagnetic calorimeters. Each of these iron wheels weighs about 1200 tons. The barrel muon stations are inserted in the gaps within the wheels. This whole barrel yoke structure is now completed and assembled in the surface hall of CMS. Work is now progressing on the assembly of the 2×3 iron discs of the end-cap yoke (structure visible in Fig. 4). The endcap muon stations are attached to these discs. The hadronic and electromagnetic calorimeters are supported by the front endcap discs. The full yoke, barrel plus endcap will be completed by summer 2002.

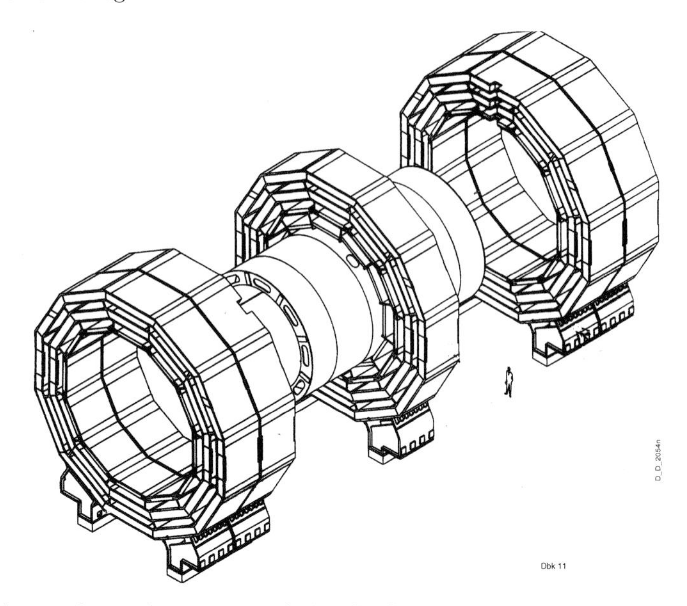

Fig. 5. The mechanical structure of the CMS barrel return yoke, with the central "wheel" carrying the cryostat for the solenoid.

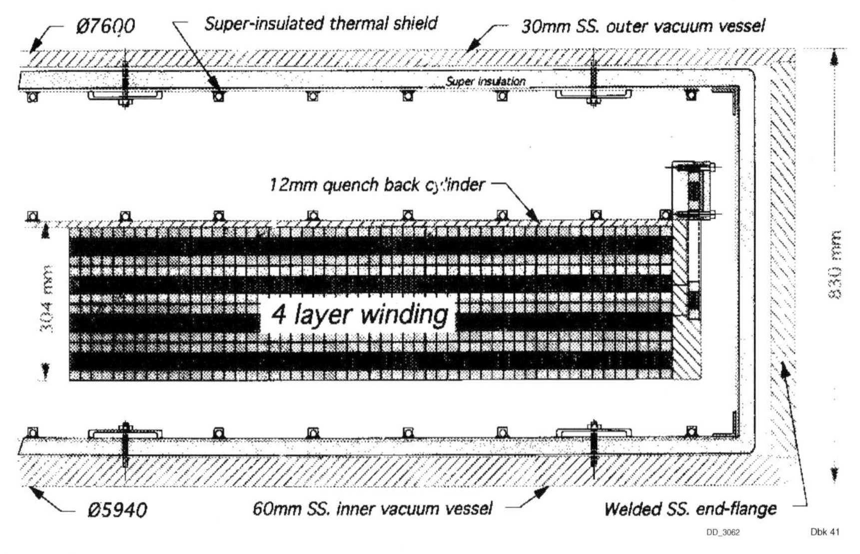

Fig. 6. Cross-section through the end of the solenoid of CMS with the 4-layers solenoid winding.

A cross section through the end of the CMS solenoid and cryostat is shown in Fig. 6 [8]. To achieve the needed number of ampere-turns for the desired 4 Tesla field, a four layers winding is required. Longitudinally the solenoid is subdivided into five sections. The cross section through the solenoid conductor is shown

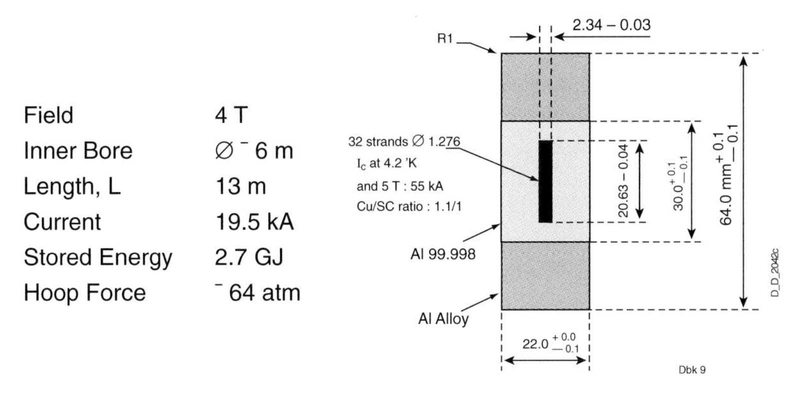

Fig. 7. Cut through the CMS solenoid conductor, with Rutherford superconducting cable and the pure Al electrical and thermal stabiliser and the Al-alloy mechanical stabiliser layers.

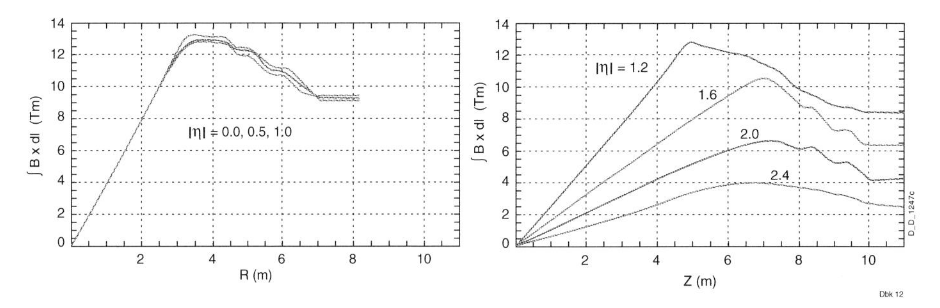

Fig. 8. The bending power of the CMS solenoid at various rapidities, including the field in the instrumented return yoke.

in Fig. 7. The conductor consists of three parts, with an inner superconducting Rutherford cable surrounded by a pure Aluminium layer for thermal and electric stabilisation and an external Aluminium alloy layer for mechanical strength. The transverse dimensions of the conductor are $22 \times 64\,\mathrm{mm}^2$. The total length of conductor needed is 53 km. It is subdivided into 20 elements of $\sim 2.8\,\mathrm{km}$ each. Cable production is now in full swing, by now two of these elements have been produced. The coil was designed by CEA-Saclay and coil winding is under the responsability of the INFN-Italy [8].

The bending power at 4 T of the CMS solenoid is shown in Fig. 8 as a function of radius and at various rapidities. An advantage of a solenoidal magnet design is that in the central part within the coil coverage, where most of the interesting hard collisions physics is, the total bending power is $\sim 13\,\mathrm{Tm}$, and including the field in the return yoke it is $\sim 16\,\mathrm{Tm}$. This large bending power allows for tracker and muon chamber techniques less demanding on measurements precisions, and on a relative muon chambers-tracker alignment system of a $\sim 100\,\mu\mathrm{m}$ precision.

Muon System

A robust muon detection system is central to the CMS concept, muons being the easiest particles to detect and measure in the crowded and hostile high luminosity LHC environment [5a]. Muons also provide the best signature for a number of physics signals, $H \to 4\mu^{\pm}$, $H_{SUSY} \to 2\mu^{\pm}$, $4\mu^{\pm}$, sparticle signals, B-physics channels, ψ, $\Upsilon \to \mu^{+}\mu^{-}$ in pp and heavy ion collisions etc.

The muon system has three purposes, to identify muons, to trigger and to measure their momenta. Muons are identified and measured in four muon stations inserted in the return yoke, Figs. 4, 5. The precision momentum measurement however comes from the inner tracking and the interconnection between inner and outer measurements. The iron thickness in the yoke decouples two consecutive stations for the debris which may accompany a muon. Precision tracking in the muon stations in the barrel yoke is made with drift tube planes and with cathode strip chambers in the endcaps [9].

A cut through a barrel muon station in its final position in the iron yoke with the support structure is shown in Fig. 9. A barrel muon station consists of drift planes for precision measurements and one or two resistive plane chambers (RPC's) on the outside for triggering [9].

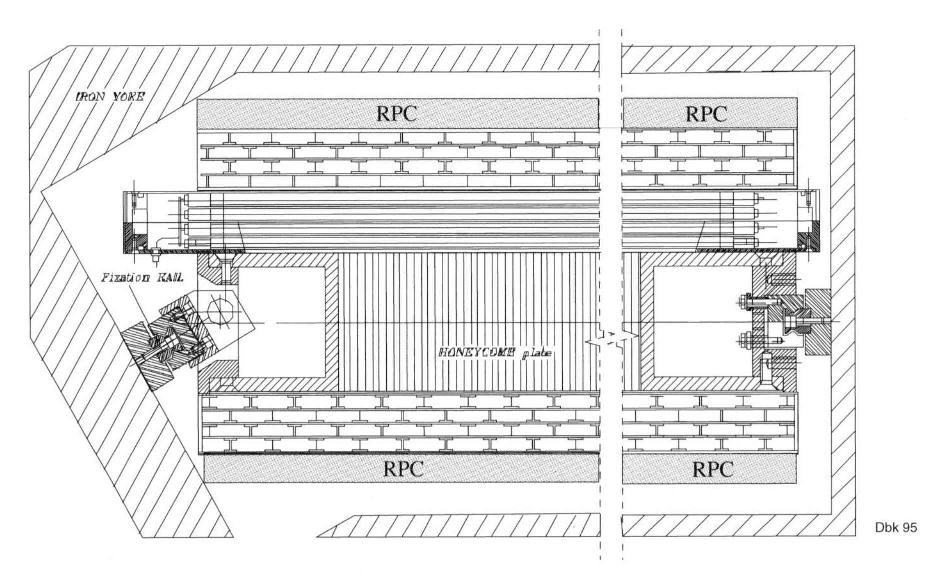

Fig. 9. Cross section of a barrel muon chamber. in the rϕ (bending) plane. The two drift tube superlayers with wires parallel to the beam line are seen face on, the z-layer has wires parallel to beam. The two RPC chambers on the outside are also visible.

There are 12 drift tube layers per barrel muon station, two quadruplets measuring in the bending plane (rϕ), one the longitudinal (z) coordinate (except in station 4). A transverse view of the basic drift-tube cell is shown in Fig. 10, with drift lines and isochrones. The choice of a drift chamber as a tracking detector

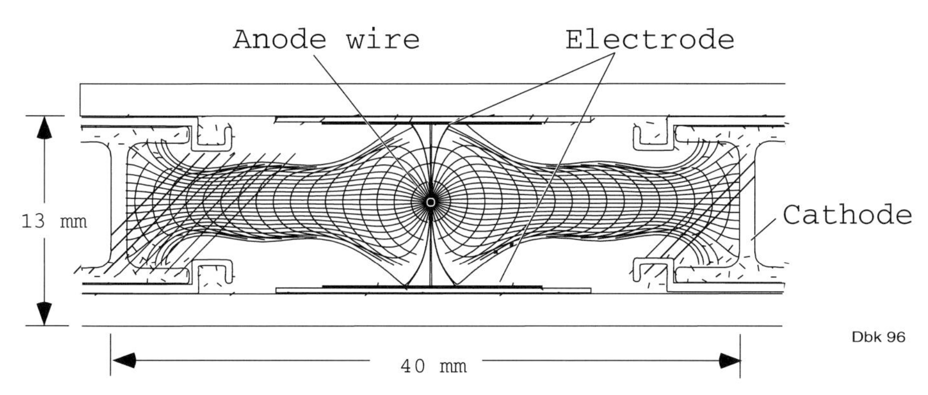

Fig. 10. Transverse view of the basic drift-tube cell, with drift lines and isochrones for a typical voltage configuration of the electrodes.

in the barrel is dictated by the low expected rate and the relatively low intensity of the local stray magnetic field in the region of chambers. The resolution per layer is 250 μm. In a test beam a space resolution of $\approx 100\,\mu$m and an angular accuracy on a local muon track segment of $\approx 1\,$mrad is achieved per station. Meantimer circuits provide a time resolution of a few nanoseconds as needed to identify the bunch crossing with a periodicity of 25 nsec [9]. About 250 such drift chambers of typically $2 \times 3\,\mathrm{m}^2$ are needed for the barrel; with some delay relative to initial expectations, mass production in the various centres, Madrid, Padova, Aachen, Torino, has recently started.

For the endcaps – Fig. 4, due to a high magnetic field and high particle rates, cathode strip chambers have been chosen for precise spatial and timing information [9]. The principle of coordinate measurements is sketched in Fig. 11. Close wire spacing allows for fast chamber response, while a track coordinate along the wires can be measured by interpolating strip charges. Each of the modules contains six layers, with strips running radially and wires for timing perpendicular to strips. The precision requirements for the endcap chambers is higher as the saggitta is smaller, tracks being harder and the magnet bending power diminishing, Fig. 8. For ME1/1 chambers (Fig. 4) in particular, a resolution of $\approx 75\,\mu$m is needed. Overall, the endcap muon system consists of 540 six-plane trapezoidal chambers roughly 1 m wide and 1.5 m long. A typical chamber has about 1000 readout channels. The mass production of components and of chambers is proceeding at full speed in the various centres (Fermilab, UCLA, U. of Florida, PNPI-St. Petersburg, IHEP-Peking, Dubna).

A muon trigger in the barrel drift tubes is generated using a meantimer to identify patterns. In the endcaps the trigger is generated from cathode strip readout patterns with timing from anode wires. For robustness, in both barrel and endcaps up to rapidity 2.1 the muon stations also include a dedicated muon trigger system based on fast resistive plate chambers (RPC) with a $\sim 3\,$nsec time resolution [9], Fig. 9. They are segmented in $\sim 1\,$cm wide strips allowing

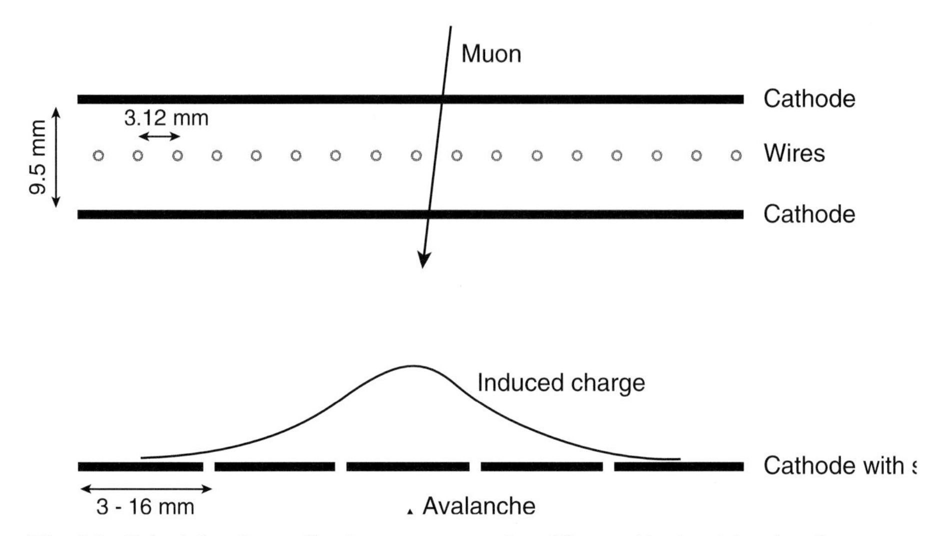

Fig. 11. Principle of coordinate measurements with a cathode strip chamber, cross section across wires (top) and across cathode strips (bottom).

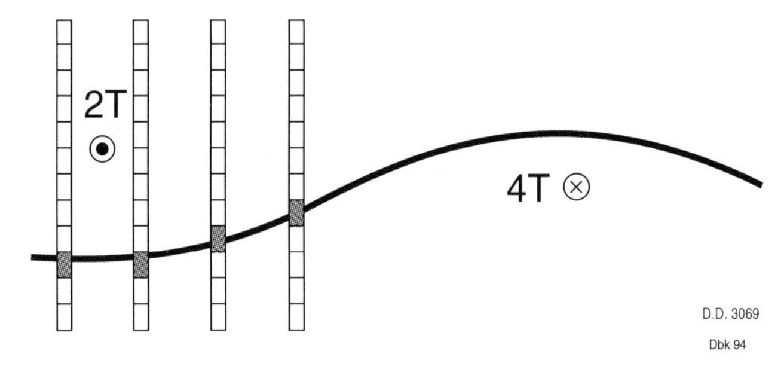

Fig. 12. The basic principle of the RPC trigger, with RPC chambers in the barrel return yoke.

a trigger level p_t measurement up to $\sim 50\,\text{GeV}$. The trigger is based on prerecorded patters of hits as sketched in Fig. 12.

Tracker

The tracker is the only CMS subdetector which has gone through major modifications from the initial concept in the letter of intent, to the technical proposal [5a] and the final design [10]. The inner tracking system of CMS is designed to reconstruct high p_t muons, isolated electrons and hadrons in $|\eta| < 2.4$ with a momentum resolution of $\Delta p_t/p_t \approx 0.15 p_t \oplus 0.5\,\%$ (p_t in TeV). Hadrons must be reconstructed down to $\sim 1\,\text{GeV}$, as lepton and photon isolation is a very important selection criterion for a number of physics signals. Hadrons must also be reconstructed within jets to allow for b-tagging through impact parameter measurements. The main problem in tracking is that of pattern recognition. As

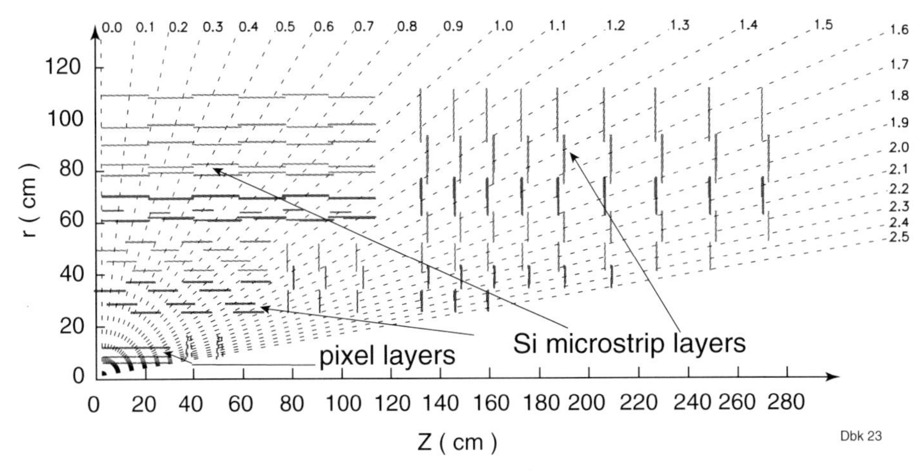

Fig. 13. Layout of one quadrant of the all- silicon CMS tracker.

at a luminosity of $10^{34}\,\mathrm{cm}^{-2}\mathrm{s}^{-1}$, interesting events will be superimposed on a background of about 500 soft charged tracks within the rapidity range considered from ~ 17 minimum bias events occurring in the same bunch crossing.

Small cell-size silicon microstrip detectors provide the required precision and granularity to maintain the cell occupancies below $\sim 1\,\%$, but the number of detection channels is large. A sketch of the final tracker layout is shown in Fig. 13. Starting from the interaction point, in the barrel part of the tracker a particle encounters first two layers (to be upgraded to three layers later on) of pixel detectors ($100\,\mu\mathrm{m} \times 150\,\mu\mathrm{m}$) followed by 10 layers of microstrip Si detectors. The pitch varies from 67 to 200 μm going outwards; strip lengths vary from 6 to 12 cm [10]. Almost every second Si microstrip layer is double to allow determination of the z coordinate of a track by a low-angle stereo measurement. The central barrel geometry is complemented at rapidities exceeding ~ 1 by the same type of Si detectors organised in a disc geometry. The inner Si μ strip sensors are $320\,\mu$m thick, the outer ones, beyond 50 cm radial distance from the beam line, are $500\,\mu$m. Altogether there are $\approx 210\,\mathrm{m}^2$ of silicon sensors with $\approx 9.6 \times 10^6$ microstrips and electronic channels and 75 k APV readout chips. Two vendors for the procurement of silicon sensors have been selected and orders are being now placed. Tests of first modules with readout electronics in the $0.25\,\mu$m technology give excellent results.

The pixel layers ($30 \cdot 10^6$ pixels of $100 \times 150\,\mu\mathrm{m}^2$), at a radius of 4, 7.7 and 11 cm in the barrel – Fig. 13, insure precise impact parameter measurements, with an asymptotic (high momentum) accuracy of $\sigma_{\mathrm{IP}} = 10$–$20\,\mu$m in the transverse plane. For the initial 'low luminosity' ($< 10^{33}\,\mathrm{cm}^{-2}\mathrm{s}^{-1}$) running period only the pixel layers at radii of 4 and 7.7 cm will be implemented. For impact parameter measurements at forward rapidities $\sim 1.5 < |\eta| < 2.4$, the tracker is equipped with two pixel discs, Fig. 13. The mechanics of the pixel system has been designed to allow independent insertion, with the beam pipe in place. There is good progress on pixel sensor and readout chip design and its adaptation to the 0.25 μm technology.

A major difficulty in the design of a tracker for an LHC experiment is the large amount of material in the tracker volume and due to sensor material, electronics, mechanical support, cooling structures etc. The material budget of the CMS tracker is shown in Fig. 14. Despite all efforts to minimise it, it exceeds 1 radiation length at rapidities ~ 1.5. This hampers significantly the reconstruction and precise measurement of photons and electrons.

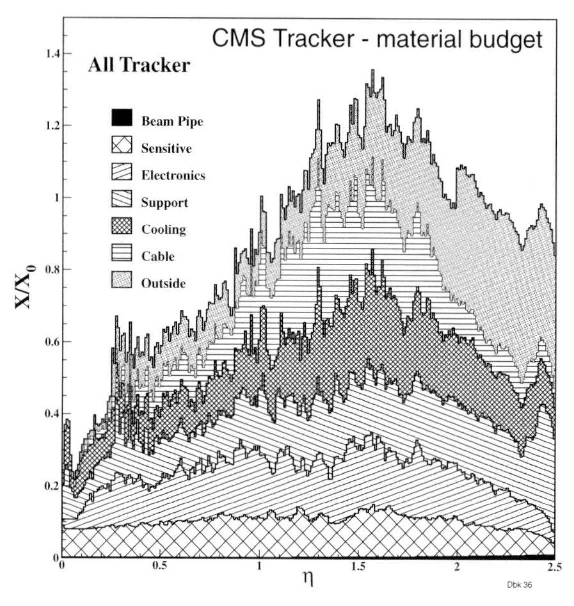

Fig. 14. The material budget in terms of radiation lengths of the various components of the CMS tracker.

Throughout the active tracker volume a track encounters always at least 12 precision measurement layers of pixels+Si microstrips. Tracks are reconstructed with good efficiency even for tracks within jets. Figure 15 shows the expected momentum resolution in the CMS tracker alone, and Fig. 16 for impact parameter measurements. This is obtained after a very detailed simulation, implementing a complete pattern recognition and track reconstruction procedure [10]. For high momentum muons, the combination of tracker and muon chamber measurements improves very much the resolution: $\Delta p_t/p_t \approx 0.07$ for a $p = 1$ TeV muon in $|\eta| < 1.5$. This requires that the relative alignment of the inner and outer systems be known to within $\sim 100\,\mu$m [9].

The expected b-tagging efficiency and sample purity has been studied in detail with several tagging algorithms [10]. Figure 17 shows the expected tagging efficiency versus mistagging rate for three jet E_t values and two possible tag-

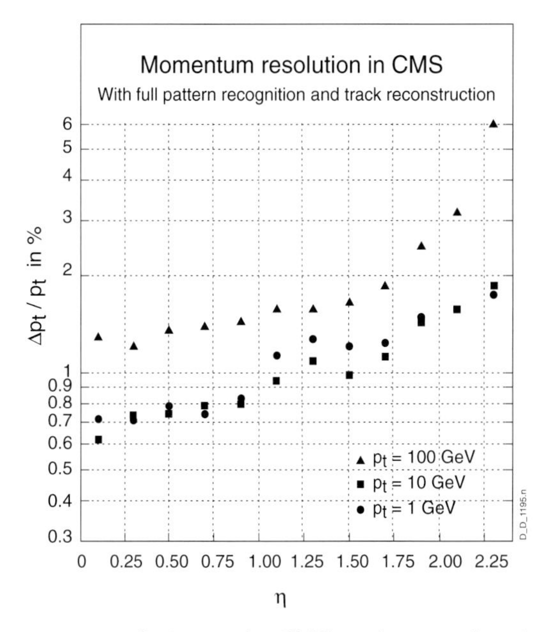

Fig. 15. Momentum resolution in the CMS tracker as a function of rapidity.

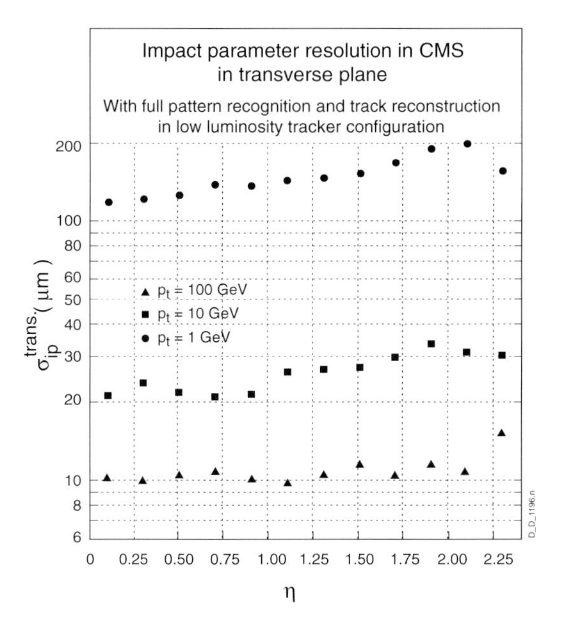

Fig. 16. Expected impact parameter resolution in transverse and longitudinal plane versus rapidity [10].

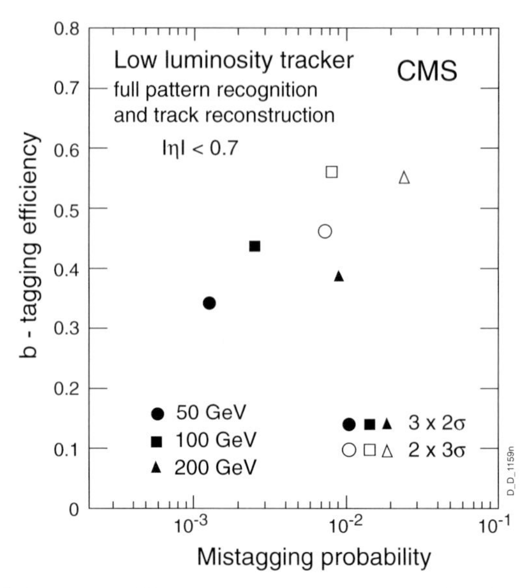

Fig. 17. Expected b-tagging efficiency vs. mistagging rate for three E_t jet values and two b-tagging algorithms.

ging algorithms obtained from detailed simulations. For b-jets in $t\bar{t}$ events the efficiency is $\sim$30–50 %, whilst the expected fake rate probability is $\sim$0.3–1 %.

Electromagnetic Calorimeter

The main task of the electromagnetic calorimeter is to measure precisely electrons and photons, and in conjunction with the hadron calorimeter to measure also jets. The primary goal is to have a $\gamma\gamma$ mass resolution $\sigma_M < 1$ GeV at $10^{34}\,\mathrm{cm^{-2}s^{-1}}$.

The electromagnetic calorimeter system of CMS is made of high resolution lead-tungstate ($PbWO_4$) scintillating crystals extending up to $|\eta| = 3.0$, and a Si preshower in the endcap region $1.5 < |\eta| < 2.5$ [11]. The high density of $PbWO_4$ crystals ($8.3\,\mathrm{g/cm^3}$), with a small Moliere radius of 2.2 cm and a short radiation length of 0.9 cm makes it also possible to have a very compact calorimeter. The arrangement of crystals is shown in Figs. 4 and 18. The geometry is slightly off-pointing (by $\sim$3 deg.). In the barrel the crystals are $26X_0$ deep, the lateral granularity is $\approx$2.2 cm $\times$ 2.2 cm i.e. $\Delta\eta \times \Delta\phi = 0.0175 \times 0.0175$. In the endcaps the granularity is $\approx$2.5 cm $\times$ 2.5 cm, and the crystals plus preshower system is $28X_0$ deep. As the intrinsic light yield of PbWO4 is low, crystals have to read out with photodetectors with gain, which are also required to operate in a high magnetic field. In the barrel crystals are read out with Si avalanche photodiodes (APD's) [11] and in the endcaps, where the radiation dose is much higher, with vacuum phototriodes (VPT) [11]. The ECAL readout chain is sketched in Fig. 19. The total number of crystals is $\approx 0.8 \times 10^5$. The residual variation of the light output with radiation dose can be followed and is corrected for through

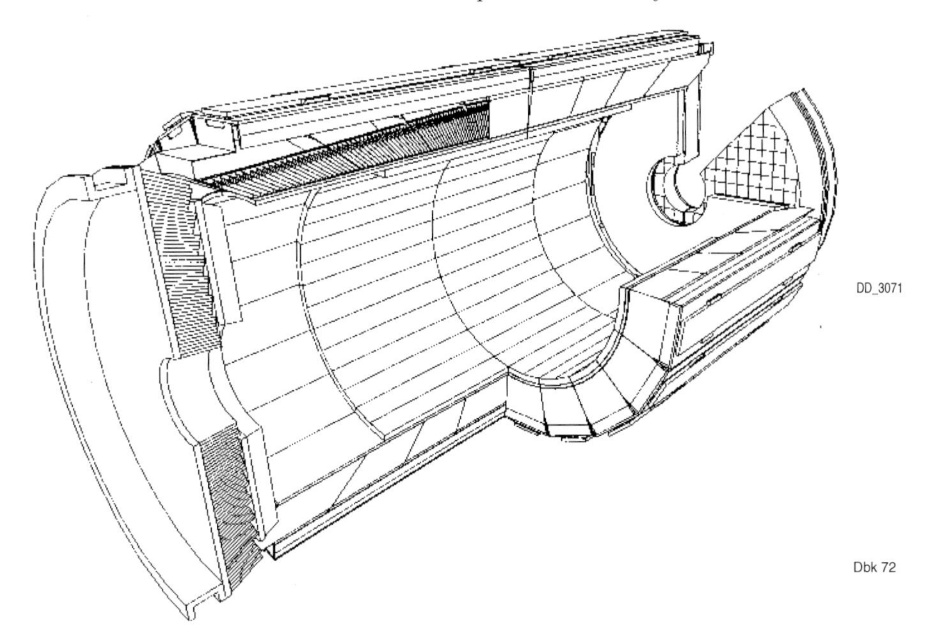

Fig. 18. A 3-D view of the electromagnetic calorimeter, with the preshower in front of the endcap crystals. The modular structure of the calorimeter is also clearly visible.

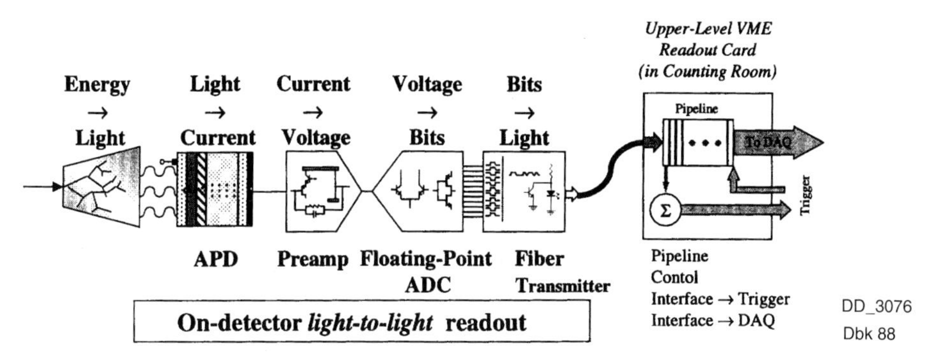

Fig. 19. Electromagnetic calorimeter readout chain.

the monitoring system [11]. With a prototype $PbWO_4$ calorimeter in a test beam an energy resolution of $\sigma_E/E \approx 0.6\,\%$ has been obtained for electrons of $E_t = 120\,GeV$.

Large scale production of crystals started in 2000. By now, end of 2001, about 10000 crystals with required mechanical, light yield and radiation hardness properties have been produced at the Bogorodijsk factory in Russia. Avalanche photodiode production (120 k APD's are needed) is also going on at full speed at Hammamatsu, about 15000 have already been delivered. Industrial pre-production of VPT's for the endcaps has also started. The regional centres for the assembly of modules and supermodules are ready. The major problem at present in the electromagnetic calorimeter is the much higher than expected cost of crystals (cost overruns and unfavourable evolution of the $US vs SF exchange rate) and the front-end electronics chain (Fig. 19). The present front-end chip (FPPA) has a ~ 4 times larger noise than expected and is being redesigned. The optical links are also significantly more expensive than expected. A major redesign of the ECAL electronics is under way.

The main task of the preshower [11] is to reduce the pizero background to photons. It is particularly important for $h/H \to \gamma\gamma$ where it brings a background reduction by about a factor of 3 at $E_t^\gamma \approx 40$ GeV. Silicon sensors for the preshower are now in production in Russia, India and Taiwan and preamplifiers have been successfully tested.

Hadron Calorimeters

Hadron calorimetry with large geometrical coverage for measurement of multi-jet final states and missing transverse energy is essential in all sparticle searches, as it is the missing E_t which provides evidence for the escaping LSP-neutralinos.

The hadron calorimeter of CMS is made of brass absorber plates interleaved with scintillator tiles read-out with embedded wavelength shifting fibres [12]. Brass is a satisfactory compact and non-magnetic absorber material of relatively low Z to minimise muon multiple scattering. The readout in the 4 Tesla field is done with hybrid photodetectors [12]. In both barrel and endcaps the tiles are organised in towers (Fig. 4) giving a lateral segmentation of $\Delta\eta \times \Delta\phi \approx 0.09 \times 0.09$. This central hadron calorimetry extends up to $|\eta| = 3.0$. Particular attention has been given to the design of the barrel/endcap transition region at $\eta \sim 1.5$ to avoid pointing cracks and dead material. The hadronic resolution obtained in a test beam is $\sigma_E/E \approx 100\,\%/\sqrt{E} \oplus 5\,\%$ for the combined ECAL-$PbWO_4$ and hadronic calorimeter system, Fig. 20. In such a mixed calorimeter - with significantly different electromagnetic to hadronic (e/h) response ratios in the two parts ($e/h \sim 1.8$ in ECAL vs. 1.2 in HCAL) – linearity is more difficult to achieve than in a more homogeneous calorimeter system. To restore linearity and to compensate for the effects of dead material in the space between the ECAL and HCAL, weighting techniques will be used. At present a deviation from linearity of $< 4\,\%$ can be obtained for test beam pions with energy from 20 to 300 GeV [12].

The central calorimetry is complemented in the forward region $3.0 < |\eta| < 5.0$ by quartz-fibre 'very forward calorimeters' (Fig. 4) [12]. Their function is to insure detector hermeticity for good missing transverse energy resolution, and to extend the forward jet detection and jet vetoing capability of CMS which is essential in Higgs, slepton, chargino and neutralino searches [13]. Detector hermeticity is particularly important for processes where the physical (real) missing E_t is on the order of few tens of GeV, as is the case in h, H, $A \to \tau\tau$, $W \to l\nu$, $t \to l\nu b$ $t \to H^\pm b \to \tau\nu b$ etc. Calorimetry extending up to $\eta \approx 5$ (i.e. with VF-CAL) reduces the fake (instrumental) E_t^{miss} by an order of magnitude in the 20–120 GeV range [12].

Despite all efforts, CMS does not achieve completely its goal of having a deep enough calorimetry fully contained within the coil. Insufficient shower contain-

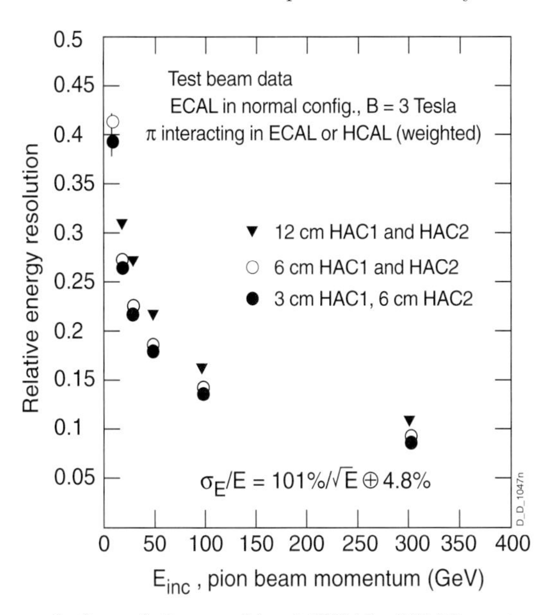

Fig. 20. Energy resolution of the combined HCAL+ECAL system from test beam data, for various thicknesses of the HCAL absorber plates.

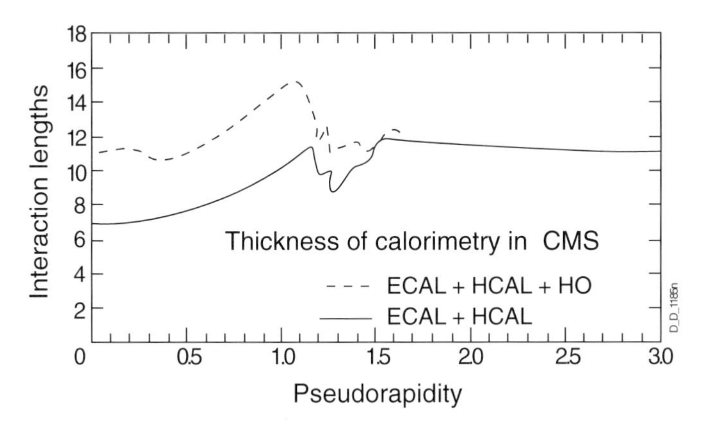

Fig. 21. Total thickness of calorimeters in CMS versus rapidity, with and without the tail catcher layers.

ment in depth is responsible for the presence of low-energy tails in hadronic (jet) energy measurements, which is one of the main sources of instrumental missing E_t. Tail-catcher scintillator layers with fiber optic readout (outer calorimeter) are inserted in the barrel region just behind the coil [12]. Inclusion of tail-catcher layers allows hadron energy measurements with at least 10.5 absorber lengths everywhere, as visible from Fig. 21.

The barrel hadron calorimeter is made of 2×18 modules (wedges). It is fixed onto the inner wall of the coil cryostat. The hadron calorimeter is the main structural element onto which is mounted the ECAL and to which is attached the tracker. All the HCAL wedges have been fabricated (in Felguera, Spain) and delivered to CERN where the first half-barrel has been assembled. The scintillator-fibre active elements ("megatiles"), produced and tested at FNAL, are now being mounted inside the brass structure. The full barrel HCAL should be finished by mid-2002. The hadron calorimeter endcaps, which are mounted on the "nose" of the first endcap yoke disc (Fig. 4), are machined in Minsk. One of the two endcaps has already been delivered at CERN and the manufacture of optical elements is proceeding well in Protvino. The mass production of forward calorimeter wedges has also started in Tcheliabinsk, Russia. The quartz fibers and photomultipliers will soon be ordered. About one third of the outer calorimeter (tail-catcher) scintillator tiles have also been already produced in India.

Trigger and Data Acquisition

The four main components of the trigger and data acquisition system of CMS, sketched in Fig.22, are the front-end electronics, the muon and calorimeter first level trigger processors, the readout network and the online event filter system. At the design luminosity of $10^{34}\,\mathrm{cm}^{-2}\mathrm{s}^{-1}$ and a bunch crossing frequency of 40 MHz and with an expected non-diffractive inelastic pp cross section of 55 mb at $\sqrt{s} = 14\,\mathrm{TeV}$, we expect on average 17 pp collisions per bunch crossing. The collision rate of 40 MHz has to be reduced by the hardwired first-level trigger system [14] to a rate of 30 to 50 kHz by selecting on candidate muons, electrons/photons, jets, missing energy or the global transverse energy flow. The fron-end electronics and first level trigger processors are pipelined with a pipeline depth of $\sim 3\,\mu$sec. The first-level trigger rate will be shared approximately equally between calorimetric and muon hardware triggers. The higher level triggers in CMS are software implemented. A typical event size is 1 Mbyte. An event builder, a large (512×512 ports) switching network with a throughput of 500 Gbits/s,

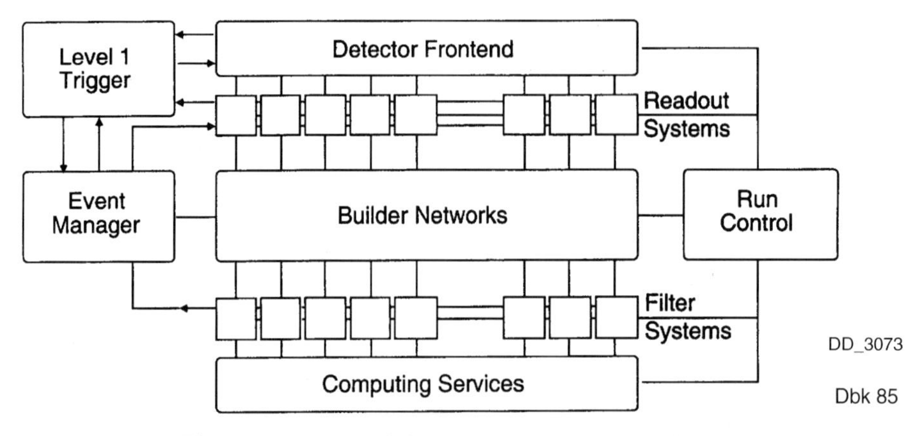

Fig. 22. Trigger and data acquisition scheme of CMS.

feeds the event filter which will be made of a large number of commercially available processors organised into farms [5a] with a total computing power of $5 \cdot 10^6$ MIPs. This event filter stage, capable of accepting from the first level trigger an input rate of 75 kHz to assure a safety margin, must reduce in several steps, executing "higher level trigger" physics algorithms, the data rate down to about 100 Hz to match the data taking capability of the mass storage devices. A single CPU processes one event. The farm of processors will be used both for on-line and off-line applications. The total data production rate will be ~ 1 Tbyte per day.

4 Higgs Expectations, SM and MSSM

SM Higgs Coverage and SUSY Higgs Reach in MSSM Parameter Space

The main task at the LHC will be to investigate the mechanism responsible for electroweak symmetry breaking. Both ATLAS and CMS have studied in great detail the observability of the SM Higgs in its various decay modes $\mathrm{H} \to \gamma\gamma$, bb, WW ($\to \mathrm{l}\nu\mathrm{l}\nu$) ZZ*/ZZ ($\to 4\mathrm{l}^{\pm}$) etc. In fact both detectors CMS and ATLAS have to a large extent been designed and optimized in view of the SM Higgs search. The outcome of these studies is summarized in Fig. 23. It gives the expected SM Higgs signal significance for 100 fb^{-1} according to ATLAS expectations, Fig. 23a, and in Fig. 23b the integrated luminosity required to see a 5σ signal according to CMS expectations [15-17]. The SM Higgs cannot escape detection over the entire expected mass range. The region $200 < \mathrm{m_H} < 600$ GeV can be explored with less

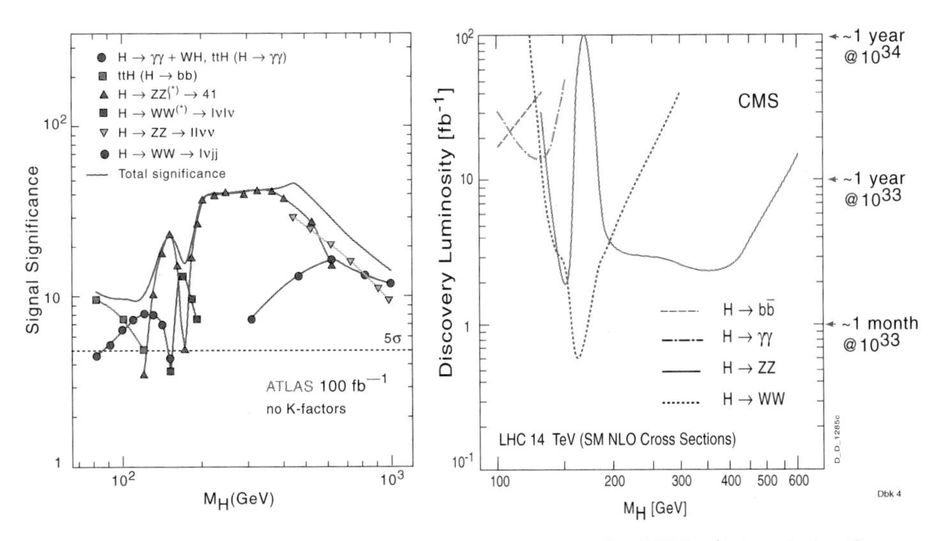

Fig. 23. Observability of the SM Higgs versus m_H at the LHC; a) signal significance in various decay modes for 100 fb^{-1} according to ATLAS expectations [15], b) required luminosity for a 5σ signal in the main decay modes according to CMS expectations.

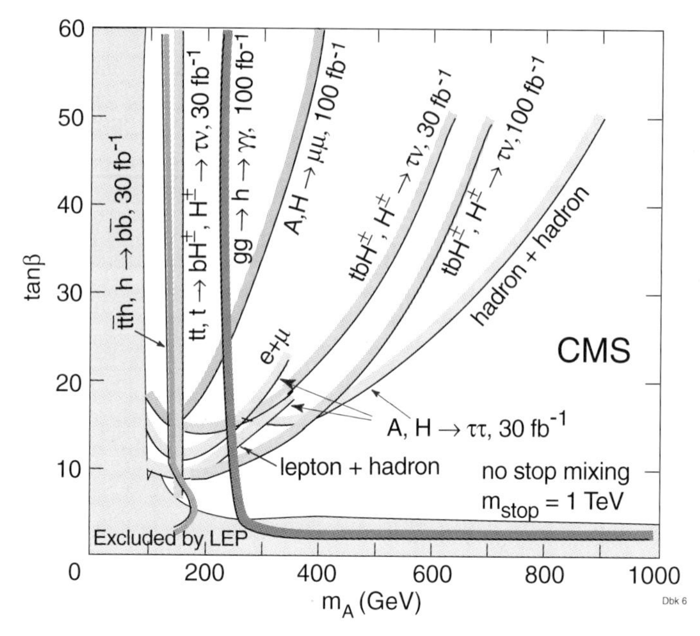

Fig. 24. Coverage of parameter space in various MSSM Higgs boson searches at luminosities of 30 and $100\,\mathrm{fb}^{-1}$ according to CMS expectations.

than $10\,\mathrm{fb}^{-1}$. Most demanding is the region $m_H < 125\,\mathrm{GeV}$ which relies on the $H \to \gamma\gamma$ and $H \to bb$ modes, and requires $> 30\,\mathrm{fb}^{-1}$.

The observability of the five MSSM Higgs bosons (h, H, A, $H^{\pm}$) throughout the parametyer space has also been studied in great detail [15-17]. For MSSM Higgses some addtional decay modes appear such as h, H, $A \to \tau\tau$ or $H^{\pm} \to \tau\nu$ requiring detection of τ's in final states, or h, H, $A \to \mu\mu$ which plays no role in SM Higgs searches. In fact the importance of τ-final states – especially with hadronic τ decays – is such that CMS had to change significantly its level-1 τ and jet triggering strategy and trigger algorithms to optimize efficiency for τ's. Another distinctive feature of MSSM Higgs production is the importance of bbH_{SUSY} associated production final states, especially at large $\tan\beta$. The need to detect and identify (b-tag) efficiently these relatively soft accompanying b-jets has also played a major role in the final optimization of the pixel/microvertexing part of the CMS tracker [10].

Coverage of parameter space in some of the MSSM Higgs boson searches at luminosities of 30 and $100\,\mathrm{fb}^{-1}$ according to CMS expectations is shown in Fig. 24 [16, 17]. The entire $\mathrm{tg}\,\beta - m_A$ MSSM parameter space can be explored with less than $30\mathrm{fb}^{-1}$ and at least one of the MSSM Higgs bosons always found. There are large parts of parameter space where several Higgs bosons can be found, especially at large $\tan\beta$ (> 10) where $A/H/H^{\pm}$ masses up to 500–800 GeV are within reach with $100\,\mathrm{fb}^{-1}$. The role and importance of τ's in h, H, $A \to \tau\tau$ or $H^{\pm} \to \tau\nu$ searches is evident, as these are the channels allowing to explore the largest portions of parameter space. These studies are continuing, more specifically on the possibilities to observe sparticle decay modes such as H, A $\rightarrow \tilde{\chi}_2^0\tilde{\chi}_2^0 \rightarrow 4l^{\pm} + E_t^{miss}$ or $H^{\pm} \rightarrow \chi^{\pm}{}_i\chi^0{}_j \rightarrow 3l^{\pm} + E_t^{miss}$ [18]. The channel H, A $\rightarrow \tilde{\chi}_2^0\tilde{\chi}_2^0$ is particularly interesting as it allows to explore a region complementary to H, A $\rightarrow \tau\tau$. Another domain of present investigations are the WW(ZZ) boson fusion production mechanism channels with H $\rightarrow \gamma\gamma$ and $\tau\tau$ or h, H $\rightarrow \tilde{\chi}_1^0\tilde{\chi}_1^0$ "invisible Higgs" decays [19].

5 Sparticle Studies in CMS

A variety of studies on sparticle reach have been performed, mostly assuming R-parity conservation, but also with R parity non-conservation and in GMSB scenarios [20, 21]. As well known, the lightest neutralino ($\tilde{\chi}_1^0$) of R-parity conserving scenarios is a good particle physics candidate for cosmological dark matter [3]. It escapes direct detection and reveals itself through missing E_t. The most extensive studies of sparticle observability have been done within the minimal supergravity-constrained MSSM model [22, 20]. The five parameters which determine all masses, couplings, production cross sections, decay branching ratios are: m_0 and $m_{1/2}$ (respectively the common scalar and gaugino masses at GUT scale), tg β, A_0 and sign(μ).

Squarks and gluinos are abundantly produced in pairs $\tilde{q}\tilde{q}$, $\tilde{g}\tilde{q}$, $\tilde{g}\tilde{g}$, but the associated production modes $\tilde{g}\tilde{\chi}$, $\tilde{q}\tilde{\chi}$ are also significant, Fig. 25. With increasing $\tilde{q}$ / $\tilde{g}$ mass, decays to charginos, neutralinos and even sleptons become important. These 'cascade decays' can lead to very complex decay schemes and final states

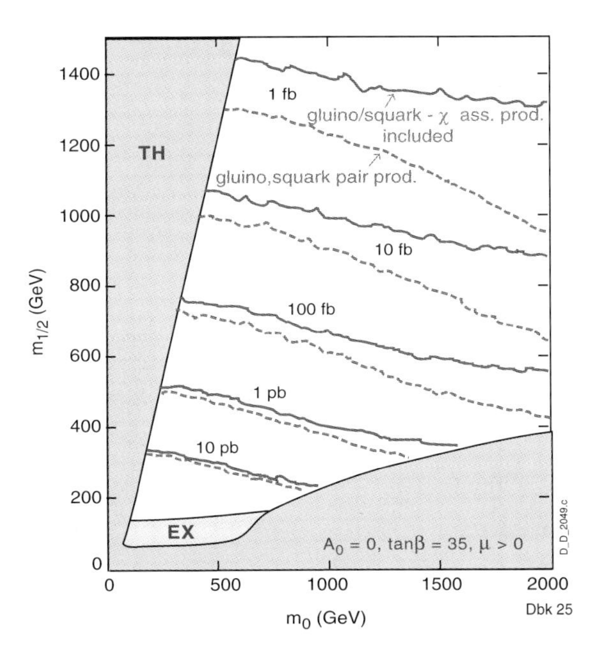

Fig. 25. Squark and gluino production cross sections at LHC at NLO in terms of mSUGRA parameters.

containing a number of leptons, jets and missing E_t. B-quarks appear with substantial branching ratios in the decays of $\tilde{g}$ or $\tilde{q}$. As there are always at least two escaping LSP's, no sparticle mass peak can be entirely reconstructed, and evidence for a SUSY signal will depend primarily on an excess of events over expected backgrounds.

Gluinos and Squarks

To evaluate the $\tilde{g}$ and $\tilde{q}$ mass reach, a systematic study of signal observability has been done in final states with a variable number of leptons, with at least two jets and a significant E_t^{miss}. The cuts have been optimised for the different topologies as a function of dominant backgrounds and as a function of the domain of parameter space probed [20]. Figure 26 shows the 5σ squark and gluino discovery contours at low $\tan\beta$ for an integrated luminosity of $10^2\,\mathrm{fb}^{-1}$ in various final states with one lepton (1l), two leptons of opposite sign (2l OS), two leptons of same sign (2l SS), three leptons (3l) and four leptons (4l) in final states, with at least two central jets and E_t^{miss} of at least 100 GeV. Squark and gluino isomass curves are also shown. The mass reach for $\tilde{q}$, $\tilde{g}$ is in excess of 2 TeV. Figure 27 shows the domain of parameter space which can be explored with integrated luminosities varying from 1 to $300\,\mathrm{fb}^{-1}$ in the inclusive topology E_t^{miss} + jets, ir-

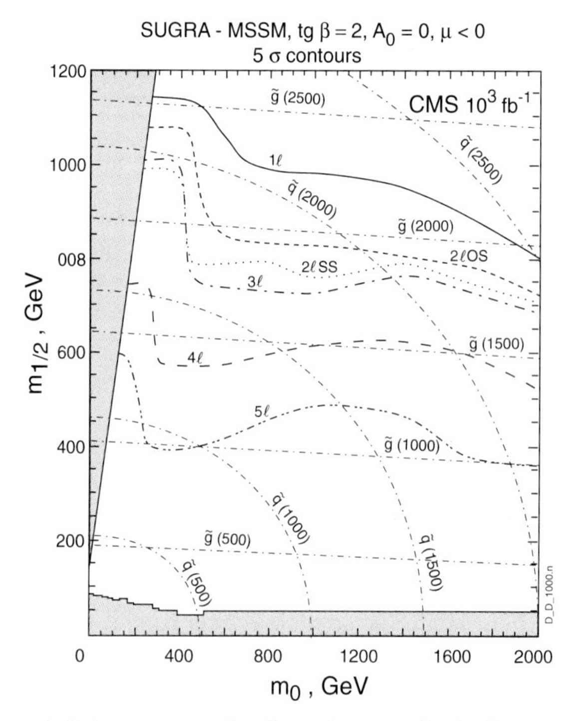

Fig. 26. Squark and gluino mass reaches in various topologies from 1 to 5 leptons with $10^2\,\mathrm{fb}^{-1}$ in mSUGRA for $\tan\beta = 2$ and $\mu < 0$.

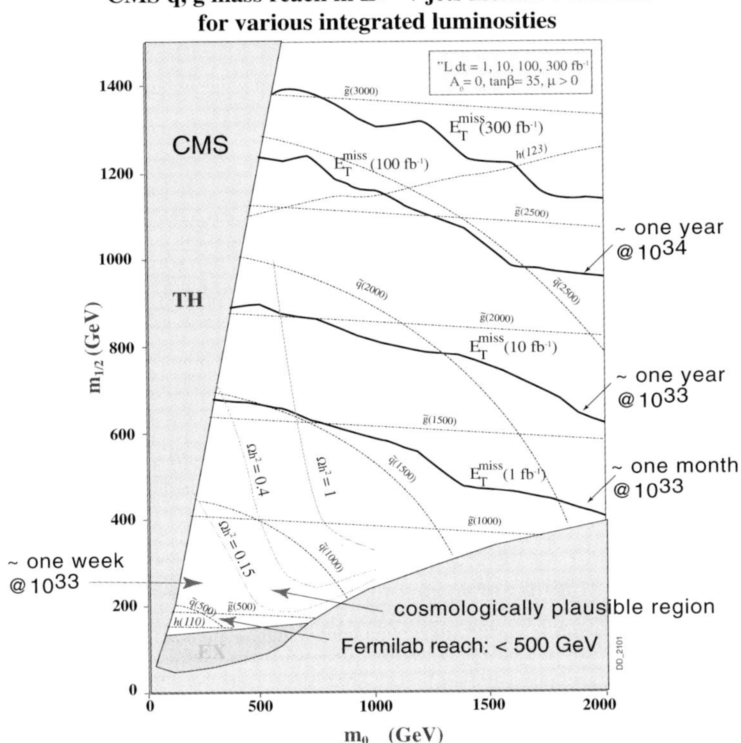

Fig. 27. Squark xand gluino discovery contours in mSUGRA for $\tan\beta = 35$, $\mu > 0$ in jets + E_t^{miss} final states for integrated luminosities from 1 to 300 fb^{-1}. The neutralino dark matter density contours of Ωh^2 equal to 0.15 and 0.4 are indicated [25].

respective of number of leptons in final state, for $\tan\beta = 35$ [20]. The squark and gluino mass reach is again in the ~ 2.5 TeV range for $10^2\,fb^{-1}$. The mass range ~ 0.5–0.7 TeV can be explored with just few weeks of data taking. The appearing as the end-product in the $\tilde{q}$, $\tilde{g}$ decay chains over the explorable $\tilde{q}$, $\tilde{g}$ mass domain spans a mass range from ~ 50 to ~ 500 GeV. The main cosmologically plausible region of $\Omega h^2 < 0.4$ can be also rapidly explored.

Mass Reconstruction

The issue of reconstructing sparticle masses in a hadron collider experiment has been discussed in a number of articles [22, 23]. Indirect $\tilde{\chi}_2^0$ production from $\tilde{q}$, $\tilde{g}$ cascades can be so abundant that the kinematical upper limit in the l^+l^- mass spectrum from $\tilde{\chi}_2^0 \to l^+l^-\tilde{\chi}_2^0$ decay is easily visible as a distinct edge even in *inclusive* isolated dilepton studies, Fig. 28a for example. This edge is the starting point of sparticle mass reconstruction, an example is given in what follows. This remains so as long as the $\tilde{\chi}_2^0 \to l^+l^-\tilde{\chi}_1^0$ decay mode is significant ie at low to moderate $\tan\beta$ [23]. At larger $\tan\beta$ values, say > 15, with the increasing role played

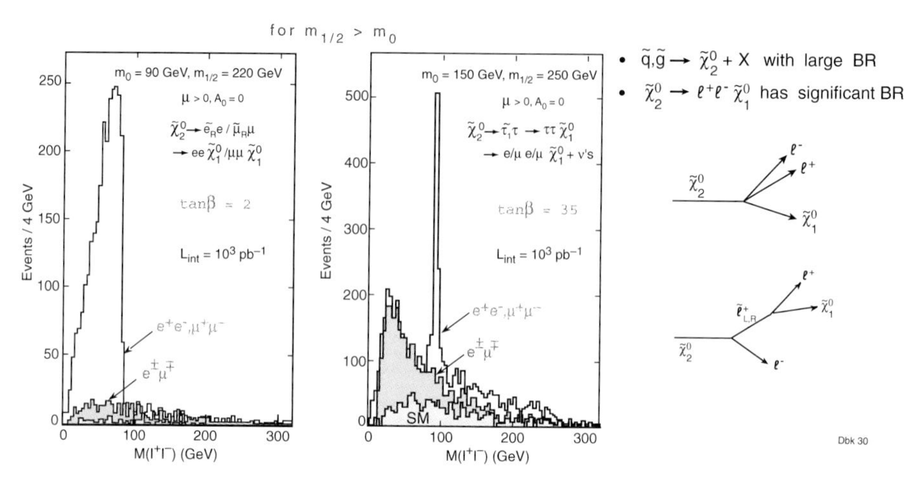

Fig. 28. Dilepton mass spectra in inclusive two leptons + E_t^{miss} final states at the indicated point in mSUGRA parameter space with $\tan\beta = 2$ and 35; there are no cuts on jets; the expected SM background is shaded.

by staus and taus in final states in both $\tilde{\chi}^0_2 \to \tau^+\tau^-\tilde{\chi}^0_1$ and $\tilde{\chi}^0_2 \to \tilde{l}l \to \tau^+\tau^-\tilde{\chi}^0_1$ regions [22, 24], the visibility of the dilepton edge diminishes. It is gradually replaced by a pronounced dilepton low mass enhancement, indicative of its $\tau\tau$ origin in the opposite-sign eμ final states. Ultimately, at large $\tan\beta$, Fig. 28b, the staus and taus dominate so much, that the dilepton spectrum shapes in same flavour and different flavour final states are practically the same, as are the production cross sections for the threshold mass enhancement. From the comparison of shapes and magnitudes of same flavour vs. different flavour dileptons it is then possible to constrain the $\tan\beta$ parameter [24].

The domain of parameter space where the dilepton (ee, $\mu\mu$) spectrum is showing this prominent edge has been mapped for several sets of mSUGRA parameters [24]. Figure 29a shows such a domain where the dilepton edge is visible at low tanβ. Figure 29b shows the domain of visibility of the low-mass enhancement in dilepton spectra expected at large $\tan\beta$ when τ (stau) dominates lepton production.

As mentioned previously, from dilepton edges it is possible, by choosing events at the edge, to reconstruct the parent $\tilde{\chi}^0_2$ momentum by the methods developed in [23]. It is then possible to reconstruct the $\tilde{q}$ mass from the $\tilde{q} \to q\tilde{\chi}^0_2$ chain by associating the $\tilde{\chi}^0_2$ four-momentum with nearby q-jets. An example of such reconstructed squark mass is shown in Fig. 30a for the decay $\tilde{b} \to b\tilde{\chi}^0_2$ [26]. The mSUGRA parameters for Fig. 30 are: $m_0 = 100$ GeV, $m_{1/2} = 250$ GeV, $\tan\beta = 10$, $A_0 = 0$, $\mu > 0$. There are significant regions of parameter space, where the chain of decays can be reconstructed and sparticle masses determined, at least for low to moderate $\tan\beta$ i.e. as long as sharp dilepton edges are observable, Fig. 29a. Figure 30b shows the next step in reconstruction of the decay chain, namely $\tilde{g}$ reconstruction from $\tilde{g} \to b\tilde{b} \to bb\tilde{\chi}_2^0 \to bb\, l^+l^-\tilde{\chi}_1^0$. For both $\tilde{b}$ and $\tilde{g}$ the masses are reconstructed with a resolution of $\approx 15\,\%$.

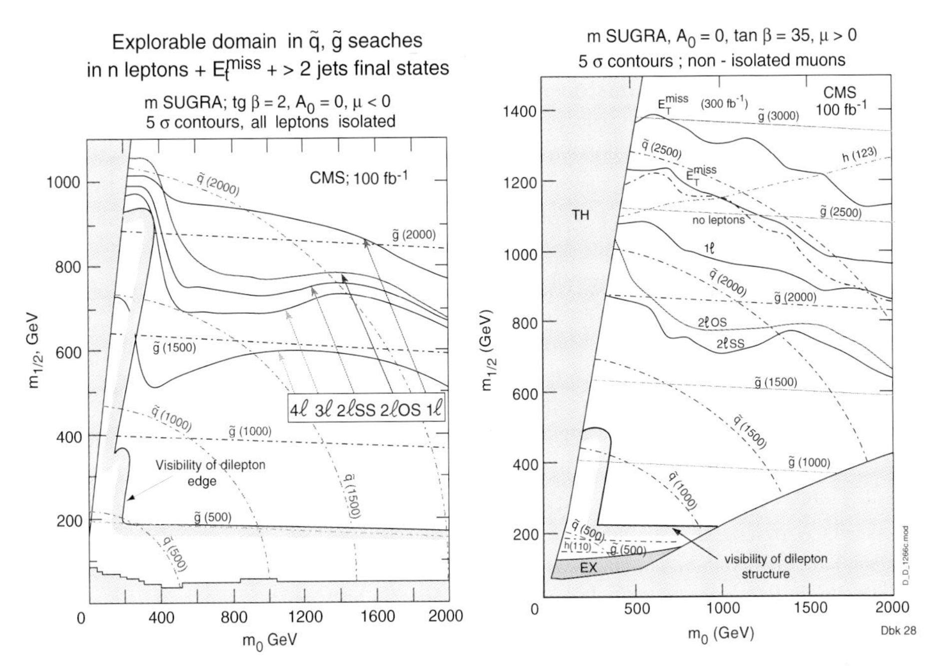

Fig. 29. Domains of mSUGRA parameter space where in inclusive $l^+l^- + \mathrm{E}_t^{\,\mathrm{miss}}$ final states the dilepton edge can be observed at low $\tan\beta$ (a) and where the dilepton low mass enhancement can be observed at large $\tan\beta$ (b).

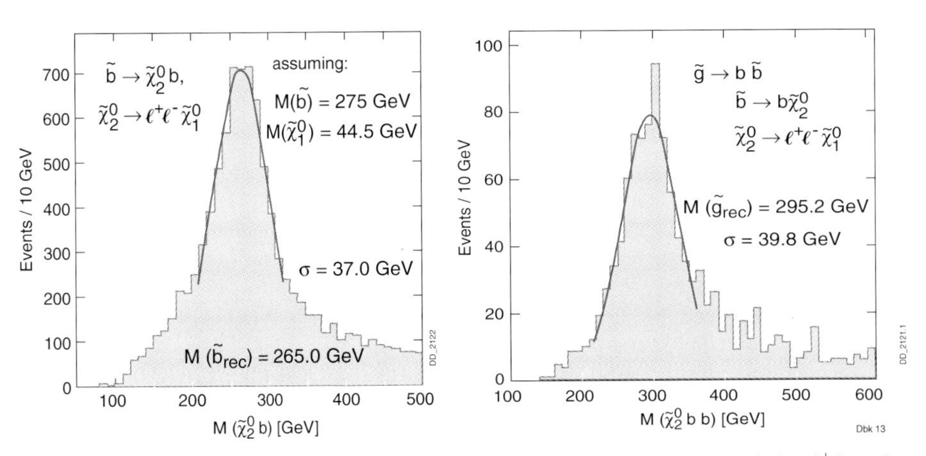

Fig. 30. (a) Reconstruction of $\tilde{q}$ mass selecting events in the region of the l^+l^- edge from $\tilde{\chi}_2^0 \to l^+l^-\tilde{\chi}_1^0$ decays; (b) Reconstruction of $\tilde{g} \to b\tilde{b}$ decay exploiting the l^+l^- edge from $\tilde{\chi}_2^0 \to l^+l^-\tilde{\chi}_1^0$ decays

6 Summary and Conclusions

Construction of the LHC is proceeding rapidly. The large underground experimental areas for ATLAS and CMS are being excavated. The last series of prototypes of main machine dipoles has very satisfactory performance. The industrial preproduction of dipoles has been launched. CMS construction is proceeding at full speed. All the major subsystems are now in the production/construction phase. The overall detector cost amounts to ≈ 500 MSF, about 50 % has already been committed and available resources cover close to 90 % of the full detector cost. LHC commissioning is expected at the end of 2006/beginning 2007 and CMS should be ready to take data.

The SM Higgs cannot escape detection over the expected mass range. The region $200 < m_H < 600$ GeV can be explored with less than $10\,fb^{-1}$. The entire $tg\,\beta - m_A$ MSSM parameter space can be explored with less than $30\,fb^{-1}$ and at least one of the MSSM Higgs bosons always found. There are large parts of parameter space where several Higgs bosons can be found. The ATLAS and CMS detector designs are flexible and powerful enough that they would also allow to explore ew symmetry breaking schemes other than the Higgs one. The LHC is also the most appropriate machine to look for squarks and gluinos. The mass reach in the 2–3 TeV range. The LHC allows also investigation of a large fraction of the SUSY sparticle spectrum. Reconstruction of the sparticle mass spectrum will however be a long and tedious task. Exploration of the 2–3 TeV squark/gluino mass range allows also to probe extensively the

cosmological dark matter scenarios. How to probe at the LHC the recent hypothesis of possibly large extra dimensions is actively investigated [4, 27].

References

[1] G. Altarelli, Proc. of LHC Workshop, Aachen, CERN 90-10, 1990 and references therein; J.F. Gunion, H.E. Haber, G.L. Kane and S. Dawson,"The Higgs Hunter's Guide", Addison-Wesley, Redwood City, CA, 1990.

[2] Z. Kunszt and F. Zwirner, Nucl. Phys. B385 (1992) 3, and references therein; H.P. Nilles, Phys. Rep. 110 (1984) 1; M. Carena et al., CERN-TH/95-45.

[3] E.W. Kolb and M.S. Turner, The Early Universe, Addison-Wesley, Redwood City, CA 1989; V. Berezinsky et al., "Neutralino dark matter: an overview", The Dark Side of the Universe, World Scientific, 1996, p. 77.

[4] N. Arkani-Hamed, S. Dimopoulos, G. Dvali, Phys. Lett. B429,(1998) 263; L. Randall, R. Sundrum, Phys. Rev. Lett.83 (1999) 3370.

[5a] Technical Proposal, CMS Collaboration, CERN/LHCC 94-38, LHCC/P1, December 1994.

[5b] Technical Proposal, ATLAS Collaboration, CERN/LHCC/94-43,LHCC/P2, December 1994.

[6] see for ex.: Proc. of the Workshop on "Standard Model Physics (and more) at the LHC", CERN yellow report 2000-004, May 2000, chapters 3 and 4.

[7] G. Gaur et al., "Heavy Ion Physics Program in CMS", CMS NOTE 2000/060, December 2000.

[8] CMS Collaboration, The Magnet Project, Technical Design Report, CERN/LHCC 97-10, CMS TDR1, 1997.
[9] CMS Collaboration, The Muon Project, Technical Design Report, CERN/LHCC 97-32, CMS TDR3, 1997.
[10] CMS Collaboration, The Tracker Project, Technical Design Report, CERN/LHCC 98-6, CMS TDR5, 1998; CMS, Addendum to CMS Tracker TDR, CERN/LHCC 2000-016.
[11] CMS Collaboration, The Electromagnetic Calorimeter, Technical Design Report, CERN/LHCC 97-33.
[12] CMS Collaboration, The Hadron Calorimeter Project, Technical Design Report, CERN/LHCC 97-31, 1977.
[13] S. Abdullin et al., CMS Coll., Discovery Potential for Supersymmetry in CMS, CMS NOTE 1998/006, April 1998; hep-ph/9806366, June 1998.
[14] CMS, The Trigger and Data Acquisition Project, Vol. 1, The Level-1 Trigger, Technical Design Report, CERN/LHCC 2000-038.
[15] ATLAS, Detector and Physics Performance, Technical Design Report, Vol. II, CERN/LHCC/99-15, ATLAS TDR 15, May 1999; J.G. Branson et al., "High transverse momentum physics at the large hadron collider", ATLAS and CMS Collaborations, hep-ph/0110021, Oct. 2001.
[16] V. Drollinger et al., "Searching for Higgs Bosons in Association with Top Quark Pairs in the $H \to b\bar{b}$ Decay Mode", CMS NOTE 2001/054, Nov. 2001, hep-ph/0111312, 23 Nov. 2001. R. Kinnunen and A. Nikitenko, "Study of $H_{SUSY} \to \tau\tau \to 2\tau$-jets in CMS", CMS NOTE 2001/031;
[17] D. Denegri et al., "Summary of the CMS discovery potential fot the MSSM SUSY Higgses", CMS NOTE 2001/032, Nov. 2001, hep-ph/0112045, 4 Dec. 2001, and references therein.
[18] F. Moortgat et al., "Observability of MSSM Higgs bosons via sparticle decay modes in CMS", CMS NOTE 2001/042, Dec. 2001, hep-ph/0112046, 4 Dec. 2001.
[19] The Higgs working group, Les Houches 2001 Summary Report, hep-ph/0203056, 5 March 2002.
[20] S. Abdullin et al.,"Discovery potential for supersymmetry in CMS", CMS NOTE 1998/006; hep-ph/9806366, June 1998; S. Abdullin and F. Charles, N.P.B 547 (1999) 60.
[21] M. Kazana, G. Wrochna, P. Zalewski, "Study of the NLSP from the GMSB models in the CMS detector at the LHC", Proc. EPS HEP99 Conf., Tampere, Finland, 15–21 July 1999; CMS CR 1999/019.
[22] H. Baer et al., Low Energy Supersymmetry Phenomenology, CERN-PPE/95-45, and refs. therein; also FSU-HEP-951215; FSU-HEP-950204; H. Baer et al., "Probing minimal Supergravity at the CERN LHC for large $\tan\beta$", hep-ph/9809223, Sept 98; FSU-HEP-980730.
[23] F. Paige, Determining SUSY Particle Masses at LHC, hep-ph/9609373; "SUSY Studies in CMS", in LHCC SUSY Workshop, CMS Document 1996-149, CERN, Oct. 29–30, 1996; I. Hinchliffe et al., Phys. Rev. D55 (1997) 5520, and references therein.
[24] D. Denegri, W. Majerotto and L. Rurua, Phys. Rev. D, Vol 58 (1998) 095010; D. Denegri, W. Majerotto and L. Rurua, Phys. Rev. D, Vol 60 (1999) 035008.
[25] H. Baer and M. Brhlik, "Cosmological Relic Density from Minimal Supergravity with Implications for Collider Physics", FSU-HEP-950818; H. Baer and M. Brhlik, Phys. Rev. D 53 (1996) 597; H. Baer and M. Brhlik, "Neutralino Dark Matter in Minimal Supergravity: Direct Detection vs. Collider Searches", hep-ph/9706509,

25 June 1997; J. Ellis, T. Falk and K. Olive, "Neutralino-Stau Coannihilation and the Cosmological Upper Limit on the Mass of the Lightest Supersymmetric Particle", hep-ph/9810360.
[26] M. Chiorboli, presentation at CMS Physics meeting, March 2002.
[27] P. Traczyk, G, Wrochna, "Search for Randall-SundrumGraviton Excitations in the CMS Experiment", CMS NOTE 2001/003, Dec. 2001, and references therein.

The ATLAS Detector and Physics Potential

Yoram Rozen

Physics Department, Technion – Israel Institute of Technology, Haifa 32000, Israel

Abstract. The ATLAS detector designed and built by a collaboration of ≈ 2000 physicists will begin taking data at the Large Hadron Collider around 2007. This document describes the physics motivation for the construction of the detector, the different components and their expected performance.

1 Introduction

The success of the SM of strong, weak and electromagnetic interactions has drawn increased attention to its limitations. In its simplest version, the model has 19 parameters, the three coupling constants of the gauge theory $SU(3) \times SU(2) \times U(1)$, three lepton and six quark masses, the mass of the Z boson which sets the scale of weak interactions, and the four parameters which describe the rotation from the weak to the mass eigenstates of the charge $-1/3$ quarks (CKM matrix). All of these parameters are known with varying errors. Of the two remaining parameters, a CP-violating parameter associated with the strong interactions must be very small. The last parameter is associated with the mechanism responsible for the breakdown of electroweak $SU(2) \times U(1)$ to $U(1)_{em}$. This can be taken as the mass of the, as yet undiscovered, Higgs boson. The couplings of the Higgs boson are determined once its mass is given.

The gauge theory part of the SM has been well tested, but there is no direct evidence either for or against the simple Higgs mechanism for electroweak symmetry breaking. All masses are tied to the mass scale of the Higgs sector. Although within the model there is no guidance about the Higgs mass itself, some constraints can be delivered from the perturbative calculations within the model requiring the Higgs couplings to remain finite and positive up to an energy scale [1]. Such calculations exists at the two-loop level for both lower and upper Higgs mass bounds. With present experimental results on the SM parameters, if the Higgs mass is in the range 160 to 170 GeV [2] then the renormalisation-group behaviour of the Standard Model is perturbative and well behaved up to Planck scale $\Lambda_{Pl} \approx 10^{19}$ GeV. For smaller or larger values of m_H new physics must set in below Λ_{Pl}. As its mass increases, the self couplings and the couplings to the W and Z bosons grow [3]. This feature has a very important consequence. Either the Higgs boson must have a mass less than about 800 GeV, or the dynamics of WW and ZZ interactions with centre-of-mass energies of order 1 TeV will reveal new structure. It is this simple argument that sets the energy scale that must be reached to guarantee that an experiment will be able to provide information on the nature of electroweak symmetry breaking.

The presence of a single elementary scalar boson is unsatisfactory to many theorists. If the theory is part of some more fundamental theory, which has some other larger mass scale (such as the scale of grand unification or the Planck scale), there is a serious 'fine tuning' or naturalness problem. Radiative corrections to the Higgs boson mass result in a value that is driven to the larger scale unless some delicate cancellation is engineered ($(m_0^2 - m_t^2) \approx m_W^2$ where m_0 and m_1 are order 10^{15} GeV or larger). There are two ways out of this problem which involve new physics on the scale of 1 TeV. New strong dynamics could enter that provides the scale of m_W, or new particles could appear so that the larger scale is still possible, but the divergences are cancelled on a much smaller scale. In any of the options, Standard Model, new dynamics or cancellations, the energy scale is the same; something must be discovered at the TeV scale.

Supersymmetry [4] is an appealing concept for which there is so far no experimental evidence. It offers the only presently known mechanism for incorporating gravity into the quantum theory of particle interactions and provides an elegant cancellation mechanism for the divergences, provided that at the electroweak scale the theory is supersymmetric. The successes of the Standard Model (such as precision electroweak predictions) are retained, while avoiding any fine tuning of the Higgs mass. Some supersymmetric models allow for the unification of gauge couplings at a high scale and a consequent reduction of the number of arbitrary parameters.

Supersymmetric models postulate the existence of superpartners for all the presently observed particles: bosonic superpartners of fermions (squarks and sleptons), and fermionic superpartners of bosons (gluinos and gauginos). There are also multiple Higgs bosons: h, H, A and $H^{\pm}$. There is thus a large spectrum of presently unobserved particles, whose exact masses, couplings and decay chains are calculable in the theory given certain parameters. Unfortunately these parameters are unknown. Nonetheless, if supersymmetry is to have anything to do with electroweak symmetry breaking, the masses should be in the region below or order of 1 TeV.

An example of the strong coupling scenario is 'technicolour' for models based on dynamical symmetry breaking [5]. Again, if the dynamics is to have anything to do with electroweak symmetry breaking we would expect new states in the region below 1 TeV; most models predict a large spectrum of such states. An elegant implementation of this appealing idea is lacking. However, all models predict structure in the WW scattering amplitude at around 1 TeV centre-of-mass energy.

There are also other possibilities for new physics that are not necessarily related to the scale of electroweak symmetry breaking. There could be new neutral or charged gauge bosons with mass larger than the Z and W; there could be new quarks, charged leptons or massive neutrinos, or quarks and leptons could turn out not to be elementary objects. While we have no definitive expectations for the masses of these objects, the LHC experiments must be able to search for them over the available energy range.

2 Motivation and Chalanges

In the initial phase at low luminosity, the experiment will function as a factory for QCD processes, heavy flavour and gauge bosons production. This will allow a large number of precision measurements in the early stages of the experiment.

A large variety of QCD related processes will be studied. These measurements are of importance as studies of QCD 'per se' in a new energy regime with high statistics. Of particular interest will be jet and photon physics, open charm and beauty production and gauge bosons production. A study of diffractive processes will present significant experimental challenges itself, given the limited angular coverage of the ATLAS detector. Several aspects of diffractive production of jets, gauge bosons, heavy flavour partons will be nevertheless studied in detail. LHC will extend the exploration of the hard partonic processes to large energy scales (of few hundred GeV^2), while reaching small fractional momentum of the proton being carried by a scattered partons. Precise constraints on the partonic distribution functions will be derived from measurements of Drell-Yan production, of W and Z bosons production, of production of direct photons and high-p_T jets, heavy flavours and gauge boson pairs. Deviation from the theoretical predictions for QCD processes themselves might indicate the onset of new physics, such as compositeness. Measurement and understanding of these QCD processes will be essential as they form the dominant background searches for new phenomena.

Even at low luminosity, LHC is a beauty factory with 10^{12} $b\bar{b}$ expected per year. The available statistics will be limited only by the rate at which data can be recorded. The proposed B-physics programme is therefore very wide. Specific B-physics topics include the search for and measurement of CP violation, of B_s^0 mixing and of rare decays. ATLAS can perform competitive high-accuracy measurements of B_s^0 mixing, covering the statistically preferred range of the Standard Model predictions. Rare B mesons such as B_c will be copiously produced at LHC. The study of B-baryon decay dynamics and spectroscopy of rare B hadrons will be also carried out.

LHC has a great potential for performing high precision top physics measurements with about eight million $t\bar{t}$ pairs expected to be produced for an integrated luminosity of $10 fb^{-1}$. It would allow not only for the precise measurements of the top-quark mass (with a precision of $\approx$2 GeV) but also for the detailed study of properties of the top-quark itself. The single top production should be observable and the high statistics will allow searches for many rare top decays. The precise knowledge of the top-quark mass places strong constraints on the mass of the Standard Model Higgs boson, while a detailed study of its properties may reveal as well new physics.

One of the challenges to the LHC experiments will be whether the precision of the W-mass measurement can be improved. Given the 300 million single W events expected in one year of data taking, the expected statistical uncertainty will be about 2 MeV. The very ambitious goal for both theory and experiment is to reduce the individual sources of systematic errors to less than 10 MeV, which would allow for the measurement of the W mass with precision of better than 20 MeV. This would ensure that the precision of the W mass is not the dominant source of errors in testing radiative corrections in the SM prediction for the Higgs mass.

The large rate of gauge boson pair production at the LHC enables ATLAS to provide critical tests of the triple gauge-boson couplings. The gauge cancellations predicted by the Standard Model will be studied and measurements of possible anomalous couplings made. These probe underlying non-standard physics. The most sensitive variables to compare with Standard Model predictions are the transverse momentum spectra of high-p_T photons or reconstructed Z bosons.

If the Higgs boson is not discovered before LHC begins operation, the searches for it and its possible supersymmetric extensions in the Minimal Supersymmetric Standard Model (MSSM) will be a main focus of activity. Search strategies presented here explore a variety of possible signatures, being accessible already at low luminosity or only at design luminosity. Although the cleanest one would lead to reconstruction of narrow mass peaks in the photonic or leptonic decay channels, very promising are the signatures which lead to multi-jet or multi-τ final states. In several cases signal-to-background ratios much smaller than one are expected, and in most cases detection of the Higgs boson will provide an experimental challenge. Nevertheless, the ATLAS experiment alone will cover the full mass range up to 1 TeV for the SM Higgs and also the full parameter space for the MSSM Higgs scenarios. It has also a large potential for searches in alternative scenarios.

Discovering SUSY at the LHC will be straightforward if it exists at the electroweak scale. Copious production of squarks and gluinos can be expected, since the cross-section should be as large as a few pb for squarks and gluinos as heavy as 1 TeV. Their cascade decays would lead to a variety of signatures involving multi-jets, leptons, photons, heavy flavours and missing energy. In several models, discussed in detail in this volume, the precision measurement of the masses of SUSY particles and the determination of the model parameters will be possible. The main challenge would be therefore not to discover SUSY itself, but to reveal its nature and determine the underlying SUSY model.

3 The ATLAS Detector

3.1 Introduction

The Large Hadron Collider (LHC) is a proton-proton collider with 14 TeV centre of mass energy and design luminosity of $10^{34} cm^{-2} s^{-1}$. Beam crossings are 25 ns apart and at design luminosity there are 23 interactions per crossing [6].

The ATLAS experiment has now entered the construction phase for many of its detector components, with a strict schedule to meet the first collisions at LHC in summer 2006.

3.2 Detector Concept

A broad spectrum of detailed physics studies led to the overall detector concept presented in the ATLAS Technical Proposal [7]. The basic design criteria of the detector include the following.

- Very good electromagnetic calorimetry for electron and photon identification and measurements, complemented by full-coverage hadronic calorimetry for accurate jet and missing transverse energy (E_T^{miss}) measurements;
- High-precision muon momentum measurements, with the capability to guarantee accurate measurements at the highest luminosity using the external muon spectrometer alone;
- Efficient tracking at high luminosity for high-p_T lepton-momentum measurements, electron and photon identification, τ lepton and heavy-flavour identification, and full event reconstruction capability at lower luminosity;
- Large acceptance in pseudorapidity (η) with almost full azimuthal angle (ϕ) coverage everywhere. The azimuthal angle is measured around the beam axis, whereas pseudorapidity relates to the polar angle (θ) where θ is the angle from the z direction.
- Triggering and measurements of particles at low-p_T thresholds, providing high efficiencies for most physics processes of interest at LHC.

The overall detector layout is shown in Fig. 1. The magnet configuration is based on an inner thin superconducting solenoid surrounding the inner detector cavity, and large superconducting air-core toroids consisting of independent coils arranged with an eight-fold symmetry outside the calorimeters.

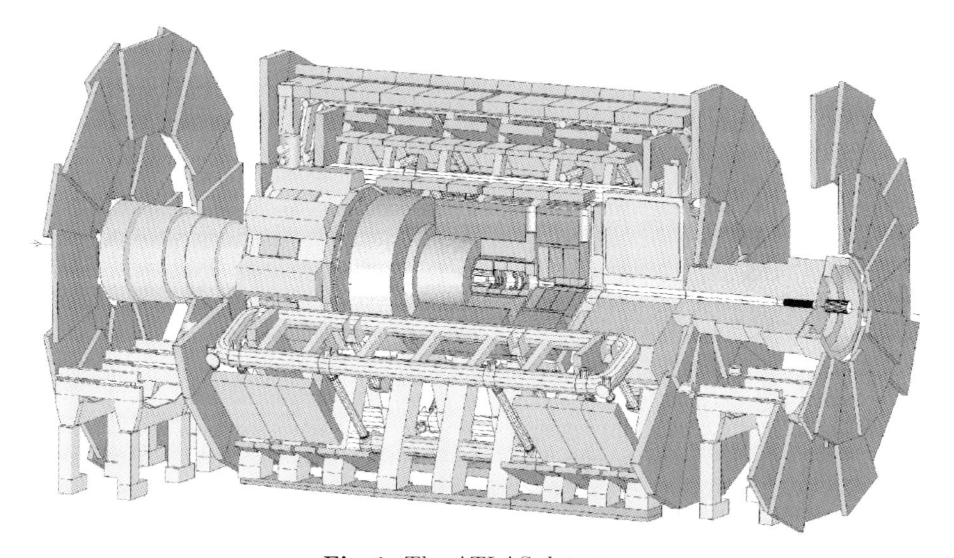

Fig. 1. The ATLAS detector

The Inner Detector (ID) is contained within a cylinder of length 7 m and a radius of 1.15 m, in a solenoidal magnetic field of 2 T. Pattern recognition, momentum and vertex measurements, and electron identification are achieved with a combination of discrete high-resolution semiconductor pixel and strip detectors in the inner part of the tracking volume, and continuous straw-tube tracking detectors with transition radiation capability in its outer part.

Highly granular liquid-argon (LAr) electromagnetic (EM) sampling calorimetry, with excellent performance in terms of energy and position resolution, covers the pseudorapidity range $|\eta| < 3.2$. In the end-caps, the LAr technology is also used for the hadronic calorimeters, which share the cryostats with the EM end-caps. The same cryostats also house the special LAr forward calorimeters which extend the pseudorapidity coverage to$|\eta| = 4.9$ The bulk of the hadronic calorimetry is provided by a novel scintillator-tile calorimeter, which is separated into a large barrel and two smaller extended barrel cylinders, one on each side of the barrel. The overall calorimeter system provides the very good jet and E_T^{miss} performance of the detector.

The LAr calorimetry is contained in a cylinder with an outer radius of 2.25 m and extends longitudinally to ±6.65 m along the beam axis. The outer radius of the scintillator-tile calorimeter is 4.25 m and its half length is 6.10 m. The total weight of the calorimeter system, including the solenoid flux-return iron yoke which is integrated into the tile calorimeter support structure, is about 4000 Tons.

The calorimeter is surrounded by the muon spectrometer. The air-core toroid system, with a long barrel and two inserted end-cap magnets, generates a large magnetic field volume with strong bending power within a light and open structure. Multiple-scattering effects are thereby minimised, and excellent muon momentum resolution is achieved with three stations of high- precision tracking chambers. The muon instrumentation also includes as a key component trigger chambers with very fast time response.

The muon spectrometer defines the overall dimensions of the ATLAS detector. The outer chambers of the barrel are at a radius of about 11 m. The half-length of the barrel toroid coils is 12.5 m, and the third layer of the forward muon chambers, mounted on the cavern wall, is located about 23 m from the interaction point. The overall weight of the ATLAS detector is about 7000 Tons.

The primary goal of the experiment is to operate at high luminosity with a detector that provides as many signatures as possible. The variety of signatures is considered to be important in the harsh environment of the LHC in order to achieve robust and redundant physics measurements with the ability of internal cross-check. The measurement of the luminosity itself will be a challenge. Precision measurements employing the total and elastic cross-sections require specialised detectors. A measurement with a precision of 5% to 10% may be obtained from the machine parameters.

3.3 Inner Detector

The Inner Detector (ID) [8] combines high-resolution detectors at the inner radii with continuous tracking elements at the outer radii, all contained in the CS which provides a nominal magnetic field of 2 T.

The momentum and vertex resolution requirements from physics call for high-precision measurements to be made with fine-granularity detectors, given the very large track density expected at the LHC. Semiconductor tracking detectors, using silicon microstrip (SCT) [9] and pixel [10] technologies offer these features. The highest granularity is achieved around the vertex region using semi-conductor pixel detectors. The total number of precision layers must be limited because of the material they introduce, and because of their high cost. Typically, three pixel layers and eight strip layers (four space points) are crossed by each track. A large number of tracking points (typically 36 per track) is provided by the straw tube tracker (TRT) [9], which provides continuous track-following with much less material per point and a lower cost. The combination of the two techniques gives very robust pattern recognition and high precision in both ϕ and z coordinates. The straw hits at the outer radius contribute significantly to the momentum measurement, since the lower precision per point compared to the silicon is compensated by the large number of measurements and the higher average radius. The relative precision of the different measurements is well matched, so that no single measurement dominates the momentum resolution. This implies that the overall performance is robust. The high density of measurements in the outer part of the tracker is also valuable for the detection of photon conversions and of V0 decays. The latter are an important element in the signature of CP violation in the B system. In addition, the electron identification capabilities of the whole experiment are enhanced by the detection of transition-radiation photons in the xenon-based gas mixture of the straw tubes.

The outer radius of the ID cavity is 115 cm, fixed by the inner dimension of the cryostat containing the LAr EM calorimeter, and the total length is 7 m, limited by the position of the end-cap calorimeters. Mechanically, the ID consists of three units: a barrel part extending over $\pm$ 80 cm, and two identical end-caps covering the rest of the cylindrical cavity. The precision tracking elements are contained within a radius of 56 cm, followed by the continuous tracking, and finally the general support and service region at the outermost radius. In order to give uniform coverage over the full acceptance, the final TRT wheels at high z extend inwards to a lower radius than the other TRT end-cap wheels.

In the barrel region, the high-precision detector layers are arranged on concentric cylinders around the beam axis, while the end-cap detectors are mounted on disks perpendicular to the beam axis. The pixel layers are segmented in R and z, while the SCT detector uses small angle (40 mrad) stereo strips to measure both coordinates, with one set of strips in each layer measuring . The barrel TRT straws are parallel to the beam direction. All the end-cap tracking elements are located in planes perpendicular to the beam axis. The strip detectors have one set of strips running radially and a set of stereo strips at an angle of 40 mrad. The continuous tracking consists of radial straws arranged into wheels. The layout provides full tracking coverage over $|\eta| < 2.5$, including impact parameter measurements and vertexing for heavy-flavour and tagging. The secondary vertex measurement performance is enhanced by the innermost layer of pixels, at a radius of about 4 cm, as close as is practical to the beam pipe. The lifetime of such a detector will be limited by radiation damage, and may need replacement after a few years, the exact time depending on the luminosity profile. A large amount of interesting physics can be done with this detector during the initial lower-luminosity running, especially in the B sector, but physics studies have demonstrated the value of good b-tagging performance during all phases of the LHC operation, for example in the case of Higgs and supersymmetry searches. It is therefore considered very important that this innermost pixel layer (or B-layer) can be replaced to maintain the highest possible performance throughout the experiment's lifetime. The mechanical design of the pixel system allows the possibility of replacing the B-layer.

3.4 Calorimeters

The ATLAS calorimeters [11] consists of an electromagnetic (EM) calorimeter covering the pseudorapidity region $|\eta| < 3.2$, a hadronic barrel calorimeter covering $|\eta| < 1.7$, hadronic end-cap calorimeters covering $1.5 < |\eta| < 3.2$, and forward calorimeters covering $3.1 < |\eta| < 4.9$. The EM calorimeter is a lead/liquid-argon (LAr) detector with accordion geometry [12]. Over the pseudorapidity range $|\eta| < 1.8$, it is preceded by a presampler detector, installed immediately behind the cryostat cold wall, and used to correct for the energy lost in the material (ID, cryostats, coil) upstream of the calorimeter.

The hadronic barrel calorimeter is a cylinder divided into three sections: the central barrel and two identical extended barrels. It is based on a sampling technique with plastic scintillator plates (tiles) embedded in an iron absorber [13]. At larger pseudorapidities, where higher ra- diation resistance is needed, the intrinsically radiation-hard LAr technology is used for all the calorimeters [12]: the hadronic end-cap calorimeter, a copper LAr detector with parallel-plate geometry, and the forward calorimeter, a dense LAr calorimeter with rod-shaped electrodes in a tungsten matrix.

The barrel EM calorimeter is contained in a barrel cryostat, which surrounds the Inner Detector cavity. The solenoid which supplies the 2 T magnetic field to the Inner Detector is integrated into the vacuum of the barrel cryostat and is placed in front of the EM calorimeter. Two end-cap cryostats house the end-cap EM and hadronic calorimeters, as well as the integrated forward calorimeter. The barrel and extended barrel tile calorimeters support the LAr cryostats and also act as the main solenoid flux return.

The approximately 200,000 signals from the LAr calorimeters leave the cryostats through cold- to-warm feedthroughs located between the barrel and the extended barrel tile calorimeters, and at the back of each end-cap. The electronics up to the digitisation stage will be contained in radial boxes attached to each feedthrough and located in the vertical gaps between the barrel and extended barrel tile calorimeters.

Electromagnetic Calorimeter. The EM calorimeter [12] is divided into a barrel part ($|\eta| < 1.475$) and two end-caps ($1.375 < |\eta| < 3.2$). The barrel calorimeter consists of two identical half-barrels, separated by a small gap (6 mm) at z = 0. Each end-cap calorimeter is mechanically divided into two coaxial wheels: an outer wheel covering the region $1.375 < |\eta| < 2.5$, and an inner wheel covering the region $2.5 < |\eta| < 3.2$.

The EM calorimeter is a lead LAr detector with accordion-shaped Kapton electrodes and lead absorber plates over its full coverage. The accordion geometry provides complete symmetry without azimuthal cracks. The lead thickness in the absorber plates has been optimised as a function of η in terms of EM calorimeter performance in energy resolution. The LAr gap has a constant thickness of 2.1 mm in the barrel. In the end-cap, the shape of the Kapton electrodes and lead converter plates is more complicated, because the amplitude of the accordion waves increases with radius. The absorbers have constant thickness, and therefore the LAr gap also increases with radius. The total thickness of the EM calorimeter is > 24 radiation lengths (X_0) in the barrel and $> 26X_0$ in the end-caps. Over the region devoted to precision physics ($|\eta| < 2.5$), the EM calorimeter is segmented into three longitudinal sections. The strip section, which has a constant thickness of $\approx$6 X_0 (upstream material included) as a function of η , is equipped with narrow strips with a pitch of 4 mm in the direction. This section acts as a 'preshower' detector, enhancing particle identification and providing a precise position measurement in . The middle section is transversally segmented into square towers of size $\Delta\eta \times \Delta\phi = 0.025 \times 0.025$. The total calorimeter thickness up to the end of the second section is 24 X_0, tapered with increasing rapidity (this includes also the upstream material). The back section has a granularity of 0.05 in η and a thickness varying between 2 X_0 and 12 X_0. For $|\eta| > 2.5$, i.e. for the end-cap inner wheel, the calorimeter is segmented in two longitudinal sections and has a coarser lateral granularity than for the rest of the acceptance. This is sufficient to satisfy the physics requirements (reconstruction of jets and measurement of E_T^{miss}). The calorimeter cells point towards the interaction region over the complete coverage. The total number of channels is $\approx$ 190 000.

The total material seen by an incident particle before the calorimeter front face is about 2.3 X_0 at $\eta = 0$, and increases with pseudorapidity in the barrel because of the particle angle. In the region where the amount of material exceeds 2 X_0, a presampler is used to correct for the energy lost by electrons and photons upstream of the calorimeter. The presampler consists of an active LAr layer of thickness 1.1 cm (0.5 cm) in the barrel (end-cap) region. At the transition between the barrel and the end-cap calorimeters, i.e. at the boundary between the two cryostats, the amount of material in front of the calorimeter reaches a localised maximum of about 7 X_0. In this region, the presampler is complemented by a scintillator slab inserted in the crack between the barrel and end-cap cryostats and covering the region $1.0 < |\eta| < 1.6$. The region $1.37 < |\eta| < 1.52$ is not used for precision physics measurements involving photons because of the large amount of material situated in front of the EM calorimeter.

Hadronic Calorimeters. The ATLAS hadronic calorimeters cover the range $|\eta| < 4.9$ using different techniques best suited for the widely varying requirements and radiation environment over the large range. Over the range $|\eta| < 1.7$, the iron scintillating-tile technique is used for the barrel and extended barrel tile calorimeters and for partially instrumenting the gap between them with the intermediate tile calorimeter (ITC). This gap provides space for cables and services from the innermost detectors. Over the range $1.5 < |\eta| < 4.9$, LAr calorimeters were chosen: the hadronic end-cap calorimeter (HEC) extends to $|\eta| < 3.2$, while the range $3.1 < |\eta| < 4.9$ is covered by the high-density forward calorimeter (FCAL). Both the HEC and the FCAL are integrated in the same cryostat as that housing the EM end-caps.

An important parameter in the design of the hadronic calorimeter is its thickness: it has to provide good containment for hadronic showers and reduce punch-through into the muon system to a minimum. The total thickness is 11 interaction lengths (λ) at $\eta = 0$, including about 1.5 λ from the outer support, which has been shown both by measurements and simulation to be sufficient to reduce the punch-through well below the irreducible level of prompt or decay muons. Close to 10 λ of active calorimeter are adequate to provide good resolution for high energy jets. Together with the large coverage, this will also guarantee a good E_T^{miss} measurement, which is important for many physics signatures and in particular for SUSY particle searches.

3.5 Muon Spectrometer

The muon spectrometer [14] is based on the magnetic deflection of muon tracks in the large superconducting air-core toroid magnets, instrumented with separate trigger and high-precision tracking chambers. Over the range $|\eta| < 1.0$, magnetic bending is provided by the large barrel toroid. For $1.4 < |\eta| < 2.7$, muon tracks are bent by two smaller end-cap magnets inserted into both ends of the barrel toroid. Over $1.0 < |\eta| < 1.4$, usually referred to as the transition region, magnetic deflection is provided by a combination of barrel and end-cap fields. This magnet configuration provides a field that is mostly orthogonal to the muon trajectories, while minimising the degradation of resolution due to multiple scattering.

The anticipated high level of particle fluxes has had a major impact on the choice and design of the spectrometer instrumentation, affecting required performance parameters such as rate capability, granularity, ageing properties and radiation hardness. Trigger and reconstruction algorithms have been optimised to cope with the difficult background conditions resulting from penetrating primary collision products and from radiation backgrounds, mostly neutrons and photons in the 1 MeV range, produced from secondary interactions in the calorimeters, shielding material, beam pipe and LHC machine elements.

In the barrel region, tracks are measured in chambers arranged in three cylindrical layers ('stations') around the beam axis; in the transition and end-cap regions, the chambers are installed vertically, also in three stations. Over most of the range, a precision measurement of the track coordinates in the principal bending direction of the magnetic field is provided by Monitored Drift Tubes (MDTs). At large pseudorapidities and close to the interaction point, Cathode Strip Chambers (CSCs) with higher granularity are used in the innermost plane over $2 < |\eta| < 2.7$, to withstand the demanding rate and background conditions. Optical alignment systems have been designed to meet the stringent requirements on the mechanical accuracy and the survey of the precision chambers.

The precision measurement of the muon tracks is made in the R–z projection, in a direction parallel to the bending direction of the magnetic field; the axial coordinate (z) is measured in the barrel and the radial coordinate (R) in the transition and end-cap regions. The MDTs provide a single-wire resolution of 80 μm when operated at high gas pressure (3 bar) together with robust and reliable operation thanks to the mechanical isolation of each sense wire from its neighbours. The construction of prototypes has demonstrated that the MDTs can be built to the required mechanical accuracy of 30 μm.

The trigger system covers the pseudorapidity range $|\eta| < 2.4$. Resistive Plate Chambers (RPCs) are used in the barrel and Thin Gap Chambers (TGCs) in the end-cap regions. The trigger chambers for the ATLAS muon spectrometer serve a threefold purpose:

- bunch crossing identification, requiring a time resolution better than the LHC bunch spacing of 25 ns;
- a trigger with well-defined pT cut-offs in moderate magnetic fields, requiring a granularity of the order of 1 cm;
- measurement of the second coordinate in a direction orthogonal to that measured by the precision chambers, with a typical resolution of 5–10 mm.

Muon Chamber Layout. The overall layout of the muon chambers in the ATLAS detector indicates the different regions in which the four chamber technologies described above are employed. The chambers are arranged such that particles from the interaction point traverse three stations of chambers. The positions of these stations are optimised for essentially full coverage and momentum resolution. In the barrel, particles are measured near the inner and outer field boundaries, and inside the field volume, in order to determine the momentum from the sagitta of the trajectory. In the end-cap regions, for $|\eta| > 1.4$, the magnet cryostats do not allow the positioning of chambers inside the field volume. Instead, the chambers are arranged to determine the momentum with the best possible resolution from a point-angle measurement (this is also the case in the barrel region in the vicinity of the coils).

The barrel chambers form three cylinders concentric with the beam axis, at radii of about 5, 7.5, and 10 m. They cover the pseudorapidity range $|\eta| < 1$. The end-cap chambers cover the range $1 < |\eta| < 2.7$ and are arranged in four disks at distances of 7, 10, 14, and 21–23 m from the interaction point, concentric with the beam axis. All chambers combined provide almost complete coverage of the pseudorapidity range $1.0 < |\eta| < 2.7$. There is an opening in the central R–ϕ plane ($\eta = 0$) for the passage of cables and services of the ID, the CS, and the calorimeters.

In the barrel, the chambers are arranged in projective towers. Particles are measured in 2×4 sensitive layers in the inner station and in $2{\times}3$ layers each in the middle and outer stations. Within a projective tower, the chambers are optically connected by alignment rays which monitor the relative chamber positions. A different alignment strategy is used in the end-caps, where the positions of complete chamber planes are monitored. No active repositioning of the chambers is foreseen.

Both in the barrel and the end-caps, a 16-fold segmentation in azimuth has been chosen that follows the eightfold azimuthal symmetry of the magnet structure. The chambers are arranged in large and small sectors. The large sectors cover the regions between the BT coils and the small sectors the azimuthal range around the BT coils. In two lower barrel sectors, the rails carrying the calorimeter and their feet require specially shaped chambers to maximise the detector acceptance.

The trigger function in the barrel is provided by three stations of RPCs. They are located on both sides of the middle MDT station, and directly inside the outer MDT station. In the end-caps, the trigger is provided by three stations of TGCs located near the middle MDT station.

Maximum standardisation and the smallest possible number of different chamber sizes have also been important goals of the detector layout. The barrel chambers are of rectangular shape with areas of 2–10 m^2. The end-cap chambers are of trapezoidal shape ('staircase' approximation) with tapering angles of 8.5^0 and 14^0 for the small and large chambers, respectively. Their areas range from 1–10 m^2 for individual chamber modules and up to 30 m^2 when several of them are preassembled for installation.

4 Summary

A large verity of tests to the SM in it's most critical energy region and an even larger potential for discovery of it's missing pieces is planned for the ATLAS detector. The construction of the detectors components is underway assuring that the needed accuracy to complete the above task is achieved. The future of high energy physics is guaranteed to be an exciting one.

Acknowledgment

The Author would like to thank the VolkswagenStiftung and the organizers for a great gathering in a beautiful setting. In particular I would like to thank J. Trampetic and I. Biljan for helping me getting back home in those worldshaking days.

References

1. L. Maiani, G. Parisi and R. Petronzio, Nucl. Phys. B **136**, 115 (1979); N. Cabbibo, L. Maiani, G. Parisi and R. Petronzio, Nucl. Phys. B **158**, 295 (1979); R. Dashen and H. Neuberger, Phys. Rev. Lett. **50**, 1897 (1983); D. J. E. Callaway, Nucl. Phys. B **233**, 189 (1984); M. A. Beg, C. Panagiatakopolus and A. Sirlin, Phys. Rev. Lett **52**, 883 (1984); M. Lindner, Z. Phys. C **31**, 295 (1986).
2. T. Hambye and K. Riesselmann, Phys. Rev. D **55**, 7255 (1997).

3. C. Quigg, B.W. Lee and H. Thacker, Phys. Rev. D **16**, 1519 (1977); M. Veltman, Acta Phys. Polon. B **8**, 475 (1977).
4. J. Wess and B. Zumino, Nucl. Phys. B **70**, 39 (1974).
5. For a review, see K.D. Lane, hep-9605257.
6. 'LHC White Book', CERN/AC/93-03; 'LHC Conceptual Design Report, CERN/AC/95-05.
7. ATLAS Collaboration, Technical Proposal for a General Purpose pp Experiment at the Large Hadron Collider at CERN, CERN/LHCC/94-43, LHCC/P2, 15 December 1994.
8. ATLAS Collaboration, Inner Detector Technical Design Report, Volume 1, CERN/LHCC/97-16, 30 April 1997.
9. ATLAS Collaboration, Inner Detector Technical Design Report, Volume 2, CERN/LHCC/97-17, 30 April 1997.
10. ATLAS Collaboration, Pixel Detector Technical Design Report, CERN/LHCC/98-13, 31 May 1998.
11. ATLAS Collaboration, Calorimeter Performance Technical Design Report, CERN/LHCC/96-40, 15 December 1996.
12. ATLAS Collaboration, Liquid Argon Calorimeter Technical Design Report, CERN/LHCC/96-41, 15 December 1996.
13. ATLAS Collaboration, Tile Calorimeter Technical Design Report, CERN/LHCC/96-42, 15 December 1996.
14. ATLAS Collaboration, Muon Spectrometer Technical Design Report, CERN/LHCC/97-22, 31 May 1997.

Flavour Oscillation and CP Violation: Experimental Results on B Mesons

Roland Waldi

Universität Rostock, D-18051 Rostock, Germany

Abstract. The field of flavour oscillation and CP violation of B mesons has received fresh input from the new B meson factories and is rapidly changing. We see a clear support for the ideas of Kobayashi and Maskawa on the origin of CP violation in the Standard Model, fitting both K and B decays. Many crosschecks will be possible of which first hints are already showing up today, and we can still hope for surprises in the near future.

1 Introduction

Flavour oscillation and CP violation of K mesons is known since many decades and is nowadays textbook physics. The observation of similar effects with quantitatively different realisation at B mesons is a rather young field, and CP violation of B mesons has been established for the first time only in the year of this symposium.

This second CP violating system has beautifully confirmed the explanation of CP violation by Kobayashi and Maskawa [1] through the physical phase in their 3×3 flavour mixing matrix

$$\mathbf{V} = \begin{pmatrix} V_{ud} & V_{us} & V_{ub} \\ V_{cd} & V_{cs} & V_{cb} \\ V_{td} & V_{ts} & V_{tb} \end{pmatrix} \approx \begin{pmatrix} 1-\frac{\lambda^2}{2} & \lambda & A\lambda^3(\rho - i\eta) \\ -\lambda & 1-\frac{\lambda^2}{2} & A\lambda^2 \\ A\lambda^3(1-\rho-i\eta) & -A\lambda^2 & 1 \end{pmatrix}$$

that occurs in charged current weak interactions and has its origin in the transition from massless to massive quark fields. While 5 of 6 phases of this unitary matrix can be absorbed in unobservable phases of the quarks, one phase cannot be eliminated and gives rise to CP violating interference effects.

At the same time, the fact that transitions between all pairs of up-type and down-type quarks are allowed leads to oscillation of several neutral mesons into their antiparticles. This flavour oscillation is observed in K^0 and B^0 mesons, and established, though not quantitatively, in B_s mesons, and is a prerequisite for large CP violating asymmetries. The available results have been recently augmented by the B factories at SLAC and KEK, motivating a new review on the topic.

2 Production of B Mesons

After the discovery of the Υ states, B meson properties have been investigated since the mid-80s at e^+e^- storage rings operating at the $\Upsilon(4S)$ resonance: DORIS

II at DESY (Gemany), with the experiment ARGUS, and CESR at Cornell (USA) with the detector CLEO. ARGUS had collected 0.25 Million $B\overline{B}$ events when it stopped in 1992, CLEO has accumulated 10 Million $B\overline{B}$ events by the end of 2000. These experiments have the advantage to investigate events with nothing else but two B mesons, which are even almost at rest since the mass of the $\Upsilon(4S)$ is only 20MeV above the $B\overline{B}$ threshold.

This source is also exploited by the asymmetric e^+e^- colliders PEP-II at SLAC (USA) and KEK-B at KEK (Japan), where the experiments BABAR and BELLE have started taking data in 1999. They both have reached record luminosities above $3 \cdot 10^{33}/\text{cm}^2/\text{s}$, and have collected 23 and 11 Million $B\overline{B}$ pairs by the end of 2000, respectively. Their data samples are increasing rapidly, and they are aiming both for over 100 Million at the end of 2002.

These B factories have different electron and positron energies to produce the $\Upsilon(4S)$ with a boost of $\beta\gamma = 0.55$ (BABAR) and 0.42 (BELLE) in order to measure the difference of the lifetime of the two B mesons. This is an essential information for the observation of time-dependent CP asymmetries, as will be discussed below.

In the 1990s the four experiments ALEPH, DELPHI, L3 and OPAL at the LEP storage ring at CERN (Switzerland) started investigating $b\overline{B}$ jets from Z^0 decays. They have each a sample of almost one Million $b\overline{B}$ events. They were joined by SLD which accumulated polarized Z^0 events at the linear collider SLC at SLAC (USA).

Hadronic production of $b\overline{B}$ jets in addition to the fragments of the original particles are the source of B mesons at the $p\bar{p}$ storage ring Tevatron at Fermilab (USA). Hadronic production of $b\overline{B}X$ at high energies is orders of magnitude higher than any other source, but the samples of triggered and detected events being only a small fraction, the exploitation of these vast amounts of data is a challenge, which has been met in the past by the CDF detector which was the first experiment to collect enough $B^0 \to J/\psi K^0_S$ decays for a meaningful exclusive CP violation analysis. Both CDF and D0 will start collecting new data this year. Hadronic production will also be the source of B mesons at the planned experiments ATLAS, CMS and the dedicated experiment LHCb at the LHC pp storage ring, and the BTeV experiment at the Tevatron. These experiments will ultimately deliver enough $b\overline{B}X$ events for high precision measurements of CP violation parameters that can be expected about ten years from now.

3 $B\overline{B}$ Oscillation

B^0 and B_s mesons are oscillating into their anti-particles much like neutral kaons do, but with different parameters. Starting with pure B^0 mesons at proper lifetime $t = 0$ is described by a wave function

$$|\psi(t)\rangle = e^{-imt-T/2}\left[\cos(x-iy)\frac{T}{2}|B^0\rangle + i\frac{q}{p}\sin(x-iy)\frac{T}{2}|\overline{B}^0\rangle\right] \qquad (1)$$

Here $p, \pm q$ are the B^0 and $\overline{B}^0$ components of the mass eigenstates, and we use dimensionless variables, $T = \Gamma t = t/\tau$ and

$$x = \frac{\Delta m}{\Gamma}, \quad y = \frac{\Delta\Gamma}{2\Gamma} = \frac{\Gamma_H - \Gamma_L}{\Gamma_H + \Gamma_L} = \frac{\Gamma - \Gamma_L}{\Gamma} = \frac{\tau_L - \tau_H}{\tau_L + \tau_H}$$

The angular oscillation frequency is Δm, the difference of the masses of the two eigenstates B_H and B_L of the Hamiltonian for the two-particle system $B^0, \overline{B}^0$. The same parameters describe the other oscillating mesons, and experimental or, lacking those, theoretically expected values of these parameteres are given in Table 1.

Table 1. Parameters of the four neutral oscillating meson pairs.

	$K^0/\overline{K}^0$	$D^0/\overline{D}^0$	$B^0/\overline{B}^0$	$B_s/\overline{B}_s$
τ [ps]	89.4 ± 0.1; 51700 ± 400	$0.413 \pm .003$	1.548 ± 0.021	1.49 ± 0.06
Γ [s^{-1}]	$5.61 \cdot 10^9$	$2.4 \cdot 10^{12}$	$(6.41 \pm 0.16) \cdot 10^{11}$	$(6.7 \pm 0.3) \cdot 10^{11}$
$y = \frac{\Delta\Gamma}{2\Gamma}$	-0.9966	$\lvert y\rvert < 0.06$	$\lvert y\rvert \lesssim 0.01^*$	$-(0.01 \ldots 0.10)^*$
Δm [s^{-1}]	$(5.300 \pm 0.012) \cdot 10^9$	$< 7 \cdot 10^{10}$	$(4.89 \pm 0.09) \cdot 10^{11}$	$> 15 \cdot 10^{12}$
Δm [eV]	$(3.49 \pm 0.01) \cdot 10^{-6}$	$< 5 \cdot 10^{-6}$	$(3.2 \pm 0.1) \cdot 10^{-4}$	$> 1.0 \cdot 10^{-2}$
$x = \frac{\Delta m}{\Gamma}$	0.945 ± 0.002	< 0.03	0.76 ± 0.02	$21 \ldots 40^*$
δ_ϵ	$(3.27 \pm 0.12) \cdot 10^{-3}$		$\sim -10^{-3\,*}$	$\lvert\delta_\epsilon\rvert < 10^{-3\,*}$
$\left\lvert\frac{q}{p}\right\rvert^2$	0.99348 ± 0.00024	$\approx 1^*$	$1 \ldots 1.002^*$	$\approx 1^*$

* Standard Model expectation

Oscillation can be described by a time-dependent asymmetry function

$$\begin{aligned} a(T) &= 1 - 2\chi(T) \\ &= \left.\frac{\dot{N}(X \to X) - \dot{N}(X \to \overline{X})}{\dot{N}(X \to X) + \dot{N}(X \to \overline{X})}\right|_T \\ &= \frac{(1 - |\frac{q}{p}|^2)\cosh yT + (1 + |\frac{q}{p}|^2)\cos xT}{(1 + |\frac{q}{p}|^2)\cosh yT + (1 - |\frac{q}{p}|^2)\cos xT} \\ &\approx \frac{\cos xT}{\cosh yT} \end{aligned} \tag{2}$$

where $X \to \overline{X}$ is shorthand for X at production time $T = 0$ and $\overline{X}$ at decay time T etc. Neglecting CP violation in the oscillation which is proportional to

$$\delta_\epsilon = \langle B_H | B_L \rangle = |p|^2 - |q|^2 = \frac{1 - |\frac{q}{p}|^2}{1 + |\frac{q}{p}|^2} = \frac{2\,\mathcal{R}e\,\epsilon}{1 + |\epsilon|^2}$$

and very small for all four oscillating mesons, we have $|\frac{q}{p}| = 1$ and the asymmetry reduces to the form shown on the last line of (2). The lifetime difference is large

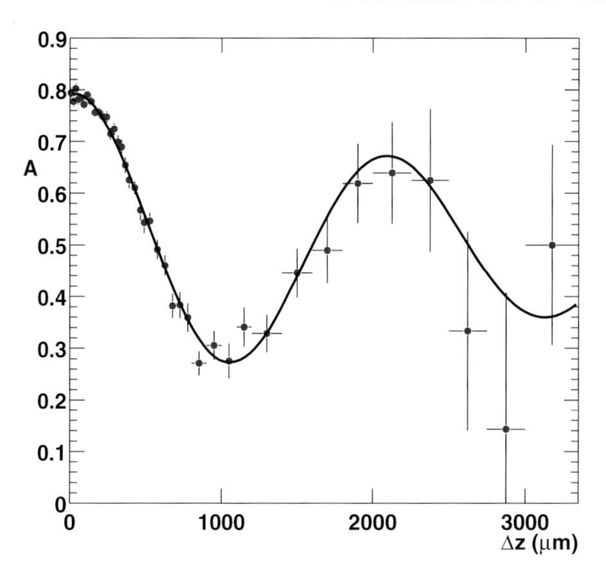

Fig. 1. The $B^0\overline{B}^0$ oscillation asymmetry seen at the PEP II B factory with the BABAR detector [3] exhibits a new quality of time-resolved measurements.

for neutral K mesons, and expected to be observable for the B_s meson, but very small for B^0 mesons. Therefore, for the latter, $y = 0$ and the asymmetry is simply

$$a(T) = \cos xT \tag{3}$$

which has first been observed as a time-integrated effect in 1987 by ARGUS [2]. The time-dependent asymmetry has been seen six years later by experiments at LEP. Recently, the asymmetric B factories KEK-B and PEP II have contributed to these measurements with unprecedented precision. An example of these measurements is the asymmetry from dilepton events shown in Fig. 1. Here, semileptonic B decays are used to tag the flavour of the two B mesons.

3.1 Oscillation at the $\Upsilon(4S)$

The $B\overline{B}$ system from strong interaction Υ(4S) decay is in a state of angular momentum $L = 1$ with an odd wavefunction. This system has to be treated as a coherent entangled quantum state. The time evolution of a state with odd symmetry is different from that of one with even symmetry. This is due to the fact, that only one antisymmetric $X\overline{X}$ state,

$$|\overline{X}(1)X(2)\rangle - |X(1)\overline{X}(2)\rangle$$

is possible, so it has to stay constant. There are, however, three symmetric states,

$$\begin{aligned}&|\overline{X}(1)X(2)\rangle + |X(1)\overline{X}(2)\rangle\\&|X(1)X(2)\rangle\\&|\overline{X}(1)\overline{X}(2)\rangle\end{aligned}$$

and their relative amplitudes may change with time.

Explicitly, for initial $B\overline{B}$ states of odd symmetry, the state evolves in time into

$$|\psi_-(T_1,T_2)\rangle = \frac{e^{-(im/\Gamma+\frac{1}{2})(T_1+T_2)}}{2pq} \cdot \left[\cos(x-iy)\frac{T_1-T_2}{2}\left(|B^0(1)\overline{B}^0(2)\rangle - |B^0(2)\overline{B}^0(1)\rangle\right) -i\sin(x-iy)\frac{T_1-T_2}{2}\left(\frac{1}{\frac{q}{p}}|B^0(1)B^0(2)\rangle - \frac{q}{p}|\overline{B}^0(1)\overline{B}^0(2)\rangle\right)\right]$$

This is observable if we associate the times T_1 and T_2 with the times of decay of the two B mesons. For $T_1 = T_2$ only the antisymmetric state is present, and mixed states, i.e. two final states indicating the same beauty flavour, will show up only at $T_1 \neq T_2$.

As a consequence, the variable T in (2) and (3) is the **difference** of the lifetime of the two B mesons, $T = (t_2 - t_1)/\tau$, for all time-dependent measurements at the $\Upsilon(4S)$. For time integrated measurements, this implies that the two B mesons do not oscillate independently and hence the mixing probability χ of a single B meson can directly be measured as the fraction of mixed $(BB, \overline{B}\overline{B})$ from all events.

3.2 Experimental Results

While only the integrated effect can be observed on the $\Upsilon(4S)$ at symmetric colliders, an observation of the oscillating behaviour was first possible at the Z^0, where the lifetime can be measured. This yields directly the frequency ν as $2\pi\nu = \Delta m$ from the asymmetry (2) equivalent to

$$a(t) = \cos \Delta m\, t$$

Results are summarized in Table 2. They have been recently augmented by results from asymmetric colliders on the $\Upsilon(4S)$.

The measurements differ mostly in the tagging method used. Reconstructing hadronic final states of $B^0/\overline{B}^0$ decays is denoted "B^0" in Table 2. This technique requires a large number of B mesons and has therefore only been used at the asymmetric B factories at SLAC and KEKB. The same holds for the channel $B^0 \to D^{*-}l^+\nu$.

The experiments at LEP use charged leptons (l), fully (D^*) or partially (π) reconstructed decays of $D^{*+} \to D^0\pi^+$, alone and associated with a lepton indicating a $B^0 \to D^{*-}(X)l^+\nu$ final state, and various definitions of a jet charge. The C and P asymmetry of polarized Z^0 decays is exploited at the Stanford Linear Collider by SLD.

There is a background fraction $f_\pm$ from other b hadrons, predominantly charged $B^\pm$ which do not oscillate. This modifies the rates and leads for B

Table 2. $B^0/\overline{B}^0$ eigenstate mass difference from the oscillation frequency. Tagging particles are given, where B^0 means a fully reconstructed B meson, Q_J is a jet charge technique, π^+ is a same-jet tag, and NN is a neural network exploiting many particles in the event.

.40 .45 .50	Δm [ps^{-1}]	tags	experiment
	$0.446 \pm 0.020 \pm 0.018$	$D^*, l/Q_J; l/l$	ALEPH 96/97 [4]
	$0.504 \pm 0.019 \pm 0.012$	$D^*, l, \pi l/Q_J; l/l$	DELPHI 97 [5]
	$0.479 \pm 0.018 \pm 0.015$	$D^*/l; D^*l/l, Q_J$	OPAL 96–00 [6,7]
	$0.444 \pm 0.028 \pm 0.028$	$l/l, Q_J$	L3 98 [8]
	$0.507 \pm 0.023 \pm 0.019$	K, Q_J, l [a]	SLD 96 prel. [9]
	$0.495 \pm 0.026 \pm 0.025$	$\pi^+, D/l;$ $D^{(*)}, D^*l/l, Q_J$	CDF 97–99 [10]
	0.490 ± 0.013	$l/l;$ $D^*l\nu, B^0/l, K, \pi$	BELLE 01 [11,12]
	0.503 ± 0.011	$l/l;$ $D^*l\nu, B^0/l, K, \text{NN}$	BABAR 01 [3,13,14]
	0.489 ± 0.009		average

[a] using the Z^0 polarization asymmetry at SLC

mesons at $T = 0$ to an asymmetry

$$a(T) = D_\pm \cos xT + I_\pm \tag{4}$$

with an effective dilution factor $D_\pm = \frac{1}{1+f_\pm/2}$ and an offset $I_\pm = \frac{f_\pm}{2+f_\pm}$. The value $a(0) = D_\pm + I_\pm = 1$ is independent of $f_\pm$.

If the flavour of one B meson is assigned wrong with probability w, the asymmetry is given by

$$a(T) = (1 - 2w)(D_\pm \cos xT + I_\pm)$$

with a tagging dilution factor $D_t = a(0) = 1 - 2w$. A limited error in vertex reconstruction leads to a smearing of the T distribution. This implies an amplitude reduction factor D_r. Assuming a Gaussian resolution in Δt of σ_t, we have $D_r = e^{-\frac{1}{2}(x\sigma_t/\tau)^2}$ which is multiplied to the tagging dilution. This approach to time smearing ignores the asymmetry entered by the exponential decay distribution. If the true distributions are convoluted with a Gaussian, and the asymmetry is calculated from the smeared distributions, there is in addition to the reduction in amplitude also a distortion.

At the b jet experiments, the distance between the primary vertex of $b\overline{B}$ jet production and the B decay vertex is measured with high precision silicon detectors. The resolution effects can be observed at the tail to negative lifetimes. In oscillation measurements at asymmetric colliders at the $\Upsilon(4S)$ energy this is not possible, since the lifetime difference has naturally negative values, and is in fact symmetric about 0. Therefore, these measurements rely on a fit of their resolution parameters, which can be verified by lifetime measurements on high statistics samples of charged and neutral B mesons.

All constant and time-dependent resolution effects differ from experiment to experiment, and depend on the method of B identification. They are convoluted with the full time dependence, including the exponential which makes the effective "smearing" of the true lifetime asymmetric. These together with a possible bias from cascade charm decays, resolution tails and background effects are modelled for each experiment and are taken into account in the fit functions (as the one shown in Fig. 1). The proper modelling of the distortions is crucial for a bias-free extraction of the oscillation frequency.

The average of all measurements in Table 2 is $\Delta m(B^0) = (0.489 \pm 0.009)/\text{ps}$ corresponding to an oscillation frequency $\nu = \Delta m/2\pi = (77.8 \pm 1.6)\text{GHz}$ and a mass difference $\Delta m = (0.322 \pm 0.007)\text{meV}$. This is a fraction of $6\text{e}{-14} \cdot m(B^0)$.

The dimensionless mixing parameter x can be calculated from the mixing probability $\chi = \frac{x^2}{2(1+x^2)}$ and from the oscillation frequency as $x = \Delta m\,\tau$ which requires also precise knowledge on the average lifetime $\tau_d = (1.548 \pm 0.021)\text{ps}$ [15] of the B^0 meson. The two independent methods agree very well and average to

$$x = 0.756 \pm 0.016$$

The common value of the scaled mass difference is dominated by the direct measurements of the oscillation frequency and has reached a precision of 2%.

There is no observation of $\Delta\Gamma$ of the B^0 mass eigenstates. In the Standard Model, a value $y \sim 10^{-3}$ is expected. CLEO finds $|y| < 0.41$ from their mixing analysis [16]. BELLE [11] gives an upper limit of $|y| < 0.08$ at 90%CL.

3.3 Experimental Results on the B_s Meson

The first hint on large B_s mixing was obtained by UA1 [17] even before $B^0\overline{B}^0$ oscillation was established. They observed an average mixing probability $\chi = 0.12 \pm 0.05$ in b jets from $\bar{p}p$ annihilation. From the same quantity measured at e^+e^- annihilation and Z^0 decay, a value χ_s close to the maximum 0.5 can be inferred with large errors due to the small B_s fraction in b jets. Direct (non-) observations of the oscillation leads to more stringent limits on the frequency. To determine a limit, the LEP Working group on B oscillations [18] combined the fit results of the B_s asymmetry amplitude A (which is 1 for true oscillations) for a series of assumed values of Δm_s from 13 measurements by the five experiments ALEPH, CDF, DELPHI, OPAL, and SLD. All measurements have been adjusted to a common set of inputs before averaging. Systematic correlations are taken into account. The average amplitudes lead to a 95%CL lower limit which is calculated using the point where the hypothesis $A = 1$ is excluded at 5% significance level in a one-sided Gaussian test. This means all values of Δm_s for which the combined amplitude A plus $1.645 \cdot \sigma(A)$ is smaller than 1 are excluded at 95%CL, where $\sigma(A)$ is the total error on A. The procedure excludes all values with $\Delta m_s < 14.9\text{ps}^{-1}$ leading to a lower limit

$$\Delta m_s > 14.9\text{ps}^{-1} \quad (95\%\text{CL}).$$

For the $B_s/\overline{B}_s$ system, using this lower limit on the oscillation frequency and the B_s lifetime value $\tau_s = (1.47 \pm 0.06)$ps [15], the present lower limit is

$$x_s > 21 \quad (95\%\text{CL})$$

which is already above the lowest expected values. It corresponds to a mixing probability $\chi_s > 0.498$.

The available results on the B_s lifetime are average lifetimes which can all have different weights of τ_H and τ_L due to the different mixture of final states used. The experimental information on the lifetime difference of the mass eigenstates is still weak and no significant lifetime difference has been observed. The combination of all results has been calculated by the LEP $\Delta\Gamma_s$ Working Group to be $\Delta\Gamma/\Gamma = 2|y_s| = 0.24 \pm {}^{0.16}_{0.13}$ [19]. They have also re-evaluated the constraint $\tau_s = \tau_0$ for individual experiments leading to $\Delta\Gamma/\Gamma = 2|y_s| = 0.16 \pm {}^{0.08}_{0.09}$.

4 CP Violation of B Mesons

CP violation is expected in the Standard Model through the physical phase in the CKM matrix [1]. A 3×3 unitary matrix fulfills 12 unitarity conditions, 6 of those are three complex numbers summing up to 0. In the complex plane, these form the six unitarity triangles which have angles that are independent of phase conventions for the matrix elements and therefore are physical observables. One of the two very similar triangles that are not collapsed to almost a single line is the one shown in Fig. 2. Its angles can all be measured via CP violating asymmtries of B mesons. These phases change sign when a process is CP-transformed (CP odd phases).

CP violation in B decays (as in K decays) occurs always via interference of (at least) two amplitudes with different CP even and CP odd phases, in three different ways:

1. Direct CP violation $\Gamma(B \to X) \neq \Gamma(\overline{B} \to \overline{X})$ can be observed by final state counting experiments. It occurs from the interference of two decay amplitudes with different phases that transform as $\text{CP}\phi = -\phi$ (CP odd phases from the CKM matrix), and with different phases that transform as $\text{CP}\delta = +\delta$ (CP even phases from the strong interaction). Direct CP violation is not restricted to neutral mesons, but may also be observed in charged meson or baryon decays.
2. CP violation induces a small asymmetry in the oscillation probability $P(B^0 \to \overline{B}^0) \neq P(\overline{B}^0 \to B^0)$ due to $|\frac{q}{p}| \neq 1$. This is due to the interference of other amplitudes with the leading box diagram of $B/\overline{B}$ mixing.
3. The interference of mixed and unmixed amplitudes leads to lifetime dependent differences $\Gamma(B^0|_{t=0} \to X|_t) \neq \Gamma(\overline{B}^0|_{t=0} \to X|_t)$ for a common final state of B and $\overline{B}$ with asymmetry amplitude modulation $\propto \sin \Delta m t = \sin xT$. This is also called "CP violation from interference of oscillation and decay", or "mixing-induced CP violation". Here, the two interfering phases

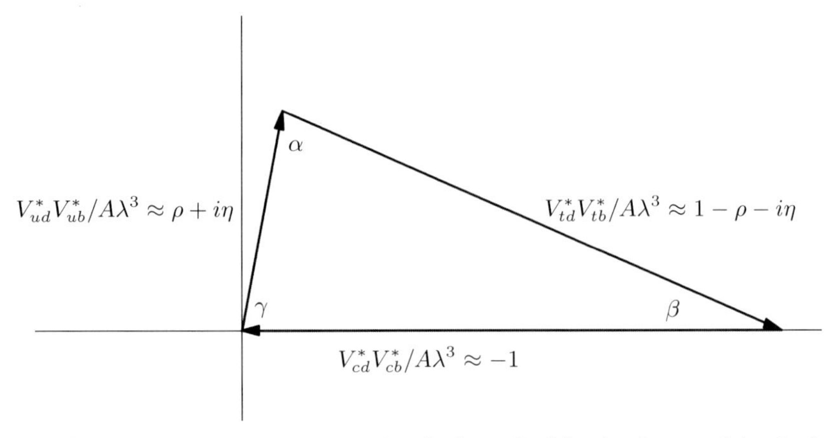

Fig. 2. The most important unitarity triangle from the *bd* unitarity condition is shown in the complex plane. Up to corrections of $\mathcal{O}(\lambda^4)$ the top point is at $(\bar{\rho} = [1 - \frac{\lambda^2}{2}]\rho, \bar{\eta} = [1 - \frac{\lambda^2}{2}]\eta)$, and the rightmost point is $(1, 0)$. Changing the phase convention for the CKM matrix will rotate all unitarity triangles in the complex plain, but their shape is invariant under those transformations. A second triangle from the *ut* unitarity condition is very similar to this one, with the same angle α and the two other angles $\beta' \approx \beta$ and $\gamma' \approx \gamma$.

are CP-odd phases from the CKM matrix and the phase $\frac{\pi}{2}$ between the coefficients $\cos x\frac{T}{2}$ and $i \sin x\frac{T}{2}$ from (1).
The final state X can be a CP eigenstate, like $J/\psi K_S^0$ (CP $= -1$) or $\pi^+\pi^-$ (CP $= +1$), or a state that can be reached from both B^0 mesons via different processes.
In the Standard Model, this CP violating interference can lead to almost maximum asymmetries. In many cases, large values are expected, and the time-dependence is a further handle to avoid misinterpretation of data. Therefore, all present and future experiments focus on these effects.

4.1 Direct CP Violation

Decays with direct CP asymmetries require in the Standard Model at least two interfering channels with different CKM phase $\phi_{1,2}$ and different strong phases $\delta_{1,2}$. This defines the amplitudes

$$\begin{aligned} A(B \to X) &= |A_1| e^{i\phi_1 + i\delta_1} + |A_2| e^{i\phi_2 + i\delta_2} \\ \bar{A}(\overline{B} \to \overline{X}) &= |A_1| e^{-i\phi_1 + i\delta_1} + |A_2| e^{-i\phi_2 + i\delta_2} \end{aligned}$$

where $|A_1| e^{i\delta_1}$ and $|A_2| e^{i\delta_2}$ is unchanged due to CP invariance of the strong interaction. They contribute to the rate for B and $\overline{B}$ decays with a different interference term, leading to

$$|\bar{A}|^2 - |A|^2 = 4|A_1||A_2| \sin(\phi_1 - \phi_2) \sin(\delta_1 - \delta_2)$$

The asymmetry is

$$a = \frac{N(\overline{B} \to \overline{X}) - N(B \to X)}{N(\overline{B} \to \overline{X}) + N(B \to X)}$$
$$= \frac{2|A_1||A_2|\sin(\phi_1 - \phi_2)\sin(\delta_1 - \delta_2)}{|A_1|^2 + |A_2|^2 + 2|A_1||A_2|\cos(\phi_1 - \phi_2)\cos(\delta_1 - \delta_2)}$$

An example is the B decay to $K\pi$, where a substantial rate asymmetry is possible due to the interference of two decay diagrams: The dominating one is a penguin diagram $b \to s$ with a t quark in the loop, with a CKM phase $\arg(V_{tb}^* V_{ts}^*)$ and a $K\pi$ state with isospin $\frac{1}{2}$. The second amplitude occurs via a tree diagram $b \to u + \bar{u}s$, with $\arg(V_{ub}^* V_{us}^*)$ and isospin $\frac{1}{2}$ and $\frac{3}{2}$ amplitudes. The interference terms are proportional to $\sin\gamma' \sin(\delta_1 - \delta_2)$. Asymmetries of this type can be observed both in neutral and charged B decays, e.g. $B^\pm \to K^\pm\pi^0$.

The asymmetry is limited, however, by the ratio of amplitudes. This is approximately $P : \lambda^2 T$, where $P : T < 1$ reflects the suppression of the loop penguin diagram (P) with respect to a tree diagram (T) of same order. This ratio has been believed to be small, $P : T \sim \lambda^2$, which would allow indeed for large asymmetries, but is now known to be $\sim \lambda$ from the measurement of the $K\pi$ and $\pi\pi$ branching fractions, as can be seen in Table 3.

Table 3. Tree (T) and leading penguin (P) contributions to two-body final states. The experimental branching fractions are averages over charged and neutral B decays (which implies that corrections from the colour suppressed contribution to the $B^+ \to \pi^+\pi^0$ amplitude are ignored).

channel	amplitude	$\mathcal{B}$	
$B \to K\pi$	$A\lambda^2 P + A\lambda^4(\rho - i\eta)T$	$(14.9 \pm 1.0) \cdot 10^{-6}$	[20–22]
$B \to \pi\pi$	$A\lambda^3(1 - \rho - i\eta)P + A\lambda^3(\rho - i\eta)T$	$(4.4 \pm 0.8) \cdot 10^{-6}$	[20–22]

CKM unitarity angles can only be extracted from those asymmetries when the strong phase difference is known. This can be obtained, however, from flavour SU(3) and isospin relations on a set of results from related channels [23]. The experimental branching fractions for individual channels from [20–22] can be used to extract information on the angle γ in the unitarity triangle [24] and provide an additional constraint.

First experimental measurements of asymmetries of this type have been reported by CLEO, BABAR and BELLE and are summarized in Table 4.

There is presently no evidence for CP violation. The experimental precision is not yet sensitive to the range of asymmetries predicted by the Standard Model, but large effects from new CP violating interactions could have shown up already and have not been observed.

Table 4. Results on direct CP violation in B meson decays: the negative asymmetry between the decay rates for the given channel and its CP (or C) conjugate is measured.

channel	asymmetry a	experiment
$B^+ \to J/\psi K^+$	$+0.018 \pm 0.043 \pm 0.004$	CLEO 00 [25]
	$-0.009 \pm 0.027 \pm 0.005$	BABAR 01 prel.
$B^+ \to \psi(2S) K^+$	$+0.020 \pm 0.091 \pm 0.010$	CLEO 00 [25]
$B^+ \to D^0_{CP+} K^+$	$+0.04 \pm ^{0.40}_{0.35} \pm 0.15$	BELLE 01 [26]
$B^0 \to K^+\pi^-$	-0.04 ± 0.16	CLEO 99 [27]
	$(-0.07 \pm 0.08 \pm 0.02)$	BABAR 01 [28]
	$-0.05 \pm 0.06 \pm 0.01$	BABAR 02 prel. [29]
	$0.04 \pm ^{0.19}_{0.17} \pm 0.02$	BELLE 01 [30]
$B^+ \to K^+\pi^0$	-0.29 ± 0.23	CLEO 99 [27]
	$0.00 \pm 0.18 \pm 0.04$	BABAR 01 [31]
	$-0.06 \pm ^{0.22}_{0.20} \pm ^{0.06}_{0.02}$	BELLE 01 [30]
$B^+ \to K^0_S\pi^+$	$+0.18 \pm 0.24$	CLEO 99 [27]
	$-0.21 \pm 0.18 \pm 0.03$	BABAR 01 [31]
	$0.10 \pm ^{0.43}_{0.34} \pm ^{0.02}_{0.06}$	BELLE 01 [30]
$B^+ \to K^+\eta$	$+0.03 \pm 0.12$	CLEO 99 [27]
$B^+ \to K^+\eta'$	$-0.11 \pm 0.11 \pm 0.02$	BABAR 01 [32]
$B^+ \to K^+\phi$	$-0.05 \pm 0.20 \pm 0.03$	BABAR 01 [32]
$B^+ \to K^{*+}\phi$	$-0.43 \pm ^{0.36}_{0.30} \pm 0.06$	BABAR 01 [32]
$B^0 \to K^{*0}\phi$	$0.00 \pm 0.27 \pm 0.03$	BABAR 01 [32]
$B^+ \to \omega\pi^+$	-0.34 ± 0.25	CLEO 99 [27]
	$-0.01 \pm ^{0.29}_{0.31} \pm 0.03$	BABAR 01 [32]
$B^0, B^+ \to K^*\gamma$	$+0.08 \pm 0.13 \pm 0.03$	CLEO 99 [33]
	$-0.035 \pm 0.076 \pm 0.012$	BABAR 01 prel.
$B^0/B^+ \to s\gamma$	$-0.079 \pm 0.108 \pm 0.022$	CLEO 01 [34]

4.2 CP Violation in the Oscillation

CP violation induces a small asymmetry in the oscillation probability $P(B^0 \to \overline{B}^0) \neq P(\overline{B}^0 \to B^0)$ due to $|\frac{q}{p}| \neq 1$. This is due to the interference of other amplitudes with the leading box diagram of $B/\overline{B}$ mixing, e.g. replacing one t quark in the loop with a c quark. The oscillation asymmetry (2) starting with an initial B^0 meson can be expanded in δ_ϵ as

$$a(T) = \left. \frac{\dot{N}(B) - \dot{N}(\overline{B})}{\dot{N}(B) + \dot{N}(\overline{B})} \right|_T = \frac{\cos xT}{\cosh yT} + \delta_\epsilon \left(1 - \frac{\cos^2 xT}{\cosh^2 yT} \right) + \mathcal{O}(\delta_\epsilon^2)$$

which is for $y = 0$

$$a(T) \approx \cos xT + \delta_\epsilon \sin^2 xT = \cos xT + \frac{\delta_\epsilon}{2}(1 - \cos 2xT)$$

In the Standard Model we have $\delta_\epsilon < 0$ and there are always slightly more $B \to \overline{B}$ than $\overline{B} \to B$ oscillations. Using leptons as flavour tag the net asymmetry can be observed at the $\Upsilon(4S)$ by counting like-sign lepton pairs which originate directly from semileptonic neutral B meson decays. They occur in final states with mixing and show an asymmetry

$$a = \frac{N(B\overline{B} \to l^+l^+) - N(B\overline{B} \to l^-l^-)}{N(B\overline{B} \to l^+l^+) + N(B\overline{B} \to l^-l^-)} = \frac{1 - |\frac{q}{p}|^4}{1 + |\frac{q}{p}|^4} = \frac{2\delta_\epsilon}{1 + \delta_\epsilon^2} \tag{5}$$

constant in time. This same parameter can also be observed as an asymmetry in the time-dependent total decay rate of B and $\overline{B}$. This rate for an initially pure B^0 sample is

$$\frac{\mathrm{d}N}{\mathrm{d}t} = N_0\,\Gamma \frac{1}{1+\delta_\epsilon} e^{-T} \left\{ \cosh yT + \delta_\epsilon \cos xT - y \sinh yT + x\delta_\epsilon \sin xT \right\}$$

In the approximation $y = 0$ expected to be good for the B^0 the rate is

$$\frac{\mathrm{d}N}{\mathrm{d}t} = N_0 \Gamma e^{-T} \left[1 + \delta_\epsilon \cdot (-1 + \cos xT + x \sin xT)\right] + \mathcal{O}(\delta_\epsilon^2)$$

An initially pure $\overline{B}^0$ sample gives the same rates with the replacement $\frac{q}{p} \longleftrightarrow 1/\frac{q}{p}$ and $\delta_\epsilon \longleftrightarrow -\delta_\epsilon$. The asymmetry (where the B flavour is understood as the one at $T = 0$) is therefore

$$\begin{aligned} a(T) &= \left. \frac{\dot{N}(\overline{B}^0 \to \text{anything}) - \dot{N}(B^0 \to \text{anything})}{\dot{N}(\overline{B}^0 \to \text{anything}) + \dot{N}(B^0 \to \text{anything})} \right|_T \\ &= \delta_\epsilon \left(1 - \cos xT - x \sin xT\right) = 2\delta_\epsilon \left(-\frac{x}{2} \sin xT + \sin^2 \frac{xT}{2} \right) \end{aligned} \tag{6}$$

In 1993 CLEO [35] has obtained an asymmetry (5) of $a = 0.031 \pm 0.096 \pm 0.032$ from lepton pairs, corresponding to $\delta_\epsilon = 0.016 \pm 0.048 \pm 0.016$. This first upper limit of $\delta_\epsilon < 0.09$ (90%CL) was still far above expectation. A recent update [36] augmented by an independent method [16] using reconstructed $B^0 \to D^{*-}\pi^+(\pi^0)$ on one side and leptons from the other B gives $\delta_\epsilon = 0.007 \pm 0.021 \pm 0.003$. A dilepton analysis by BABAR [37] gives $a = 0.005 \pm 0.012 \pm 0.014$ and dominates the average in Table 5.

With B^0 mesons from jets, the interpretation of any charge asymmetry in lepton production is more difficult. OPAL at LEP gives a value $\delta_\epsilon = 0.004 \pm 0.014 \pm 0.006$ [7]. A measurement by ALEPH using semileptonic B^0 decays [40] gave the result $a = -0.037 \pm 0.032 \pm 0.007$ corresponding to $\delta_\epsilon = -0.018 \pm 0.016 \pm 0.003$.

A first measurement of the total asymmetry (6) at LEP gave the preliminary result $\delta_\epsilon = -0.011 \pm 0.015 \pm 0.005$ [38]. Results from OPAL ($a = 0.005 \pm 0.055 \pm 0.013$) [39] and ALEPH ($a = 0.016 \pm 0.034 \pm 0.009$) [40] are also given in Table 5. This is an alternative way to measure CP violation in $B^0/\overline{B}^0$ oscillation, but it implies the danger of a possible bias due to the B^0 event selection.

Table 5. Results on $\delta_\epsilon(B^0)$.

−0.02 0.00 0.02	δ_ϵ $[10^{-3}]$	method	experiment
	$-11 \pm 15 \pm 5$	incl.	DELPHI prel. 97 [38]
	$4 \pm 14 \pm 6$	*ll*	OPAL 97 [7]
	$3 \pm 28 \pm 6$	incl.	OPAL 99 [39]
	$-18 \pm 16 \pm 3$	*ll*	ALEPH 00 [40]
	$8 \pm 17 \pm 4$	incl.	ALEPH 00 [40]
	$7 \pm 21 \pm 3$	*ll*, *Bl*	CLEO 01 [35,36]
	$2.4 \pm 5.8 \pm 7.2$	*ll*	BABAR 02 [37]
	-0.7 ± 5.7		average

CPT conservation is assumed in these results. Averaging gives

$$\delta_\epsilon = -0.0007 \pm 0.0057$$

or $-0.012 < \delta_\epsilon < 0.010$ at 95%CL.

4.3 CP Violation in Common Final States of B^0 and $\bar{B}^0$

The most pronounced manifestation of CP violation in the $B^0/\overline{B}^0$ system is expected in interference of oscillation and decay to final states common to B^0 and $\overline{B}^0$ [41]. The effect is largest for CP eigenstates, but may occur at any final state where the amplitudes of the mixed and unmixed decay can interfere:

$$\begin{array}{ccc} & \overline{B}^0 & \\ \nearrow & & \searrow \\ B^0 & \longrightarrow & X \end{array} \tag{7}$$

The decay rate of an initial B^0 meson is proportional to the matrix element

$$|\mathcal{M}|^2 = e^{-T}|A|^2 \frac{1+|\rho|^2}{2} \Big\{ \cosh yT - \Theta_0 \cos xT + \Omega_0 \sinh yT - \Lambda_0 \sin xT \Big\} \tag{8}$$

where the coefficients are

$$\Omega_0 := \frac{2\,\mathcal{R}e\,\rho}{1+|\rho|^2}, \quad \Lambda_0 := \frac{2\,\mathcal{I}m\,\rho}{1+|\rho|^2}, \quad \Theta_0 := \frac{|\rho|^2-1}{|\rho|^2+1}$$

which are given by the two real numbers $\mathcal{R}e\,\rho$ and $\mathcal{I}m\,\rho$ and related via $\Omega_0^2 + \Lambda_0^2 + \Theta_0^2 = 1$, and where the ratio of the upper and lower path's amplitudes in (7) is

$$\rho := \frac{q}{p}\frac{\bar{A}}{A} = \frac{\langle \overline{B}^0|B_L\rangle}{\langle B^0|B_L\rangle}\frac{\langle X|\mathbf{H}|\overline{B}^0\rangle}{\langle X|\mathbf{H}|B^0\rangle}$$

The matrix element (8) and its antiparticle counterpart lead to a general asymmetry

$$a(T) = \left.\frac{\dot N(\overline{B} \to X) - \dot N(B \to X)}{\dot N(\overline{B} \to X) + \dot N(B \to X)}\right|_T = \frac{|\overline{\mathcal{M}}|^2 - |\mathcal{M}|^2}{|\overline{\mathcal{M}}|^2 + |\mathcal{M}|^2}$$
$$= \frac{\Theta_0 \cos xT + \Lambda_0 \sin xT + \delta_\epsilon(\cosh yT + \Omega_0 \sinh yT)}{\cosh yT + \Omega_0 \sinh yT + \delta_\epsilon(\Theta_0 \cos xT + \Lambda_0 \sin xT)}$$

At the presently accessible precision, where $\delta_\epsilon = 0$ (corresponding to $|\frac{q}{p}| = 1$) is a very good approximation, and with $y \approx 0$ for B^0 mesons, we obtain the simplified asymmetry function

$$a(T) = \Theta_0 \cos xT + \Lambda_0 \sin xT \tag{9}$$

If X is a CP eigenstate, the ratio $\bar{A}/A$ is often just a phase, which includes the sign of the CP eigenvalue of X. The phase of the product ρ is independent of conventions, and is in fact an observable. More general, A and $\bar{A}$ can have also different magnitudes, which in the absence of oscillation would still imply an asymmetry and corresponds to the direct CP violation of non-oscillating particles. This direct CP violation is responsible for the cos-term in (8) for final states that can be reached by both B^0 and $\overline{B}^0$.

4.4 Experimental Data on $\sin 2\beta$

If we assume that the Standard Model explanation for CP violation is the correct and only one, then the channel $B \to J/\psi\, K^0_S$ where the K^0_S decays to a CP $= +1$ eigenstate, offers a clean measurement of $\sin 2\beta$. Results have become available since 1999. A summary of all present experimental information on $\sin 2\beta$ is given in Table 6. The statistical error is limited by $\sigma(\Lambda_0) > 1.7/\sqrt{N_{\text{sig}} \cdot (\epsilon_t D_t^2)_{\text{eff}}}$, and typically larger due to background, time-resolution and the statistical error on the dilution.

The first measurement with non-zero information content on the CP-violating parameter $\sin 2\beta$ was performed by the CDF experiment and published 1999 [44]. The golden final state $\mu^+\mu^-\pi^+\pi^-$ was reconstructed. A total sample of 395 ± 31 signal events is divided into a subsample of 202 ± 18 events with both muons measured in the silicon vertex detector, and the remainder without lifetime information. This latter sample is used to measure the time integrated asymmetry, while the subsample with vertex information is submitted to a fit of the asymmetry function

$$a(T) = \Lambda_0 \sin xT \tag{10}$$

The flavour was tagged combining information from soft leptons and a jet charge variable of the opposite jet, and same-side tagging with a charged particle (taken as pion) close to the reconstructed B^0 meson in direction and rapidity. The opposite-jet tags have been investigated using fully reconstructed charged B mesons, and the same-jet tags using $D^{(*)}lX$ candidates. The overall effective performance is $(\epsilon_t D_t^2)_{\text{eff}} = 0.063 \pm 0.017$.

Table 6. Results on the asymmetry amplitude $A_0 = \sin 2\beta$ from $B \to J/\psi K^0_S$ and related channels. The number of signal events and the effective number $(\epsilon_t D_t^2)_{\text{eff}} N_{\text{sig}}$ is given to compare the sensitivity of the measurements.

0.0 0.5 1.0	A_0	$N_{\text{sig}} \cdot (\epsilon_t D_t^2)_{\text{eff}} / N_{\text{sig}}$	experiment	
	$(3.2 \pm ^{1.8}_{2.0} \pm 0.5)$	$1.6/14 \pm 2$	OPAL 98 [42]	
	$(1.8 \pm 1.1 \pm 0.3)$	$12/198 \pm 17$	CDF 98 [43]	
	$0.79 \pm ^{0.38}_{0.41} \pm 0.16$	$25/397 \pm 31$	CDF 99 [44]	*
	$(0.84 \pm ^{0.82}_{1.04} \pm 0.16)$	4.4/17	ALEPH 00 [45]	
	$0.34 \pm 0.20 \pm 0.05$	140/520	BABAR 01 [46]	
	$0.58 \pm ^{0.32}_{0.34} \pm ^{0.09}_{0.10}$	70/260	BELLE 01 [47]	
	$0.59 \pm 0.14 \pm 0.05$	280/1080	BABAR 01 [48]	
	$0.99 \pm 0.14 \pm 0.06$	280/1030	BELLE 01 [49]	*
	$0.75 \pm 0.09 \pm 0.04$	540/2200	BABAR 02 prel. [50]	*
	$0.82 \pm 0.12 \pm 0.05$		BELLE 02 prel. [51]	
	0.82 ± 0.08	$\chi^2 = 1.8/2df$	average of (*)	

The year 2000 saw first results from the e^+e^- B factories at SLAC and KEK which produce $B^0\overline{B}^0$ pairs in $\Upsilon(4S)$ decays. These experiments have much lower background than experiments using B^0 mesons in b jets, and can therefore reconstruct many additional decay channels with lower branching fraction or worse resolution.

At the end of 2000, the BABAR sample [46] was the largest sample of B mesons, comprising 23 Million B meson pairs from $\Upsilon(4S)$ decays. Both BABAR and BELLE were accumulating more data rapidly. They could significantly establish CP violation in B mesons in 2001.

Signals are reconstructed for $(J/\psi \to l^+l^-) + (K^0_S \to \pi^+\pi^-, \pi^0\pi^0)$, $(\psi(2S) \to l^+l^-, J/\psi\pi^+\pi^-) + (K^0_S \to \pi^+\pi^-)$, $(\chi_{c1} \to J/\psi\gamma) + (K^0_S \to \pi^+\pi^-)$, $(J/\psi \to l^+l^-) + (K^{*0} \to K^0_S\pi^0)$, and $(J/\psi \to l^+l^-) + K^0_L$. Each event with a signal candidate is assigned a B^0 or $\overline{B}^0$ tag if the rest of the event (with the daughter tracks of the signal B removed) satisfies the criteria for one of several tagging categories. The BABAR experiment uses only events with a minimum tag discrimination, thus reducing the number of tagged events to 68% of all reconstructed events. The total statistical tagging performance is $(\epsilon_t D_t^2)_{\text{eff}} = 0.261 \pm 0.012$.

BELLE [49] has reconstructed additional final states, $(\eta_c \to K^+K^-\pi^0, K^0_S K^\pm\pi^\mp) + (K^0_S \to \pi^+\pi^-)$, from a sample about the same size as the BABAR sample. The channel $J/\psi\pi^0$ has been used in their first analysis [47], but has been dropped in [49] since it is a $c\bar{c}d\bar{d}$ final state without intermediate neutral kaon, and may have some additional phases from penguin diagrams. They tag all events including those with little or no flavour discrimination. Their total statistical tagging performance is $(\epsilon_t D_t^2)_{\text{eff}} = 0.270 \pm 0.012$ and agrees with that of BABAR.

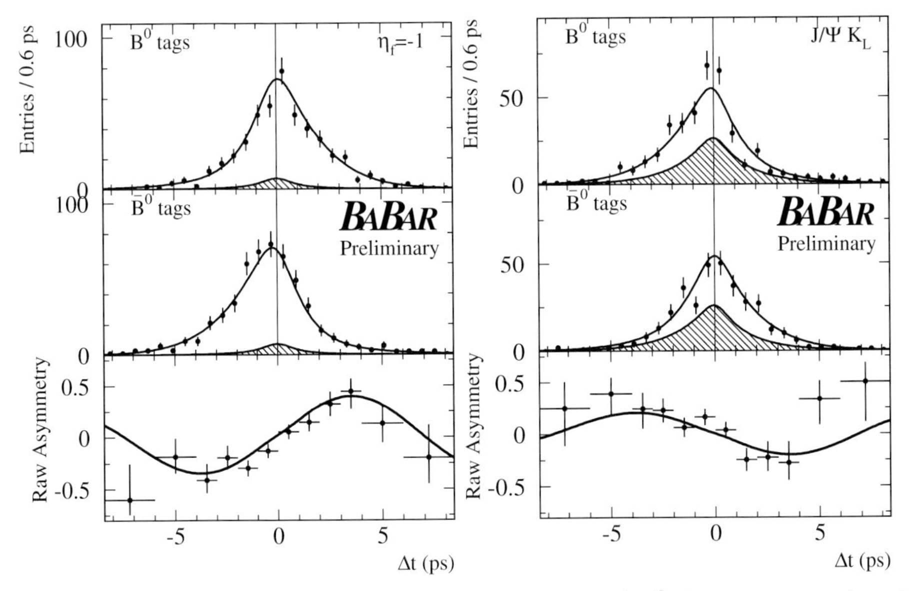

Fig. 3. Number of candidates from the $\{J/\psi, \psi(2S), \chi_{c1}\}K^0_S$ (left, CP $= -1$) and $J/\psi K^0_L$ (right, CP $= +1$) final states in the signal region with B^0 tag (top) and with $\overline{B}^0$ tag (middle), together with the raw asymmetry (bottom). Overlaid is the result of the fit for the diluted asymmtry, assuming equal numbers of B^0 and $\overline{B}^0$ tags [50].

Both experiments determine the tagging dilution D_t and the resolution function from a fit of (3) convoluted with these effects to fully reconstructed $B^0/\overline{B}^0$ mesons in self-tagging final states, where the second B-meson is tagged via exactly the same procedure as the CP sample. They fix the B^0 lifetime $\tau_{B^0} =$ 1.548ps and the oscillation frequency $\Delta m = 0.472\text{ps}^{-1}$ to the 2000 world average [15]. There is a slight correlation between Λ_0 and these data that is absorbed in the systematic error.

The $J/\psi K^{*0}$ CP eigenstates have contributions with CP $= +1$ $(L = 0, 2)$ and CP $= -1$ $(L = 1)$ eigenvalues. Therefore, a partial wave analysis has been performed to disentangle both contributions. While BABAR use their result [52] to apply a physical dilution factor $D_P = 0.65 \pm 0.07$, BELLE [49] and an updated BABAR analysis [50] include the transversity angle [53] of these events in the fit to improve the separation of the two eigenstates.

The raw asymmetries varying as $\sin xT = \sin \Delta m \Delta t$ are clearly visible in Fig. 3, which shows the analysis with the largest sample (at present).

The $J/\psi K^0_L$ channel suffers from high background, including crosstalk from $B \to J/\psi K^{*0} \to J/\psi K^0_L \pi^0$. This component can be included with an average CP eigenvalue in the fit, and a conservative range for this parameter has been taken to estimate its systematic error. Other systematic errors arise from uncertainties in input parameters to the maximum likelihood fit, incomplete knowledge of the time resolution function, uncertainties in the mistag fractions, and possible limitations in the analysis procedure. The sum of all systematic uncertainties is still considerably smaller than the statistical error (see Table 6).

A two-parameter fit according to (9) of the subset of channels with low background has been performed by BABAR [48,50]. They find $|\rho| = 0.92 \pm 0.06 \pm 0.02$, compatible with the value of 1 if there were no direct CP violation present. The value of Λ_0 is not changed by the two-parameter fit in the subsample. The fit results correspond to

$$\Lambda_0 = 0.76 \pm 0.10, \quad \Theta_0 = -0.08 \pm 0.07$$

The average value from Table 6 is $\Lambda_0 = 0.82 \pm 0.08$. In the Standard Model, this parameter is related to the tree-level ratio

$$\rho_{c\bar{c}d\bar{d}} = \frac{q}{p} \frac{A(\overline{B}^0 \to c\bar{c}d\bar{d})}{A(B^0 \to c\bar{c}d\bar{d})}$$

Using the assumption $|\rho_{c\bar{c}d\bar{d}}| = 1$ or equivalently $\mathcal{Re}\, \eta_{c\bar{c}d\bar{d}} = 0$, which is only justified within the Standard Model, we have

$$\mathcal{Im}\, \rho_{c\bar{c}d\bar{d}} = \sin 2\beta = 0.82 \pm 0.08$$

The corresponding "ϵ"-parameter is

$$\eta_{c\bar{c}d\bar{d}} = \frac{A(B_H \to c\bar{c}d\bar{d})}{A(B_L \to c\bar{c}d\bar{d})} = (0.05 \pm 0.04) - (0.52 \pm 0.09)i$$

The experimental result establishes the existence of CP violation in the B^0 system and is in good agreement with the Standard Model prediction. This is demonstrated in an update of a recent overall fit to the CKM matrix including the $\sin 2\beta$ measurements [54]. The allowed region for the tip of the unitarity triangle in the $\bar{\rho}, \bar{\eta}$-plane is shown in Fig. 4.

However, if the Standard Model description of this CP asymmetry is not complete, we have to consider different Λ_0 and Θ_0 parameters for different classes of channels, since e.g. $B \to J/\psi \pi^0$ may suffer more from loop diagrams than $B \to J/\psi K_S^0$. For cross checks like this, more final states are presently investigated by BABAR and BELLE. Among these are several with the same asymmetry as $J/\psi K_S^0$ from the leading tree diagrams. One class is $B^0 \to D^+ D^-$ or $D^{*+} D^{*-}$. Here, penguin diagrams contribute with different phases and a different asymmetry, including possibly a $\cos xT$ term from direct CP violation, is expected.

Another class is $B^0 \to \eta' K_S^0$ or ϕK_S^0, where the angle β' of the *tu* unitarity triangle is measured. Since this is very similar to the *bd* triangle shown in Fig. 2, again a similar asymmetry as for $J/\psi K_S^0$ is expected in the Standard Model.

All experimental results [29,51] on these states are, unfortunately, at present not yet sensitive to observe differences.

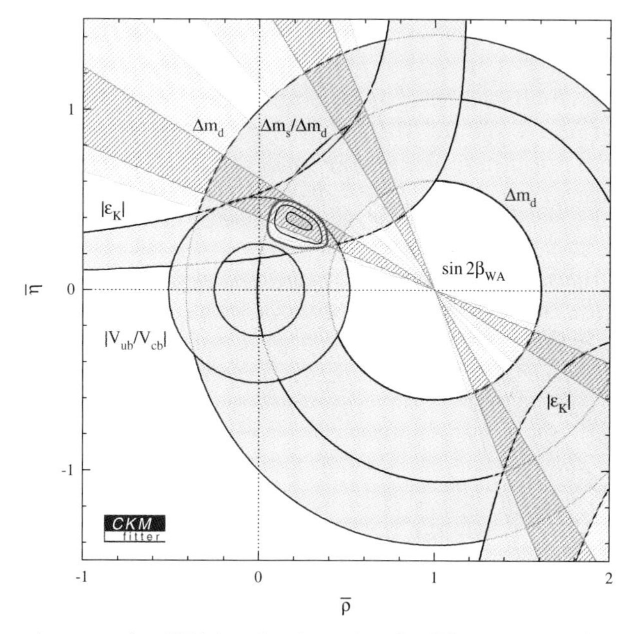

Fig. 4. Constraints on the CKM unitarity triangle. The contours for the tip of the triangle show the parameters at 90%, 32% and 5% significance level of a combined chisquare test to all measurements. The theoretical parameters are taken freely within allowed ranges to minimize this χ^2 without assigning any weight (probability) to them [54].

4.5 Experimental Data on $B \to \pi\pi$

The difficulties of the interpretation of the $B^0 \to \pi^+\pi^-$ arise from the fact that there are two competing diagrams, a penguin (a) and a dominating tree (b):

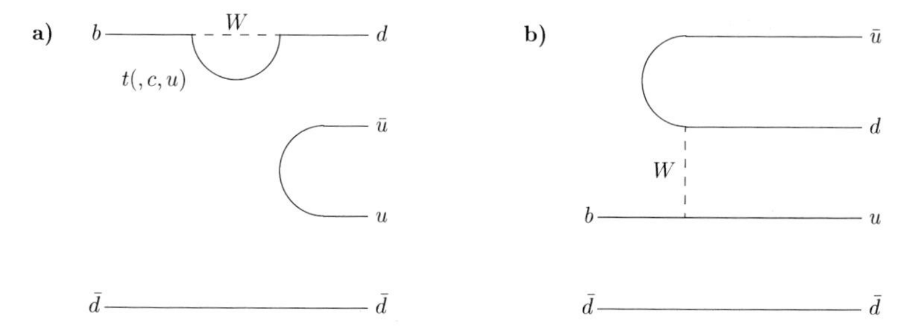

Due to crossfeed between the channels, it is always investigated experimentally together with the penguin-dominated channel $B \to K^+\pi^-$. The asymmetry in the latter, self-tagging final state is a sign of direct CP violation; experimental limits are given in Table 4.

Experimental results on branching fractions are given in Table 3. They strongly indicate a substantial penguin contribution to the $\pi^+\pi^-$ final state. Therefore, as long as this channel alone is investigated, experiments can only determine the coefficients Λ_0 and Θ_0 which allow no direct translation into the angle α (the tree alone would yield $\Lambda_0 = \sin 2\alpha$). The results are given in Table 7. The latest numbers from BABAR and BELLE for both coefficients differ by more than two standard deviations: while the BELLE result suggests a substantial CP asymmetry, BABAR's result indicates a very small deviation from the completely CP symmetric decay.

Table 7. Experimental results on the coefficients of the CP violating asymmetry $a(T) = \Theta_0 \cos xT + \Lambda_0 \sin xT$ for the $B \to \pi^+\pi^-$ decay.

Λ_0	Θ_0	experiment
$0.03 \pm^{0.53}_{0.56} \pm 0.11$	$0.25 \pm^{0.45}_{0.47} \pm 0.14$	BABAR 01 [28]
$-0.01 \pm 0.37 \pm 0.07$	$0.02 \pm 0.29 \pm 0.07$	BABAR 02 prel. [29]
$-1.21 \pm^{0.38}_{0.27} \pm^{0.16}_{0.13}$	$0.94 \pm^{0.25}_{0.31} \pm 0.09$	BELLE 02 prel. [51]

A full analysis requires a flavour-tagged measurement of the decay $B \to \pi^0\pi^0$. For this channel, up to now no significant signal has been observed.

5 Summary and Outlook

The field of flavour oscillation and CP violation of B mesons has received fresh input from the new B meson factories and is rapidly changing, as the reader may see by comparing the updated results in this writeup with the ones presented last year at Dubrovnik. By now we see a clear support for the ideas of Kobayashi and Maskawa on the origin of CP violation in the Standard Model, fitting both K and B decays. Many crosschecks will be possible of which first hints are already showing up today, and we can still hope for surprises in the near future.

This paper has profited from fruitful discussions with many colleagues from inside and outside the BABAR collaboration. Most notably, I enjoyed an interesting conference at the inspiring surroundings of Dubrovnik.

References

1. M Kobayashi, T Maskawa, Prog. Theor. Phys. **49**, 652 (1973).
2. ARGUS Collab., Phys. Lett. B **192**, 245 (1987).
3. BABAR Collab., SLAC-PUB-9096 (2001).
4. ALEPH Collab., Z. Phys. C **75**, 397 (1997).
5. DELPHI Collab., Z. Phys. C **76**, 579 (1997).
6. OPAL Collab., Z. Phys. C **72**, 377 (1996); OPAL Collab., Z. Phys. C **76**, 417 (1997); OPAL Collab., Phys. Lett. B **493**, 266 (2000).
7. OPAL Collab., Z. Phys. C **76**, 401 (1997).
8. L3 Collab., Eur. Phys. J C **5**, 195 (1998).

9. SLD Collab., SLAC-PUB-7230 (1996); SLD Collab., SLAC-PUB-7228 (1996); SLD Collab., SLAC-PUB-7229 (1996).
10. CDF Collab., Phys. Rev. Lett. **80**, 2057 (1998); CDF Collab., Phys. Rev. D **59**, 032001 (1999).
11. BELLE Collab., Phys. Rev. Lett. **86**, 3228 (2001).
12. H Tajima (BELLE), Proc. of the 36th Recontres de Moriond on QCD and High Energy Hadronic Interactions, Bourg-Saint-Maurice, 2001.
13. BABAR Collab., SLAC-PUB-8530 (2000).
14. BABAR Collab., SLAC-PUB-9061 (2001).
15. The Particle Data Group, Eur. Phys. J C **15**, 1 (2000).
16. CLEO Collab., Phys. Lett. B **490**, 36 (2001).
17. UA1 Collab., Phys. Lett. B **186**, 247 (1987).
18. The LEP B Oscillation Working Group, CERN-EP-2000-096, and updates on the World Wide Web, `URL=http://www.cern.ch/LEPBOSC/`.
19. The LEP $\Delta\Gamma_s$ Working Group, P Coyle et al., on the World Wide Web, `URL=http://lepbosc.web.cern.ch/LEPBOSC/deltagamma_s/`, Apr. 2001.
20. CLEO Collab., CLNS 99/1650.
21. BABAR Collab., Phys. Rev. Lett. **87**, 151802 (2001).
22. BELLE Collab., Phys. Rev. Lett. **87**, 101801 (2001).
23. M Gronau et al., Phys. Rev. D **50**, 4529 (1994); Phys. Rev. D **52**, 6374 (1994); Phys. Lett. B **333**, 500 (1994); Phys. Rev. Lett. **73**, 21 (1994); L Wolfenstein, Phys. Rev. D **52**, 537 (1995); N G Deshpande, X G He, Phys. Rev. Lett. **75**, 1705 (1995); Phys. Rev. Lett. **75**, 3064 (1995); A J Buras, R Fleischer, Phys. Lett. B **360**, 138 (1995); Phys. Lett. B **365**, 390 (1996); G Kramer, W F Palmer, Phys. Rev. D **52**, 6411 (1995); B Grinstein, R F Lebed, Phys. Rev. D **53**, 6344 (1996).
24. M Beneke, G Buchalla, M Neubert, C T Sachrajda, Nucl. Phys. B **606**, 245 (2001).
25. CLEO Collab., Phys. Rev. Lett. **84**, 5940 (2000).
26. BELLE Collab., BELLE-CONF-0108, contributed to the 20th International Symposium on Lepton and Photon Interactions at High Energies, Rome, Italy, 23-28 Jul 2001.
27. CLEO Collab., Phys. Rev. Lett. **85**, 525 (2000).
28. BABAR Collab., Phys. Rev. D **65**, 051502 (2002).
29. BABAR Collab., presented at the XXXVII Recontres de Moriond, March 2002.
30. BELLE Collab., Phys. Rev. D **64**, 071101 (2001).
31. BABAR Collab., Phys. Rev. Lett. **87**, 151802 (2001)
32. BABAR Collab., Phys. Rev. D **65**, 051101 (2002).
33. CLEO Collab., Phys. Rev. Lett. **84**, 5283 (2000).
34. CLEO Collab., Phys. Rev. Lett. **86**, 5661 (2001).
35. CLEO Collab., Phys. Rev. Lett. **71**, 1680 (1993).
36. CLEO Collab., Phys. Rev. Lett. **86**, 5000 (2001).
37. BABAR Collab., SLAC-PUB-9149, subm. to Phys. Rev. Lett. (2002).
38. DELPHI Collab., M Feindt et al., DELPHI 97-98 CONF 80, contributed paper to the 18th Int. Symp. on Lepton-Photon Interactions, Hamburg 1997.
39. OPAL Collab., Eur. Phys. J C **12**, 609 (2000).
40. ALEPH Collab., Eur. Phys. J C **20**, 431 (2001).
41. A B Carter, A I Sanda, Phys. Rev. Lett. **45**, 952 (1980); Phys. Rev. D **23**, 1567 (1981); I I Bigi, A I Sanda, Nucl. Phys. B **193**, 85 (1981); Nucl. Phys. B **281**, 41 (1987).
42. OPAL Collab., Eur. Phys. J C **5**, 379 (1998).
43. CDF Collab., Phys. Rev. Lett. **81**, 5513 (1998).

44. CDF Collab., Phys. Rev. D **61**, 72005 (2000).
45. ALEPH Collab., Phys. Lett. B **492**, 259 (2000).
46. BABAR Collab., Phys. Rev. Lett. **86**, 2515 (2001).
47. BELLE Collab., Phys. Rev. Lett. **86**, 2509 (2001).
48. BABAR Collab., Phys. Rev. Lett. **87**, 091801 (2001).
49. BELLE Collab., Phys. Rev. Lett. **87**, 091802 (2001).
50. BABAR Collab., SLAC-PUB-9153 (2002).
51. BELLE Collab., presented at the XXXVII Recontres de Moriond, March 2002.
52. BABAR Collab., Phys. Rev. Lett. **87**, 241801 (2001).
53. I Dunietz et al., Phys. Rev. D **43**, 2193 (1993).
54. A Höcker, H Lacker, S Laplace and F Le Diberder, Eur. Phys. J C **21**, 225 (2001).

Part IV

Noncommutative Field Theories

A Short Review of Noncommutative Field Theory

Loriano Bonora

[1] International School for Advanced Studies (SISSA/ISAS), Via Beirut 2, 34014 Trieste, Italy
[2] INFN, Sezione di Trieste

Abstract. Some properties of noncommutative field theories are reviewed. The emphasis is in particular on renormalization and anomalies.

1 Introduction

Theories defined in noncommutative spaces have been considered over the years both by mathematicians and by physicists. Recently there has been a renewal of interest in this subject due to the realization of physical models of noncommutative spaces based on open strings on D–branes, see [1] and references therein for early contributions to the subject. The possibility of pursuing this research with two different languages, that of (noncommutative) field theory and the one of string theory, with the possibility to compare the results, has naturally attracted a lot of activity. One could mention several lines of research. Among others an extremely interesting one, which however will not be covered in this review, is the search for classical solutions in a noncommutative field theory, i.e. solitons and instantons. Noncommutative solitons in particular turn out to be particularly interesting because of their mimicking solutions of String Field Theory and their connection with tachyon condensation. The present short review concerns instead the properties of noncommutative gauge theories. Open strings attached to D–branes contain in their spectrum a massless vector field. It is a standard matter to find the amplitudes of the corresponding vertex operators. In the field theory limit ($\alpha' \to 0$) these amplitudes coincide with those of an ordinary $U(1)$ gauge field theory. If, however, we switch on a constant $B_{\mu\nu}$ field with nonzero components only in the space directions parallel to the D–brane, and repeat the above calculation, we find that, in the field theory limit, the amplitudes have changed. They are not the amplitudes of an ordinary gauge field theory, rather they correspond to the amplitudes of a *noncommutative field theory*, in which the noncommutative parameter is precisely related to the value of the B–field.

It is natural to ask whether this new field theory, which is nonlocal in the ordinary sense, enjoys the same properties as the ordinary field theories. In particular, is it renormalizable and unitary? Does it have the same chiral anomalies as ordinary gauge theories? There has been intense research on these subjects and the previous questions have been answered at least in part. This is a short review of such results.

2 Noncommutative $U(N)$ Gauge Field Theories

In the introduction it was pointed out that noncommutative gauge field theories can be embedded in string theory. However these theories can also be defined on their own, without reference to string theory. Let us consider the Euclidean space $\mathbf{R}^d$ and define on it the complex algebra $\mathcal{A}_\theta$ endowed with the Moyal–Weyl product

$$f \star g(x) \equiv f(x) e^{\frac{i}{2}\theta^{\mu\nu}\overleftarrow{\partial}_\mu \overrightarrow{\partial}_\nu} g(x), \tag{1}$$

where $\theta^{\mu\nu}$ is the deformation parameter. This implies in particular noncommutativity in $\mathbf{R}^d$

$$x^\mu \star x^\nu - x^\nu \star x^\mu = i\theta^{\mu\nu} \tag{2}$$

It is natural to try to define a field theory in which the ordinary product is replaced by the Moyal product. In particular for a gauge theory we will assume the existence of a noncommutative connection A_μ with curvature

$$F_{\mu\nu} = \partial_\mu A_\nu - \partial_\nu A_\mu + iA_\nu \star A_\mu - iA_\mu \star A_\nu \tag{3}$$

and gauge transformation

$$\delta A_\mu = \partial_\mu \lambda + i\lambda \star A_\mu - iA_\mu \star \lambda \tag{4}$$

It is understood that both A and λ are hermitean: $A_\mu = A_\mu^B t^B$ and $\lambda = \lambda^B t^B$, where t^B is a complete set of $N \times N$ hermitean matrices: $(t^B)^\dagger = t^B$. In other words the connection A and the infinitesimal gauge transformations λ are $u(n)$–valued functions on $\mathbf{R}^d$. The theory we are introducing can therefore be called noncommutative $U(N)$ ($NCU(N)$) gauge theory. Its action is

$$S = -\frac{1}{4g^2} \int \mathrm{Tr}\,(F \star F) \tag{5}$$

We notice that if we expand the integrand in this action in power series of θ, we obtain an infinite series in the ordinary field $A_\mu(x)$ and its derivatives. The presence of higher and higher derivative terms would render an ordinary field theory non–local and untreatable. The remarkable thing of noncommutative theories is that, although they are non–local, the Moyal–Weyl product organizes such infinite series of terms so that they often behave like ordinary local theories.

For instance, the Feynman rules for the theory (5) can be extracted in the usual way. The propagator is the same as in the ordinary theory, but the vertices are different. As an example, the vertex for three gluons with momenta p_1, p_2, p_3 and polarizations ξ_1, ξ_2, ξ_3 is given, in the $NCU(1)$ case, by

$$(\xi_1 \cdot \xi_2 \, p_2 \cdot \xi_3 + \xi_1 \cdot \xi_3 \, p_1 \cdot \xi_2 + \xi_2 \cdot \xi_3 \, p_3 \cdot \xi_1) \, e^{-\frac{i}{2} p_1 \theta p_2} \tag{6}$$

where $p\theta q$ means $p_\mu \theta^{\mu\nu} q_\nu$.

As a matter of notation, we will use a basis of hermitean matrices $t^A = (t^A)^j{}_i$, (capital letters $A, B, ... = 0, \dots N^2-1$ will denote indices in the Lie algebra $u(N)$, while $i, j = 1, \dots, N$ are the indices in the fundamental representation), with the normalization

$$\mathrm{Tr}(t^A t^B) = \frac{1}{2}\,\delta^{AB} \ . \qquad (7)$$

This can be done, for example, by using a basis of hermitean matrices for the Lie algebra of $SU(N)$, t^a, (whenever necessary, lower case letters $a, b, ... = 1, \dots N^2-1$ will denote indices in the adjoint of $su(N)$), and adjoining $t^0 = \frac{1}{\sqrt{2N}}\mathbf{1}_N$. The basis t^A satisfies

$$[t^A, t^B] = i f_{ABC} t^C, \quad \{t^A, t^B\} = d_{ABC} t^C \qquad (8)$$

where f_{ABC} is completely antisymmetric, f_{abc} is the same as for $su(N)$ and $f_{0BC} = 0$, while d_{ABC} is completely symmetric; d_{abc} is the same as for $su(N)$, $d_{0BC} = \sqrt{\frac{2}{N}}\delta_{BC}$, $d_{00c} = 0$ and $d_{000} = \sqrt{\frac{2}{N}}$, see [5].

3 String Theory Embedding

As explained in the introduction, the noncommutative gauge theory introduced in the previous section (let us consider for the time being the $NCU(1)$ theory) can be immersed in a string theory. Let us follow the approach of N.Seiberg and E.Witten, [1]. Think of a closed string theory in the presence of a D–brane. The closed string theory contains in the gravity spectrum an antisymmetric massless B–field, which always appears in the equations of motion under the differentiation symbol. Therefore, it is always possible to add a constant part to B without affecting the field equations. In particular the vacuum configuration of such a theory is always defined up to a constant B field. In the absence of any D–brane, a constant B field can always be gauged away. But if the vacuum configuration contains a D–brane, this operation is not possible anymore along the D–brane world–volume. The reason is that in the D–brane world volume there exists a $U(1)$ gauge field A which together with B form a gauge invariant combination $B - dA$. Therefore a constant B–field can be eliminated by means of a gauge transformation on the bulk, but not along the D–brane world–volume. The upshot of this discussion is that when we are in presence of such a configuration we should allow for a constant B–field, rather than put it to zero.

Therefore let us suppose that we have a constant B field along a D–brane[1]. Now let us see the consequences of having a nonvanishing B–field. The D–brane dynamics is represented by the open strings attached to it. These open strings

[1] We suppose throughout the review that only $B_{\mu\nu}$ space components are nonvanishing. A nonvanishing time component would lead not to noncommutative field theory, but rather to an open string theory in a noncommutative ambient space, the so–called NCOS theories, which will not be considered in this review.

interact with B by the endpoints. This can be seen by looking at the sigma model action for such open strings:

$$S = \frac{1}{4\pi\alpha'}\int_\Sigma (g_{\mu\nu}\partial_a X^\mu \partial^a X^\nu - 2\pi\alpha' B_{\mu\nu}\epsilon^{ab}\partial_a X^\mu \partial_b X^\nu)$$
$$= \frac{1}{4\pi\alpha'}\int_\Sigma (g_{\mu\nu}\partial_a X^\mu \partial^a X^\nu) - \frac{i}{2}\int_{\partial\Sigma} B_{\mu\nu} X^\mu \partial_t X^\nu \tag{9}$$

after partial integration. Here $g_{\mu\nu}$ is the closed string metric, i.e. the metric background of the ambient space where the closed string lives, Σ is the world–sheet of the open string and $\partial\Sigma$ its boundary. The boundary conditions for the theory (9) are

$$g_{\mu\nu}\,\partial_n X^\nu + 2\pi\alpha'\, B_{\mu\nu}\,\partial_t X^\nu|_{\partial\Sigma} = 0 \tag{10}$$

where ∂_n, ∂_t are the normal and tangential derivatives to $\partial\Sigma$. We see that B interpolates between the Neumann boundary conditions ($B = 0$) and the Dirichlet ones ($B = \infty$).

Now, at tree level, the relevant world–sheet is the disk or the upper half plane. The string propagator in the upper half plane is

$$\langle X^\mu(z) X^\nu(z')\rangle$$
$$= -\alpha'\left(g^{\mu\nu}\ln|z-z'| - g^{\mu\nu}\ln|z-\bar{z}'| + G^{\mu\nu}\ln|z-\bar{z}'|^2 + \frac{1}{2\pi\alpha'}\theta^{\mu\nu}\ln\frac{z-\bar{z}'}{\bar{z}-z'}\right)$$

where

$$G^{\mu\nu} = \left(\frac{1}{g+2\pi\alpha' B}\,g\,\frac{1}{g-2\pi\alpha' B}\right)^{\mu\nu},$$
$$\theta^{\mu\nu} = -(2\pi\alpha')^2\left(\frac{1}{g+2\pi\alpha' B}\,B\,\frac{1}{g-2\pi\alpha' B}\right)^{\mu\nu}, \tag{11}$$

For open string amplitudes the relevant propagator is evaluated at the insertion points, i.e. on the real axis. This is

$$\langle X^\mu(\tau) X^\nu(\tau')\rangle = -\alpha' G^{\mu\nu}\ln(\tau-\tau')^2 + \frac{i}{2}\,\theta^{\mu\nu}\,\epsilon(\tau-\tau') \tag{12}$$

where τ, τ' are the real part of z, z', respectively.

Now, if we take $\alpha' \to 0$ keeping B and θ fixed we get

$$G^{\mu\nu} = -\frac{1}{(2\pi\alpha')^2}\left(\frac{1}{B}\,g\,\frac{1}{B}\right)^{\mu\nu}, \qquad \theta^{\mu\nu} = \left(\frac{1}{B}\right)^{\mu\nu}, \tag{13}$$

and

$$\langle X^\mu(\tau) X^\nu(\tau')\rangle = \frac{i}{2}\,\theta^{\mu\nu}\epsilon(\tau-\tau') \tag{14}$$

Now we can move on to compute amplitudes for the string modes. In particular the gluon is represented by the vertex operator

$$V(\xi, p) = \int_{\partial\Sigma} \xi \cdot \partial X \, e^{ip\cdot X} \tag{15}$$

where the momentum p and polarization ξ satisfy the relations: $p^2 = p \cdot \xi = 0$. The vertices are inserted on the real axis (if Σ is the upper half plane) in a definite order and then their positions are integrated over the entire real axis. The calculation from the string theory point of view is completely standard, and, once the limit $\alpha' \to 0$ is taken, the result is

$$\begin{aligned} &\langle V(\xi_1, p_1) V(\xi_2, p_2) V(\xi_3, p_3) \rangle \\ &= (\xi_1 \cdot \xi_2 \, p_2 \cdot \xi_3 + \xi_1 \cdot \xi_3 \, p_1 \cdot \xi_2 + \xi_2 \cdot \xi_3 \, p_3 \cdot \xi_1) \, e^{-\frac{i}{2} p_1 \theta p_2} \end{aligned} \tag{16}$$

which coincides with the vertex (6) above, which was obtained from the $NCU(1)$ gauge field theory.

It takes more work, but it can be proven that the generic n–gluon amplitudes obtained in the same way from string theory coincide with the tree level vertices derived from the $NCU(1)$ gauge theory. The generalization to $U(N)$ is straightforward. String amplitudes in this case are simply multiplied by the appropriate Chan–Paton (CP) factors. For instance, in the 3–gluon case, the CP factor is $\mathrm{Tr}(t^{A_1}\, t^{A_2} t^{A_3})$, where t^{A_i} belong to the basis of $N \times N$ hermitean matrices introduced above. It is easy to see that this make the amplitudes coming from strings coincide with those derived from $NCU(N)$ gauge theory.

In conclusion, at the tree level, there is a perfect correspondence between the gluon amplitudes obtained via string theory in the $\alpha' \to 0$ limit, and the analogous amplitudes obtained from $NCU(N)$ gauge theory. Then a question arises immediately: is this pattern going to persist also at one–loop? This means, on the field theory side, renormalizing $NCU(N)$ gauge theory at one loop and calculating the relevant renormalized amplitudes. On the string theory side it means calculating the string theory one–loop corrected amplitudes after taking the field theory limit of the latter. Finally one has to compare the two results and see whether they coincide.

This is what we are going to see in the next section.

3.1 Renormalization of $NCU(N)$

In this section we study one–loop renormalization of $NCU(N)$ theory. The theory we have to renormalize is specified by the gauge–fixed action

$$S = \int d^4x \, \mathrm{Tr} \left(-\frac{1}{4} F_{\mu\nu} F^{\mu\nu} - \frac{1}{2\alpha} (\partial_\mu A^\mu)^2 + \frac{1}{2} (i\bar{c} * \partial_\mu D^\mu c \; - \; i\partial_\mu D^\mu c * \bar{c}) \right) \tag{17}$$

where $c = c^A t^A$ is the Faddeev–Popov ghost field. The notation is as in the previous sections. To simplify the calculations we will choose $\alpha = 1$. The Feynman rules for this theory are given in the Appendix of [5], see also [2–4]. For instance, the propagators are the same as in the corresponding ordinary gauge theory, while the 3–gluon vertex is given by

$$-g\Big(f_{ABC}\,\cos(p\times q)+d_{ABC}\,\sin(p\times q)\Big)$$
$$\cdot\,(g_{\mu\nu}\,(p-q)_\lambda+g_{\nu\lambda}\,(q-k)_\mu+g_{\lambda\mu}(k-p)_\nu) \tag{18}$$

where the external gluons carry labels (A,μ,p), (B,ν,q) and (C,λ,k) for the Lie algebra, momentum and Lorentz indices and are ordered in anticlockwise sense. Moreover we use the notation $p\times q=\frac{1}{2}p_\mu\theta^{\mu\nu}q_\nu$.

Evaluating the one–loop contributions is lengthy but straightforward. The contributions split into two distinguished sets: *planar* and *nonplanar*. The first are characterized by the fact that the noncommutative factors (which are quadratic exponentials of the momenta) contain only external momenta, while in the nonplanar ones the noncommutative factors contain also the momentum running along the loop. Since eventually we integrate over the running momentum, it follows that in the latter case the noncommutative factors become smoothing factors for ultraviolet singularities. Therefore we should not expect ultraviolet divergences from nonplanar diagram contributions. As a consequence in the following we limit ourselves only to planar diagrams. In [5] the planar part of the 2–, 3– and 4–point functions were evaluated adopting the dimensional regularization ($\epsilon=4-D$, as usual). Here we write down some of the results. For instance for the 2–point function we have two nonvanishing contribution to the UV divergent part:

– gluons circulating inside the loop:

$$i\frac{1}{(4\pi)^2}\frac{2}{\epsilon}\delta_{AB}N\left[\frac{19}{12}g_{\mu\rho}p^2-\frac{11}{6}p_\mu p_\nu\right] \tag{19}$$

– ghosts circulating inside the loop:

$$i\frac{1}{(4\pi)^2}\frac{2}{\epsilon}\delta_{AB}N\left[\frac{1}{12}g_{\mu\rho}p^2+\frac{1}{6}p_\mu p_\nu\right] \tag{20}$$

Their sum is:

$$i\frac{1}{(4\pi)^2}\frac{2}{\epsilon}\delta_{AB}N\frac{5}{3}\left[g_{\mu\rho}p^2-p_\mu p_\nu\right] \tag{21}$$

which entails the usual renormalization constant

$$Z_3=1+\frac{5}{3}g^2N\frac{1}{(4\pi)^2}\frac{2}{\epsilon}\quad. \tag{22}$$

The same happens for the 3– and 4–point functions. They give rise to renormalization constants

$$Z_1=1+\frac{2}{3}g^2N\frac{1}{(4\pi)^2}\frac{2}{\epsilon}\quad. \tag{23}$$

and

$$Z_4 = 1 - \frac{1}{3}\, g^2\, N\, \frac{1}{(4\pi)^2}\, \frac{2}{\epsilon} \quad . \tag{24}$$

These are the same renormalization constant that occur in ordinary $U(N)$ Yang–Mills theories. Therefore, *the noncommutative* $U(N)$ *Yang–Mills theories are one–loop renormalizable.*

Now that we have examined the field theory side let us have a look at the string theory side. The one–loop calculation from the string theory point of view is much more complicated and we will limit ourselves to a short summary. The calculation of the one–loop corrections to the amplitudes considered in the previous section are the annulus amplitudes. I.e. the relevant world–sheet is the annulus. The latter can be represented as a unit disk from which a smaller disk centered at the origin has been cut out. This two dimensional surface has one modulus, which can be chosen to be the radius q of the smaller disk. The vertices relevant for the amplitudes in question are inserted at the border of the annulus. So they can be inserted either on the same (internal or external) boundary circle, giving rise to *planar* amplitudes, or are inserted on both boundary circles, giving rise to *non-planar* amplitudes. The field theory limit, in this situation, is not simply the limit $\alpha' \to 0$, because we have to put a condition also on the modulus. As intuition suggests, this corresponds to $q \to 1$, i.e. to the annulus being squeezed to a circle, in which case the string diagrams look as skinny as Feynman diagrams.

In the field theory limit the string diagrams that may give rise to divergences are only the planar ones. Since we are interested in a comparison with the divergent parts that appear in the one–loop renormalization just considered, we will limit ourselves to planar diagrams. Now the modification of the latter when a constant B–field is switched on is particularly simple. If we denote by $A_{(1)}(p_1, \ldots, p_n)$ and $\mathcal{A}_{(1)}(p_1, \ldots, p_n)$ the n–point one–loop planar amplitudes with and without B field, respectively, the relation is, [7],

$$\mathcal{A}_{(1)}(p_1, \ldots, p_n) = \prod_{i<j} e^{p_i \theta p_j}\, A_{(1)}(p_1, \ldots, p_n) \tag{25}$$

This result was extended to higher loops in [8]. Now what we have to do is find the field theory limit in the theory without B field and multiply it by the noncommutative factor as in (25). Finding a field theory limit of a one–loop amplitude is a nontrivial exercise, but the result can be found in the literature, see [6] and references therein. Once multiplied by the relevant noncommutative factor it reproduces exactly the divergences found in the renormalization of the $NCU(N)$ gauge theory above.

In conclusion we can therefore quote the striking result that the $NCU(N)$ gauge theory and the field theory limit of string theory with $U(N)$ CP factors in the presence of a B field do exactly correspond, at least up to one–loop.

3.2 Other Noncommutative Gauge Field Theories

In addition to $NCU(N)$ gauge theories there have been some attempts at defining noncommutative gauge field theories based on subalgebras of $u(N)$. Up to tree level there seems to be no obstruction to defining field theories of orthogonal or symplectic type, [14]. These theories can also be embedded in some particular configurations of string theory, that is they can be viewed as effective field theories living on D–branes immersed in some kind of string theory in the presence of a B–field. However the one–loop situation is still not clear, [16]. On the one hand, a consistent one–loop renormalization procedure has not yet been found; on the other hand, the string theory one–loop calculations are not entirely unambiguous in the presence of a B–field. Until these question marks have been removed one cannot safely rely on these new theories.

For a different approach to noncommutative gauge theories based on a generic Lie algebra, see [15].

4 Consistency Problems in Noncommutive Field Theories

Renormalizability is a first test of quantum consistency of a theory, but there are others. After considering renormalization of noncommutative gauge theories in the previous section, a fundamental consistency test is the absence of chiral anomalies. The presence of chiral anomalies is a fatal disease for a theory because it prevents us from defining the fermion functional integral. It is therefore of utmost importance to find out whether a theory is plagued by chiral anomalies. What we are interested in in the following is of course whether noncommutativity brings in any difference as far as anomalies are concerned. The answer will be that it does.

4.1 Chiral Anomalies

Let us couple a $NCU(N)$ gauge theory to fermionic matter and let us start from the simplest situation in which chiral anomalies are relevant, i.e. an action with chiral spinor in the fundamental representation in D dimensions:

$$S = \int d^D x \left(\bar{\psi}_i \star \gamma^\mu (i\partial_\mu \psi^i + A^i_{\mu\ k} \star P_+ \psi^k) \right) \tag{26}$$

Here $P_\pm = \frac{1}{2}(1 \pm \hat{\gamma})$ and $\hat{\gamma} = \gamma_0 \gamma_1 \dots \gamma_{D-1}$, and we have used the correspondence

$$A^j_{\mu i} \equiv A^B_\mu (t^B)^j{}_i, \qquad A_\mu = A^B_\mu t^B, \qquad A^B_\mu = 2\,\mathrm{tr}(t^B A_\mu)$$

where tr denotes the trace in the fundamental representation.

There are basically two ways to calculate chiral anomalies. One is based on Feynman diagram techniques [9,10,12], the second on the WZ consistency conditions, [11,13]. The method we review here is the second. It relies on the concept of *nc-locality*, which means that the space of cochains (i.e. field theory monomials such as action terms) we consider is the same as in ordinary local field theories with the ordinary product replaced by the Weyl–Moyal product. This principle of nc–locality is suggested by one–loop renormalization of noncommutative field theories, where counterterms are precisely of the above type, and, in the cases in which the noncommutative field theories can be embedded in string theory in the presence of a B field, can be traced back to the properties of string amplitudes, precisely to the fact that such string amplitudes factorize into noncommutative factors and ordinary string amplitudes (see previous section). The advantage of using this method is that, once the formalism is established, many conclusions are evident without resorting to explicit Feynman diagram calculations.

To write down the WZ consistency conditions we consider a matrix–valued one–form $A = A_\mu dx^\mu$, with gauge field strength two–form $F = dA + iA \star A$ and gauge transformation parameter c (it is the same as λ above but we take it to be a Grassmann–odd, i.e. the Faddeev–Popov ghost with ghost number 1). All these quantities are valued in the Lie algebra generated by the t^A. They are therefore hermitean matrices. The gauge (BRST) transformations are:

$$sA = dc - iA \star c + ic \star A, \qquad sc = -\, c \star c \tag{27}$$

d and s are assumed to commute. As a consequence the transformations (27) are nilpotent like in the ordinary case.

Now, following [11,13], we can write down the descent equations relevant to $D = 2n$ dimensions, starting from a closed and BRST invariant $2n+2$–form Ω_{2n+2}, constructed as a polynomial of F and referred to as the *top form*:

$$\begin{aligned} \Omega_{2n+2} &= d\Omega^0_{2n+1} \\ s\Omega^0_{2n+1} &= d\Omega^1_{2n} \\ s\Omega^1_{2n} &= d\Omega^2_{2n-1} \end{aligned} \tag{28}$$

where the upper index is the ghost number and the lower index is the form order. Ω^1_{2n} is the (unintegrated) anomaly. Upon integrating the last equation, one gets

$$s\left(\int d^D x\, \Omega^1_{2n}\right) = 0 \tag{29}$$

which precisely says that $\int d^D x\, \Omega^1_{2n}$ satisfies the Wess–Zumino consistency conditions. The virtue of the discent equations formalism is that it provides explicit expressions for anomalies and one is spared the details of the complicated verification that Ω^1_{2n} does indeed satisfy the Wess–Zumino consistency conditions. The latter is an automatic consequence of the top form Ω_{2n+2} being closed (and nontrivial).

In noncommutative gauge theories there is however a complication. This method does not work straightforwardly, because there exists no closed invariant polynomial that can be built with the noncommutative curvature F. There is however a way out that was pointed out in [11]: the differential space of cochains must be constituted by forms that are defined up to an overall cyclic permutations of the Moyal product factors involved. So, keeping this specification in mind, we can easily obtain the anomaly expression from the top form $\Omega_{2n+2} = \mathrm{tr}(F \star F \star F \star \ldots \star F)$ even though the latter, strictly speaking, is neither closed nor invariant. The anomaly is

$$\Omega^2_{2n-1} = n \int_0^1 dt \frac{(t-1)^2}{2} \mathrm{Tr}(dc * dc * A * F_t * ... * F_t + \ldots) \tag{30}$$

where the dots represent $(n-1)(n-2)-1$ terms obtained from the first by permuting in all distinct ways dc, A and $F_t = tdA + it^2 A \star A$, keeping track of the grading and keeping dc fixed in the first position.

In four dimensions the anomaly takes the form

$$\Omega^1_4 = -\frac{1}{2}\mathrm{Tr}(dc * A * dA + dc * dA * A + dc * A * A * A) \tag{31}$$

It we integrate it over space–time, it coincides with the result obtained via Feynman diagram methods, see [10], (24).

Before we discuss this equation let us consider other situations which may be potentially anomalous. There are not many. In fact it is well–known by now that, beside the fundamental representation of $U(N)$, only the anti–fundamental and the adjoint representations of $u(N)$ extend to linear representations of the Lie algebra of noncommutative $u(N)$ gauge transformations. So we can build noncommutative gauge theories only with the latter representations (or direct sums of them). Now as long as we stick to D=4, in ordinary gauge theories with chiral fermions in the adjoint representation the chiral anomaly identically vanishes. In such theories one can resort to the well–known argument of reality of the adjoint representation to reach the relevant conclusion. However this conclusion can be easily extended to noncommutative theories, [12,13]. Therefore, in $D = 4$, the only (potentially) anomalous action is the one specified by (26). The anomaly corresponding to it is given by (31). Let us discuss this anomaly in detail.

The main question is of course whether this anomaly may vanish under specific conditions. The first two terms in (31) are proportional to $\mathrm{Tr}(T^A T^B T^C)$. Therefore the anomaly (31) vanishes only if $\mathrm{Tr}(T^A T^B T^C) = 0$. Notice that in ordinary theories the anomaly is proportional to $\mathrm{Tr}(T^A[T^B, T^C])$, which vanishes for instance in the case of $SU(2)$. Is this possible for $NCU(N)$? The answer is no. In fact $\mathrm{Tr}(T^A T^B T^C) = \frac{1}{2}\mathrm{Tr}(T^A\{T^B, T^C\}) + \frac{1}{2}\mathrm{Tr}(T^C[T^B, T^C])$. The first term in the RHS is the usual symmetric ad–invariant third order tensor; the second term, which is absent in the commutative case, is proportional to the structure constant and vanishes only when all the structure constants do. Therefore we see that (31) cannot vanish. Therefore the only possibility for a noncommutative theory to be chiral anomaly free is to be non–chiral.

This is a very drastic conclusion, which can be extended to any even dimension (see [13]), and prevents, for instance, a simple extension of the Standard Model to the noncommutative case. Whether we can get around it and define anomaly free chiral noncommutative gauge theories is not a minor problem we have to face in the construction of noncommutative theories.

4.2 Unitarity and IR/UV Mixing

The previous subsection tells us that consistent noncommutative gauge theories can be found in the realm of theories without fermions or with nonchiral fermionic matter. Let us limit ourselves to such theories. However even in these theories there are new problems (as compared to ordinary theories) due to the overlap between IR and UV properties, [17].

We have noticed above that nonplanar diagrams do not give rise to UV singularities. This is the beneficial effect of IR/UV mixing, and it is due to the smoothing effect of the noncommutative factors in the high momentum region. This can be seen most clearly in the simple case of a noncommutative ϕ^4 theory in 4D, with mass m and coupling constant g. The two–point vertex at one loop can be easily evaluated. It splits, as usual, into a planar and a nonplanar part, which, after introducing a regulator Λ, are respectively given by

$$\Gamma^{(2)}_{planar} = \frac{g^2}{48\pi^2}\left(\Lambda^2 - m^2 \ln(\frac{\Lambda^2}{\mathrm{m}^2}) + \dots\right) \tag{32}$$

$$\Gamma^{(2)}_{nonplanar} = \frac{g^2}{96\pi^2}\left(\Lambda^2_{eff} - m^2 \ln(\frac{\Lambda^2_{\mathrm{eff}}}{\mathrm{m}^2}) + \dots\right) \tag{33}$$

where

$$\Lambda^2_{eff} = \frac{1}{\frac{1}{\Lambda^2} + p \circ p}, \qquad p \circ p = -p_\mu \theta^{\mu\lambda} \theta_\lambda{}^\nu p_\nu$$

the metric signature being $(-1, 1, \dots, 1)$.

From (32) above, we see that, as long as $p \circ p \neq 0$ we can safely remove the regulator Λ ($\Lambda \to \infty$).

However this beneficial effect in the UV is compensated by an increasing singularity pattern in the IR. To see this it is enough to look at the one–loop 1PI quadratic effective action, which takes the form (forgetting logarithms for simplicity)

$$S^{(2)}_{1PI} = \int d^4p \, \frac{1}{2}, \phi(p) \left(p^2 + M^2 + \frac{g^2}{96\pi^2(\frac{1}{\Lambda^2} + p \circ p)} + \dots\right) \phi(-p) \tag{34}$$

where $M^2 = m^2 + \frac{g^2 \Lambda^2}{48\pi^2} + \dots$ is the renormalized mass. From (34) we see that besides the usual propagator pole in p^2 we have another pole in $p \circ p$. This is not in the original spectrum of the theory.

The feature shown in this simple example is actually characteristic of all the noncommutative field theories, including the gauge theories. These modes, which appear dynamically in the spectrum, have been recognized as due to the lack of decoupling between the modes of the field theory leaving on the D–brane (i.e. the massless open string modes) and the closed string modes that live on the bulk.

4.3 Comments

The above–mentioned new modes, which are not present in the original spectrum of the theory, represent a problem for noncommutative field theories. They may affect renormalization and unitarity of noncommutative field theories at higher loops. But they also raise more radical questions. In fact it is clear that in general noncommutative field theories represent an intermediate species between ordinary field and string theory. They are interesting in themselves, but they might represent a serious opportunity as a new method to extract string theory results in a simplified way, without having to resort to full–fledged string theory . However the new modes, as well as the chiral anomaly problem we have seen above, represent an obstruction in this direction. The questions we would like to be able to answer are: can noncommutative field theories have an autonomous formulation, much in the same way ordinary field theories do? or, at some point, do we have to resort to our knowledge of string theory to understand otherwise incomprehensible facts about noncommutative field theories? Some answers to these issues have already been put forward, but the discussion is open.

Acknowledgements

These notes originate from talks given at the Conference on Symmetry Methods in Physics (Yerevan, Armenia, July 3-8, 2001) and at the 8th Adriatic Meeting on Particle Physics in the New Millennium, (Dubrovnik, Croatia, 4 - 14 September 2001). I would like to thank the organizers for their invitation and hospitality.

References

1. N. Seiberg and E. Witten, JHEP **09**, 032 (1999).
2. M. Sheikh–Jabbari, JHEP **06**, 015 (1999).
3. T. Krajewski and R. Wulkenhaar, Int. J. Mod. Phys. A **15**, 1011 (2000).
4. A. Armoni, Nucl. Phys. B **593**, 229 (2001).
5. L. Bonora and M. Salizzoni, Phys. Lett. B **504**, 80 (2001).
6. P. Di Vecchia et al, Nucl. Phys. B **469**, 235 (1996).
7. A. Bilal, C.-S. Chu, and R. Russo, Nucl. Phys. B **582**, 65 (2000).
8. C.-S. Chu, R. Russo, and S. Sciuto, Nucl. Phys. B **585**, 193 (2000).
9. F. Ardalan and N. Sadooghi, Int. Jour. Mod. Phys. A **16**, 3151 (2001).
10. J. Gracia–Bondia and C. P. Martin, Phys. Lett. B **479**, 321 (2000).
11. L. Bonora, M. Schnabl, and A. Tomasiello, Phys. Lett. B **485**, 311 (2000).
12. C. P. Martin, J. Phys. A **34**, 9037 (2001).
13. L. Bonora and A. Sorin, Phys. Lett. B **521**, 421 (2001).
14. L. Bonora et al, Nucl. Phys. B **589**, 461 (2000).
15. J. Madore et al, Eur. Phys. J. C **16**, 161 (2000).
16. L. Bonora and M. Salizzoni, hepth/0105307.
17. S. Minwalla, M. Van Raamsdonk, and N. Seiberg, JHEP **02**, 020 (2000); M. Van Raamsdonk and N. Seiberg, JHEP **03**, 035 (2000).

Regularization and Renormalization of Quantum Field Theories on Noncommutative Spaces

Harald Grosse

University of Vienna, Austria

Abstract. We first review regularization methods based on matrix geometry and show how an ultraviolet cut-off for scalar fields respecting symmetries results. Sections of bundles over the sphere can be quantized too. This procedure even allows to regularize supersymmetry without violating it. This work was extended recently to include quantum group covariant regularizations.

In a second part recent attempts to renormalize fourdimensional deformed quantum field theory models is reviewed. For scalar models the well-known IR-UV mixing does not allow to use standard techniques. The same applies to the Yang-Mills model in four dimensions. Only additional symmetry, as it occurs in the Wess-Zumino model, allows to avoid this problem.

Nevertheless there is some hope that the Yang-Mills model can be handled too. We used the Seiberg-Witten map to transform the noncommutative gauge field to a commutative one and used the degree of freedom of this map to obtain counter terms for the renormalization procedure. Finally a derivation of the Seiberg-Witten map from natural requirements is skechted.

1 Introduction

Almost ten years ago, I learnt from John Madore what he called the Fuzzy sphere. I was immediately enthusiastic: If you work within a geometry which does not have points, singularities of quantum field theory models may be cured. We applied these ideas first to simple two-dimensional models [1].

Lateron, together with Klimcik and Presnajder we treated all kind of models from scalar fields to spinor and gauge fields, even supersymmetric models were successfully treated. Lateron we dealt with models defined on $CP(2)$, on a cylinder respectively on an hyperboloid. All this was within what is now called the Lie-algebra-deformation.

Very recently, together with John Madore and Harold Steinacker we went to q-deformed models. We obtained a cut-off procedure based on a sequence of embedded Podles spheres and were able to formulate a differential calculus which led to the formulation of gauge models too. This way we connected the Yang-Mills models and the Chern-Simons model to those obtained from string theory.

Despite many attempts four dimensional quantum field theory on commutative space-time is still in bad shape. For two dimensional models constructive methods as well as new algebraic methods led to enormous insights. Even three-dimensional models are quite well understood. In four dimensions we have to rely on renormalized perturbation theory, and attempts to cure the diseases go from unification ideas to adding additional dimensions, strings etc. It is natural therefore to ask for alternatives.

One idea is to change the geometry. The final goal would be to include full quantized gravity, but on the way we may well study quantum field theoretic models on quantized space-time. This led recently to some surprises and new problems have shown up.

Nevertheless there is some hope that the Yang-Mills model can be handled and the first steps have been done in joined work with Wulkenhaar and the group at the Technical University of Vienna. Following the Muniche group we used the Seiberg-Witten map to expand the models on noncommutative space-time into a commutative gauge field. The idea is to use the degree of freedom of the SW-map to include the necessary counter terms for loop renormalization. This program is under development.

There is an old and simple argument that a smooth space-time manifold contradict quantum physics. If one localizes an event within a region of extension l an energy of the order hc/l is transfered. This energy generates a gravitational field. A strong gravity field prevents on the other hand signals to reach an observer. If one inserts the energy density to the rhs of Einsteins equation and puts a length r characterizing the curvature of space-time we get

$$1/r^2 = (G/c^4)(hc/l)(1/l^3) \tag{1}$$

Next we put the two length scales to be equal, since it is certainly operational impossible to localize an event beyond this resulting Planck length. To the best of our knowledge the first time this argument was put to precise mathematics was in the work by Doplicher, Fredenhagen and Roberts. [2] They obtained what is now called the canonical deformation but averaged over two-spheres. I believe that this averaging will not improve the divergences problems, although it seems that nobody did do explicite calculations.

We used in our work part of the ideas of noncommutative geometry. There are a number of books giving an overview [3], [4], [5]. We replace first the algebra of functions over a manifold by a noncommutative deformed algebra. Three kinds of deformations are treated. If the commutator of coordinates is put equal to a constant antisymmetric tensor, we call it the canonical deformation. If it is put to be a linear function of coordinates we call it a Liealgebra deformation, and if it is put to be a quadratic expression we call it a quantum group deformation. In the last case the Yang-Baxter equation has to be fulfilled.

Next step concerns the differential calculus. We replace vector fields by derivations on the algebra, we define differential forms by duality. Although it is somewhat tricky to do this for the q-deformed sphere, in all three cases one succeeds. Hodge duality, Lie-derivatives, the Laplace operator etc., in summary most of the steps of ordinary differential geometry can be simulated.

An essential step concerns fields. Classically they are sections of bundles, but also modules over the algebra. This last notion can be taken over to the noncommutative situation. We study finitely generated projective modules over the deformed algebra and are able to quantize scalar, spinor and gauge fields.

Finally we write down actions and integrate over the algebra, so we use certain trace functionals. Next steps concerns cohomology problems, the formulation of the spectral triples with the help of a Dirac operator and the use of cyclic cocyles. The final version of Alain Connes ideas concerns the spectral action which allows to unify all the interactions within one principle.

2 Regularization of Quantum Field Theory

2.1 Scalar Fields

The ideas of deforming or quantizing the commutative algebra of functions over a manifold can best be explained for the simple example of the two-sphere and leads to the Fuzzy Sphere. The euclidean action of a scalar field is given by

$$S[\Phi] = \frac{1}{4\pi} \int_{S^2} d\Omega [(J_i \Phi)^2 + \mu^2 (\Phi)^2 + POL(\Phi)] \tag{2}$$

and is invariant under isometies or rotations of the sphere. The generators are angular momentum operators, they close under $su(2)$. $\phi(x)$ can be expanded according to the infinite set of irreducible representations of $su(2)$.

Next we truncate this tower: Consider vector spaces transforming according to the first N representations. They can be identified with mappings from the representation space $N/2$ to itself. These mappings are $N+1$ times $N+1$ dimensional matrices, the noncommutative product of these is taken as the product within the algebra. In addition we have to give a precise description of the embedding of these algebras for different N, which gives a precise meaning also to the limit. For details see [6]. The Lie-algebra of the generators of this algebra is easy to describe. They form the $su(2)$ Lie- algebra with suitable rescaling, such that the Casimir operator still fulfills the defining relation for the two-sphere as an operator.

$$[\hat{X}_i, \hat{X}_j] = i\lambda \varepsilon_{ijk} \hat{X}_k \ , \quad \sum_{i=1}^{3} \hat{X}_i^2 = R^2 \ , \tag{3}$$

$$R\lambda^{-1} = \sqrt{\frac{N}{2}\left(\frac{N}{2} + 1\right)} \tag{4}$$

Since we work on a matrix algebra it is easy to introduce a differential calculus. All derivations are inner, they are given by the adjoint action with the generators themself. For the two-sphere we take the generators introduced above and can develop a derivation based differetial calculus. Our aim next is to make sense out of a functional integral of the type

$$\langle F[\Phi] \rangle = \frac{\int D\Phi e^{-S[\Phi]} F[\Phi]}{\int D\Phi e^{-S[\Phi]}} \tag{5}$$

The action (2) will be replaced by

$$S[\phi] = 1/NTr([\hat{X}_i, \phi][\hat{X}_i, \phi] + POL(\phi)) \tag{6}$$

where everything depends on N. As for the measure we take just the product measure for the finite number of degrees of freedom. This makes the functional integral well-defined.

In the limit where N tends to infinity, the old divergences show up. Feynman rules can be developed, a tadpole graph for the ϕ^4 model diverges logarithmically in the limit.

In the same spirit it is possible to quantize Kähler manifolds like $CP(n)$, models on the cylinder and on hyperboloids have been treated too.

In addition it is possible to quantize sections of line bundles over S^2, which are characterized by the Chern number. This way we construct projective modules which lead in a certain limit to the sections of the line bundles over S^2. We start from the Hopf-fibration of S^3 over S^2 with fiber $U(1)$. Let χ_1 and χ_2 be components of a spinor. Ristrict the sum of the squares of these two complex numbers to be R. This defines the sphere S^3. We study next expansions in terms of these two complex coordinpr.tex ates and there complex conjugates. Define $X_i = \chi^\dagger \sigma_i \chi$. The squares of X_i sum up to R^2 and define therefore S^2. If in an expansion an equal number of χ's and $\chi^\dagger$'s occur, it becomes a well-defined function on S^2. Otherwise it is a section of a bundle.

We next quantize this scheme: Replace the complex quantities by creation and annihilation operators of two bosons:

$$[A_i, A_j^\dagger] = \delta_{ij}, i, j = 1, 2 \tag{7}$$

This means we use the Jordan-Schwinger representation of $su(2)$. Next we define N-dimensional subspaces F_N of the Fockspace given by a fixed number of $N-1$ creation operators. They form an irreducible representation of $su(2)$. We study now maps from one such subspace to another one. If they are of equal dimension, square matrices will map one to the other and the sequence of them forms again the Fuzzy Sphere. If they are of different dimensions and their difference equals twice the topological charge, we quantize in a certain sence sections of the bundles. This way we obtain a sequence of embedded modules formed from nonsquare matrices of special size.

The Dirac operator and spinors were obtained from a supersymmetric treatment. We extended $su(2)$ to the supergroup $osp(2/1)$ and obtained a quantization of superfields through an embedded sequence of graded matrix algebras.

2.2 The Fuzzy Q-Sphere

The above described Fuzzy Sphere is invariant under the action of $SO(3)$, or equivalently under the action of $U(so(3))$. Together with John Madore and Harold Steinacker we defined a sequence of finite algebras, which have analogous properties, but which are covariant under the quantized universal enveloping algebra $U_q(su(2))$ [7]. This has been done for real q as well as for q being a root of unity. In the later case certain restrictions have to be obeyed. Covariance of an algebra A under $U_q(su(2))$ means that there exists an action

$$U_q(su(2)) \times A \to A, (u,a) \mapsto u \triangleright a \tag{8}$$

such that $u \triangleright (ab) = (u_{(1)} \triangleright a)(u_{(2)} \triangleright b)$ for $a, b \in A$. Here $\Delta(u) = u_{(1)} \otimes u_{(2)}$ is the Sweedler notation for the coproduct.

We may follow now the undeformed scheme and define q-deformed creation and annihilation operators.

$$A^{\dagger i} A_j = \delta^i_j + q\hat{R}^{ik}_{jl} A_k A^{\dagger l} \tag{9}$$

Next one considers again N dimensional subspaces of Fock space which form irreducible representations of $U_q(su(2))$. It is possible to define the q-deformed spheres in terms of $Z_i = A^{\dagger\alpha}\epsilon_{\alpha\beta}\sigma_i^{\beta\gamma}A_\gamma$ by

$$\epsilon^{ij}_k Z_i Z_j = constant Z_k \tag{10}$$

and to prove that the full matrix algebra is generated and to map the generators from the universal enveloping algebra to the others. After a study of the reality structure we introduced an invariant integral and studied a differential calculus. There exists a unique threedimensional module of 1-forms. As opposed to the classical case, an additional radial one-form shows up. This leads to the addition of a scalar field.

Next it is possible to write down actions for scalar fields as well as for gauge fields. Three possible actions can be formed for the later.

$$S_1 = \int A * A + 2A\Theta \tag{11}$$

$$S_2 = 2\int A^3 + 3(AdA + A * A) + 6A\Theta + 2\Theta^3 \tag{12}$$

$$S_3 = \int (dA + A^2) * (dA + A^2) \tag{13}$$

were * denotes the Hodge star operation and Θ a special one-form which plays the role of the Maurer-Cartan form. The commutator of algebra elements with this special form gives one-forms.

$$df = [\Theta, f] \tag{14}$$

As for the step from one-forms to two-forms one has to take the commutator but subtract the Hodge star of the one form in order to get two-forms.

$$d\alpha = [\Theta, \alpha]_+ - *\alpha \tag{15}$$

where the anticommutator enters. The step to three-forms is given again by the commutator with this special form.

It is interesting to note that special linear combinations of S_i 's correspond to the Yang-Mills action and to the Chern-Simons action. A special combination of both resulted from string theory after taking a particular limit [8]. Very recently we studied second quantization of a field theory on the q-deformed fuzzy sphere for q real. For this case it was necessary to perform a path integral over modes, which generate a quasiassociative algebra. This way we kept the symmetry and obtained a smooth limit for q going to one.

3 Renormalization of Deformed Quantum Field Theory

3.1 IR-UV Mixing

Recently after the work of Seiberg and Witten [9] many attempts have been done to study field theory on noncommutative space-time. In analogy to quantum mechanics one assumes that the four dimensional coordinates obey a commutator relation like

$$[x^\mu, x^\nu] = -2i\theta^{\mu\nu}, \tag{16}$$

This leads to the simplest noncommutative space-time and allows explicite calculations. Plane waves $u_p = e^{ip_\mu x^\mu}$ obey the algebra

$$u_p u_q = e^{i\theta^{\mu\nu} p_\mu q_\nu} u_{p+q} \tag{17}$$

were p and q denote commutative d-dimensional momenta. Derivations on this algebra are defined through the multiplication with these momenta, an integral on this algebra is defined by mapping to the $p = 0$ part. For the noncommutative torus p and q runs over an integer lattice. The later occured through compactification of M-theory on noncommutative 2-tori [10] and corresponds to turning on a background three-form.

Superpositions of plane waves with smooth functions $f(x)$ gives the Weyl operators $W[f]$. A standard question is which product of functions allows to encode this algebra. The answer is given by the Moyal-Weyl product

$$W[f]W[g] = W[f * g], (f * g)(x) = e^{i\theta^{\mu\nu}\partial_\mu\partial_\nu} f(x)g(y) \tag{18}$$

with x equal to y.

Since divergences come from singularities due to point like interactions, there was hope that fields on these deformed algebras may be better behaved. Unfortunately divergences still persist. Loop integrals in nc theories differ only by phase factors $e^{i\theta^{\mu\nu}p_\mu q_\nu}$ [11]. For $p = q$, which corresponds to planar diagrams, the same integral with the same divergences occurs as in undeformed models.

Next question concerns renormalizability: It turns out that the Yang-Mills model is one-loop renormalizable, divergences can be absorbed such that Ward-Slavnov identities are fulfilled.

For higher loop contributions a phenomenon similar to the IR-UV mixing of the ϕ^4 theory occurs. Although the one-loop integrals for nonplanar diagrams is finite for generic external momenta, it diverges for zero external momenta.

This causes infrared problems even in massive theories. Inserted in higher loop contributions it gives rise to divergences which cannot be absorbed by standard procedures [12]. They would give rise to counter terms which are of different structure as the initial action was. We therefore conclude that the regulating phase becomes inefficient for vanishing momenta, which manifests itself as IR divergences. Higer loop contributions have been analysed by Chepelev and Roiban [13]. They represented Feynman graphs by ribbons, have drawn them on a genus-g Rieman surface with boundary and established a convergence theorem except for two kind of dangerous cases. So called rings stacked on the came cycle led to singularities (which may be summed-up). Commutants occur if exceptional momenta lead to a vanishing of the oscillating phase factor. This leads to non-local counterterms.

One possible way out concerns additional symmetries. We used a superfield formulation and established a proof, that the Wess-Zumino model does not have the above mentioned disease [14]. Since only logarithmically divergent diagrams occur, and we showed earlier that even iterated integrals of this type are harmless, the standard renormalizability proof can be adapted.

3.2 Yang-Mills Fields

Recently nc Yang-Mills theory attracted many people. Especially Seiberg and Witten argued that there should exist an equivalence of regularization schemes (point splitting versus Pauli-Villars), and that there should exist a map which relates the nc gauge field with nc gauge parameter to counterparts on ordinary space-time.

Denote by $\hat{\lambda}$, $\hat{A}$ and $\hat{F}$ gauge parameter, gauge connection and field strength for the deformed model, and by λ, A and F the appropriate quantities for the undeformed case. For the Seiberg-Witten map one requires that

$$\delta_{\lambda[\hat{\lambda},A]}\hat{A}[A] = \hat{A}[\delta_\lambda A] \tag{19}$$

which means that the gauge transformed nc field should be equal to the nc field of the gauge transformed commutative one. This requirement leaves a great degree of freedom open. Various methods are avalable to get the mapping from A to $\hat{A}$, or better to get the expansion of the later in terms of an expansion of A.

This map is meant to be given as a formal power expansion series in $\theta^{\mu\nu}$. The Muniche group [15] [10] used these ideas and argued that this is the way to obtain a finite number of degrees of freedom in non-Abelian nc Yang-Mills theories.

If one inserts the Seiberg-Witten map to the Yang-Mills action it leads to a gauge field theory with an infinite number of vertices and Feynman graphs with unbounded degree of divergences, which seems to rule out a perturbative renormalization. An explicite quantum field theoretical investigation using this map for nc Maxwell theory led to the surprising result that the one-loop photon self-energy is gauge-invariant and gauge independent. It was not renormalizable.

However, it was possible to absorb the divergences into gauge invariant extension terms added to the classical action involving θ, which was interpreted as coming from a modified scalar product.

It turns out, that the extended action is actually a part of the freedom which exists in the Seiberg-Witten map. It means that a renormalization of constants which are free in that map can be used to absorb the one-loop divergences. This extends to a complete proof of all-order renormalizability of the photon self-energy. This freedom in the Seiberg-Witten map can be regarded as a field redefinition.

3.3 Observer versus Particle Lorentz Transformations

In general one should distinguish between two kinds of Lorentz (or more general, conformal) transformations. Lorentz transformations in special relativity relate physical observations made in two inertial reference frames characterized by different velocities and orientations. These transformations can be implemented as coordinate changes, known as *observer Lorentz transformations*. Alternatively one considers transformations which relate physical properties of two particles with different helicities or momenta within one specific inertial frame. These are known as *particle Lorentz transformations*. Usually (without background) these two approaches are equivalent. However, in the presence of a background tensor field this equivalence fails, because the background field will transform as a tensor under observer Lorentz transformation and as a set of scalars under particle Lorentz transformations.

Thirdly, having a background tensor field one may consider the transformations of *all* fields within a specific inertial frame simultaneously, including the background field. These transformations are known as *(inverse) active Lorentz transformations* and are equivalent to observer Lorentz transformations.

What kind of 'field' is $\theta^{\alpha\beta}$? Since we are considering the case of a constant θ, it certainly is a background field. Therefore we will refer always to 'observer' transformations. This also matches to the setting of noncommutative field theory appearing in string theory. Here θ is related to the inverse of a 'magnetic field' (mostly taken to be constant). In this sense, Lorentz invariance of the action means that its value is the same for observers in different inertial reference frames. Since invariance of the action always involves the sum of conformal transformations of $\hat{A}$ and θ, one can however take the 'particle' point of view and regard our 'observer' invariance as the quantitative amount of 'particle' symmetry breaking due to the presence of θ.

In the rest of the review we will simply refer to conformal transformations, leaving out the 'observer' prefix.

3.4 Rigid Conformal Symmetries

The Lie algebra of the rigid conformal transformations is generated by momentum, Lorentz generators and dilatation $P_\tau, M_{\alpha\beta}, D$ and obey the standard commutation relations.

A particular representation is given by infinitesimal rigid conformal transformations of the coordinates x^μ, which is given for example for the Lorentz transformation by

$$(x^\mu)^R = (1 + \omega^{\alpha\beta}\rho_x(M_{\alpha\beta}))x^\mu + \mathcal{O}(\omega^2), \rho_x(M_{\alpha\beta}) = x_\beta\partial_\alpha - x_\alpha\partial_\beta \qquad (20)$$

for constant parameters $\omega^{\alpha\beta}$.

A field is by definition an irreducible representation of the Lie algebra. In view of the noncommutative generalization we are interested in the Yang-Mills field A_μ (and the constant antisymmetric two-tensor field $\theta^{\mu\nu}$) for which a representation of the generator of Lorentz transformation is given by

$$W^R_{A;\alpha\beta} := \int d^4x\, \mathrm{tr}\Big(\big(g_{\mu\alpha}A_\beta - g_{\mu\beta}A_\alpha + x_\alpha\partial_\beta A_\mu - x_\beta\partial_\alpha A_\mu\big)\frac{\delta}{\delta A_\mu}\Big), \qquad (21)$$

We use the following differentiation rule for an antisymmetric two-tensor field:

$$\frac{\partial\theta^{\mu\nu}}{\partial\theta^{\rho\sigma}} := \frac{1}{2}\left(\delta^\mu_\rho\delta^\nu_\sigma - \delta^\mu_\sigma\delta^\nu_\rho\right) . \qquad (22)$$

and observe that the Yang-Mills action

$$\Sigma = -\frac{1}{4g^2}\int d^4x\, \mathrm{tr}\big(F_{\mu\nu}F^{\mu\nu}\big) , \qquad (23)$$

where $F_{\mu\nu} = \partial_\mu A_\nu - \partial_\nu A_\mu - \mathrm{i}[A_\mu, A_\nu]$ is the Yang-Mills field strength and g a coupling constant, is invariant under all our transformations. Moreover the action (23) is invariant under gauge transformations

$$W^G_{A;\lambda} = \int d^4x\, \mathrm{tr}\Big(D_\mu\lambda\frac{\delta}{\delta A_\mu}\Big) , \qquad D_\mu\bullet = \partial_\mu\bullet - \mathrm{i}[A_\mu, \bullet] , \qquad (24)$$

with a possibly field-dependent transformation parameter λ.

Next we were able to show that the noncommutative gauge field forms an irreducible representation of *the same undeformed* Lie algebra of rigid conformal transformations. To obtain the representation we took the symmetric product when going to the noncommutative realm: $AB \to \frac{1}{2}\{A, B\}_\star$. Compatibility with gauge transformations implies that only the sum of the conformal transformations of gauge field $\hat{A}$ and θ has a meaning. A covariant splitting of this sum allows a θ-expansion into a commutative gauge theory.

3.5 Conformal Transformations of the Noncommutative Gauge Field

We generalize the (rigid) conformal transformations to noncommutative Yang-Mills theory, i.e. a gauge theory for the field $\hat{A}_\mu$ transforming according to (21):

$$W^R_{\hat{A};\alpha\beta} := \int d^4x\, \mathrm{tr}\Big(\Big(\frac{1}{2}\{x_\alpha, \partial_\beta\hat{A}_\mu\}_\star - \frac{1}{2}\{x_\beta, \partial_\alpha\hat{A}_\mu\}_\star + g_{\mu\alpha}\hat{A}_\beta - g_{\mu\beta}\hat{A}_\alpha\Big)\frac{\delta}{\delta\hat{A}_\mu}\Big), \qquad (25)$$

where $\{U,V\}_\star := U\star V + V\star U$ is the $\star$-anticommutator. It is important to take the symmetric product in the "quantization" $x_\alpha \partial_\beta A_\mu \mapsto \frac{1}{2}\{x_\alpha, \partial_\beta \hat{A}_\mu\}_\star$. Let us introduce the convenient abbreviation $W^?_{\hat{A}}$ standing for one of the operators $\{W^T_{\hat{A};\tau}, W^R_{\hat{A};\alpha\beta}, W^D_{\hat{A}}\}$ and similarly for $W^?_\theta$.

Applying $W^R_{\hat{A};\alpha\beta}$ to the noncommutative Yang-Mills field strength $\hat{F}_{\mu\nu} = \partial_\mu \hat{A}_\nu - \partial_\nu \hat{A}_\mu - \mathrm{i}[\hat{A}_\mu, \hat{A}_\nu]_\star$ one obtains $W^R_{\hat{A};\alpha\beta}\hat{F}_{\mu\nu}$ which is not the expected Lorentz transformation of the field strength. However, we must also take the θ-transformation into account, which acts on the $\star$-product in the $\hat{A}$-bilinear part of $\hat{F}_{\mu\nu}$. Using the differentiation rule for the $\star$-product

$$W^?_\theta(U\star V) = \left(W^?_\theta U\right)\star V + U\star\left(W^?_\theta V\right) + \frac{\mathrm{i}}{2}\left(W^?_\theta \theta^{\mu\nu}\right)(\partial_\mu U)\star(\partial_\nu V)\ , \tag{26}$$

which is a consequence of (22), together with

$$W^?_\theta \hat{A}_\mu = 0\ , \tag{27}$$

one finds that $W^R_{\theta;\alpha\beta}\hat{F}_{\mu\nu}$ cancels exactly part of the above expression. It is then easy to verify that the noncommutative Yang-Mills (NCYM) action

$$\hat{\Sigma} = -\frac{1}{4g^2}\int d^4x\, \mathrm{tr}(\hat{F}^{\mu\nu}\star\hat{F}_{\mu\nu}) \tag{28}$$

is invariant under noncommutative translations, rotations and dilatations[1]:

$$W^T_{\hat{A}+\theta;\tau}\hat{\Sigma} = 0, W^R_{\hat{A}+\theta;\alpha\beta}\hat{\Sigma} = 0, W^D_{\hat{A}+\theta}\hat{\Sigma} = 0, \tag{29}$$

with the general notation

$$W^?_{A;C} + W^?_{B;C} = W^?_{A+B;C}. \tag{30}$$

Computing the various commutators between $W^?_{\hat{A}}$ one convinces oneself that the noncommutative gauge field $\hat{A}_\mu$ forms an irreducible representation of the conformal Lie algebra. It is remarkable that the conformal group remains the same and should not be deformed when passing from a commutative space to a noncommutative one whereas the gauge groups are very different in both cases. This shows that the fundamentals of quantum field theory—Lorentz covariance, locality, unitarity—have good chances to survive in the noncommutative framework.

In particular, the Wigner theorem that a field is classified by mass and spin holds. The conformal Lie algebra is of rank 2, hence its irreducible representations ρ are (in nondegenerate cases) classified by two Casimir operators,

$$m^2 = -g^{\tau\sigma}\rho(P_\tau)\rho(P_\sigma)\ , \qquad s(s+1)m^2 = -g_{\mu\nu}W^{PL;\mu}W^{PL;\nu}\ , \tag{31}$$

[1] In [20] we have shown that an identity like $W^D_\phi\hat{\Sigma} - 2\theta^{\mu\nu}(\partial\hat{\Sigma}/\partial\theta^{\mu\nu}) = 0$ exists for dilatation in the case of noncommutative ϕ^4 theory.

where

$$W^{PL;\mu} = -\frac{1}{2}\epsilon^{\mu\tau\alpha\beta}\rho(P_\tau)\rho(M_{\alpha\beta}) \tag{32}$$

is the Pauli-Ljubanski vector and m and s mass and spin of the particle, respectively. In our case where $\rho(?)$ is given by the action of $W^?_{\hat{A}+\theta}$ on $\hat{A}_\mu$ we find

$$m^2\hat{A}_\mu = -\partial^\tau\partial_\tau\hat{A}_\mu\ , g_{\rho\sigma}W^{PL;\rho}_{\hat{A}}W^{PL;\sigma}_{\hat{A}}\hat{A}_\mu = 2(g_{\mu\tau}\partial^\sigma\partial_\sigma - \partial_\mu\partial_\tau)\hat{A}^\tau + 0\,\partial_\mu\partial_\tau\hat{A}^\tau\ , \tag{33}$$

which means that the transverse components of $\hat{A}_\mu$ have spin $s = 1$ and the longitudinal component spin $s = 0$.

3.6 Compatibility with Gauge Symmetry

The NCYM action (9) is additionally invariant under noncommutative gauge transformations

$$W^G_{\hat{A};\hat{\lambda}} = \int d^4x\,\mathrm{tr}\Big(\big(\partial_\mu\lambda - \mathrm{i}\big[\hat{A}_\mu,\hat{\lambda}\big]_\star\big)\frac{\delta}{\delta\hat{A}_\mu}\Big), \tag{34}$$

where $\hat{\lambda}$ is a possibly $\hat{A}$-dependent gauge parameter. This means that the symmetry algebra of the NCYM action is at least[2] given by the Lie algebra

$$\mathcal{L} = \mathcal{G} \rtimes \mathcal{C} \tag{35}$$

of Ward identity operators, which is the semidirect product of the Lie algebra $\mathcal{G}$ of possibly field-dependent gauge transformations $W^G_{\hat{A};\hat{\lambda}}$ with the Lie algebra $\mathcal{C}$ of rigid conformal transformations $W^{\{T,R,D\}}_{\hat{A}+\theta}$. The commutator relations of $\mathcal{L}$ can now be computed.

It is crucial to use the sum of the individual transformations $W^{\{R,D\}}_{\hat{A}}$ and $W^{\{R,D\}}_\theta$ because the individual commutators do not preserve the Lie algebra $\mathcal{L}$:

One may ask whether there exists a 'rotation' in $(\hat{A},\theta)$ space so that the 'rotated fields' preserve individually the mixed commutators. To be concrete, what we look for is a splitting

$$W^?_{\hat{A}+\theta} \equiv W^?_{\hat{A}} + W^?_\theta = \tilde{W}^?_{\hat{A}} + \tilde{W}^?_\theta\ , [\tilde{W}^?_{\hat{A}}, W^G_{\hat{A};\hat{\lambda}}] = W^G_{\hat{A};\hat{\lambda}^?_{\hat{A}}}\ , [\tilde{W}^?_\theta, W^G_{\hat{A};\hat{\lambda}}] = W^G_{\hat{A};\hat{\lambda}^?_\theta}\ , \tag{36}$$

for appropriate field-dependent gauge parameters $\hat{\lambda}^?_{\hat{A}}$ and $\hat{\lambda}^?_\theta$. Each of the two relations in (36) is of course the consequence of the other relation. Furthermore, we impose the condition that the splitting should be universal in the sense $\tilde{W}^?_\theta = W^?_\theta(\theta^{\rho\sigma})\frac{d}{d\theta^{\rho\sigma}}$:

$$\tilde{W}^?_{\hat{A}} = W^?_{\hat{A}} - W^?_\theta(\theta^{\rho\sigma})\int d^4x\,\mathrm{tr}\Big(\frac{d\hat{A}_\mu}{d\theta^{\rho\sigma}}\frac{\delta}{\delta\hat{A}_\mu}\Big)\ , \tag{37}$$

[2] Renormlizability seems to require that the symmetry algebra of the NCYM action is actually bigger than $\mathcal{L}$.

$$\tilde{W}^?_\theta = W^?_\theta + W^?_\theta(\theta^{\rho\sigma}) \int d^4x \,\mathrm{tr}\Big(\frac{d\hat{A}_\mu}{d\theta^{\rho\sigma}} \frac{\delta}{\delta \hat{A}_\mu}\Big) \equiv W^?_\theta(\theta^{\rho\sigma}) \frac{d}{d\theta^{\rho\sigma}} \,. \tag{38}$$

The notation $\frac{d\hat{A}_\mu}{d\theta^{\rho\sigma}}$ is for the time being just a symbol for a field-dependent quantity with three Lorentz indices and power-counting dimension 3. Inserted into (36) one gets the *equivalent conditions*

$$-\mathrm{i}\big[\tilde{W}^?_{\hat{A}} \hat{A}_\mu, \hat{\lambda}\big]_\star - W^G_{\hat{A};\hat{\lambda}}(\tilde{W}^?_{\hat{A}}(\hat{A}_\mu)) = \hat{D}_\mu(\hat{\lambda}^?_{\hat{A}} - \tilde{W}^?_{\hat{A}}(\hat{\lambda}))\,, \tag{39}$$

$$W^?_\theta(\theta^{\rho\sigma})\Big(-\mathrm{i}\Big[\frac{d\hat{A}_\mu}{d\theta^{\rho\sigma}}, \hat{\lambda}\Big]_\star + \frac{1}{2}\{\partial_\rho \hat{A}_\mu, \partial_\sigma \hat{\lambda}\}_\star - W^G_{\hat{A};\hat{\lambda}}\Big(\frac{d\hat{A}_\mu}{d\theta^{\rho\sigma}}\Big)\Big) = \hat{D}_\mu(\hat{\lambda}^?_\theta - \tilde{W}^?_\theta(\hat{\lambda}))\,. \tag{40}$$

Whereas (40) cannot be solved without prior knowledge of the result[3], we can trivially solve (39) by a covariance ansatz were $\hat{\Omega}_{\rho\sigma\mu}$, a polynomial in the covariant quantities $\theta, \hat{X}, \hat{F}, \hat{D} \dots \hat{D}\hat{F}$ which is antisymmetric in ρ, σ and of power-counting dimension 3 shows up. For physical reasons (e.g. quantization) an $\hat{X}$-dependence of $\hat{\Omega}_{\rho\sigma\mu}$ should be excluded. We denote these transformations as *covariant transformations* of the noncommutative gauge field $\hat{A}$, because they reduce in the commutative case to the 'gauge-covariant conformal transformations' of Jackiw [18,19].

It follows that $\tilde{W}^?_\theta$ and thus $\frac{d\hat{A}_\mu}{d\theta^{\rho\sigma}}$ are (up to a gauge transformation) precisely the missing piece to complete an invariance of the action,

$$(\tilde{W}^R_{\hat{A};\alpha\beta} + \tilde{W}^R_{\theta;\alpha\beta})\hat{\Sigma} = 0\,, \qquad (\tilde{W}^D_{\hat{A};\alpha\beta} + \tilde{W}^D_{\theta;\alpha\beta})\hat{\Sigma} = 0\,. \tag{41}$$

Applying to the NCYM action (9) we obtain for $\hat{\Omega}_{\rho\sigma\mu} = 0$

$$\tilde{W}^T_{\hat{A};\tau}\hat{\Sigma} = 0\,, \tilde{W}^R_{\hat{A};\alpha\beta}\hat{\Sigma} = \frac{1}{g^2}\int d^4x \,\mathrm{tr}\Big(\theta_{\alpha\rho}\hat{F}^{\rho\sigma} \star \hat{T}_{\beta\sigma} - \theta_{\beta\rho}\hat{F}^{\rho\sigma} \star \hat{T}_{\alpha\sigma}\Big)\,, \tag{42}$$

$$\tilde{W}^D_{\hat{A}}\hat{\Sigma} = \frac{1}{g^2}\int d^4x \,\mathrm{tr}\Big(\theta^\delta{}_\rho \hat{F}^{\rho\sigma} \star \hat{T}_{\delta\sigma}\Big)\,, \tag{43}$$

where the quantity

$$\hat{T}_{\mu\nu} = \frac{1}{2}\hat{F}_{\mu\rho} \star \hat{F}_\nu{}^\rho + \frac{1}{2}\hat{F}_{\nu\rho} \star \hat{F}_\mu{}^\rho - \frac{1}{4} g_{\mu\nu}\hat{F}_{\rho\sigma} \star \hat{F}^{\rho\sigma} \tag{44}$$

resembles (but is not) the energy-momentum tensor. The calculation uses however the symmetry $\hat{T}_{\mu\nu} = \hat{T}_{\nu\mu}$ and tracelessness $g^{\mu\nu}\hat{T}_{\mu\nu} = 0$. As we have shown the first (rotational) condition has, reinserting $\hat{\Omega}_{\rho\sigma\mu}$, the solution

$$\frac{d\hat{A}_\mu}{d\theta^{\rho\sigma}} = -\frac{1}{8}\{\hat{A}_\rho, \partial_\sigma \hat{A}_\mu + \hat{F}_{\sigma\mu}\}_\star + \frac{1}{8}\{\hat{A}_\sigma, \partial_\rho \hat{A}_\mu + \hat{F}_{\rho\mu}\}_\star + \hat{\Omega}_{\rho\sigma\mu}\,, \tag{45}$$

---

[3] One can make of course an ansatz for $\frac{d\hat{A}_\mu}{d\theta^{\rho\sigma}}$ with free coefficients to be determined by (40).

which is also compatible with the second (dilatational) condition in (41). The solution (10a) is for $\hat{\Omega}_{\rho\sigma\mu} = 0$ known as the Seiberg-Witten differential equation It is now straightforward to check (40) for an arbitrary field-dependent gauge parameter $\hat{\lambda}$. The gauge parameters in (38) are

$$\hat{\lambda}^T_\tau = \hat{A}_\tau \,, \hat{\lambda}^R_{\alpha\beta} = \frac{1}{4}\{2x_\alpha + \theta_\alpha{}^\rho \hat{A}_\rho, \hat{A}_\beta\}_\star - \frac{1}{4}\{2x_\beta + \theta_\beta{}^\rho \hat{A}_\rho, \hat{A}_\alpha\}_\star \,, \hat{\lambda}^D = \frac{1}{2}\{x^\delta, \hat{A}_\delta\}_\star \,. \tag{46}$$

3.7 Summary and Outlook

We have established rigid conformal transformations for the noncommutative Yang-Mills field $\hat{A}$. Our results related to these transformations can be summarized as follows.

The (classical) noncommutative Yang-Mills action (9) is invariant under the Lie algebra $\mathcal{L}$ of gauge transformations $W^G_{\hat{A};\hat{\lambda}}$ and the sum $W^?_{\hat{A}} + W^?_\theta$ of conformal transformations of $\hat{A}$ and θ. The commutation relations $[W^?_{\hat{A}} + W^?_\theta, W^G_{\hat{A};\hat{\lambda}}] = W^G_{\hat{A};\hat{\lambda}'}$ in $\mathcal{L}$ suggest a *covariant* splitting $W^?_{\hat{A}} + W^?_\theta = \tilde{W}^?_{\hat{A}} + \tilde{W}^?_\theta$. The relation $[\tilde{W}^?_{\hat{A}}, W^G_{\hat{A};\hat{\lambda}}] = W^G_{\hat{A};\hat{\lambda}''}$ is trivially solved by a covariance ansatz. Then, the covariant complement $\tilde{W}^?_\theta$ is simply obtained from invariance of the NCYM action under $\tilde{W}^?_{\hat{A}} + \tilde{W}^?_\theta$ transformation. The solution for $\tilde{W}^?_\theta$ is given by the Seiberg-Witten differential equation. What we have thus achieved is a more transparent—and less restrictive—derivation of the Seiberg-Witten differential equation which does not require the usual ansatz of gauge equivalence.

Interpreting the Seiberg-Witten differential equation as an evolution equation we can express the noncommutative Yang-Mills field $\hat{A}$ in terms of its initial value A. The resulting θ-expansion of the NCYM action is invariant under *commutative* gauge transformations. Moreover, noncommutative conformal transformations reduce after θ-expansion to commutative conformal transformations. In this way we associate to the NCYM theory a gauge theory on commutative space-time for a commutative gauge field A coupled to a translation-invariant external field θ. Both gauge theories can be quantized by adding appropriate gauge-fixing terms and yield two quantum field theories respectively. It is unclear in which sense these two quantum field theories are equivalent. At least on a perturbative level the quantum field theories are completely different.

Loop calculations and power-counting analysis [13] reveal a new type of infrared singularities which so far could not be treated. Loop calculations [21] are free of infrared problems but lead apparently to an enormous amount of ultraviolet singularities. This is not necessarily a problem. For instance, all UV-singularities in the photon selfenergy are *field redefinitions* [22] which are possible in presence of a field $\theta^{\mu\nu}$ of negative power-counting dimension. For higher N-point Green's functions the situation becomes more and more involved and a renormalization seems to be impossible without a symmetry for the θ-expanded NCYM-action. We did hope at the beginning of the work that this symmetry searched for could be the Seiberg-Witten expansion of the noncommutative conformal symmetries. As we have seen this is not the case and the complete renormalization of NCYM theory remains an open problem.

We have proved that the noncommutative gauge field is an irreducible representation of the *undeformed* conformal Lie algebra. The noncommutative spin-$\frac{1}{2}$ representations for fermions have been worked out too. This shows that classical concepts of particles and fields extend without modification to a noncommutative space-time. We believe this makes life in a noncommutative world more comfortable.

Of course much work remains to be done. First we have considered a very special noncommutative geometry of a constant $\theta^{\mu\nu}$. This assumption should finally be relaxed; at least the treatment of those non-constant $\theta^{\mu\nu}$ which are Poisson bivectors as in [12] seems to be possible. The influence of the modified concept of locality on causality and unitarity of the S-matrix must be studied. Previous results with different consequences according to whether the electrical components of $\theta^{\mu\nu}$ are zero must be invariantly formulated in terms of the signs of the two invariants $\theta^{\mu\nu}\theta_{\mu\nu}$ and $\epsilon_{\mu\nu\rho\sigma}\theta^{\mu\nu}\theta^{\rho\sigma}$. Eventually the renormalization puzzle for noncommutative Yang-Mills theory ought to be solved.

Acknowledgement

I would like to thank J. Madore, H. Steinacker, R. Wulkenhaar, M. Schweda and his group for enjoyable collaborations and the organizers of this school for invitation.

References

1. Grosse, H., and Madore, J., *Phys. Lett.*, **B283**, 218 (1992)
2. Doplicher, S., Fredenhagen, K., and Roberts, J.E., *Commun. Math. Phys.*, **172**, 187 (1995).
3. Connes, A., *Noncommutative Geometry*, Academic Press, San Diego 1994.
4. J. Madore, *An introduction to noncommutative differential geometry and its physical applications*, Cambridge University Press, 1999.
5. J. M. Gracia-Bondia, J. C. Varilly, and H. Figueria, *Elements of noncommutative geometry*, Birkhäuser, Boston 2000.
6. H. Grosse, C. Klimcik, and P. Presnajder, Int. J. Theor. Phys. **35**, 231 (1996), Commun. Math. Phys. **178**, 507 (1996), Commun. Math. Phys. **185**, 155 (1997).
7. H. Grosse, J. Madore, and H. Steinacker, J. Geom. Phys. **38**, 208 (2001), hep-th/0005273, hep-th/0103164.
8. A. Yu. Alekseev, A. Recknagel, and V. Schomerus, JHEP **09**, 023 (1999).
9. N. Seiberg and E. Witten, JHEP **09**, 032 (1999).
10. A. Connes, M. R. Douglas, and A. Schwarz, JHEP **02**, 003 (1998).
11. T. Filk, Phys. Lett B **376**, 53 (1996).
12. S. Minwalla, M. Van Raamsdonk, and N. Seiberg, JHEP **02**, 002 (2000).
13. I. Chepelev and R. Roiban, JHEP **03**, 001 (2001).
14. A. Bichl et al, JHEP **10**, 046 (2000).
15. B. Jurčo, S. Schraml, P. Schupp and J. Wess, Eur. Phys. J. C **17**, 521 (2000).
16. J. Madore, S. Schraml, P. Schupp and J. Wess, Eur. Phys. J. C **16**, 161 (2000).

17. B. Jurčo, P. Schupp and J. Wess, Nucl. Phys. B **604**, 148 (2001).
18. R. Jackiw, Phys. Rev. Lett. **41**, 1635 (1978).
19. R. Jackiw, Acta Physica Austr. Suppl. XXII (1980) 383.
20. A. Gerhold, J. Grimstrup, H. Grosse, L. Popp, M. Schweda and R. Wulkenhaar, hep-th/0012112.
21. A. Bichl et al, JHEP **06**, 013 (2001).
22. A. Bichl et al, Eur. Phys. J. C **24**, 165 (2002).

Physical Instances of Noncommuting Coordinates

Roman Jackiw

Center for Theoretical Physics, Massachusetts Institute of Technology, Cambridge MA 02139-4307, USA

Abstract. Noncommuting spatial coordinates and fields can be realized in actual physical situations. Plane wave solutions to noncommuting photodynamics exhibit violaton of Lorentz invariance (special relativity).

1 Introduction

These days, investigators are probing the validity of Lorentz invariance (special relativity). This activity is documented by the papers presented at the Indiana meeting and submitted to the (recently postponed) Harvard meeting. Experimental and theoretical studies are pursued: experimentalists measure limits on Lorentz-violating processes; theorists build plausible Lorentz-violating extensions of the standard model.

When selecting Lorentz-violating terms, for possible inclusion in a modified standard model, we prefer to use structures that have a preexisting role in physics or mathematics. Thus our old proposal to add to the Maxwell Lagrangian the Lorentz-noninvariant quantity $\frac{m}{2}\int \mathrm{d}^3r\,\boldsymbol{A}\cdot\boldsymbol{B} = \frac{m}{2}\int \mathrm{d}^3r\,\boldsymbol{A}\cdot(\boldsymbol{\nabla}\times\boldsymbol{A})$, which leads to birefringence of the vacuum and to a Faraday-like rotation for the polarization of light propagating through the vacuum, makes use of the $\int \mathrm{d}^3r\,\boldsymbol{A}\cdot\boldsymbol{B}$ quantity, which was previously known in magnetohydrodynamics as the "magnetic helicity", in fluid mechanics (with the fluid velocity $\boldsymbol{v}$ replacing the electromagnetic vector potential $\boldsymbol{A}$) as the "kinetic vorticity", and in mathematics as the "Chern-Simons term". While the inclusion of this term in an electrodynamical theory leads to Lorentz, parity, and CTP violation, experiment conclusively rules out such a modification in Nature [1].

Another mechanism for Lorentz invariance breaking has become the focus of recent research: the suggestion is made that spatial coordinates need not commute. While present attention to this idea derives from string theory, we shall place this mechanism in the more familiar context of quantum mechanics and quantum field-theory.

Like many interesting quantal ideas, the notion that spatial coordinates may not commute can be traced to Heisenberg who, in a letter to Peierls, suggested that a coordinate uncertainty principle may ameliorate the problem of infinite self-energies. We shall describe later the physical application that Peierls made with Heisenberg's idea. Evidently, Peierls also described it to Pauli, who told it to Oppenheimer, who told it to Snyder, who wrote the first paper on the subject [2].

Let us begin with a physical application of the idea that goes back to Peierls.

2 Noncommutativity in the Presence of Strong Magnetic Fields

2.1 Particle Noncommutativity in the Lowest Landau Level

We are interested in a point-particle moving on a plane, with an external magnetic field $\mathbf{b}$ perpendicular to the plane. The equation for the 2-vector $\boldsymbol{r} = (x, y)$ is

$$m\dot{v}^i = \frac{e}{c}\varepsilon^{ij} v^j b + f^i(\boldsymbol{r}) \tag{1}$$

where $\mathbf{v}$ is the velocity $\dot{\boldsymbol{r}}$, and $\mathbf{f}$ represents other forces, which we take to be derived from a potential V: $\boldsymbol{f} = -\boldsymbol{\nabla} V$. Absent additional forces, the quantized theory gives rise to the well-known Landau levels, with separations $O(b/m)$. The limit of large b effectively projects onto the lowest Landau level, and is equivalent to small m. Setting the mass to zero in (1) leaves a first order equation

$$\dot{r}^i = \frac{c}{eb}\varepsilon^{ij} f^j(\mathbf{r}) \ . \tag{2}$$

This may be obtained by taking Poisson brackets of $\mathbf{r}$ with the Hamiltonian

$$H_0 = V \tag{3}$$

provided the fundamental brackets describe noncommuting coordinates,

$$\{r^i, r^j\} = \frac{c}{eb}\varepsilon^{ij} \tag{4}$$

so that

$$\dot{r}^i = \{H_0, r^i\} = \{r^j, r^i\}\partial_j V = \frac{c}{eb}\varepsilon^{ij} f^j(\mathbf{r}) \ . \tag{5}$$

The noncommutative algebra (4) and the associated dynamics can be derived in the following manner. The Lagrangian for the equation of motion (1) is

$$L = \tfrac{1}{2}mv^2 + \frac{e}{c}\boldsymbol{v}\cdot\boldsymbol{A} - V \tag{6}$$

where we choose the gauge $\boldsymbol{A} = (0, bx)$. Setting m to zero leaves

$$L_0 = \frac{eb}{c}x\dot{y} - V(x, y). \tag{7}$$

which is of the form $p\dot{q} - h(p, q)$, and one sees that $(\frac{eb}{c}x, y)$ form a canonical pair. This implies (4), and identifies V as the Hamiltonian.

Finally, we give a canonical derivation of noncommutativity in the $m \to 0$ limit, starting with the Hamiltonian

$$H = \frac{\pi^2}{2m} + V \ . \tag{8}$$

H gives (1) upon bracketing with $\boldsymbol{r}$ and $\boldsymbol{\pi}$, provided the following brackets hold:

$$\{r^i, r^j\} = 0 \tag{9}$$

$$\{r^i, \pi^j\} = \delta^{ij} \tag{10}$$

$$\{\pi^i, \pi^j\} = -\frac{eb}{c}\varepsilon^{ij} \ . \tag{11}$$

Here π is the kinematical (noncanonical) momentum, $m\dot{\boldsymbol{r}}$, related to the canonical momentum $\boldsymbol{p}$ by $\pi = \boldsymbol{p} - \frac{e}{c}\boldsymbol{A}$.

We wish to set m to zero in (8). This can only be done provided π vanishes, and we impose $\pi = 0$ as a constraint. But according to (11), the bracket of the constraints is nonzero, and the constraints are recognized to be "second-class" in Dirac's terminology. To proceed with the canonical formalism, we must introduce Dirac brackets. We omit the details of that technology, but merely record the resulting Dirac bracket:

$$\{r^i, r^j\}_D = \frac{c}{eb}\varepsilon^{ij} \ . \tag{12}$$

In this approach, noncommuting coordinates arise as Dirac brackets in a system constrained to lie in the lowest Landau level. Notice that the coordinate noncommutativity is already established at the classical level in that the Poisson bracket of coordinates is nonvanishing. Later we shall discuss the quantum version [3].

Peierls observed that when an impurity in the electron system is described by V, one can obtain the first-order energy shift of the lowest Landau level by taking the coordinates of (x, y) on which V depends to be noncommuting [4].

A further interesting subject, which is not discussed here, concerns the behavior of the wave function in the phase-space reductive, $m \to 0$, limit that projects onto the lowest Landau level. Before the reduction, the wave function is a normalized expression depending on the two coordinates. After the reduction, the wave function can depend only on one coordinate, because the other is a conjugate variable. How all this comes about is explained in the literature [3].

2.2 Field Noncommutativity in the Lowest Landau Level

The above demonstrates that spatial coordinates of particles in an intense magnetic field do not (Poisson) commute. But we are interested in fields. To find an example of noncommuting fields, we turn to the equations of a charged fluid, moving on a plane in an external magnetic field perpendicular to the plane. The fluid is described by a density ρ and velocity $\boldsymbol{v}$, both defined on the two-dimensional plane. A mass parameter m is introduced for dimensional reasons, so that the mass density is $m\rho$. The fields ρ and $\boldsymbol{v}$ are functions of t and $\boldsymbol{r}$ and give an Eulerian description of the fluid. The equations that are satisfied are the continuity equation

$$\dot{\rho} + \boldsymbol{\nabla} \cdot (\rho \boldsymbol{v}) = 0 \tag{13}$$

which expresses matter conservation, and the Euler equation

$$m\dot{v}^i + m\boldsymbol{v} \cdot \boldsymbol{\nabla} v^i = \frac{e}{c}\varepsilon^{ij} v^j b + f^i \tag{14}$$

which is the force equation. Here f^i describes additional forces, e.g., $-\frac{1}{\rho}\boldsymbol{\nabla}P$ where P is pressure. We shall take the force to be derived from a potential of the form

$$\boldsymbol{f}(\boldsymbol{r}) = -\boldsymbol{\nabla}\frac{\delta}{\delta\rho(\boldsymbol{r})}\int \mathrm{d}\boldsymbol{r}\, V. \tag{15}$$

[For isentropic systems, the pressure is only a function of ρ; (15) holds with V a function of ρ, related to the pressure by $P(\rho) = \rho V'(\rho) - V(\rho)$. Here we allow more general dependence of V on ρ (e.g., nonlocality or dependence on derivatives of ρ) and also translation noninvariant, explicit dependence on $\boldsymbol{r}$ [5].]

Equations (13)–(15) follow by bracketing ρ and $\boldsymbol{\pi}$ with the Hamiltonian

$$H = \int \mathrm{d}^2 r \Big(\rho\frac{\pi^2}{2m} + V\Big) \tag{16}$$

provided that fundamental brackets are taken as

$$\{\rho(\boldsymbol{r}), \rho(\boldsymbol{r}')\} \quad = 0 \tag{17}$$

$$\{\pi(\boldsymbol{r}), \rho(\boldsymbol{r}')\} \quad = \boldsymbol{\nabla}\delta(\boldsymbol{r} - \boldsymbol{r}') \tag{18}$$

$$\{\pi^i(\boldsymbol{r}), \pi^j(\boldsymbol{r}')\} = -\varepsilon^{ij}\frac{1}{\rho(\boldsymbol{r})}\Big(m\omega(\boldsymbol{r}) + \frac{eb}{c}\Big)\delta(\boldsymbol{r} - \boldsymbol{r}') \tag{19}$$

where $\varepsilon^{ij}\omega(\boldsymbol{r})$ is the vorticity $\partial_i v^j - \partial_j v^i$, and $\pi = m\boldsymbol{v}$.

We now consider a strong magnetic field and take the limit $m \to 0$, which is equivalent to large b. Equations (14) and (15) reduce to

$$v^i = -\frac{c}{eb}\varepsilon^{ij}\frac{\partial}{\partial r^j}\frac{\delta}{\delta\rho(\boldsymbol{r})}\int \mathrm{d}^2 r\, V\ . \tag{20}$$

Combining this with the continuity equation (13) gives the equation for the density "in the lowest Landau level":

$$\dot{\rho}(\boldsymbol{r}) = \frac{c}{eb}\frac{\partial}{\partial r^i}\rho(\boldsymbol{r})\varepsilon^{ij}\frac{\partial}{\partial r^j}\frac{\delta}{\delta\rho(\boldsymbol{r})}\int \mathrm{d}^2 r\, V \tag{21}$$

(For the right-hand side not to vanish, V must not be solely a function of ρ.)

The equation of motion (21) can be obtained by bracketing with the Hamiltonian

$$H_0 = \int \mathrm{d}^2 r\, V \tag{22}$$

provided the charge density bracket is nonvanishing, showing noncommutativity of the ρ's [6]:

$$\{\rho(\boldsymbol{r}), \rho(\boldsymbol{r}')\} = -\frac{c}{eb}\varepsilon^{ij}\partial_i\rho(\boldsymbol{r})\partial_j\delta(\boldsymbol{r}-\boldsymbol{r}') \ . \tag{23}$$

H_0 and this bracket may be obtained from (16) and (17) – (19) with the same Dirac procedure presented for the particle case: We wish to set m to zero in (16); this is possible only if π is constrained to vanish. But the bracket of the π's (19) is nonvanishing, even at $m=0$, because $b \neq 0$. Thus at $m=0$ we are dealing with a second-class constraint that leads to a nonvanishing Dirac bracket of densities as in (23):

$$\{\rho(\boldsymbol{r}), \rho(\boldsymbol{r}')\}_D = -\frac{c}{eb}\varepsilon^{ij}\partial_i\rho(\boldsymbol{r})\partial_j\delta(\boldsymbol{r}-\boldsymbol{r}') \ . \tag{24}$$

The ρ-bracket (23), (24) enjoys a more appealing expression in momentum space. Upon defining

$$\tilde{\rho}(\boldsymbol{p}) = \int \mathrm{d}^2 r\, e^{i\boldsymbol{p}\cdot\boldsymbol{r}}\rho(\boldsymbol{r}) \tag{25}$$

we find

$$\{\tilde{\rho}(\boldsymbol{p}), \tilde{\rho}(\boldsymbol{q})\} = -\frac{c}{eb}\varepsilon^{ij}p^i q^j \tilde{\rho}(\boldsymbol{p}+\boldsymbol{q}). \tag{26}$$

The form of the charge density bracket (23), (24), (26) can be understood by reference to the particle substructure for the fluid. Take

$$\rho(\boldsymbol{r}) = \sum_n \delta(\boldsymbol{r}-\boldsymbol{r}_n) \tag{27}$$

where n labels the individual particles. When the coordinates of each particle satisfy the nonvanishing bracket (4), (12), the $\{\rho(\boldsymbol{r}), \rho(\boldsymbol{r}')\}$ bracket takes the form (23), (24), (26).

2.3 Quantization

Quantization before the reduction to the lowest Landau level is straightforward. For the particle case (9)–(11) and for the fluid case (17)–(19) we replace brackets with $i/\hbar$ times commutators. After reduction to the lowest Landau level we do the same for the particle case thereby arriving at the "Peierls substitution," which (as mentioned previously) states that the effect of an impurity [V in (6)] on the lowest Landau energy level can be evaluated to lowest order by viewing the (x,y) arguments of V as noncommuting variables [4].

For the fluid, quantization presents a choice. On the one hand, we can simply promote the bracket (23), (24), (26) to a commutator by multiplying by $i/\hbar$.

$$[\rho(\boldsymbol{r}), \rho(\boldsymbol{r}')] = i\hbar\frac{c}{eb}\varepsilon^{ij}\partial_i\rho(\boldsymbol{r}')\partial_j\delta(\boldsymbol{r}-\boldsymbol{r}') \tag{28}$$

$$[\tilde{\rho}(\boldsymbol{p}), \tilde{\rho}(\boldsymbol{q})] = i\hbar\frac{c}{eb}\varepsilon^{ij}p^i q^j \tilde{\rho}(\boldsymbol{p}+\boldsymbol{q}) \tag{29}$$

Alternatively we can adopt the expression (27), for the operator $\rho(\boldsymbol{r})$, where the $\boldsymbol{r}_n$ now satisfy the noncommutative algebra

$$\left[r_n^i, r_{n'}^j\right] = -i\hbar\frac{c}{eb}\varepsilon^{ij}\delta_{nn'} \tag{30}$$

and calculate the ρ commutator as a derived quantity.

However, once $\boldsymbol{r}_n$ is a noncommuting operator, functions of $\boldsymbol{r}_n$, even δ-functions, have to be ordered. We choose the Weyl ordering, which is equivalent to defining the Fourier transform as

$$\tilde{\rho}(\boldsymbol{p}) = \sum_n e^{i\boldsymbol{p}\cdot\boldsymbol{r}_n} \; . \tag{31}$$

With the help of (30) and the Baker-Hausdorff lemma, we arrive at the "trigonometric algebra"

$$[\tilde{\rho}(\boldsymbol{p}), \tilde{\rho}(\boldsymbol{q})] = 2i\sin\Big(\frac{\hbar c}{2eb}\varepsilon^{ij}p^i q^j\Big)\tilde{\rho}(\boldsymbol{p}+\boldsymbol{q}) \; . \tag{32}$$

This reduces to (29) for small $\hbar$.

This form for the commutator, (32), is connected to a Moyal star product in the following fashion. For an arbitrary c-number function $f(\boldsymbol{r})$ define

$$\langle f\rangle = \int \mathrm{d}^2 r\, \rho(\boldsymbol{r}) f(\boldsymbol{r}) = \frac{1}{(2\pi)^2}\int \mathrm{d}^2 p\, \tilde{\rho}(\boldsymbol{p})\tilde{f}(-\boldsymbol{p}) \; . \tag{33}$$

Multiplying (32) by $\tilde{f}(-\boldsymbol{p})\tilde{g}(-\boldsymbol{q})$ and integrating gives

$$[\langle f\rangle, \langle g\rangle] = \langle h\rangle \tag{34}$$

with

$$h(\boldsymbol{r}) = (f\star g)(\boldsymbol{r}) - (g\star f)(\boldsymbol{r}) \tag{35}$$

where the "$\star$" product is defined as

$$(f\star g)(\boldsymbol{r}) = e^{\frac{i}{2}\frac{\hbar c}{eb}\varepsilon^{ij}\partial_i\partial'_j} f(\boldsymbol{r})g(\boldsymbol{r}')|_{\boldsymbol{r}'=\boldsymbol{r}} . \tag{36}$$

Note however that only the commutator is mapped into the star commutator. The product $\langle f\rangle\langle g\rangle$ is not equal to $\langle f\star g\rangle$.

The lack of consilience between (29) and (32) is an instance of the Groenwald-VanHove theorem, which establishes the impossibility of taking over into quantum mechanics all classical brackets [7]. Equations (30)–(36) explicitly exhibit the physical occurrence of the star product for fields in a strong magnetic background.

3 Various Algebras

Before proceeding with our construction of a noncommutative Maxwell field theory, let us summarize here the various (nontrivial) algebras that we have encountered in the above development.

The fluid velocity algebra (19) at $b = 0$ and $m = 1$ reads in any spatial dimension

$$\{v^i(\boldsymbol{r}), v^j(\boldsymbol{r}')\} = -\frac{1}{\rho(\boldsymbol{r})}\big(\partial_i v^j(\boldsymbol{r}) - \partial_j v^i(\boldsymbol{r})\big)\delta(\boldsymbol{r}-\boldsymbol{r}') \ . \tag{37}$$

This was first given by Landau [8]. In spite of the awkward appearance, the algebra in fact takes a familiar form when we define the momentum density $\boldsymbol{\mathcal{P}} = \rho\boldsymbol{v}$, and use (17), (18) for the ρ brackets. Then (37), with (17) and (18) implies

$$\{\mathcal{P}^i(\boldsymbol{r}), \mathcal{P}^j(\boldsymbol{r}')\} = \Big(\mathcal{P}^j(\boldsymbol{r})\frac{\partial}{\partial r^i} + \mathcal{P}^i(\boldsymbol{r}')\frac{\partial}{\partial r^j}\Big)\delta(\boldsymbol{r}-\boldsymbol{r}') \ . \tag{38}$$

This is the usual momentum density algebra, which also describes diffeomorphisms of space in the following fashion. If an infinitesimal coordinate transformation is given by

$$\delta r^i = -f^i(\boldsymbol{r}) \tag{39}$$

we define the average $\langle f\rangle$ of f^i by integrating with $\mathcal{P}^i$

$$\langle f\rangle \equiv \int \mathrm{d}\boldsymbol{r}\, f^i(\boldsymbol{r})P^i(\boldsymbol{r}) \tag{40}$$

then (38) has the consequence that for two such functions f and g we have

$$\{\langle f\rangle, \langle g\rangle\} = -\langle h\rangle \tag{41}$$

where h is the Lie bracket of f and g:

$$h^i = g^j\partial_j f^i - f^j\partial_j g^i \ . \tag{42}$$

By scaling ρ the noncommutative density algebra (23), (24) may be presented as

$$\{\rho(\boldsymbol{r}), \rho(\boldsymbol{r}')\} = \varepsilon^{ij}\partial_i\rho(\boldsymbol{r})\partial_j\delta(\boldsymbol{r}-\boldsymbol{r}') \ . \tag{43}$$

This intrinsically two-dimensional structure is the area-preserving algebra, studied by Arnold [9]. Area-preserving coordinate transformations (volume preserving in arbitrary dimensionality) possess unit Jacobian. For the infinitesimal form of the transformation (39) this means that f^i is transverse: $\partial_i f^i = 0$. Therefore, in two dimensions, an area-preserving transformation is generated by a scalar:

$$f^i = \varepsilon^{ij}\partial_j f \ . \tag{44}$$

When an average $\langle f\rangle$ is defined by

$$\langle f\rangle = \int \mathrm{d}^2 r\, f(\boldsymbol{r})\rho(\boldsymbol{r}) \tag{45}$$

equation (43) again implies (41), but now we have

$$h = \varepsilon^{ij}\partial_i f\partial_j g \tag{46}$$

which also follows from (42) when all three functions take the form (44).

Finally the algebra (32)

$$\{\tilde{\rho}(\boldsymbol{p}), \tilde{\rho}(\boldsymbol{q})\} = -\frac{2}{\hbar}\sin\Big(\frac{\hbar c}{2eb}\varepsilon^{ij}p^i q^j\Big)\tilde{\rho}(\boldsymbol{p}+\boldsymbol{q})$$

which also leads to the Moyal-star product (36) for averages (45), is called a trigonometric algebra, which was introduced by D. Fairlie, P. Fletcher, and C. Zachos [10].

4 Noncommutative Electrodynamics

Stimulated by the occurrence of the star product in the discussion of charged fluids in an intense magnetic field, we abstract the idea and use it in the new setting of noncommutative Maxwell theory. This theory is described by the vector potential $\hat{A}_\mu$ (the caret denotes noncommuting quantities) and the theory is built on a gauge-invariance principle. Gauge transformations act on $\hat{A}_\mu$ according to

$$\hat{A}_\mu \to \hat{A}_\mu^\lambda = (e^{i\lambda}) \star (\hat{A}_\mu + i\partial_\mu) \star (e^{i\lambda})^{-1} \ . \tag{47}$$

The star ($\star$) product of two quantities is defined by

$$(O_1 \star O_2)(\boldsymbol{r}) = e^{\frac{i}{2}\theta^{\mu\nu}\frac{\partial}{\partial r^\mu}\frac{\partial}{\partial r'^\nu}} O_1(\boldsymbol{r})O_2(\boldsymbol{r}')\big|_{\boldsymbol{r}=\boldsymbol{r}'} \tag{48}$$

and we take $\theta^{\mu\nu}$ to have no time components ($\theta^{0i} = 0$, $\theta^{ij} = \varepsilon^{ijk}\theta^k$). The field strength $\hat{F}_{\mu\nu}$ is constructed from $\hat{A}_\mu$ in a manner such that the gauge transformation (47) effects a covariant transformation:

$$\hat{F}_{\mu\nu} \to \hat{F}_{\mu\nu}^\lambda = (e^{i\lambda}) \star F_{\mu\nu} \star (e^{i\lambda})^{-1} \ . \tag{49}$$

This requirement is met, provided $\hat{F}_{\mu\nu}$ is given by

$$\hat{F}_{\mu\nu} = \partial_\mu \hat{A}_\nu - \partial_\nu \hat{A}_\mu - i[\hat{A}_\mu, \hat{A}_\nu]_\star \tag{50}$$

where $[\hat{A}_\mu, \hat{A}_\nu]_\star = \hat{A}_\mu \star \hat{A}_\nu - \hat{A}_\nu \star \hat{A}_\mu$. Finally, the action is taken to be

$$\begin{aligned} \hat{I} &= -\tfrac{1}{4}\int \mathrm{d}^4x\, \hat{F}^{\mu\nu} \star \hat{F}_{\mu\nu} \\ &= -\tfrac{1}{4}\int \mathrm{d}^4x\, \hat{F}^{\mu\nu}\hat{F}_{\mu\nu} \ . \end{aligned} \tag{51}$$

One would like to find the equations of motion, calculate physically interesting quantities, and compare them to corresponding quantities in the Maxwell theory. In this way one could assess the effect of noncommutativity and perhaps place experimental limits on it. However, a problem arises: local quantities in noncommutative electrodynamics are gauge variant and no invariant meaning can be assigned to their profiles. Nonlocal, integrated, expressions can be gauge invariant, (for example, the action (51) is gauge invariant) but in the ordinary Maxwell theory we deal with local quantities (like profiles of electromagnetic waves) and we would like to compare these classical local disturbances to corresponding quantities in the noncommutative theory.

A way out of this difficulty is provided by Seiberg and Witten's observation that the noncommuting gauge theory may be equivalently described by a commuting gauge theory that is formulated in terms of ordinary (not star) products of a commuting vector potential A_μ, together with an explicit dependence on $\theta^{\alpha\beta}$, which acts as a constant "background". This equivalence is established by expressing the noncommuting vector potential $\hat{A}_\mu$ as a function of A_μ and $\theta^{\alpha\beta}$ that solves the Seiberg-Witten equation [11]

$$\frac{\partial \hat{A}_\mu}{\partial \theta^{\alpha\beta}} = -\tfrac{1}{8}\{\hat{A}_\alpha, \partial_\beta \hat{A}_\mu + \hat{F}_{\beta\mu}\}_\star - (\alpha \leftrightarrow \beta) \tag{52}$$

where the bracketed expression denotes the "star" anticommutator. Solutions of this equation are expressed in terms of $\theta^{\alpha\beta}$ and the "initial condition" $\hat{A}_\mu\big|_{\theta^{\alpha\beta}=0}$; the latter quantity being just the commuting A_μ.

We work to lowest order in θ and find

$$\hat{A}_\mu = A_\mu - \tfrac{1}{2}\theta^{\alpha\beta} A_\alpha\big(\partial_\beta A_\mu + F_{\beta\mu}\big) \; . \tag{53}$$

The noncommuting action, expressed in terms of the commuting quantities A_μ, $F_{\mu\nu} = \partial_\mu A_\nu - \partial_\nu A_\mu$, and $\theta^{\alpha\beta}$, now reads [12]

$$\hat{I} = -\tfrac{1}{4}\int \mathrm{d}^4x \Big((1 - \tfrac{1}{2}\theta^{\alpha\beta}F_{\alpha\beta})F^{\mu\nu}F_{\mu\nu} + 2\theta^{\alpha\beta}F_{\mu\alpha}F_{\nu\beta}F^{\mu\nu}\Big) \; . \tag{54}$$

This is gauge invariant in the conventional sense, and from the equations of motion that are implied by $\hat{I}$ we can determine the gauge-invariant electric ($E^i = F^{i0}$) and magnetic fields ($B^i = -\varepsilon^{ijk}F_{jk}$).

These fields satisfy the equations, which maintain a Maxwell form.

$$\begin{aligned} \frac{1}{c}\frac{\partial}{\partial t}\boldsymbol{B} + \boldsymbol{\nabla}\times\boldsymbol{E} &= 0 \\ \boldsymbol{\nabla}\cdot\boldsymbol{B} &= 0 \end{aligned} \tag{55}$$

$$\begin{aligned} \frac{1}{c}\frac{\partial}{\partial t}\boldsymbol{D} - \boldsymbol{\nabla}\times\boldsymbol{H} &= 0 \\ \boldsymbol{\nabla}\cdot\boldsymbol{D} &= 0 \end{aligned} \tag{56}$$

The first set (55) reflects the gauge invariance of the system, namely, that $\boldsymbol{E}$ ant $\boldsymbol{B}$ are given in terms of potentials. The second set (56) is a consequence of the nonlinear dynamics implied by (54). The constitutive relations relating $\boldsymbol{D}$ and $\boldsymbol{H}$ to $\boldsymbol{E}$ and $\boldsymbol{B}$ follow from (54):

$$\boldsymbol{D} = (1 - \boldsymbol{\theta}\cdot\boldsymbol{B})\boldsymbol{E} + (\boldsymbol{\theta}\cdot\boldsymbol{E})\boldsymbol{B} + (\boldsymbol{E}\cdot\boldsymbol{B})\boldsymbol{\theta} \tag{57}$$

$$\boldsymbol{H} = (1 - \boldsymbol{\theta}\cdot\boldsymbol{B})\boldsymbol{B} - (\boldsymbol{\theta}\cdot\boldsymbol{E})\boldsymbol{E} + \tfrac{1}{2}(\boldsymbol{E}^2 - \boldsymbol{B}^2)\boldsymbol{\theta} \; . \tag{58}$$

Note that parity is preserved – coordinate reflection leaves the constant vector $\boldsymbol{\theta}$ unchanged, hence $\boldsymbol{\theta} \cdot \boldsymbol{B}$ transforms as a scalar field and $\boldsymbol{\theta} \cdot \boldsymbol{E}$ as a pseudoscalar field. Similarly, $(\boldsymbol{E} \cdot \boldsymbol{B})\boldsymbol{\theta}$ behaves as a vector field, while $(\boldsymbol{E}^2 - \boldsymbol{B}^2)\boldsymbol{\theta}$ as a pseudovector.

We seek plane-wave solutions to (55)–(57) — functions of $\omega t - \boldsymbol{k} \cdot \boldsymbol{r}$ — keeping terms to lowest order in θ. Such solutions indeed exist provided the dispersion relation, relating $\boldsymbol{k}$ and ω, takes the following form. In the absence of an external magnetic field the dispersion relation is conventional, $\omega = ck$. However, plane wave solutions to our system of equations exist even in the presence of a constant background magnetic induction $\boldsymbol{b}$. Then the dispersion relation is modified to

$$\omega = ck(1 - \boldsymbol{\theta}_T \cdot \boldsymbol{b}_T) \tag{59}$$

where $\boldsymbol{\theta}_T$ and $\boldsymbol{b}_T$ are components transverse to $\boldsymbol{k}$, the direction of propagation $\boldsymbol{k} \cdot \boldsymbol{\theta}_T = \boldsymbol{k} \cdot \boldsymbol{b}_T = 0$ [6].

The result (59) puts into evidence an explicit violation of Lorentz invariance. Conservation of parity, which we remarked on previously, ensures that both polarizations travel at the same velocity, which generically differs from c by the factor $(1 - \boldsymbol{\theta}_T \cdot \boldsymbol{b}_T)$, and there is no Faraday rotation. Let us also observe that the effective Lagrange density in (54) possesses two interaction terms proportional to θ, with definite numerical constants. Owing to the freedom of rescaling θ, only their ratio is significant. It is straightforward to verify that if the ratio is different from what is written in (54), the two linear polarizations travel at different velocities. Thus the noncommutative theory is unique in affecting the two polarizations equally, at least to $O(\theta)$.

The change in velocity for motion relative to an external magnetic induction $\boldsymbol{b}$ allows searching for the effect with a Michelson-Morley experiment. In a conventional apparatus with two legs of length ℓ_1 and ℓ_2 at right angles to each other, a light beam of wavelength λ is split in two, and one ray travels along $\boldsymbol{b}$ (where there is no effect), while the other, perpendicular to $\boldsymbol{b}$, feels the change of velocity and interferes with the the first. After rotating the apparatus by 90°, the interference pattern will shift by $2(\ell_1 + \ell_2)\theta_T \cdot \boldsymbol{b}_T/\lambda$ fringes. Taking light in the visible range, $\lambda \sim 10^{-5}$cm, a field strength $b \sim 1$ tesla, and using the current bound on $\theta \leq (10\text{TeV})^{-2}$ obtained in [13], one finds that a length $\ell_1 + \ell_2 \geq 10^{18}$ cm ~ 1 parsec would be required for a shift of one fringe. Galactic magnetic fields are neither that strong nor coherent over such large distances, so another experimental setting needs to found to test for noncommutativity.

Finally, what about Heisenberg's intuition that noncommuting coordinates will ammeliorate divergences in relativistic field theory? It turns out that that is indeed true as far as ultraviolet divergences are concerned. However, novel infrared divergences appear, so the problem of divergences remains, albeit in another form. Indeed, these infrared effects associated with noncommutative coordinates provide another obstacle to physical applicatons of this idea.

References

1. S. Carroll, G.B. Field, and R. Jackiw, Phys. Rev. D **41**, 1231 (1990); see also J. Harvey and S. Naculich, Phys. Lett. **B217**, 231 (1989).
2. H. Snyder, Phys. Rev. **71**, 38 (1947). This story was documented and told by J. Wess.
3. G. Dunne, R. Jackiw and C. Trugenberger, Phys. Rev. D **41**, 661 (1990); G. Dunne and R. Jackiw, Nucl. Phys. (Proc. Suppl.) **33C**, 114 (1993).
4. R. Peierls, Z. Phys. **80**, 763 (1933).
5. For a review of fluid mechanical field theory, see, for example, R. Jackiw, physics/0010042.
6. Z. Guralnik, R. Jackiw, S.-Y. Pi, and A.P. Polychronakos, Phys. Lett. B **517**, 450 (2001).
7. For a review, see G. Dunne, J. Phys. A **21**, 2321 (1988).
8. L. Landau, Zh. Eksp. Teor. Fiz. **11**, 592 (1941) [English translation: J. Phys. USSR, **5**, 71 (1941)]; for a review see [5].
9. V. Arnold, *Mathematical Methods in Classical Mechanics*, (Springer, New York 1978).
10. D.B. Fairlie, P. Fletcher and C.K. Zachos, Phys. Lett. B **218**, 203 (1989).
11. N. Seiberg and E. Witten, JHEP **9909**, 032 (1999)
12. A. Bichl, J. Grimstrup, L. Popp, M. Schweda, and R. Wulkenhaar, hep-th/0102044.
13. Bounds on θ from experimental limits on modifications to a fermion sector are $\theta = O(10\,\mathrm{TeV})^{-2}$; M. Chaichian, M. Sheikh-Jabbari, and A. Tureanu, Phys. Rev. Lett. **86**, 2716 (2001); S. Carroll, J. Harvey, V. A. Kostelecky, C. Lane and T. Okamoto, Phys. Rev. Lett. **87**, 141601 (2001).

Particle Physics on Noncommutative Space-Time

Peter Schupp

University of Munich, Theoretical Physics, Theresienstr. 37, 80333 München, Germany

Abstract. This is a concise overview of the construction of Yang-Mills theories with realistic gauge groups on noncommutative generalizations of space-time and of the construction of the noncommutative standard model. We discuss the relevant mathematical tools including tensor products, indicate some exciting physical consequences and conclude with a short-list of challenging open problems.

1 Introduction

Noncommutativity of position and momentum operators in quantum mechanics is the mathematical manifestation of Heisenberg's uncertainty relations that state that we cannot localize a particle with arbitrary high precision and at the same time know its exact momentum. This idea can be extended to space-time itself by promoting coordinate functions to operators with non-trivial commutation relations and corresponding uncertainty relations [1]. Such noncommutative space-time structures could play a role in a reformulation of Einstein's theory of gravity that takes the microscopic structure of space-time into consideration. To probe space-time at very short distances we need high energy transfers, but we know that high energy densities also induce strong gravitational fields that distort space time. This leads to the picture of a foamy space-time structure with a fundamental limit to the precision with which we can localize space-time events. In this way the interplay of gravity and quantum physics can lead to a natural ultraviolet cut-off and there are in fact indications that quantum field theory on certain noncommutative space-time structures is less divergent. As we have argued, we have compelling reasons for drastic changes to the picture of a continuous, commutative description of space-time near the Planck scale. It is important to realize, however, that the noncommutativity scale is in principle independent of the Planck-scale. The appearance of noncommutative effects at energies as low as a few TeV is possible and experimentally not excluded.[1]

Our present goal is an effective formulation of particle physics on noncommutative space-time with realistic gauge group and with realistic particle content. We want to describe the scattering of particles taking effects due to space-time noncommutativity in the interaction region into account. We shall assume that space-time has an asymptotically commutative description. This is in accordance

[1] That is the limit from accelerator experiments [2–5]. Much higher limits exist for low-energy physics [6]; these should, however, be interpreted with care. See also [7].

with the philosophy that the quantum structure of space-time is relevant only at very short distances; there may even be a phase transition. A future goal is the construction of models that take the dynamics of noncommutative space-time structures into account and incorporate gravity – possibly within the framework of string/M-theory.

1.1 Models of Noncommutative Space-Time

The basic strategy of noncommutative geometry is to focus on the algebra of functions on a space-time manifold rather than on the set of points that it is made of. In the noncommutative realm this algebra is then replaced by an arbitrary associative algebra while the notion of "points" looses its meaning. Noncommutative geometry is therefore sometimes also called "point-less" geometry. An impressive collection of mathematical tools has been developed to deal with noncommutative geometries [8,9]. For the purpose of this article, however, it will often suffice to consider some of the following concrete examples of noncommutative space-time structures, see [10]. In the *canonical structure* the usual coordinate functions x^μ are promoted to operators $\hat{x}^\mu$ with commutation relations

$$[\hat{x}^\mu, \hat{x}^\nu] = i\theta^{\mu\nu}, \quad [\hat{x}^\xi, \theta^{\mu\nu}] = 0. \tag{1}$$

We can take $\theta^{\mu\nu}$ to be an antisymmetric matrix with real constant entries. Another interesting case is that of a *Lie structures* with commutation relations that close linearly in the $\hat{x}^\xi$,

$$[\hat{x}^\mu, \hat{x}^\nu] = iC^{\mu\nu}{}_\xi \hat{x}^\xi, \tag{2}$$

where the structure constants $C^{\mu\nu}{}_\xi$ satisfy the Jacobi identity. Such a noncommutative structure is important for fuzzy spheres [15] (where one considers representations of $SU(2)$ up to a certain spin). It can also be used to define noncommutative versions of $\mathbb{R}^n$. An interesting special case of the Lie structure is $[\hat{x}^\mu, \hat{x}^\nu] = i(v^\mu \hat{x}^\nu - v^\nu \hat{x}^\mu)$ with a constant vector v^μ. More complicated commutation relations are of course possible. An example are *quantum space structures* that can be written in terms of generators modulo relations that are typically quadratic in the generators. These quantum spaces are covariant under generalizations of the usual Lorentz and Poincaré symmetries called quantum groups. An interesting aspect of quantum spaces is the existence of lattice-like space time-structures, see Fig. 1, that result when one considers Hilbert space representations of the deformed Heisenberg algebra over a quantum space [16].

1.2 Star Products

The noncommutative algebras that we have just introduced and many other examples can be written as star products on the space of functions over a suitable manifold M [17]. The canonical structure (1) is, e.g., realized by the Moyal-Weyl

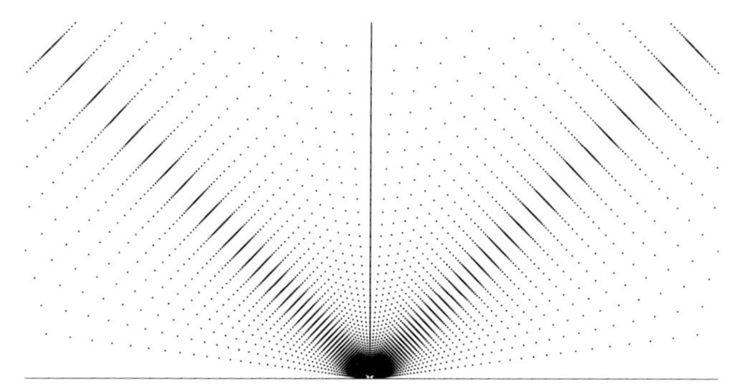

Fig. 1. Space-time lattice structure of quantum Minkowski space: Shown are eigenvalues of r versus t in spacelike as well as forward and backward timelike regions [16].

star product

$$f \star g = \sum_{n=0}^{\infty} \frac{1}{n!} \left(\frac{i\hbar}{2} \right)^n \theta^{\mu_1 \nu_1} \cdots \theta^{\mu_n \nu_n} \partial_{\mu_1} \cdots \partial_{\mu_n} f \cdot \partial_{\nu_1} \cdots \partial_{\nu_n} g, \tag{3}$$

where f and g are ordinary C^∞ functions on $\mathbb{R}^n$ (flat space-time) and $\hbar$ is a formal deformation parameter.[2] We find $x^\mu \star x^\nu - x^\nu \star x^\mu = i\hbar\theta^{\mu\nu}$ for coordinate functions on $\mathbb{R}^n$. More generally a local star product $\star$ is an associative $\mathbb{C}[[\hbar]]$-bilinear product of C^∞ functions on a suitable manifold M, written as a formal power series in bidifferential operators starting with the pointwise product of functions. The star product is a quantization of a given Poisson structure $\{f, g\} = \theta^{ij} \partial_i f \partial_j g$:

$$[f \stackrel{\star}{,} g] = i\hbar\theta^{ij} \partial_i f \partial_j g + \mathcal{O}(\hbar^2). \tag{4}$$

Two star products $\star$, $\star'$ are called equivalent if they are related by a formal linear differential operator $\mathcal{D}$ starting with the identity, such that $\mathcal{D}(f \star' g) = \mathcal{D}f \star \mathcal{D}g$. $\mathcal{D}$ is formally invertible and associativity of $\star'$ follows from associativity of $\star$.

Star products provide a convenient language to formulate noncommutative field theory and the scattering of particles that enter from an asymptotically commutative region. The semi-classical Poisson limit and the classical "commutative" limit $\hbar \to 0$ are directly build into the definition of a star product.

2 Gauge Theory On Noncommutative Space-Time

The construction of a gauge theory on a given non-commutative space rests on few basic ideas: the concept of covariant coordinates and functions, the requirement of locality, and gauge equivalence and consistency conditions [10–14].

[2] In the following sections we will not write $\hbar$ explicitly.

2.1 Covariant Coordinates

Let us consider an infinitesimal non-commutative local gauge transformation $\hat{\delta}$ of a fundamental matter field that carries a representation ρ_Ψ

$$\hat{\delta}\widehat{\Psi} = i\rho_\Psi(\widehat{\Lambda}) \star \widehat{\Psi}. \tag{5}$$

In the non-Abelian case $\widehat{\Psi}$ is a vector, $\rho_\Psi(\widehat{\Lambda})$ is a matrix and $\star$ includes matrix multiplication. The product of a field and a coordinate, $\widehat{\Psi} \star x^\mu$, yields a new field that transforms just like $\widehat{\Psi}$. The opposite product $x^\mu \star \widehat{\Psi}$, however, is not a covariant object because the gauge parameter does not commute with x^μ. In analogy to the covariant derivatives of ordinary gauge theory we thus need to introduce covariant coordinates [10] $X^\mu = x^\mu + \theta^{\mu\nu}\widehat{A}_\nu$, where $\widehat{A}_\nu$ is a non-commutative analog of the gauge potential with the following transformation property:[3]

$$\hat{\delta}\widehat{A}_\mu = \partial_\mu\widehat{\Lambda} + i[\widehat{\Lambda} \stackrel{\star}{,} \widehat{A}_\mu]. \tag{6}$$

More generally we also need covariant functions $\mathcal{D}(f)$ [12], where f is an ordinary function (or a matrix-valued function) and $\mathcal{D}$ is an invertible linear differential operator that transforms under gauge transformations: $\delta\mathcal{D}(f) = i[\widehat{\Lambda} \stackrel{\star}{,} \mathcal{D}(f)]$. The product of a covariant coordinate or function and a field $\widehat{\Psi}$ *is* a covariant object that transforms like $\widehat{\Psi}$. Noting that the covariantizing map $\mathcal{D}$ defines an equivalent star product $\star'$ we see that the space of fields (sections) is naturally a left $\star'$-module and a right $\star$-module. From the covariant coordinates one can construct further covariant objects: The covariant derivative

$$\widehat{D}_\mu\widehat{\Psi} = \partial_\mu\widehat{\Psi} - i\rho_\Psi(\widehat{A}_\mu) \star \widehat{\Psi}, \tag{7}$$

is related to the covariant expression $\rho_\Psi(X^\mu) \star \widehat{\Psi} - \widehat{\Psi} \star x^\mu$ which is simply the difference between the left and right actions of x^μ on $\widehat{\Psi}$. The corresponding non-commutative field strength

$$\widehat{F}_{\mu\nu} = \partial_\mu\widehat{A}_\nu - \partial_\nu\widehat{A}_\mu - i[\widehat{A}_\mu \stackrel{\star}{,} \widehat{A}_\nu], \qquad \hat{\delta}\widehat{F}_{\mu\nu} = i[\widehat{\Lambda} \stackrel{\star}{,} \widehat{F}_{\mu\nu}], \tag{8}$$

is related to the commutator of covariant coordinates X^μ. (In the following we shall often omit the symbol ρ_Ψ, when its presence is obvious.)

We can now write a noncommutative generalization of the Yang-Mills action:

$$\widehat{S} = \int d^4x \, \frac{-1}{2g^2}\, Tr(\widehat{F}_{\mu\nu} \star \widehat{F}^{\mu\nu}) + \overline{\widehat{\Psi}} \star i\,\widehat{\not{D}}\widehat{\Psi}. \tag{9}$$

This action is invariant under noncommutative gauge transformations, because of the trace property of the integral with respect to the Moyal-Weyl star product. Applying the usual QFT techniques directly to this action we quickly run into several difficulties that include the following apparent problems:

[3] Here and in the following we use $\theta^{\mu\nu}$ to lower indices, yielding expressions that are more convenient to work with. We should note that this is in general only possible in the case of constant $\theta^{\mu\nu}$. Otherwise one should work with $\tilde{A}^\mu$, where $X^\mu = x^\mu + \tilde{A}^\mu$.

- The choice of gauge groups seems to be restricted to $U(N)$ in the fundamental representation. Related to this even for Abelian gauge theory charge can apparently only take on the discrete values $Q/e = -1, 0, +1$.
- It is not clear a priori how to write covariant Yukawa terms.
- Tensor products do not appear to be well-defined.
- The meaning of noncommutative "in" and "out"-states is opaque.

2.2 Locality, Gauge Equivalence and Consistency Conditions

So far it has not been important that the noncommutative structures that we considered can be realized as star products. We shall now introduce another important concept that will help us overcome the apparent problems mentioned at the end of the previous section. That concept is the principle of locality. A star product is a formal power series starting with the ordinary product plus higher order terms that are chosen so as to yield an associative product. The star product can be pictured as a tower build upon the leading Poisson tensor $\theta^{\mu\nu}$. It is a natural to ask whether it is possible to express also the non-commutative fields $\hat{A}$, $\widehat{\Psi}$ and non-commutative gauge parameter $\hat{\Lambda}$ in a similar fashion as towers build upon the corresponding ordinary fields A, Ψ and ordinary gauge parameter Λ. This is indeed the case; the non-commutative fields and parameter can be expressed as local functions of their commutative counterparts

$$\widehat{A}_\mu[A] = A_\mu + \frac{1}{4}\theta^{\xi\nu}\Big\{A_\nu, \partial_\xi A_\mu + F_{\xi\mu}\Big\} + \dots\,, \tag{10a}$$

$$\widehat{\Psi}[\Psi, A] = \Psi + \frac{1}{2}\theta^{\mu\nu}\rho_\Psi(A_\nu)\partial_\mu\Psi + \frac{1}{4}\theta^{\mu\nu}\rho_\Psi(\partial_\mu A_\nu)\Psi + \dots\,, \tag{10b}$$

$$\hat{\Lambda}_\Lambda[A] = \Lambda + \frac{1}{4}\theta^{\mu\nu}\{A_\nu, \partial_\mu\Lambda\} + \dots\,, \tag{10c}$$

where $F_{\mu\nu} = \partial_\mu A_\nu - \partial_\nu A_\mu - i[A_\mu, A_\nu]$ is the ordinary field strength. A local function of a field is a formal series that at each order in θ depends on the field and a finite number of derivatives of the field. The expressions (10a)–(10c) have the remarkable property that ordinary gauge transformations $\delta_\Lambda A_\mu = \partial_\mu\Lambda + i[\Lambda, A_\mu]$ and $\delta_\Lambda\Psi = i\Lambda\cdot\Psi$ induce noncommutative gauge transformations (5), (6) of $\hat{A}$ and $\widehat{\Psi}$ with parameter $\hat{\Lambda}$: We have gauge equivalence conditions

$$\delta_\Lambda\widehat{A}_\mu[A] = \partial_\mu\widehat{\Lambda}_\Lambda[A] + i[\widehat{\Lambda}_\Lambda[A] \stackrel{\star}{,} \widehat{A}_\mu[A]]\,, \tag{11a}$$

$$\delta_\Lambda\widehat{\Psi}[\Psi, A] = i\widehat{\Lambda}_\Lambda[A] \star \widehat{\Psi}[\Psi, A]. \tag{11b}$$

Here is an example how this works in the abelian case for a matter field ψ:

$$\begin{aligned}\delta_\lambda\hat{\psi} &= i\lambda\psi + \tfrac{1}{2}\theta^{ij}(\partial_j\lambda)\partial_i\psi + \tfrac{1}{2}\theta^{ij}a_j\partial_i(i\lambda\psi) + \dots \\ &= i(\lambda + \tfrac{1}{2}\theta^{ij}a_j\partial_i\lambda + \dots) \star (\psi + \tfrac{1}{2}\theta^{ij}a_j\partial_i\psi + \dots) \\ &= i\hat{\lambda} \star \hat{\psi}\,.\end{aligned}$$

The expressions (10a), (10c) for the noncommutative field strength and noncommutative gauge parameter in terms of their classical counterparts were introduced by Seiberg and Witten as field redefinitions in the context of string theory [18]; they are called Seiberg-Witten (SW) maps.

Any pair of non-commutative gauge parameters $\hat{\Lambda}_\alpha[a]$, $\hat{\Lambda}_\beta[a]$ has to satisfies the following consistency condition (cocycle condition) [13]

$$[\hat{\Lambda}_\alpha[A] \stackrel{\star}{,} \hat{\Lambda}_\beta[A]] + i\delta_\alpha \hat{\Lambda}_\beta[A] - i\delta_\beta \hat{\Lambda}_\alpha[A] = \hat{\Lambda}_{[\alpha,\beta]}[A], \tag{12}$$

which follows from $[\delta_\alpha, \delta_\beta]\widehat{\Psi}[\Psi, A] = \delta_{-i[\alpha,\beta]}\widehat{\Psi}[\Psi, A]$. The infinitesimal gauge transformations can be exponentiated to yield finite gauge transformations with "NC group elements" $\widehat{G}_g[A]$ corresponding to $g = \exp(i\Lambda)$. Care has to be taken, however, because $\widehat{\Lambda} = \widehat{\Lambda}_\Lambda[A]$ depends on the gauge potential:

$$\widehat{G}_g[A] \star \widehat{\Psi} = \exp(\delta_\Lambda)\widehat{\Psi} = \Big(1 + i\widehat{\Lambda} + \frac{i}{2}(\delta_\Lambda \widehat{\Lambda}) - \frac{1}{2}\widehat{\Lambda} \star \widehat{\Lambda} + \dots\Big) \star \widehat{\Psi}\,. \tag{13}$$

In terms of the $\widehat{G}_g[A]$ the consistency condition becomes

$$\widehat{G}_{g_1}[A_{g_2}] \star \widehat{G}_{g_2}[A] = \widehat{G}_{g_1 \cdot g_2}[A], \qquad A_{g_2} \equiv g_2 \cdot A \cdot g_2^{-1} + i g_2 \mathrm{d} g_2^{-1}\,. \tag{14}$$

This "NC group law" is the starting point for the construction of noncommutative vector bundles with nontrivial gauge fields [22]. The consistency condition (12) is important for the practical calculation of SW maps, since it involves only the gauge parameter and because it reduces the task to a cohomological problem [13,19–21]. The gauge equivalence and consistency conditions do not uniquely determine SW maps. To first order in θ we have the freedom of classical field redefinitions and noncommutative gauge transformations. We have used that freedom to choose maps with Hermitean $\hat{A}_\mu$ and $\hat{\Lambda}$. The constants that parametrize the freedom in the SW map receive quantum corrections in noncommutative gauge theory [30]. The freedom in the Seiberg-Witten map is also important in the context of tensor products of fields and gauge groups.

2.3 θ-Expanded Noncommutative Yang-Mills Action

Using the SW maps (10a) and (10b) to expand the fields $\widehat{A}$ and $\widehat{\Psi}$ in the action (9) yields an action that is written in ordinary fields. Expanding to first order in θ and integrating by parts we find:

$$\begin{aligned} S = \int d^4x \Big(&-\frac{1}{2g^2}\mathrm{tr} F_{ij} \star F^{ij} + \frac{1}{4g^2}\theta^{ij}\mathrm{tr} F_{ij} F_{kl} F^{kl} - \frac{1}{g^2}\theta^{ij}\mathrm{tr} F_{ik} F_{jl} F^{jl} \\ &+ \overline{\Psi}(i\not{D} - m)\Psi - \frac{1}{4}\theta^{ij}\overline{\Psi} F_{ij}(i\not{D} - m)\Psi - \frac{i}{2}\theta^{ij}\overline{\Psi}\gamma^k F_{ki} D_j \Psi \Big) \end{aligned}$$

with $D_i\Psi \equiv \partial_i\Psi - iA_i\Psi$ and $F_{ij} = \partial_i A_j - \partial_j A_i - i[A_i, A_j]$. This formulation of noncommutative gauge theory [10–13,23] has the following advantages:

- We can freely choose the gauge group, and its representations.
- Related to this, there is no problem with charges.
- Tensor products of fields and of gauge groups exist.
- Nontrivial gauge fields (and vector bundles) are possible.

We shall now discuss some of these points in more detail.

3 Enveloping Algebra Valued Gauge Fields

With the SW maps we can freely choose a structure group G (gauge group). The fields A, F and the gauge parameter Λ are valued in the Lie algebra of G while their noncommutative counterparts $\widehat{A}$, $\widehat{F}$, $\hat{\Lambda}$ are valued in the enveloping algebra of the Lie algebra, i.e., they contain polynomials in the Lie algebra generators

$$\hat{\Lambda} = \Lambda_a(x)T^a + \Lambda^1_{ab}(x) : T^aT^b : + \Lambda^2_{abc}(x) : T^aT^bT^c : + \ldots, \tag{15}$$

where : : denotes symmetric ordering [11]. This is a general feature of space-time noncommutativity as can be seen by considering the commutator of two Lie algebra-valued noncommutative gauge parameters:

$$[\hat{\Lambda} \stackrel{\star}{,} \hat{\Lambda}'] = \frac{1}{2}\{\Lambda_a(x) \stackrel{\star}{,} \Lambda'_b(x)\}[T^a, T^b] + \frac{1}{2}[\Lambda_a(x) \stackrel{\star}{,} \Lambda'_b(x)]\{T^a, T^b\}. \tag{16}$$

It is enveloping algebra valued, because the coefficient of $\{T^a, T^b\}$ is in general non-zero since the functions $\Lambda_a(x)$, $\Lambda'_b(x)$ do not commute.

We can of course consider a matrix representation of the Lie algebra; then $\widehat{A}$, $\widehat{F}$ and $\hat{\Lambda}$ will also be matrix-valued. For $U(N)$ *in the fundamental representation* this solves the problem, but, e.g., for $SU(N)$ A, F and Λ are traceless but $\widehat{A}$, $\widehat{F}$, $\hat{\Lambda}$ are not, see [10]. The noncommutative fields nevertheless have the correct number of degrees of freedom since they are functions of the classical fields via the SW map. That is one of the important reasons for the use of the SW map in the formulation of noncommutative gauge theory. See [24] for a recent careful study of θ-expanded perturbation theory for subgroups of $U(N)$.

3.1 Charge Quantization Problem and Its Resolution

In noncommutative QED one faces the problem that the theory can apparently accommodate only charges $\pm q$ or zero for one fixed q [25]. That is so because the only couplings of the non-commutative gauge boson $\widehat{A}_\mu$ to a matter field $\widehat{\Psi}$ compatible with the non-commutative gauge transformation (6) are

$$+i\widehat{\Psi} \star \widehat{A}_\mu, \quad -i[\widehat{A}_\mu \stackrel{\star}{,} \widehat{\Psi}], \quad \text{and } -i\widehat{A}_\mu \star \widehat{\Psi}. \tag{17}$$

For the Standard Model on NC space-time this would be a major problem because of the existence of fractional charges and hypercharges, see Table 1.

The problem can be traced back to the fact that the gauge fields are enveloping algebra valued. The charge quantization problem should really be seen as a

Table 1. Charges of the Standard Model fields

	$SU(3)_C$	$SU(2)_L$	$U(1)_Y$	$U(1)_Q$
e_R	**1**	**1**	-1	-1
$\begin{pmatrix}\nu_L\\ e_L\end{pmatrix}$	**1**	**2**	$-1/2$	$\begin{pmatrix}0\\ -1\end{pmatrix}$
u_R	**3**	**1**	$2/3$	$2/3$
d_R	**3**	**1**	$-1/3$	$-1/3$
$\begin{pmatrix}u_L\\ d_L\end{pmatrix}$	**3**	**2**	$1/6$	$\begin{pmatrix}2/3\\ -1/3\end{pmatrix}$
$\begin{pmatrix}\phi^+\\ \phi^0\end{pmatrix}$	**1**	**2**	$1/2$	$\begin{pmatrix}1\\ 0\end{pmatrix}$

problem with the degrees of freedom (more precisely the number of fundamental fields). In the θ-expanded approach based on Seiberg-Witten maps the fundamental field $a_\mu(x)$ enters as $A_\mu = eQa_\mu(x)$, where e is the coupling constant and Q is the charge operator. The resulting $\widehat{A}_\mu$ is a highly nonlinear function of Q, so that for differently charged particles we will also get different noncommutative gauge fields $\widehat{A}^q_\mu = eq\widehat{a}^q_\mu(x)$. The important point is that these fields are not independent. They are all functions of the physical field $a_\mu(x)$ and consequently there is no restriction on the allowed charges in θ-expanded NC QED [14,26].

4 Tensor Products

The gauge parameter for a product $G \times G'$ of two groups is $\Lambda = \lambda + \lambda'$, where λ and λ' are valued in the respective Lie algebras of the groups G and G'. Similarly the gauge potential is $A_\mu = a_\mu + a'_\mu$. The noncommutative gauge parameter $\hat{\mathbf{\Lambda}}$ depends on λ, λ', a, a' and can also be written as the sum of two terms:

$$\hat{\mathbf{\Lambda}}_{(\lambda,\lambda')}[a,a'] = \hat{\mathbf{\Lambda}}_\lambda[a,a'] + \hat{\mathbf{\Lambda}}'_{\lambda'}[a,a']. \tag{18}$$

The consistency relation (12) for $\hat{\mathbf{\Lambda}}_{(\lambda,\lambda')}[a,a']$ implies a total of three relations for $\hat{\mathbf{\Lambda}}_\lambda[a,a']$ and $\hat{\mathbf{\Lambda}}'_{\lambda'}[a,a']$: $\hat{\mathbf{\Lambda}}_\lambda[a,a']$ and $\hat{\mathbf{\Lambda}}'_{\lambda'}[a,a']$ satisfy CR's separately

$$[\hat{\mathbf{\Lambda}}_\alpha \stackrel{\star}{,} \hat{\mathbf{\Lambda}}_\beta] + i\delta_\alpha \hat{\mathbf{\Lambda}}_\beta - i\delta_\beta \hat{\mathbf{\Lambda}}_\alpha = \hat{\mathbf{\Lambda}}_{[\alpha,\beta]}, \tag{19a}$$

$$[\hat{\mathbf{\Lambda}}'_{\alpha'} \stackrel{\star}{,} \hat{\mathbf{\Lambda}}'_{\beta'}] + i\delta_{\alpha'} \hat{\mathbf{\Lambda}}'_{\beta'} - i\delta_{\beta'} \hat{\mathbf{\Lambda}}'_{\alpha'} = \hat{\mathbf{\Lambda}}'_{[\alpha',\beta']} \tag{19b}$$

and there is a new mixed consistency relation

$$\boxed{[\hat{\mathbf{\Lambda}}_\alpha \stackrel{\star}{,} \hat{\mathbf{\Lambda}}'_{\beta'}] + i\delta_\alpha \hat{\mathbf{\Lambda}}'_{\beta'} - i\delta_{\beta'} \hat{\mathbf{\Lambda}}_\alpha = 0.} \tag{20}$$

Note that there is no inhomogeneous term on the RHS because $[\alpha, \beta'] = 0$.

It is not hard to find solutions to these equations [27]. A particularly simple **symmetric solution** is obtained by plugging $\Lambda = \lambda + \lambda'$ and $A_\mu = a_\mu + a'_\mu$ into the formula for the SW map (10c):

$$\hat{\mathbf{\Lambda}}_{(\lambda,\lambda')} = \hat{\Lambda}_{\lambda+\lambda'}[a_\mu + a'_\mu]. \tag{21}$$

A completely **assymetric solution** is found by setting $\delta_{\beta'} \hat{\mathbf{\Lambda}}_\alpha \equiv 0$. This implies $\delta_\alpha \hat{\mathbf{\Lambda}}'_{\beta'} = i[\hat{\mathbf{\Lambda}}_\alpha \stackrel{\star}{,} \hat{\mathbf{\Lambda}}'_{\beta'}]$, i.e., $\hat{\mathbf{\Lambda}}'_{\beta'}$ is a covariant function.

$$\hat{\mathbf{\Lambda}}_\lambda[a,a'] = \Lambda_\lambda[a], \qquad \hat{\mathbf{\Lambda}}'_{\lambda'}[a,a'] = \mathcal{D}_{[a]}(\Lambda'_{\lambda'}[a']). \tag{22}$$

Here $\mathcal{D}_{[a]}$ is a differential operator that is a local function of a and θ Finally, there is an **interpolating solution** with real parameter k. To order θ:

$$\hat{\mathbf{\Lambda}}_\lambda[a,a'] = \lambda + \frac{\theta^{\mu\nu}}{2}\left\{\partial_\mu\lambda, \frac{a_\nu}{2} + k a'_\nu\right\}, \quad \hat{\mathbf{\Lambda}}'_{\lambda'}[a,a'] = \lambda' + \frac{\theta^{\mu\nu}}{2}\left\{\partial_\mu\lambda', \frac{a'_\nu}{2} + (1-k)a_\nu\right\}. \tag{23}$$

Similar expressions exist for the noncommutative gauge potential $\widehat{\mathbf{A}}_\mu[a,a']$.

4.1 Hybrid Seiberg-Witten Map

For the product of fields $\psi\psi'$, where ψ transforms under G and ψ' transforms under G' we could try to construct a SW map $\widehat{\mathbf{\Psi}}[\psi,\psi',a,a']$ directly.[4] It is, however, more convenient to follow a different strategy. For this we need a hybrid SW map [14,26] that interpolates between the SW map for the gauge field (10a) and the SW map for the matter field (10b): Consider a field Φ that transforms on the left and on the right under two arbitrary gauge groups G_L and G_R, respectively: $\delta\Phi = i\Lambda \cdot \Phi - i\Phi \cdot \Lambda'$, where Λ is valued in (a representation of) Lie(G_L) and Λ is valued in (a representation of) Lie(G_R). We shall also assume that there are gauge fields A and A' corresponding to the respective gauge groups, that transform as $\delta A_\nu = \partial_\nu \Lambda + i[\Lambda, A_\nu]$ and $\delta A'_\nu = \partial_\nu \Lambda' + i[\Lambda', A'_\nu]$. The following hybrid Seiberg-Witten map (given to order θ)

$$\widehat{\Phi}^H[\Phi,A,A'] = \Phi + \frac{1}{2}\theta^{\mu\nu}\left(A_\nu\partial_\mu\Phi + \frac{1}{2}\partial_\mu A_\nu \Phi + \partial_\mu\Phi A'_\nu + \frac{1}{2}\Phi\partial_\mu A'_\nu + i A_\nu \Phi A'_\mu\right) \tag{24}$$

has the property that the transformations of Φ, A and A' that were given above induce the following transformation of $\widehat{\Phi}^H$:[5]

$$\delta\widehat{\Phi}^H = i\widehat{\Lambda} \star \widehat{\Phi}^H - i\widehat{\Phi}^H \star \widehat{\Lambda}'. \tag{25}$$

Here $\widehat{\Lambda} = \widehat{\Lambda}_\Lambda[A]$ and $\widehat{\Lambda}' = \widehat{\Lambda}'_{\Lambda'}[A']$ according to the usual SW map (10c). There is also a corresponding formula for the covariant derivative of $\delta\widehat{\Phi}^H$:

$$\widehat{D}_\mu\widehat{\Phi}^H = \partial_\mu\widehat{\Phi}^H - i\widehat{A}_\mu \star \widehat{\Phi}^H + i\widehat{\Phi}^H \star \widehat{A'_\mu}. \tag{26}$$

---

[4] The naive choice $\widehat{\psi} \star \widehat{\psi}'$ does of course not work, because the gauge parameter $\widehat{\lambda}'$ does not $\star$-commute with $\widehat{\psi}$ in the second term of $\delta\widehat{\psi} \star \widehat{\psi}' = i\widehat{\lambda} \star \widehat{\psi} \star \widehat{\psi}' + i\widehat{\psi} \star \widehat{\lambda}' \star \widehat{\psi}'$.

[5] A similar formula with a + sign in front of the second term would require the use of the opposite SW map (with $\theta \to -\theta$) for Λ'.

The SW map for the product of fields ψ and ψ' can now be written as

$$\widehat{\boldsymbol{\Psi}}[\psi,\psi',a,a'] = \widehat{\psi}^{H}[\psi,a+a',a'] \star \widehat{\psi}'[\psi',a']. \tag{27}$$

The gauge transformation $\delta\psi = i\lambda\psi$, $\delta\psi' = i\lambda'\psi'$, $\delta a_\mu = \partial_\mu\lambda + i[\lambda,a_\mu]$, $\delta a'_\mu = \partial_\mu\lambda' + i[\lambda',a'_\mu]$ now induces the desired transformation

$$\delta\widehat{\boldsymbol{\Psi}}[\psi,\psi',a,a'] = i\widehat{\Lambda}_{(\lambda+\lambda')}[a+a'] \star \widehat{\boldsymbol{\Psi}}[\psi,\psi',a,a'], \tag{28}$$

where we have used that ψ and λ' commute. We see that the given version of the hybrid SW map corresponds to the symmetric solution for the gauge parameter.

5 Noncommutative Standard Model

The structure group of the Standard Model is $G_{SM} = SU(3)_C \times SU(2)_L \times U(1)_Y$. The gauge potential A_μ and gauge parameter Λ are valued in Lie(G_{SM}):

$$A_\nu = g'\mathcal{A}_\nu(x)Y + g\sum_{a=1}^{3} B_{\nu a}(x)T_L^a + g_S\sum_{b=1}^{8} G_{\nu b}(x)T_S^b \tag{29a}$$

$$\Lambda = g'\alpha(x)Y + g\sum_{a=1}^{3} \alpha_a^L(x)T_L^a + g_S\sum_{b=1}^{8} \alpha_b^S(x)T_S^b, \tag{29b}$$

where Y, T_L^a, T_S^b are the generators of $u(1)_Y$, $su(2)_L$ and $su(3)_C$ respectively. In addition to the gauge bosons we have three families of left- and right-handed fermions and a Higgs doublet

$$\Psi_L^{(i)} = \begin{pmatrix} L_L^{(i)} \\ Q_L^{(i)} \end{pmatrix}, \qquad \Psi_R^{(i)} = \begin{pmatrix} e_R^{(i)} \\ u_R^{(i)} \\ d_R^{(i)} \end{pmatrix}, \qquad \Phi = \begin{pmatrix} \phi^+ \\ \phi^0 \end{pmatrix} \tag{30}$$

where i = 1,2,3 is the generation index and ϕ^+, ϕ^0 are complex scalar fields. We shall now apply the appropriate SW maps to the fields A_μ, $\Psi^{(i)}$, Φ, expand to first order in θ and write the corresponding NC Yang-Mills action [26]. We should note that this corresponds to the symmetric solution (21) for the tensor products. Special care must be taken in the definition of the trace in the gauge kinetic terms and in the construction of covariant Yukawa terms.

5.1 Noncommutative Yukawa Terms

The classical Higgs field $\Phi(x)$ commutes with the generators of the $U(1)$ and $SU(3)$ gauge transformations. It also commutes with the corresponding gauge parameters. The latter is no longer true in the noncommutative setting: The coefficients $\alpha(x)$ and $\alpha_b^S(x)$ of the $U(1)$ and $SU(3)$ generators in the gauge parameter are functions and therefore do not $\star$-commute with the Higgs field.

This makes it hard to write down covariant Yukawa terms. The solution to the problem is the hybrid SW map (24). By choosing appropriate representations it allows us to assign separate left and right charges to the noncommutative Higgs field $\widehat{\Phi}^H$ that add up to its usual charge [26]. Here are two examples:

$$\begin{array}{rccccccc} & \overline{\widehat{L}}_L \star & \rho_L(\widehat{\Phi}) & \star\, \widehat{e}_R & \overline{\widehat{Q}}_L \star & \rho_Q(\widehat{\Phi}) & \star\, \widehat{d}_R \\ Y = & 1/2 & \underbrace{-1/2+1}_{1/2} & -1 & -1/6 & \underbrace{1/6+1/3}_{1/2} & -1/3 \end{array} \tag{31}$$

We see here two instances of a general rule: The gauge fields in the SW maps and in the covariant derivatives inherit their representation (charge for Y, trivial or fundamental representation for T_L^a, T_S^b) from the fermion fields $\Psi^{(i)}$ to their left and to their right.

5.2 The Minimal Noncommutative Standard Model

The trace in the kinetic terms for the gauge bosons is not unique, it depends on the choice of representation. This would not matter if the gauge fields were Lie algebra valued, but in the noncommutative case they live in the enveloping algebra. The simplest choice is a sum of three traces over the $U(1)$, $SU(2)$, $SU(3)$ sectors with $Y = \frac{1}{2}\begin{pmatrix} 1 & 0 \\ 0 & -1 \end{pmatrix}$ in the definition of $\mathbf{tr_1}$ and the fundamental representation for $\mathbf{tr_2}$ and $\mathbf{tr_3}$. This leads to the following gauge kinetic terms

$$\begin{aligned} S_{\text{gauge}} = & -\frac{1}{4}\int d^4x\, f_{\mu\nu}f^{\mu\nu} - \frac{1}{2}\,\mathrm{Tr}\int d^4x\, F^L_{\mu\nu}F^{L\mu\nu} \\ & -\frac{1}{2}\,\mathrm{Tr}\int d^4x\, F^S_{\mu\nu}F^{S\mu\nu} + \frac{1}{4}g_S\,\theta^{\mu\nu}\,\mathrm{Tr}\int d^4x\, F^S_{\mu\nu}F^S_{\rho\sigma}F^{S\rho\sigma} \\ & -g_S\,\theta^{\mu\nu}\,\mathrm{Tr}\int d^4x\, F^S_{\mu\rho}F^S_{\nu\sigma}F^{S\rho\sigma} + \mathcal{O}(\theta^2)\,. \end{aligned} \tag{32}$$

We note, that there are neither fff nor $F^LF^LF^L$-terms. Contrary to common believe triple photon couplings are not *required* in noncommutative gauge theories! This model is minimal in the sense that it deviates as little as possible from the Standard Model on commutative space-time. The full action of the Minimal Noncommutative Standard Model is [26]:

$$\begin{aligned} S_{NCSM} = & \int d^4x \sum_{i=1}^{3} \overline{\widehat{\Psi}}{}^{(i)}_L \star i\widehat{\not{D}}\widehat{\Psi}^{(i)}_L + \int d^4x \sum_{i=1}^{3} \overline{\widehat{\Psi}}{}^{(i)}_R \star i\widehat{\not{D}}\widehat{\Psi}^{(i)}_R \\ & - \int d^4x \frac{1}{2g'}\mathbf{tr_1}\widehat{F}_{\mu\nu}\star\widehat{F}^{\mu\nu} - \int d^4x \frac{1}{2g}\mathbf{tr_2}\widehat{F}_{\mu\nu}\star\widehat{F}^{\mu\nu} \\ & - \int d^4x \frac{1}{2g_S}\mathbf{tr_3}\widehat{F}_{\mu\nu}\star\widehat{F}^{\mu\nu} + \int d^4x \Big(\rho_0(\widehat{D}_\mu\widehat{\Phi})^\dagger \star \rho_0(\widehat{D}^\mu\widehat{\Phi}) \\ & -\mu^2\rho_0(\widehat{\Phi})^\dagger\star\rho_0(\widehat{\Phi}) - \lambda\rho_0(\widehat{\Phi})^\dagger\star\rho_0(\widehat{\Phi})\star\rho_0(\widehat{\Phi})^\dagger\star\rho_0(\widehat{\Phi})\Big) \end{aligned}$$

$$
\begin{aligned}
&- \int d^4x \left(\sum_{i,j=1}^{3} W^{ij} \Big((\bar{\widehat{L}}_L^{(i)} \star \rho_L(\widehat{\Phi})) \star \widehat{e}_R^{(j)} + \bar{\widehat{e}}_R^{(i)} \star (\rho_L(\widehat{\Phi})^\dagger \star \widehat{L}_L^{(j)}) \Big) \right. \\
&+ \sum_{i,j=1}^{3} G_u^{ij} \Big((\bar{\widehat{Q}}_L^{(i)} \star \rho_{\bar{Q}}(\widehat{\bar{\Phi}})) \star \widehat{u}_R^{(j)} + \bar{\widehat{u}}_R^{(i)} \star (\rho_{\bar{Q}}(\widehat{\bar{\Phi}})^\dagger \star \widehat{Q}_L^{(j)}) \Big) \\
&\left. + \sum_{i,j=1}^{3} G_d^{ij} \Big((\bar{\widehat{Q}}_L^{(i)} \star \rho_Q(\widehat{\Phi})) \star \widehat{d}_R^{(j)} + \bar{\widehat{d}}_R^{(i)} \star (\rho_Q(\widehat{\Phi})^\dagger \star \widehat{Q}_L^{(j)}) \Big) \right) \qquad (33)
\end{aligned}
$$

where W^{ij}, G_u^{ij}, G_d^{ij} are Yukawa couplings, $\bar{\Phi} = i\tau_2 \Phi^*$ and we have omitted the superscript ^{H} in the hybrid SW map of the Higgs.

5.3 Non-minimal Versions of the NCSM

We can use the freedom in the choice of traces in kinetic terms for the gauge fields to construct non-minimal versions of the NCSM. The general form of the gauge kinetic terms is [26][Appendix C], [27]

$$
S_{\text{gauge}} = -\frac{1}{2} \int d^4x \sum_\rho \kappa_\rho \mathrm{Tr}\Big(\rho(\widehat{F}_{\mu\nu}) \star \rho(\widehat{F}^{\mu\nu})\Big), \qquad (34)
$$

where the sum is over all unitary irreducible inequivalent representations ρ of the gauge group G. The freedom in the kinetic terms is parametrized by real coefficients κ_ρ that are subject to the constraints

$$
\frac{1}{g_I^2} = \sum_\rho \kappa_\rho \mathrm{Tr}\Big(\rho(T_I^a)\rho(T_I^a)\Big), \qquad (35)
$$

where g_I and T_I^a are the usual "commutative" coupling constants and generators of $U(1)_Y$, $SU(2)_L$, $SU(3)_C$, respectively. Both formulas can also be written more compactly as

$$
S_{\text{gauge}} = -\frac{1}{2} \int d^4x\, \mathbf{Tr} \frac{1}{\mathbf{G}^2} \widehat{F}_{\mu\nu} \star \widehat{F}^{\mu\nu}, \qquad \frac{1}{g_I^2} = \mathbf{Tr} \frac{1}{\mathbf{G}^2} T_I^a T_I^a, \qquad (36)
$$

where the trace $\mathbf{Tr}$ is again over all representations and $\mathbf{G}$ is an operator that commutes with all generators T_I^a and encodes the coupling constants. The possibility of new parameters in gauge theories on noncommutative space-time is a consequence of the fact that the gauge fields are in general valued in the enveloping algebra of the gauge group.

The expansion in θ is at the same time an expansion in the momenta. The θ-expanded action can thus be interpreted as a low energy effective action. In such an effective low energy description it is natural to expect that all representations that appear in the commutative theory (matter multiplets and adjoint representation) are important. We should therefore consider the non-minimal version of the NCSM with non-zero coefficients κ_ρ at least for these representations. The number of new parameters in the non-minimal NCSM can be restricted by considering GUTs on noncommutative space-time [28,27].

6 Noncommutative Physics

A general feature of gauge theories on noncommutative space-time is the appearance of many new interactions including Standard Model-forbidden processes. The origin of these new interactions is two-fold: One source are the star products that let abelian gauge theory on NC space-time resemble Yang-Mills theory with the possibility of triple and quadruple gauge boson vertices. The other source are the gauge fields in the Seiberg-Witten maps for the gauge and matter fields. These can be pictured as a cloud of gauge bosons that dress the original 'commutative' fields and that have their origin in the interaction between gauge fields and the NC structure of space-time. One of the perhaps most striking effects and a possible signature of space-time noncommutativity is the spontaneous breaking of continuous and discrete space-time symmetries: The actions that we have written are invariant under the usual space-time symmetries *if we transform* $\theta^{\mu\nu}$ *as a tensor*. If we however consider $\theta^{\mu\nu}$ as a spectator or as the vacuum expectation of some background field, then we do find processes that violate certain space-time symmetries and the corresponding conservation laws. These typically include spin conservation, and CP; they can also include CPT, momentum and energy conservation.

Let us conclude with a shortlist of challenging problems and interesting projects: It is important to get a better understanding of the *quantization* of field theories on noncommutative space-time. In the case of the lowest order noncommutative correction to ordinary field theories that we have discussed here, quantization is straightforward. Feynman rules can be obtained either in the canonical formalism or straight from the action in a path-integral approach. That is no longer true when one considers higher orders in θ or even sums the whole series: There are subtle issues related to time-ordering that can lead to apparent violation of unitarity, if one naively uses Feynman rules that have been directly read off the Lagrangian density, see [29]. A more careful canonical approach does, however, lead to a well-defined theory. The issue of *renormalizability* of the type of noncommutative gauge theories that we have discussed here is still open. We should recall that the primary goal of the models that we have studied is to provide an effective description of particle physics on a given noncommutative space. We should not a priori expect nor require such a theory to be renormalizable. The full θ-expanded action does in fact contain infinitely many power-counting non-renormalizable terms. Nevertheless these theories do appear to be (almost) renormalizable in the following sense: Noncommutative and commutative gauge invariance alone do not uniquely single out a specific action. There is in fact quite some freedom which includes also the freedom in the choice of Seiberg-Witten map. In quantum theory the constants that parametrize the freedom become running coupling constants. This issue has been carefully studied in the case of noncommutative QED [30] and it has been found that at first order in θ at the one-loop level the quantum action has to contain (only) one additional term [31]. It is important to find more interesting processes that could serve as *experimental signatures of space-time noncommutativity* in accelerator physics, astrophysics and cosmology. Good candidates are processes that appear to violate continuous and/or discrete space-time symmetries. An example is the $Z \to \gamma\gamma$ decay depicted in Fig. 2 that violates spin conservation [5]. See, e.g., [32,33] for other examples. In this context we would also like to refer to a recent review [7] of the phenomenology of noncommutative geometry with an extensive list of references that also includes other approaches [34] than the one presented here. On the theoretical side it would be interesting to study *more complicated noncommutative structures* as models of the microscopic structure of space-time including models where the noncommutative structures are dynamical and possibly include gravity. Here we have the intriguing possibility of processes that appear to violate energy and/or momentum conservation because four-momentum can be transfered to the noncommutative space-time structure.

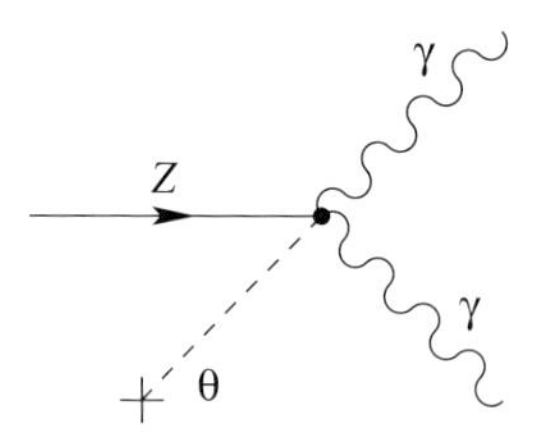

Fig. 2. $Z \to \gamma\gamma$ decay

Acknowledgements

I thank the organizers for the invitation to such a nice conference. I am grateful to Paolo Aschieri, Branislav Jurčo and Julius Wess for many helpful discussions.

References

1. H. S. Snyder, "Quantized Space-Time," Phys. Rev. **71**, 38 (1947).
2. H. Arfaei and M. H. Yavartanoo, arXiv:hep-th/0010244.
3. J. L. Hewett, F. J. Petriello and T. G. Rizzo, Phys. Rev. D **64**, 075012 (2001).
4. I. Hinchliffe and N. Kersting, Phys. Rev. D **64**, 116007 (2001).
5. W. Behr, N. G. Deshpande, G. Duplancic, P. Schupp, J. Trampetic and J. Wess, arXiv:hep-ph/0202121.
6. C. E. Carlson, C. D. Carone and R. F. Lebed, Phys. Lett. B **518**, 201 (2001).
7. I. Hinchliffe and N. Kersting, arXiv:hep-ph/0205040.
8. A. Connes, *Non-commutative geometry* (Academic Press, London 1994).
9. J. Madore, *An Introduction to Noncommutative Differential Geometry and its Physical Applications*, 2nd edn. (Cambridge University Press 1999).
10. J. Madore, S. Schraml, P. Schupp and J. Wess, Eur. Phys. J. C **16**, 161 (2000).
11. B. Jurco, S. Schraml, P. Schupp and J. Wess, Eur. Phys. J. C **17**, 521 (2000).
12. B. Jurco, P. Schupp and J. Wess, Nucl. Phys. B **604**, 148 (2001).
13. B. Jurco, L. Moller, S. Schraml, P. Schupp and J. Wess, Eur. Phys. J. C **21**, 383 (2001).
14. P. Schupp, hep-th/0111038.

15. J. Madore, Class. and Quant. Grav. **9**, 69 (1992).
16. B. L. Cerchiai and J. Wess, Eur. Phys. J. C **5**, 553 (1998).
17. F. Bayen, M. Flato, C. Fronsdal, A. Lichnerowicz, D. Sternheimer, Ann. Phys. 111, 61 (1978).
18. N. Seiberg and E. Witten, J. High Energy Phys. **9909**, 032 (1999).
19. R. Stora, privat communication.
20. D. Brace, B. L. Cerchiai, A. F. Pasqua, U. Varadarajan and B. Zumino, J. High Energy Phys. **0106**, 047 (2001); D. Brace, B. L. Cerchiai and B. Zumino, hep-th/0107225.
21. G. Barnich, F. Brandt and M. Grigoriev, hep-th/0206003.
22. B. Jurco, P. Schupp and J. Wess, Lett. Math. Phys. (to appear), hep-th/0106110.
23. A. A. Bichl, J. M. Grimstrup, L. Popp, M. Schweda and R. Wulkenhaar, hep-th/0102103.
24. H. Dorn and C. Sieg, hep-th/0205286.
25. M. Hayakawa, hep-th/9912167.
26. X. Calmet, B. Jurco, P. Schupp, J. Wess and M. Wohlgenannt, Eur. Phys. J. C **23**, 363 (2002).
27. P. Aschieri, B. Jurco, P. Schupp and J. Wess, hep-th/0205214.
28. N. G. Deshpande and X. G. He, Phys. Lett. B **533**, 116 (2002).
29. D. Bahns, S. Doplicher, K. Fredenhagen and G. Piacitelli, Phys. Lett. B **533**, 178 (2002).
30. A. Bichl, J. Grimstrup, H. Grosse, L. Popp, M. Schweda and R. Wulkenhaar, J. High Energy Phys.**0106**, 013 (2001); R. Wulkenhaar, J. High Energy Phys. **0203**, 024 (2002).
31. J. M. Grimstrup and R. Wulkenhaar, hep-th/0205153.
32. E. O. Iltan, hep-ph/0204332; hep-ph/0204129.
33. Z. Chang and Z. z. Xing, hep-ph/0204255.
34. M. Chaichian, P. Presnajder, M. M. Sheikh-Jabbari and A. Tureanu, hep-th/0107055.

Gauge Theories on Noncommutative Spaces

Julius Wess

[1] Sektion Physik der Ludwig-Maximilians-Universität, Theresienstr. 37, D-80333 München
[2] Max-Planck-Institut für Physik (Werner-Heisenberg-Institut), Föhringer Ring 6, D-80805 München

Abstract. A formalism is presented where gauge theories for nonabelian groups can be constructed on a noncommutative algebra.

1 Introduction

The idea of noncommutative coordinates ist not new. In a letter to Ehrenfest Heisenberg proposed an uncertainty relation for coordinates based on noncommuting coordinates in the year 1930 [1]. At that time Heisenberg could not formulate this idea mathematically, but the idea was propagated and in 1943 H.S.Snyder published a paper on "Quantized Space Time" [2]. Pauli in a letter to Bohr mentioned this work, calling it "mathematically ingeneous" but a "failure for reasons of physics".

Recently such ideas have been popular again, now based on more sophisticated mathematics. Starting from a gauge theory, the noncommutative structure of space time can be implemented and it leads to additional couplings of the gauge field.

To have an example in mind we choose the coordinate for the algebra

$$[x^\mu, x^\nu] = i\theta^{\mu\nu} \tag{1}$$

with constant $\theta^{\mu\nu}$, because this example has found most attention lately. This is called the canonical space time structure for coordinates and momenta. In this persentation I will always deal with this example as an algebra for coordinates.

The parameter $\theta^{\mu\nu}$ that parametrizes the noncommutativity will now enter the Lagrangian as a coupling constant. To first order in θ the Lagrangian changes as follows:

$$-\frac{1}{4}\mathrm{Tr}\int F_0^{\mu\nu} F_{0\mu\nu}\,\mathrm{d}x \tag{2}$$

$$\Rightarrow -\frac{1}{4}\mathrm{Tr}\int F_0^{\mu\nu} F_{0\mu\nu}\,\mathrm{d}x \tag{3}$$

$$+\frac{1}{8}\theta^{\kappa\lambda}\mathrm{Tr}\int F_{0\kappa\lambda} F^0_{\mu\nu} F^{0\mu\nu}\,\mathrm{d}x \tag{4}$$

$$-\frac{1}{2}\theta^{\kappa\lambda}\mathrm{Tr}\int F^0_{\mu\kappa} F^0_{\nu\lambda} F^{0\mu\nu}\mathrm{d}x \tag{5}$$

$F_{0\mu\nu}$ is the usual field strength in a nonabelian gauge theory. We see that to first order triple vertices of the gauge field occur, giving rise to new phenomenological predictions. The trace in the enveloping algebra depends on the representation. This leads to an ambiguity in the interaction terms. We do not know for what type of interactions, if any, the model is renormalizable. Thus we do not have the notion of a "minimal coupling". We consider the Lagrangians as effective Lagrangians, in this case all gauge invariant couplings are allowed and should be considered.

For the coupling of the gague fileds to the matter fields we obtain

$$\int \bar{\Psi}^0(\gamma^\mu D_\mu - m)\Psi^0 \, \mathrm{d}x \tag{6}$$

$$\Rightarrow \quad \int \bar{\Psi}^0(\gamma^\mu D_\mu - m)\Psi^0 \, \mathrm{d}x \tag{7}$$

$$- \frac{1}{2}\theta^{\kappa\lambda} \int \bar{\Psi}^0 F_{0\kappa\lambda}(\gamma^\mu D_\mu - m)\Psi^0 \, \mathrm{d}x \tag{8}$$

$$- \frac{1}{4}\theta^{\kappa\lambda} \int \bar{\Psi}^0 \gamma^\mu F^0_{\mu\kappa} D_\lambda \Psi^0 \, \mathrm{d}x \tag{9}$$

Again new couplings of the gauge field to the matter fields occur.

This formalism can be applied to the standard model with the gauge group $SU(3) \times SU(2) \times U(1)$.

The particle content is the same as in the ordinary standard model. For $\theta = 0$ the usual standard model is reproduced.

In general the new interactions will contribute to interactions already present in the usual standard model. Due to the tensorial character of the "coupling constant" $\theta^{\mu\nu}$ Lorentz invariance will be violated for the new interactions. This makes these contributions different from the usual ones and it leads to signatures by which these interactions can be identified. The most interesting case is when the new terms lead to a process that is not allowed in a usual local and Lorentz invariant quantum field theory. An example is the $Z^0 \to \gamma\gamma$ decay. Forbidden in conventional QFT, it can be obtained to first order in θ:

$$\begin{aligned} L_{Z^0\to\gamma\gamma} = \frac{e}{8}\sin 2\theta_W \left(g'^2\kappa_1 + (g'^2 - 2g^2)\kappa_2\right) \theta^{\rho\tau} \Big[2Z^{\mu\nu}\left(2A_{\mu\rho}A_{\nu\tau} - A_{\mu\nu}A_{\rho\tau}\right) \\ +8Z_{\mu\rho}A^{\mu\nu}A_{\nu\tau} - Z_{\rho\tau}A_{\mu\nu}A^{\mu\nu}\Big] , \end{aligned} \tag{10}$$

where the constants κ_1, κ_2 are defined in reference [14]. The discovery of this process would pose a problem to usual gauge theories. The same would be true for processes of the type

$$\Upsilon \to \gamma\gamma, \quad B \to K + \gamma, \quad K \to \pi\gamma \tag{11}$$

as well as a violation of the CPT theorem.

It is already interesting to see that there is a possible deformation of the usual theories where all these forbidden processes would not cause a problem.

Thus if CPT violation were discovered we still could be very close to our usual QFT.

To produce this new class of deformed QFTs a few new ideas have to be incorporated.

1. Noncommutative coordinates have to be accepted. The canonical case is the simplest example. A possible scenario would start from an x-dependent $\theta^{\mu\nu}$. Mathematically we are only able to deal with examples where $\theta^{\mu\nu}$ is linear or quadratic.

2. A star product as it is known in the deformation quantization can be used to realize the algebra. In this formalism the objects of the algebra are functions of commuting variables, the noncommutativity is present in the noncommutative star product. For the canonical case this is

$$f \star g(x) = e^{\frac{i}{2} \frac{\partial}{\partial x^\mu} \theta^{\nu\mu} \frac{\partial}{\partial y^\mu}} f(x) \cdot g(x) \Big|_{y \to x} \tag{12}$$

From this star product follows immediately

$$[x^\mu \overset{\star}{,} x^\nu] = x^\mu \star x^\nu - x^\nu \star x^\mu = i\theta^{\mu\nu} \tag{13}$$

The star product originally defined for polynomials, extended to formal power series, has been extended to functions. These functions are the objects in physics, they are identified with fields.

3. Enveloping algebra valued gauge transformations have to be used. A gauge theory is based on a Lie algebra:

$$[L^i, L^j] = i f_k^{ij} L^k \tag{14}$$

For the gauge transformation of fields the star product has to be used as well, or, which is the same, the gauge parameters depend on the noncommutative variables:

$$\delta\Psi = i\Lambda(x) \star \Psi \tag{15}$$

The commutator of two such transformations will only close if Λ is enveloping algebra valued

$$\Lambda(x) = \alpha_i^{(0)}(x) L^i + \alpha_{ij}^{(1)}(x) :L^i L^j: + \ldots + \alpha_{i_1 \ldots i_n}^{(n-1)} :L^{i_1} \cdots L^{i_n}: \tag{16}$$

The two dots indicate that a basis in the algebra is used. Completely symmetrical polynomials form such a basis.

For commuting variables Lie algebra valued transformations close, it is consistent to put α^n for $n \neq 1$. The transformation depends on L parameters $\alpha_1^0(x) \ldots \alpha_L^0(x)$. The enveloping algebra valued transformations seem to depend on an infinite set of parameters. It is, however, possible to express these parameters in terms of the parameters $\alpha_i^0(x)$ and the usual Lie algebra valued gauge field $A_\mu(x) = A_\mu^i(x) T^i$ with the transformation properties

$$\delta A_\mu = \partial_\mu \alpha^0 + i[\alpha^0, A_\mu] \tag{17}$$

$$\alpha_{i_1 \ldots i_r}^{(r-1)}(x) = \Lambda_{i_1 \ldots i_r}^{r-1} \{\alpha_i^0(x), A_\mu^i, \partial\} \tag{18}$$

This is part of the Seiberg-Witten map. We now call a transformation that uses these $\alpha^{(r)}$

$$\delta_{\alpha^0}\Psi = \alpha(x) \star \Psi \tag{19}$$

If we apply a second transformation we have to transform A_μ as well. It is now possible to find expressions $\alpha^{(r)}$ in terms of $\alpha^{(0)}$, A_μ and their derivatives such that

$$(\delta_{\alpha_0}\delta_{\beta_0} - \delta_{\beta_0}\delta_{\alpha_0})\Psi = \delta_{[\alpha_0,\beta_0]}\Psi \tag{20}$$

This equation determines $\alpha^{(r)}$ to a large extent. The problem is formulated in power series expansion in θ. To first order in θ we find

$$\alpha = \alpha_0 + \frac{1}{2}\theta^{\mu\nu} :\partial_\mu\alpha^0(x)A_\nu(x): + \ldots \tag{21}$$

4. Seiberg-Witten map. This is the most important new idea. One part is contained in the construction of the transformation parameter α, but it is also possible to express the new gauge field and the matter fileds that transform as follows:

$$\delta\hat{\Psi} \;\; = i\alpha \star \hat{\Psi} \tag{22}$$

$$\delta\hat{A}^\mu = -i[x^\mu \overset{\star}{,} \alpha] + i[\alpha \overset{\star}{,} \hat{A}^\mu] \tag{23}$$

in terms of fields that transform as usual

$$\delta\Psi \;\; = i\alpha_0\Psi \tag{24}$$

$$\delta A^\mu = \partial_\mu\alpha_0 + i[\alpha_0, A^\mu] \tag{25}$$

The new gauge potential is enveloping algebra valued.

$$\bar{\Psi} \;\; = \Psi + \frac{1}{2}\theta_{\mu\nu}A_\nu\partial_\mu\Psi + \frac{1}{4}\theta^{\mu\nu}\partial_\mu A_\nu\Psi + \ldots \tag{26}$$

$$\hat{A}_\mu = A_\mu + \frac{1}{4}\theta^{\rho\nu}\{A_\nu(\partial_\rho A_\mu + F_{\rho\mu})\} + \ldots \tag{27}$$

This opens the way to construct gauge field theories with the star product without changing the particle content.

5. Covariant coordinates. The transformation law of $\hat{A}$ allows the construction of covariant coordinates. Due to the noncommutativity coordinates will not commute with the gauge transformation, a situation similar to derivatives in a usual gauge theory. Taking the idea from covariant derivatives one can try the following Ansatz

$$\hat{X}^\mu \qquad\quad = \xi^\mu + \hat{A}^\mu \tag{28}$$

$$\delta_{\alpha_0}\hat{X}^\mu \star \hat{\Psi} = i\alpha \star \xi^\mu \star \hat{\Psi} \tag{29}$$

This defines the transformation law for $\hat{A}^\mu$.

A field strength can be introduced in complete analogy:

$$\hat{X}^\mu\hat{X}^\nu - \hat{X}^\nu\hat{X}^\mu = i\theta^{\mu\nu} + \hat{F}^{\mu\nu} \tag{30}$$

the field strength transforms tensorially

$$\delta_\alpha \hat{F}^{\mu\nu} = [\alpha \star \hat{F}^{\mu\nu}] \tag{31}$$

and can be used to construct invariant Lagrangians. This leads to the theories discussed at the beginning.

Let me now go into more detail.

2 The Algebra

Let me first exhibit the algebraic structure of $\mathbb{R}^n$ and then generalise to noncommutative coordinates. The coordinates $x^1 \dots x^n \in \mathbb{R}^n$ are considered as elements of an associative algebra over $\mathbb{C}$. The algebra, freely generated by these elements, will be denoted by $\mathbb{C}\left[[x^1, \dots, x^n]\right]$. The two brackets indicate that formal power series are allowed in the algebra.

The elements of this algebra are then subject to relations that make them commutative:

$$\mathcal{R}: \quad x^i x^j - x^j x^i = 0. \tag{32}$$

These relations generate a two-sided ideal I_R; it consists of all the elements of the algebra $\mathbb{C}\left[[x^1, \dots, x^n]\right]$ that can be obtained from the relation (32) by multiplying (32) from the left and the right by all possible products of the coordinates. We factor out this ideal and obtain the desired algebra:

$$\mathcal{A}_x = \frac{\mathbb{C}\left[[x^1, \dots, x^n]\right]}{I_{\mathcal{R}}}. \tag{33}$$

The elements of this algebra are the polynomials and the formal power series in the commuting variables $x^1, \dots, x^n \in \mathbb{R}$.

$$\begin{aligned} &f(x^1, \dots, x^n) \in \mathcal{A}_x, \\ &f(x^1, \dots, x^n) = \sum_{r_i=0}^{\infty} f_{r_1 \dots r_n} (x^1)^{r_1} \cdot \ldots \cdot (x^n)^{r_n}. \end{aligned} \tag{34}$$

Multiplication in this algebra is the pointwise multiplication of these functions.

This algebraic concept can be easily generalized to noncommutative coordinates. We consider algebras, freely generated by elements $\hat{x}^1, \dots \hat{x}^n$, again we call these elements coordinates, but now they are supposed to satisfy relations that make them noncommutative:

$$\mathcal{R}_{\hat{x},\hat{x}}: \quad [\hat{x}^i, \hat{x}^j] = i\theta^{ij}(\hat{x}). \tag{35}$$

Following L.Landau, noncommutativity carries a hat. Again the relations (35) generate an ideal and we define our algebra $\hat{\mathcal{A}}_{\hat{x}}$ as follows:

$$\begin{aligned} \mathcal{A}_{\hat{x}} &= \frac{\mathbb{C}[[\hat{x^1}, \dots, \hat{x^n}]]}{I_{\mathcal{R}_{\hat{x},\hat{x}}}}, \\ \hat{f} &\in \hat{\mathcal{A}}_{\hat{x}}. \end{aligned} \tag{36}$$

We impose one more condition on the algebra. The vectorspace of the homogeneous polynomials of degree m, $\hat{V}_{\hat{x}}^m$ should have the same dimension as V_x^m. Algebras of this type are said to have the Poincare-Birkhoff-Witt property. In the following we shall consider such algebras only.

3 The $\star$ Product

The vectorspaces V_x^m and $\hat{V}_{\hat{x}}^m$ are finite-dimensional, thus they are isomorphic. To establish an isomorphism we map a given basis of one space into a given basis of the other space. This then defines a vectorspace isomorphism between the vectorspaces $\hat{V}_{\hat{x}}$ and V_x.

We now change the algebra $\mathcal{A}_x$ to extend the above vector space isomorphism to an algebra isomorphism. For this purpose we have to change the multiplication law in $\mathcal{A}_x$. When we multiply two elements in $\hat{\mathcal{A}}_{\hat{x}}$ we can compute from the multiplication law in $\hat{\mathcal{A}}_{\hat{x}}$ the coefficient function of the product in a given basis. We define the product in the vectorspace V_x to be the element with the same coefficient function as it was calculated in $\hat{\mathcal{A}}_{\hat{x}}$. This multiplication rule we call $\star$ (star) product and this defines the algebra $\star\mathcal{A}_x$. The algebras $\hat{\mathcal{A}}_{\hat{x}}$ and $\star\mathcal{A}_x$ are isomorphic.

It is natural to use the elements of $\star\mathcal{A}_x$ as objects in physics. The pointwise product has to be replaced by the $\star$ product. In all the cases of interest the $\star$ product can be expressed with the help of a differential operator. This makes it possible to extend the $\star$ product to functions without referring to power series expansion. Thus we treat the elements $\mathcal{A}_x$ like ordinary fields but replace the pointwise product by the $\star$ product. This would be the starting point of deformation quantization. As we have based the concept on associative algebras, associativity of the $\star$ product is guaranteed.

4 Gauge Theory

In this context it is possible to formulate a gauge theory [4–6]. We start from a Lie algebra:

$$[T^a, T^b] = i f_c^{ab} T^c. \tag{37}$$

In a usual gauge theory on commutative spaces the fields will span a representation of this Lie algebra and they will transform under the usual gauge transformation with Lie algebra valued parameters:

$$\delta_{\alpha^0}\psi(x) = i\alpha^0(x)\psi(x). \tag{38}$$

$$\begin{aligned}
(\delta_{\alpha^0}\delta_{\beta^0} - \delta_{\beta^0}\delta_{\alpha^0})\psi &= -(\beta^0\alpha^0 - \alpha^0\beta^0)\psi \\
&= i(\alpha^0 \times \beta^0)\psi = \delta_{\alpha^0\times\beta^0}\psi, \\
\alpha^0 \times \beta^0 &\equiv \alpha_a^0 \beta_b^0 f_c^{ab} T^c.
\end{aligned} \tag{39}$$

The commutator of two such transformations remains Lie algebra valued.

For a theory on non-commutative spaces we start with fields that are elements of $\star\mathcal{A}_x$. Gauge transformations have to be defined with the $\star$ product:

$$\delta_\alpha \psi(x) = i\alpha(x) \star \psi(x) \tag{40}$$

The star product of functions is not commutative. The commutator of two Lie algebra valued transformations does not reproduce a Lie algebra valued parameter. Thus we shall assume that the infinitesimal transformation parameters are enveloping algebra valued [3]:

$$\alpha(x) = \alpha^0_a(x) T^a + \alpha^1_{ab}(x) : T^a T^b : + \ldots + \alpha^{n-1}_{a_1 \ldots a_n}(x) : T^{a_1} \cdot \ldots \cdot T^{a_n} : + \cdots \tag{41}$$

We have adopted the :: notation for a basis in the enveloping algebra of the Lie algebra. Completely symmetrized products could serve as a basis:

$$\begin{aligned} : T^a : \quad &= T^a, \\ : T^a T^b : &= \frac{1}{2}(T^a T^b + T^b T^a) \text{ etc.} \end{aligned} \tag{42}$$

The commutator of two transformations is certainly enveloping algebra valued.

$$(\delta_\alpha \delta_\beta - \delta_\beta \delta_\alpha)\psi = [\alpha \stackrel{\star}{,} \beta] \star \psi. \tag{43}$$

The disadvantage of this approach is that infinitely many parameters $\alpha^n(x)$ have to be introduced.

It is a surprise that it is possible to define gauge transformations where all the parameters $\alpha^n(x)$ depend on the finite set of parameters $\alpha^0(x)$ (Lie algebra valued) and in addition on the gauge potential $a(x)$ of a usual gauge theory and on their derivatives. The gauge potential $a(x)$ has the usual transformation properties:

$$\begin{aligned} \delta a_i &= \partial_i \alpha^0 + i[\alpha^0, a_i], \\ \delta a_{i,a} &= \partial_i \alpha^0_a - \alpha^0_b f_a^{bc} a_{i,c}. \end{aligned} \tag{44}$$

We will call the new type of transformation parameters $\Lambda_{\alpha^0}(x)$. The new transformations are supposed to close under a commutator into a transformation characterized by $(\alpha^0 \times \beta^0)$:

$$\begin{aligned} \delta_{\alpha^0} \psi(x) \quad &= i\Lambda_{\alpha^0(x)}(x) \star \psi(x), \\ (\delta_{\alpha^0}\delta_{\beta^0} - \delta_{\beta^0}\delta_{\alpha^0})\psi &= \delta_{\alpha^0 \times \beta^0}\psi, \\ (\alpha^0 \times \beta^0)_a \quad &= \alpha^0_b \beta^0_c f_a^{bc}. \end{aligned} \tag{45}$$

These equations define $\Lambda_{\alpha^0}(x)$. We shall see that all $\alpha_n(x)$ in (41) can be defined in terms of $\alpha^0(x)$ and the gauge potential $a(x)$. The transformation property (45) then holds as a consequence of (44). The solution of this problem, however, is not unique, this will be seen in the following.

As a consequence of the a dependence of Λ^0_α we have to transform Λ^0_α under the second variation in the commutator. This changes equation (43) and this is the reason why the new approach works.

5 Constant θ

To illustrate this approach we restrict it to the algebra where $\theta^{\mu\nu}$ is a constant. In this case we obtain in a fully symmetrized basis the following $\star$ product [11,12]:

$$(f \star g)(x) = e^{\frac{i}{2}\frac{\partial}{\partial x^i}\theta^{ij}\frac{\partial}{\partial y^j}} f(x)g(y)\Big|_{y\Rightarrow x} \tag{46}$$
$$= \int d^n y \; \delta^n(x-y) e^{\frac{i}{2}\frac{\partial}{\partial x^i}\theta^{ij}\frac{\partial}{\partial y^j}} f(x)g(y).$$

We expand in θ.

$$\Lambda_{\alpha^0} = \alpha_a^0 T^a + \theta^{ij}\Lambda^1_{\alpha^0,ij} + \ldots, \tag{47}$$

The $\star$ product has to be expanded as well. Finally we expand the defining equation for Λ_{α^0}

$$(\delta_{\alpha^0}\delta_{\beta^0} - \delta_{\beta^0}\delta_{\alpha^0})\psi = i(\delta_{\alpha^0}\Lambda_{\beta^0} - \delta_{\beta^0}\Lambda_{\alpha^0}) \star \psi + [\Lambda_{\alpha^0} \stackrel{\star}{,} \Lambda_{\beta_0}] \star \psi, \tag{48}$$
$$= \delta_{\alpha^0\times\beta^0}\psi = i\Lambda_{\alpha^0\times\beta^0} \star \psi$$

The zeroeth order in the expansion of (48) defines α^0 as Lie algebra valued.

In first order we obtain

$$\theta^{ij}\big((\delta_{\alpha^0}\Lambda^1_{\beta^0,ij} - \delta_{\beta^0}\Lambda^1_{\alpha^0,ij}) - i([\alpha^0, \Lambda^1_{\beta^0,ij}] - \tag{49}$$
$$- [\beta^0, \Lambda^1_{\alpha^0,ij}])\big) + \frac{1}{2}\partial_i\alpha_a^0\partial_j\beta_b^0 : T^aT^b := \theta^{ij}\Lambda^1_{\alpha^0\times\beta^0,ij}.$$

A closer look shows that this is an inhomogeneous linear equation for Λ^1. The inhomogeneous term is known, it contains α^0 and β^0 only. A particular solution of (49) is:

$$\theta^{ij}\Lambda^1_{\alpha^0,ij} = \frac{1}{2}\theta^{ij}(\partial_i\alpha_a^0)a_{j,b} : T^aT^b : . \tag{50}$$

Any solution of the homogeneous part of (49) can be added to (50).

We can proceed order by order in θ, the structure of the equations will always be the same. It will be an inhomogeneous linear equation, the homogeneous remains the same, the inhomogeneous part will contain known quantities only. This way we obtain Λ_{α^0} in a θ expansion.

$$\Lambda_{\alpha^0} = \alpha_a^0 T^a + \frac{1}{2}\theta^{ij}(\partial_i\alpha_a^0)a_{j,b} : T^aT^b : + \ldots \tag{51}$$

Such a construction of the transformation parameter first occured in the context of the Seiberg-Witten map [7,19,16].

6 Covariant Coordinates

In a usual gauge theory we would procede with the definition of covariant derivatives. Derivatives, however, are not a natural concept for algebras. It is more natural to introduce covariant coordinates. Based on such a concept gauge theories can be developed as well.

It is obvious that coordinates do not commute with gauge transformations, it is also natural to introduce covariant coordinates in analogy to covariant derivatives:

$$X^i = x^i + A^i(x), \qquad (52)$$
$$\delta_{\alpha^0} X^i \star \psi = i\Lambda_{\alpha^0} \star X^i \star \psi.$$

This leads to the following transformation law for the gauge potential:

$$\delta A^i = -i[x^i \stackrel{\star}{,} \Lambda_{\alpha^0}] + i[\Lambda_{\alpha^0} \stackrel{\star}{,} A^i]. \qquad (53)$$

To satisfy such a transformation law we have again to assume that $A(x)$ is enveloping algebra valued. In general, this would imply infinitely many gauge fields. For the restricted gauge transformations Λ_{α^0} it is possible to construct a gauge potential that depends on the Lie algebra valued potential $a(x)$ and its derivatives only. The transformation law (44) for $a(x)$ will imply the transformation law for $A(x)$. This is the main achievement of the Seiberg-Witten map (53).

The construction of gauge fields that transform to tensorial follows the usual concept as we know it from covariant derivatives. An obvious definition is

$$X^\mu X^\nu - X^\nu X^\mu - i\theta^{\mu\nu}(X) = \hat{F}^{\mu\nu} \qquad (54)$$

It is chosen in such a way that $\tilde{F}^{\mu\nu}$ vanishes for a vanishing gauge potential A^μ.

The tensorial transformation law of $\tilde{F}^{\mu\nu}$ follows directly from (53):

$$\delta_\alpha \tilde{F}^{\mu\nu} = [\Delta_{\alpha 0} \star \tilde{F}^{\mu\nu}] \qquad (55)$$

It should be noted, however, that the trace in the representation space of the Lie algebra of a tensor is not an invariant because the star product is not commutative.

7 The Integral

An invariant action can be constructed only if the integral has its trace property:

$$\int f \star g = \int g \star f. \qquad (56)$$

Integration is not a natural concept in an algebra. It is supposed to be a linear map from $\mathcal{A}_{\hat{x}}$ into $\mathbb{C}$.

$$\int \qquad : \mathcal{A}_{\hat{x}} \to \mathbb{C}, \qquad (57)$$
$$\int (c_1 \hat{f} + c_2 \hat{g}) = c_1 \int \hat{f} + c_2 \int \hat{g},$$

In addition the trace property is required:

$$\int \hat{f}\hat{g} = \int \hat{g}\hat{f}. \qquad (58)$$

This is equivalent to (56).

8 Gauge Theory for Constant θ

For constant θ the usual integral in x-space will have the trace property. This can be shown by a direct calculation.

Let us have a look at this formalism for constant $\theta^{\mu\nu}$:

The Seiberg-Witten map:

$$\begin{aligned}A^i(x) &= \theta^{ij}V_j,\\ V_j(x) &= a_{j,a}T^a - \frac{1}{2}\theta^{ln}a_{l,a}(\partial_n a_{j,b} + F_{nj,b} : T^aT^b : + \ldots,\\ F_{nj,b} &= \partial_n a_{j,b} - \partial_j a_{n,b} + f_b^{cd}a_{n,c}a_{j,d}.\end{aligned} \tag{59}$$

The field strength:

$$\tilde{F}_{ij} = F_{ij,a}T^a + \theta^{ln}(F_{il,a}F_{jn,l} - \frac{1}{2}a_{l,a}(2\partial_n F_{ij,b} + a_{n,c}F_{ij,d}f_e^{cd})) : T^aT^b : + \ldots. \tag{60}$$

The Lagrangian:

$$L = \frac{1}{4}\mathrm{Tr}F_{ij} \star F^{ij}. \tag{61}$$

The invariant action:

$$\begin{aligned}W &= \frac{1}{4}\int \mathrm{Tr}F_{ij} \star F^{ij}\\ &= \frac{1}{4}\int \mathrm{Tr}F_{ij}F^{ij}.\end{aligned} \tag{62}$$

New coupling terms arise, $\theta^{\mu\nu}$ appears as a coupling constant, it is a Lorentz tensor and the interaction term breaks Lorentz invariance. This was to be expected because the defining relation (32) already breaks Lorentz invariance.

These new terms in the Lagrangian will give rise to new interactions. Due to the breaking of Lorentz invariance interaction terms will occur that are forbidden in a Lorentz invariant theory. A good example is the $Z^0 \to \gamma\gamma$ decay. From (62) we find the following interaction terms that contribute to this decay if the gauge theory is based on the standard model.

$$\begin{aligned}\mathcal{L}_{Z_{\gamma\gamma}} &= \frac{e}{8}\sin 2\theta_W \left(g'^2\kappa_1 + (g'^2 - 2g^2)\kappa_2\right)\theta^{kl}\\ &\times \Big(2(-\partial_i Z_k + \partial_k Z_i)\partial_j A_l(\partial^i A^j - \partial^j A^i)\\ &+ (\partial_i A_k \partial_j A_l + \partial_k A_i \partial_l A_j - 2\partial_k A_i \partial_j A_l)(-\partial^i Z^j + \partial^j Z^i)\\ &+ (-2\partial_k Z_i \partial_l A_j + 2\partial_j Z_l \partial_k A_i + 2\partial_i Z_j \partial_k A_l + \partial_k Z_l \partial_i A_j)(\partial^i A^j - \partial^j A^i)\Big)\end{aligned} \tag{63}$$

This expression is gauge invariant under the usual Lie algebra valued gauge transformation. It contributes to the branching ratio of the Z^0 decay.

We still have to learn how the gauge potential couples to the matter fields. This will be done via covariant derivatives.

$$\begin{aligned}\mathcal{D}_i \star \psi &= (\partial_i - iV_i) \star \psi,\\ \delta_{\alpha^0}\mathcal{D}_i \star \psi &= i\Lambda_{\alpha^0} \star \mathcal{D}_i \star \psi.\end{aligned} \tag{64}$$

9 Derivatives

First we have to define derivatives. In general, the star product will depend on the coordinates, when we differentiate it the coordinate dependence of the $\star$ product will contribute as well. Nevertheless, we demand a Leibniz rule of the type

$$\partial_\mu \star (f \star g) = (\partial_\mu f) \star g + \mathcal{O}^\nu_\mu(f) \star \partial_\nu g. \tag{65}$$

From the associativity of the $\star$ product follows that $O^\nu_\mu(f)$ has to be an algebra homomorphism.

It is easier to define derivatives for $\hat{\mathcal{A}}_{\hat{x}}$. A general procedure was outlined in [13]. We first extend the algebra by algebraic elements $\hat{\partial}$ and consider the algebra $\mathbb{C}\left[\left[\hat{x}^1, \ldots, \hat{x}^n, \ldots, \hat{\partial}^1, \ldots, \hat{\partial}^n\right]\right]$. This algebra has to be divided by the ideal $I_{\hat{x},\hat{x}}$ as before. Then we have to construct a derivative, based on a Leibniz rule that is a map in $\mathbb{C}\left[\left[\hat{x}^1, \ldots, \hat{x}^n, \ldots, \hat{\partial}^1, \ldots, \hat{\partial}^n\right]\right]/I_{\hat{x},\hat{x}}$. This leads to consistency relations for the Leibniz rule The Leibniz rule can now be interpreted as a relation and the respective ideals can be constructed and factored out. Finally this has to be supplemented by $\hat{\partial}, \hat{\partial}$ relations. We treat these relations as usual and after dividing by the respective ideal we arrive at an algebra that we call $\hat{\mathcal{A}}_{\hat{x},\hat{\partial}}$.

In more detail the generalized Leibniz rule is supposed to have the form:

$$\hat{\partial}_i(\hat{f}\hat{g}) = (\hat{\partial}_i \hat{f})\hat{g} + O^l_i(\hat{f})\hat{\partial}_l \hat{g}. \tag{66}$$

From the law of associativity in $\hat{\mathcal{A}}_{\hat{x}}$ follows that the map 0 has to be an algebra homomorphism

$$O^i_j(\hat{f}\hat{g}) = O^i_l(\hat{f})O^l_j(\hat{g}). \tag{67}$$

If we define the Leibniz rule on the linear coordinates we can generalize it to all elements.

In the ${}^\star\mathcal{A}_x$ version of the algebra the Leibniz rule takes the form of equation (69). This rule can be found as follows: $\hat{\partial}$ introduces a map on the basis of $\hat{\mathcal{A}}_{\hat{x}}$, this map defines a map in ${}^\star\mathcal{A}_x$. This map has finally to be expressed with ordinary x-derivatives. This then leads to (69).

For constant $\theta^{\mu\nu}$ where the $\star$ product does not depend on x the $\partial\star$ derivatives are just the ordinary x-derivatives.

Covariant derivatives are then defined as usual:

$$\begin{aligned} \mathcal{D}_i \star \psi &= (\partial_i - iV_i) \star \psi, \\ \delta_{\alpha^0} \mathcal{D}_i \star \psi &= i\Lambda_{\alpha^0} \star \mathcal{D}_i \star \psi. \end{aligned} \tag{68}$$

The vector potential has to be enveloping algebra valued. Again, it can be expressed in terms of a_μ by a Seiberg-Witten map. Therefore we expect that A_μ and V_μ are related.

For constant $\theta^{\mu\nu}$ we find:

$$A^i(x) = \theta^{ij} V_j \tag{69}$$

Covariant derivatives exist for $\theta^{\mu\nu} = 0$. From (69) follows that A^μ vanishes in this case, coordinates are already covariant.

10 Gauge Couplings to Matter Fields

The matter field ψ that transforms like

$$\delta_{\alpha^0}\psi(x) = i\Lambda_{\alpha^0(x)}(x) \star \psi(x) \tag{70}$$

can be expressed in terms of a field ψ^0 that transforms with a Lie algebra valued parameter and the Lie algebra valued vector potential a. The transformation property (70) will be a consequence of (38) and (44).

For constant θ we find:

$$\psi = \psi^0 - \frac{1}{2}\theta^{\mu\nu} a^l_\mu T^l \partial_\nu \psi^0 + \ldots \tag{71}$$

This now leads to the Lagrangian

$$\begin{aligned} &\int \bar{\psi} \star (\gamma^\mu D_\mu \star - m)\, \psi d^4 x \\ &\quad = \int \bar{\psi^0}\, (\gamma^\mu D_\mu - m)\, \psi^0 d^4\, x - \frac{1}{4}\theta^{\mu\lambda} \int \bar{\psi^0} F^0_{\mu\lambda}\, (\gamma^\mu D_\mu - m)\, \psi^0 \\ &\qquad - \frac{1}{4}\theta^{\sigma\lambda} \int \bar{\psi^0} \gamma^\mu F^0_{\mu\sigma} D_\lambda \psi^0 d^4 x + \cdots \end{aligned} \tag{72}$$

The fields ψ^0 and $F^{\mu\nu 0}$ transform like the usual gauge fields with a Lie algebra valued parameter. $F^{\mu\nu 0}$ is just the usual field strength of a gauge theory. Accordingly, $D_\mu \psi^0$ is the usual covariant derivative with the field a_μ as a gauge potential.

11 Conclusion

Such a theory based on noncommutative coordinates should only be relevant for a region with very high energy density, thus for very short distances, i.e. well inside the confinement range. For larger distances we know that physics is described very well with commuting coordinates. $\theta^{\mu\nu}(x)$ will be a complicated function, we treat this function in a power series expansion and start with constant $\theta^{\mu\nu}$. This has a chance to be relevant for processes that take place at very short distances where the constant $\theta^{\mu\nu}$ might be dominant. The higher order contribution on the expansion become relevant at distances where the process has already occured. Such a process will not be sensitive to the functional behaviour of $\theta^{\mu\nu}(x)$ and the constant $\theta^{\mu\nu}$ approximation might be a good approximation [15,14]. To find such a process demands physical intuition.

References

1. W.Heisenberg, *Gesammelte Werke, Teil II*, Springer-Verlag.
2. H.S.Snyder, *Quantized Space-Time*, Phys.Rev. **71**, 38 (1947).

3. B. Jurčo, S. Schraml, P. Schupp and J. Wess, Eur. Phys. J. C **17**, 521 (2000).
4. J. Madore, S. Schraml, P. Schupp and J. Wess, Eur. Phys. J. C **16**, 161 (2000).
5. B. Jurčo, P. Schupp, Eur. Phys. J. C **14**, 367 (2000).
6. B. Jurčo, P. Schupp and J. Wess, Nucl. Phys. B **584**, 784 (2000).
7. N. Seiberg and E. Witten, J. High Energy Phys. **9909**, 032 (1999).
8. A. Dimakis, J. Madore, J. Math. Phys. **37**, 4647 (1996).
9. J. Madore, *An Introduction to Noncommutative Differential Geometry and it Physical Applications*, 2nd Edition (Cambridge University Press, 1999).
10. J. Wess, *q-deformed Heisenberg Algebras*, in H. Gausterer, H. Grosse and L. Pittner, eds., Proceedings of the 38. Internationale Universitätswochen für Kern- und Teilchenphysik, no. 543 in Lect. Notes in Phys., Springer-Verlag, 2000, Schladming, January 1999, math-ph/9910013.
11. H. Weyl, Z. Physik **46**, 1 (1927), *Quantenmechanik und Gruppentheorie*; *The theory of groups and quantum mechanics*, Dover, New-York (1931), translated from *Gruppentheorie und Quantenmechanik*, Hirzel Verlag, Leipzig (1928).
12. J. E. Moyal, Proc. Cambridge Phil. Soc. **45**, 99 (1949).
13. J. Wess and B. Zumino, Nucl. Phys. Proc. Suppl. **18B**, 302 (1991).
14. W.Behr, N.G.Deshpande, G.Duplancic, P.Schupp, J.Trampetic and J. Wess, hep-ph/0202121.
15. X.Calmet, B.Jurco, P.Schupp, J.Wess, M.Wohlgenannt, Eur. Phys. J. C **23**, 363 (2002).
16. B.Jurco, P.Schupp and J.Wess, hep-th/0106110.
17. B.Jurco, L.Möller, S.Schraml, P.Schupp and J.Wess, Eur. Phys. J. C **21**, 383 (2001).
18. B.Jurco, P.Schupp, J.Wess, Nucl. Phys. B **604**, 148 (2001).
19. B.Jurco, P.Schupp, J.Wess, Mod. Phys. Lett. A **16**, 343 (2001).
20. I.Hinchliffe, N.Kersting, hep-ph/0205040.
21. D.Brace, B.L.Cerchiai, A.F.Pasqua, U.Varadarajan, B.Zumino, J. High Energy Phys. **0106**, 047 (2001).
22. A.A.Bichl, J.M.Grimstrup, L.Popp, M.Schweda, R.Wulkenhaar, hep-th/0102103.
23. A.A.Bichl, J.M.Grimstrup, H.Grosse, L.Popp, M.Schweda, R.Wulkenhaar, J. High Energy Phys. **0106**, 013 (2001).
24. J.M.Grimstrup, R.Wulkenhaar, hep-th/0205153.

Part V

Diverse Topics in Theoretical Physics

QCD2 with Massless Quarks in Terms of Currents

Adi Armoni[1], Yitzhak Frishman[2], and Jacob Sonnenschein[3]

[1] Theory Division, CERN, CH-1211 Geneva 23, Switzerland
[2] Department of Particle Physics, Weizmann Institute of Science, 76100 Rehovot, Israel
[3] School of Physics and Astronomy, Beverly and Raymond Sackler Faculty of Exact Sciences, Tel Aviv University, Ramat Aviv, 69978, Israel

Abstract. We discuss the spectra of multi-flavor massless QCD_2. An approximation in which the Hilbert space is truncated to two currents states is used. We write down a 't Hooft like equation for the wave function of the two currents states. We solve this equation for the lowest massive state and find an excellent agreement with the DLCQ results. The issue of the non-hermiticity of the truncated Hamiltonian is also discussed. Talk based mainly on [1].

1 Introduction

Two dimensional quantum chromodynamics (QCD_2) is a useful model for the real world QCD. The model at the large N_c limit was solved by 't Hooft[2] showing confinement of quarks with an approximately linear Regge trajectory of states. Other issues, such as the baryonic spectrum at strong coupling[3] and questions of screening versus confinement[4,5] can also be addressed in this framework.

The two-dimensional model with fermions in the adjoint representation is also interesting and attracted a lot of attention in recent years[4–14]. In particular it was shown in [9] that the adjoint fermions model is equivalent to QCD_2 model with level $N_f = N_c$, for the massive part of the spectrum, in the case of massless fermions. Another attempt to address the adjoint fermions model, was by using the currents as building blocks of the spectrum [10].

The idea of the present work is to study the spectrum of QCD_2 at arbitrary level N_f using states built from two currents, for the case of massless fermions.

Whereas the 't Hooft model $N_f = 1$ is exactly solvable, the multi-flavor case $N_f > 1$ is not solvable model even in the Veneziano limit when both N_c and N_f are taken to infinity (with a fixed ratio), since pair creation and annihilation is not suppressed. In the present work, we use an approximation in which we restrict ourselves to two currents states. We cannot justify a-priori such an approximation for arbitrary level. However the numerical solutions for the lowest massive state admit a very close resemblance to the DLCQ results where such a truncation was not used [12–14]. A justification can be given for $\frac{N_f}{N_c}$ very small or very large.

The obtained equation suggests the following picture: the underlying degrees of freedom in the problem are interacting "gluons" with mass $\frac{e^2 N_f}{\pi}$. Actually, these are really quanta of the the color currents. As it is well known, there are no independent gluon degrees of freedom in two dimensions.

2 Massless QCD_2 and Bosonization

Massless multi-flavor QCD_2 with fermions in the fundamental representation of $SU(N_c)$ is described by the action

$$S = \int d^2x \text{ tr } (-\frac{1}{2e^2} F^2_{\mu\nu} + i\bar{\Psi} \not{D} \Psi) \tag{1}$$

where $\Psi = \Psi^i_a$, $i = 1 \dots N_c$, $a = 1 \dots N_f$.

It is natural to bosonize this system, since bosonization in the $SU(N_c) \times SU(N_f) \times U_B(1)$ scheme decouples color and flavor degrees of freedom (in the massless case). The bosonized form of the action of this theory is given by[15]

$$\begin{aligned} S_{bosonized} = & \quad (2) \\ & N_f S_{WZW}(h) + N_c S_{WZW}(g) + \int d^2x \, \frac{1}{2} \partial_\mu \phi \partial^\mu \phi - \int d^2x \text{ tr } \frac{1}{2e^2} F_{\mu\nu} F^{\mu\nu} \\ & - \frac{N_f}{2\pi} \int d^2x \text{ tr } (ih^\dagger \partial_+ h A_- + ih \partial_- h^\dagger A_+ + A_+ h A_- h^\dagger - A_+ A_-) \end{aligned}$$

where $h \in SU(N_c)$, $g \in SU(N_f)$, ϕ is the baryon number and S_{WZW} stands for the Wess-Zumino-Witten action, which for complex fermions reads

$$\begin{aligned} S_{WZW}(g) = & \frac{1}{8\pi} \int_\Sigma d^2x \text{ tr } (\partial_\mu g \partial^\mu g^{-1}) + \\ & \frac{1}{12\pi} \int_B d^3y \epsilon^{ijk} \text{ tr } (g^{-1} \partial_i g)(g^{-1} \partial_j g)(g^{-1} \partial_k g), \end{aligned}$$

Since we are interested in the massive spectrum of the theory and the flavor degrees of freedom are entirely decoupled from the system and they are massless, we can put aside the g and ϕ fields (There is a residual interaction of the zero modes of the g, h and ϕ fields, but it is not important to our discussion[9]).

Upon choosing the light cone gauge $A_- = 0$ and integrating A_+ we arrive to the effective action

$$S_{eff} = N_f S_{WZW}(h) - \frac{1}{2} e^2 \int d^2x \text{ tr } (\frac{1}{\partial_-} J^+)^2, \tag{3}$$

where $J^+ = \frac{iN_f}{2\pi} h \partial_- h^\dagger$. In terms of $J = \sqrt{\pi} J^+$, the light-cone momentum operators P^μ take the following simple form

$$P^+ = \frac{1}{N_c + N_f} \int dx^- : J^a(x^-, x^+ = 0) J^a(x^-, x^+ = 0) : \tag{4}$$

namely the Sugawara form, and

$$P^{-} = -\frac{e^{2}}{2\pi}\int dx^{-} : J^{a}(x^{-},x^{+}=0)\frac{1}{\partial_{-}^{2}}J^{a}(x^{-},x^{+}=0) : . \tag{5}$$

Our task will be to solve the eigenvalue equation

$$2P^{+}P^{-}\,|\psi\rangle = M^{2}\,|\psi\rangle. \tag{6}$$

We write P^{+} and P^{-} in terms of the Fourier transform of $J(x^{-})$ defined by $J(p^{+}) = \int \frac{dx^{-}}{\sqrt{2\pi}}e^{-ip^{+}x^{-}}J(x^{-},x^{+}=0)$. Normal ordering in the expression of P^{+} and P^{-} are naturally with respect to p, where $p<0$ denotes a creation operator. To simplify the notation we write from now on p instead of p^{+}. In terms of these variables the momenta generators are

$$P^{+} = \frac{1}{N_c+N_f}\int_0^{\infty} dp J^{a}(-p)J^{a}(p) \tag{7}$$

$$P^{-} = \frac{e^{2}}{\pi}\int_0^{\infty} dp\frac{1}{p^{2}}J^{a}(-p)J^{a}(p) \tag{8}$$

Recall that the light-cone currents $J^{a}(p)$ obey a level N_f, $SU(N_c)$ affine Lie algebra

$$[J^{a}(p),J^{b}(p')] = \frac{1}{2}N_f\;p\;\delta^{ab}\delta(p+p') + if^{abc}J^{c}(p+p') \tag{9}$$

We can now construct the Hilbert space. The vacuum $|0,R\rangle$ is defined by the annihilation property:

$$\forall p>0,\;\; J(p)\;|0,R\rangle = 0 \tag{10}$$

Where R is an "allowed" representations depending on the level. Therefore, a physical state in Hilbert space is $|\psi\rangle = \mathrm{tr}\;J(-p_1)\ldots J(-p_n)\;|0,R\rangle$. Note that this basis is not orthogonal.

3 't Hooft-Like Equation for the Two Currents Wave-Function

Let us restrict ourselves to the 2-currents sector of the Hilbert space

$$|\Phi\rangle = \frac{2}{\sqrt{N_c^{2}-1N_f}}\int_0^{1} dk\;\Phi(k)J^{a}(-k)J^{a}(k-1)\;|0\rangle, \tag{11}$$

namely to states which are color singlets of two currents with total $P^{+}=1$ momentum and a distribution of P^{-} momentum $\Phi(k)$. Note that Φ is a symmetric function

$$\Phi(k) = \Phi(1-k). \tag{12}$$

Our task now is to find the eigenvalue (Schrödinger) equation for the wave-function $\Phi(k)$. Let us act with the "Hamiltonian" P^{-} on the state $|\Phi\rangle$.

The commutator of P^- with a current $J^b(-k)$ yields the following result

$$
\begin{aligned}
&[\int_0^\infty \frac{dp}{p^2} J^a(-p)J^a(p), J^b(-k)] = \\
&\left((\frac{1}{2}N_f - N_c)\frac{1}{k} + N_c\frac{1}{\epsilon} \right) J^b(-k) + \\
&\int_k^\infty dp \left(\frac{1}{p^2} - \frac{1}{(p-k)^2} \right) i f^{abc} J^a(-p)J^c(p-k) + \\
&\int_0^k \frac{dp}{p^2} i f^{abc} J^c(p-k)J^a(-p).
\end{aligned} \tag{13}
$$

The above expression (13) contains 3 terms on the R.H.S. The first term contains a single creation operator. The second term contains an annihilation current and therefore should be commuted with $J^b(k-1)$. The third term contains two creation currents and it would lead to a 3-currents state. Thus, the affine Lie algebra created a higher state. This is a manifestation of the fact that pair creation is, generically, not suppressed in multi-flavor QCD_2, as expected in general in QFT.

Note that while deriving (13) we get an "infinite" contribution $N_c\frac{1}{\epsilon}J^b(-k)$, where ϵ represents a regularization away from $p=0$ or $p=k$ integration. This contribution will be canceled by a counter contribution which comes from the regime $p \sim k$ in the first integral on the R.H.S. of (13), as below.

The commutator of the second term in the R.H.S of (13) with $J^b(k-1)$ yields

$$
\begin{aligned}
&[\int_k^\infty dp \left(\frac{1}{p^2} - \frac{1}{(p-k)^2} \right) i f^{abc} J^a(-p)J^c(p-k), J^b(k-1)] = \\
&N_c \int_k^\infty dp \left(\frac{1}{p^2} - \frac{1}{(p-k)^2} \right) (J^a(-p)J^a(p-1) - J^a(p-k)J^a(k-p-1)).
\end{aligned} \tag{14}
$$

Our results can be summarized by the following set of equations

$$
\begin{aligned}
&M^2 \left|\Phi\right\rangle = \\
&\frac{1}{N_c N_f} \int_0^1 dk\ \tilde{\Phi}(k) J^a(-k)J^a(k-1) \left|0\right\rangle + \\
&\frac{1}{(N_c N_f)^{\frac{3}{2}}} \int_0^1 dk\ dp\ dl\ \delta(k+p+l-1)\Psi(k,p,l) i f^{abc} J^a(-k)J^b(-p)J^c(-l) \left|0\right\rangle
\end{aligned} \tag{15}
$$

with

$$
\Psi(k,p,l) = \frac{2e^2 (N_c N_f)^{\frac{1}{2}}}{\pi} \left(\frac{\Phi(l) - \Phi(k)}{p^2} \right) \tag{16}
$$

and

$$
\begin{aligned}
\tilde{\Phi}(k) = \frac{e^2}{\pi} \Bigg(&(N_f - N_c)\left(\frac{1}{k} + \frac{1}{1-k} \right)\Phi(k) + \frac{2N_c}{\epsilon}\Phi(k) - N_c \int_0^{k-\epsilon} dp \frac{\Phi(p)}{(p-k)^2} \\
&- N_c \int_{k+\epsilon}^1 dp \frac{\Phi(p)}{(p-k)^2} + N_c \left(\frac{1}{k^2} - \frac{1}{(1-k)^2} \right) \int_0^k dp\ \Phi(p) \Bigg).
\end{aligned}
$$

Ignoring the 3-currents term (see below), we get that $\Phi(k)$ obeys the following eigenvalue equation

$$\frac{M^2}{e^2/\pi}\Phi(k) = (N_f - N_c)\left(\frac{1}{k} + \frac{1}{1-k}\right)\Phi(k)$$
$$-N_c\mathcal{P}\int_0^1 dp\frac{\Phi(p)}{(p-k)^2} + N_c\left(\frac{1}{k^2} - \frac{1}{(1-k)^2}\right)\int_0^k dp\;\Phi(p). \qquad (17)$$

For general N_c and N_f discarding the 3-currents term is unjustified. However, since the length of Ψ is $|\Psi(k,p,l)|\sim e^2(N_cN_f)^{\frac{1}{2}}$, in the limit of large N_c with fixed e^2N_c and fixed N_f, or large N_f with fixed e^2N_f and fixed N_c, the 3-currents contribution is indeed negligible, as compared with the 2-currents term, the later being of order 1. Note also that while deriving (17) we assumed that $\int_0^1 dp\;\Phi(p) = 0$. We will justify this assumption in the following. Another remark is that the first integral in (17) should be calculated as a principal value integral. The divergent part of this integral (arising from the regime $p \sim k$) cancels the previously mentioned infinity.

In order to make contact with the ordinary 't Hooft equation, it is useful to integrate equation (17) with respect to k and rewrite the equation in terms of $\varphi(k) \equiv \int_0^k dp\;\Phi(p)$.

$$\frac{M^2}{e^2/\pi}\varphi(k) = (N_f - N_c)\left(\frac{1}{k} + \frac{1}{1-k}\right)\varphi(k)$$
$$-N_c\mathcal{P}\int_0^1 dp\frac{\varphi(p)}{(p-k)^2} + N_f\int_0^k dp\frac{\varphi(p)}{p^2} + N_f\int_k^1 dp\frac{\varphi(p)}{(1-p)^2} \qquad (18)$$

The derivation goes as follows. First, integrating (12) we get $\varphi(k) = -\varphi(1-k) +$ const. Taking $\varphi(1) = 0$ we get

$$\varphi(k) = -\varphi(1-k). \qquad (19)$$

Now $\varphi(1) = 0$ implies $\int_0^1 dk\Phi(k) = 0$, which was our assumption above. Then, differentiating (18) we do get (17), and by (19) we also get that there is no extra integration constant.

We would like to comment on the issue of the Hermiticity of the 'Hamiltonian' M^2. Naively, it seems that M^2 is not Hermitian with respect to the scalar product $\langle\psi|\varphi\rangle = \int_0^1 dk\psi^\star(k)\varphi(k)$, since the Hermitian conjugate of (18) is

$$\left(\frac{M^2}{e^2/\pi}\right)^\dagger\varphi(k) = (N_f - N_c)\left(\frac{1}{k} + \frac{1}{1-k}\right)\varphi(k)$$
$$-N_c\mathcal{P}\int_0^1 dp\frac{\varphi(p)}{(p-k)^2} - N_f\frac{1}{k^2}\int_0^k dp\varphi(p) - N_f\frac{1}{(1-k)^2}\int_k^1 dp\varphi(p) \quad (20)$$

However, as we shall see in the next section, the numerical solution yields real eigenvalues and eigenfunctions, though (20) is non-Hermitian. Due to its non-Hermiticity we expect non-orthogonal solutions. Note that (20) is "more regular" than (18), as in (18) it is $\varphi(p)/p^2$ that appears in the integration from zero.

Equation (18) is similar to 't Hooft equation for a massive single flavor large N_c QCD_2, with $m^2 = \frac{e^2 N_f}{\pi}$. It differs from 't Hooft's equation by having two additional terms (two last terms in (18)). It suggests that the dynamics which governs the lowest state of the multi-flavor model is given, approximately, by a model of massive "glueball" with an $SU(N_c)$ gauge interaction and additional terms which are proportional to N_f.

Before we present our solution of (18) it is important to note that it is only an approximated solution. We neglected the 3-currents state with, a-priori, no justification. We shall see, however, that the restriction to the truncated 2-currents sector is an excellent approximation for the lowest massive meson.

4 The Spectrum – Numerical Results

The most convenient way to solve (18) is to expand $\varphi(k)$ in the following basis (see, however [16], a different interesting choice of basis)

$$\varphi(k) = \sum_{i=0}^{\infty} A_i (k - \frac{1}{2}) \left(k(1-k)\right)^{\beta+i} \tag{21}$$

The value of β chosen such that the Hamiltonian will not be singular near $k \to 0$ (or $k \to 1$)[2],[17]. This consideration leads to the following equation

$$(\frac{N_f}{N_c} - 1) - \frac{N_f/N_c}{\beta+1} + \beta\pi \cot \beta\pi = 0. \tag{22}$$

This comes from (20). Had we started with (18), it would have been $-\beta$ replacing β in (22), and constrained to β larger than 1. Upon truncating the infinite sum in (21) to a finite sum, the eigenvalue problem reduces to a diagonalization of a matrix. So, the problem can be reformulated as follows

$$\lambda N_{ij} A_j = H_{ij} A_j, \tag{23}$$

with

$$N_{ij} = \int_0^1 dk (k - \frac{1}{2})^2 \left(k(1-k)\right)^{2\beta+i+j}, \tag{24}$$

and

$$\begin{aligned} H_{ij} &= (\frac{N_f}{N_c} - 1) \int_0^1 dk (k - \frac{1}{2})^2 \left(k(1-k)\right)^{2\beta+i+j-1} \\ &- \frac{N_f}{N_c} \int_0^1 dk (k - \frac{1}{2}) \left(k(1-k)\right)^{\beta+i} \frac{1}{k^2} \int_0^k (p - \frac{1}{2}) \left(p(1-p)\right)^{\beta+j} \\ &- \frac{N_f}{N_c} \int_0^1 dk (k - \frac{1}{2}) \left(k(1-k)\right)^{\beta+i} \frac{1}{(1-k)^2} \int_k^1 (p - \frac{1}{2}) \left(p(1-p)\right)^{\beta+j} \\ &- \int_0^1 dk dp \frac{(k-\frac{1}{2}) \left(k(1-k)\right)^{\beta+i} (p-\frac{1}{2}) \left(p(1-p)\right)^{\beta+j}}{(k-p)^2} \end{aligned} \tag{25}$$

Hence

$$N_{ij} = \frac{B(2\beta+i+j+2, 2\beta+i+j+2)}{2(2\beta+i+j+1)}, \tag{26}$$

and

$$\begin{aligned} H_{ij} = {} & (\frac{N_f}{N_c}-1)\frac{B(2\beta+i+j+1, 2\beta+i+j+1)}{2(2\beta+i+j)} \\ & -\frac{N_f}{N_c}\frac{B(2\beta+i+j+1, 2\beta+i+j+1)}{2(2\beta+i+j)(\beta+j+1)} \\ & +\frac{(\beta+i)(\beta+j)B(\beta+i,\beta+i)B(\beta+j,\beta+j)}{8(2\beta+i+j)(2\beta+i+j+1)}, \end{aligned} \tag{27}$$

where $B(x,y)$ is the Beta function

$$B(x,y) = \frac{\Gamma(x)\Gamma(y)}{\Gamma(x+y)}. \tag{28}$$

In practice, the process converges rapidly and a 5×5 matrix yields the 'continuum' results. Working with higher dimensional matrices should, in principle, improve the accuracy of our results. Surprisingly, for large matrices, complex eigenvalues appear[1]. It can be interpreted in two possible ways: (i) either it is related to the non-hermiticity of the truncated Hamiltonian, or (ii) it is an artifact of Maple. The first possibility can be justified by the following example. Consider an Harmonic oscillator with a non-Hermitian perturbation ga^n. For states $|k\rangle$, with $k < n$, the perturbation is effective only when applied on the 'ket' and not on the 'bra', yielding different results and possibly also complex eigenvalues for $k \geq n$. It is very similar to the present case. On the other hand, the accuracy of our results suggest that the problems are an artifact of Maple, possibly also related to the bad behavior of the basis functions near the boundaries. We postpone this crucial issue for future work.

The lowest eigenvalues of (18) as a function of the ratio $\frac{N_f}{N_c}$ are listed in Table 1 below (see also Fig. 1). Note that by $\beta = 0, N_f/N_c = 0$ we mean the limit $\beta \to 0, N_f/N_c \to 0$.

These values are in excellent agreement with recent DLCQ calculations. For comparison see [12], [13] and especially [14] for a recent work. The typical error is less than 0.1% !

An interesting observation is that the eigenvalues depends linearly on N_f (see Fig. 1). The dependence is

$$M^2 = \frac{e^2 N_c}{\pi}(5.88 + 5\frac{N_f}{N_c}). \tag{29}$$

We do not have a good understanding of this observation. It is not clear why the lowest eigenvalue sits on a straight line. It is not clear even why, as an eigenvalue equation (18) exhibits such a behavior.

A detailed discussion of some special cases, such as the 't Hooft model, adjoint QCD and the large N_f limit is given in [1].

---

[1] We thank Ariel Abrashkin for discussions on this issue.

Table 1. The mass of the lowest massive meson, in units of $\frac{e^2 N_c}{\pi}$, as a function of N_f/N_c and β.

β	N_f/N_c	M^2
0.0000	0	5.88
0.0573	0.2	6.91
0.1088	0.4	7.91
0.1552	0.6	8.91
0.1978	0.8	9.89
0.2366	1.0	10.86
0.2725	1.2	11.83
0.3050	1.4	12.77
0.3360	1.6	13.73
0.3645	1.8	14.67

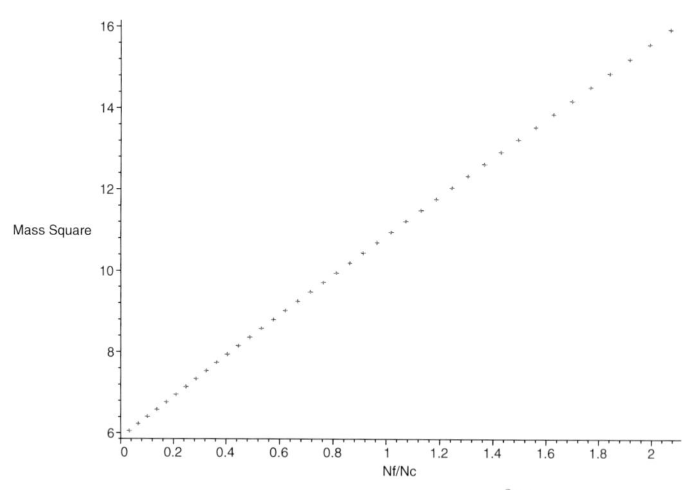

Fig. 1. The mass of the lowest massive meson, in units of $\frac{e^2 N_c}{\pi}$, as a function of N_f/N_c.

Acknowledgements

A.A. thanks the department of particle physics at the Weizmann institute of science for the warm hospitality while part of this work was done. The research of J.S was supported in part by the US–Israeli Binational Science Foundation, the Israeli Science Foundation, the German–Israeli Foundation for Scientific Research (GIF) and the Einstein Center for theoretical Physics at the Weizmann Institute. A.A would also like to thank B. Van de Sande for help with the numerical solution of 't Hooft equation. We would also like to thank A. Abrashkin for fruithful discussions.

References

1. A. Armoni, Y. Frishman and J. Sonnenschein, Nucl. Phys. B **596**, 459 (2001).
2. G. 't Hooft, Nucl. Phys. B **75**, 461 (1974).
3. G.D. Date, Y. Frishman and J. Sonnenschein, Nucl. Phys. B **283**, 365 (1987).
4. D.J. Gross, I.R. Klebanov,A.V. Matytsin and A.V. Smilga, Nucl. Phys. B **461**, 109 (1996).
5. A. Armoni, Y. Frishman and J. Sonnenschein, Phys. Rev. Lett. **80**, 430 (1998), Phys. Lett. B **449**, 76 (1999).
6. S. Dalley and I.R. Klebanov, Phys. Rev. D **47**, 2517 (1993).
7. D. Kutasov, Nucl. Phys. B **414**, 33 (1994).
8. G. Bhanot, K. Demeterfi and I.R. Klebanov, Phys. Rev. D **48**, 4980 (1993).
9. D. Kutasov and A. Schwimmer, Nucl. Phys. B **442**, 447 (1995).
10. A. Armoni and J. Sonnenschein, Nucl. Phys. B **457**, 81 (1995).
11. S. Dalley, Phys. Lett. B **418**, 160 (1998).
12. D. J. Gross, A. Hashimoto and I. R. Klebanov, Phys. Rev. D **57**, 6420 (1998).
13. F. Antonuccio and S. Pinsky, Phys. Lett. B **439**, 142 (1998).
14. U. Trittmann, hep-th/0005075.
15. Y. Frishman and J. Sonnenschein, Phys. Rept. **223**, 309 (1993).
16. O. Abe, Phys. Rev. D **60**, 105040 (1999).
17. B. Van de Sande, Phys. Rev. D **54**, 6347 (1996).

Physics of the Monopoles in QCD

Valentine I. Zakharov

Max-Planck Institut für Physik, Föhringer Ring 6, 80805 München, Germany

Abstract. We discuss implications of the recent measurements of the non-Abelian action density associated with the monopoles condensed in the confining phase of gluodynamics. The radius of the monopole determined in terms of the action was found to be small numerically. As far as the condensation of the monopoles is described in terms of a scalar field, a fine tuning is then implied. In other words, a hierarchy exists between the self energy of the monopole and the temperature of the confinement-deconfinement phase transition. The ratio of the two scales is no less than a factor of 10. Moreover, we argue that the hierarchy scale can well eventually extend to a few hundred GeV on the ultraviolet side. The corresponding phenomenology is discussed, mostly within the polymer picture of the monopole condensation.

1 Introduction

The monopole condensation is one of the most favored mechanisms [1] of the confinement, for review see, e.g., [2]. In the field theoretical language, one usually thinks in terms of a Higgs-type model:

$$S_{eff} \;=\; \int d^4x\big(|D_\mu\phi|^2+\frac{1}{4}F^2_{\mu\nu}+\;V(|\phi|^2)\big) \tag{1}$$

where ϕ is a scalar field with a non-zero magnetic charge, $F_{\mu\nu}$ is the field strength tensor constructed on the dual-gluon field B_μ, D_μ is the covariant derivative with respect to the dual gluon. Finally, $V(|\phi|^2)$ is the potential energy ensuring that $\langle\phi\rangle\neq 0$ in the vacuum. Relation of the "effective" fields ϕ, B_μ to the fundamental QCD fields is one of the basic problems of the approach considered but here we would simply refer the reader to [3] for further discussion of this problem. At this moment, it suffices to say that the "dual-superconductor" mechanism of confinement assumes formation of an Abrikosov-type tube between the heavy quarks introduced into the vacuum via the Wilson loop while the tube itself is a classical solution of the equations of motion corresponding to the effective Lagrangian (1).

By introducing scalar fields, one opens a door to the standard questions on the consistency, on the quantum level, of a $\lambda\phi^4$ theory. Here, we mean primarily the problem of the quadratic divergence in the scalar mass. At first sight, these problems are not serious in our case since (1) apparently represents an effective theory presumably valid for a limited range of mass scales.

However, if we ask ourselves, what are the actual limitations on the use of the effective theory (1) we should admit that there is no way at the moment to answer this question on pure theoretical grounds and we should turn instead to the experimental data, that is lattice measurements. This lack of understanding concerns first of all the nature of the non-perturbative field configurations that are defined as monopoles. First, it is not clear apriori which $U(1)$ subgroup of the $SU(2)$ [1] is to be picked up for the classification of the monopoles. Even if we make this choice on pure pragmatic basis and concentrate on the most successful scheme of the monopoles in the maximal Abelian projection [2] we still get very little understanding of the field configurations underlying the objects defined as monopoles in this projection, for discussion see, e.g., [4]. In particular, nothing can be said on the size of the monopole which presumably limits application of (1) on the ultraviolet side.

Direct measurements of the monopole size were reported recently [5] and brought an unexpectedly small value of the monopole radius:

$$R_{mon} \approx 0.06 \text{ fm}, \tag{2}$$

where the monopole radius is defined here in terms of the full non-Abelian action associated with the monopole and not in terms of the projected action. If we compare the radius (2) with the temperature of the confinement-deconfinement transition:

$$T_{deconf} \approx 300 \text{ MeV} \tag{3}$$

then we would come to the conclusion that there are different mass scales co-existing within the effective scalar-field theory (1). And the question, how this mass hierarchy is maintained is becoming legitimate.

Although comparison of (2) and (3) is instructive by itself, we will argue that the actual hierarchy mass scale can be much higher on the ultraviolet side. Namely, we will emphasize later that even at the size (2) the monopoles are very "hot", i.e. have action comparable to the action of the zero-point fluctuations. For physical interpretation, it is natural to understand by the radius such distances where the non-perturbative fields die away on the scale of pure perturbative fluctuations. And this radius is to be considerably smaller than (2).

Also, estimate (2) means that the asymptotic freedom is not yet reached at quite small distances and the question arises as to how reconcile this observation with such phenomena as the precocious scaling.

We cannot claim at all understanding answers to these questions but feel that it is important to start discussing them. Our approach is mostly phenomenological and we are trying to formulate which measurements could help to find answers to the puzzles outlined above. The theoretical framework which we are using is mainly the polymer approach to the scalar field theory, see, e.g., [6–8].

2 Monopole Condensation: Overview of the Theory

2.1 Compact $U(1)$

The show case of the monopole condensation is the compact $U(1)$ [9]. The crucial role of the compactness is to ensure that the Dirac string does not cost energy

[1] for simplicity we will confine ourselves to the case of $SU(2)$ as the color group.

(for a review see, e.g., [4]). The monopole self energy reduces then to the energy associated with the radial magnetic field $\mathbf{B}$. The self energy is readily seen to diverge linearly in the ultraviolet:

$$M_{mon}(a) \;=\; \frac{1}{8\pi}\int \mathbf{B}^2 d^3r \;\sim\; \frac{c}{8e^2}\frac{1}{a}\,, \tag{4}$$

where c is a constant, a is the lattice spacing, e is the electric charge and the magnetic charge is [2] $g_m = 1/2e$. Thus, the monopoles are infinitely heavy and, at first sight, this precludes any condensation since the probability to find a monopole trajectory of the length L is suppressed as

$$\exp(-S) \;=\; \exp\left(-\frac{c}{e^2}\cdot\frac{L}{a}\right)\,. \tag{5}$$

Note that the constant c depends on the details of the lattice regularization but can be found explicitly in any particular case.

However, there is an exponentially large enhancement factor due to the entropy. Namely, trajectory of the length L can be realized on a cubic lattice in $N_L = 7^{L/a}$ various ways. Indeed, the monopole occupies center of a cube and the trajectory consists of L/a steps. At each step the trajectory can be continued to an adjacent cube. In four dimensions there are 8 such cubes. However, one of them has to be excluded since the monopole trajectory is non-backtracking. Thus the entropy factor,

$$N_L \;=\; \exp\left(\ln 7\cdot\frac{\mathrm{L}}{\mathrm{a}}\right)\,, \tag{6}$$

cancels the suppression due to the action (5) if the coupling e^2 satisfies the condition

$$e^2_{crit} \;=\; c/\ln 7 \;\approx\; 1\,, \tag{7}$$

where we quote the numerical value of e^2_{crit} for the Wilson action and cubic lattice. At e^2_{crit} any monopole trajectory length L is allowed and the monopoles condense.

This simple theory works within about one percent accuracy in terms of e^2_{crit} [10]. Note that the energy-entropy balance above does not account for interaction with the neighboring monopoles.

2.2 Monopole Cluster in the Field-Theoretical Language

The derivation of the previous subsection implies that the monopole condensation occurs when the monopole action is ultraviolet divergent. On the other hand, the onset of the condensation in the standard field theoretical language corresponds to the zero mass of the magnetically charged field ϕ. It is important

[2] The notation g is reserved for the non-Abelian coupling, the magnetic coupling is denoted as g_m.

to emphasize that this apparent mismatch between the two languages is not specific for the monopoles at all. Actually, there is a general kinematic relation between the physical mass of a scalar field m^2_{phys} and the mass M defined in terms of the (Euclidean) action, $M \equiv S/L$ where L is the length of the trajectory and S is the corresponding action [3]:

$$m^2_{phys} \cdot a \ \approx \ M - \frac{\ln 7}{\mathrm{a}}, \tag{8}$$

where terms of higher order in ma are omitted. Here by m^2_{phys} we understand the mass entering the propagator of a free particle,

$$D(p^2, m^2_{phys}) \ \sim \ (p^2 + m^2_{phys})^{-1} \ ,$$

where p^2 is either Euclidean or Minkowskian momentum squared.

In view of the crucial role of the (8) for our discussion, let us reiterate the statement. We consider propagator of a free scalar particle in terms of the path integral:

$$D(x_i, x_f) \ \sim \ \Sigma_{paths} exp(\ - S_{cl}(path)), \tag{9}$$

where for the classical action associated with the path we would like to substitute simply the action of a point-like classical particle, $S_{cl} = M \cdot L$ where M is the mass of the particle and L is the length of the path. Then we learn that there is no such representation (with replacement of S_{cl} by iS_{cl})) for the propagator of a relativistic particle in the Minkowski space because of the backward-in-time motions [4]. However, in the Euclidean space the representation (9) works. The physical mass is, however, gets renormalized compared to M according to (8).

Derivation of the Eq (8) is in textbooks [5] , see, e.g., [12]. The central point is that the action for a point-like particle in the Euclidean space looks exactly the same as that of a non-interacting polymer with a non-vanishing chemical potential for the constituent atoms. The transition from the polymer to the field theoretical language is common in the statistical physics (see, e.g., [13]). The first applications to the monopole physics are due to the authors in [7]. For the sake of completeness we reproduce here the main points crucial for our discussion later. Mostly, we follow the second paper in [7].

The scalar particle trajectory represented as a random walk and the corresponding partition function is:

$$Z = \int \mathrm{d}^4 x \sum_{N=1}^{\infty} \frac{1}{N} \, e^{-\mu N} \, Z_N(x,x) \, , \tag{10}$$

---

[3] It is worth emphasizing that the results of the lattice measurements are commonly expressed in terms of Higgs masses and interaction constants, see [11]. However, these masses are obtained without subtracting the ln7 term (compare Eq (8)) and, to our belief, are not the physical mass for this reason. Where by the physical masses we understand the masses in the continuum limit. In particular, the physical masses determine the shape of the Abrikosov-like string confining the heavy quarks.

[4] I am indebted to L. Stodolsky for an illuminating discussions on this topic.

[5] Actually, one finds mostly $\ln 2D \equiv \ln 8$ instead of ln7. We do think that ln7 is the correct number but in fact this difference is not important for further discussion.

where μ is the chemical potential and $Z_N(x_0,x_f)$ is the partition function of a polymer broken into N segments:

$$Z_N(x_0,x_f)=\left[\prod_{i=1}^{N-1}\int \mathrm{d}^4x_i\right]\prod_{i=1}^{N}\left[\frac{\delta(|x_i-x_{i-1}|-a)}{2\pi^2a^3}\right]\exp\Big\{-\sum_{i=1}^{N}gV(x_i)\Big\}. \quad (11)$$

This partition function represents a summation over all atoms of the polymer weighted by the Boltzmann factors. The δ–functions in (11) ensure that each bond in the polymer has length a. The starting point of the polymer (11) is x_0 and the ending point is $x_f\equiv x_N$.

In the limit $a\to 0$ the partition function (11) can be treated analogously to a Feynman integral. The crucial step is the coarse–graining: the N–sized polymer is divided into m units by n atoms ($N=mn$), and the limit is considered when both m and n are large while a and $\sqrt{n}a$ are small. We get,

$$\prod_{i=\nu n}^{(\nu+1)n-1}\frac{1}{2\pi^2a^3}\delta(|x_i-x_{i+1}|-a)\to\left(\frac{2}{\pi n a^2}\right)^2\exp\Big\{-\frac{2}{n\,a^2}(x_{(\nu+1)n}-x_{\nu n})^2\Big\}, \quad (12)$$

where the index i, $i=\nu n\cdots(\nu+1)n-1$, labels the atoms in ν^{th} unit. The polymer partition function becomes [7]:

$$Z_N(x_0,x_f)=\mathrm{const}\cdot\left[\prod_{\nu=1}^{m-1}\mathrm{d}^4x\right]\left[\left(\frac{2}{\pi n a^2}\right)^{2m}\exp\Big\{\sum_{\nu=1}^{m}\frac{(x_\nu-x_{\nu-1})^2}{na^2}\Big\}\right]$$
$$\cdot\exp\Big\{-\sum_{\nu=1}^{m}n(\mu+V(x_\nu))\Big\}. \quad (13)$$

The x_i's have been re-labeled so that x_ν is the average value of x in at the coarser cell. Using the variables:

$$s=\frac{1}{8}na^2\nu,\quad \tau=\frac{1}{8}a^2N,\quad m_0^2=\frac{8\mu}{a^2}, \quad (14)$$

one can rewrite the partition function (10) as

$$Z=\mathrm{const}\cdot\int_0^\infty\frac{\mathrm{d}\tau}{\tau}\int_{x(0)=x(\tau)=x}Dx\,\exp\left\{-\int_0^\tau\left[\frac{1}{4}\dot{x}_\mu^2(s)+m_0^2+g_0V(x(s))\right]\mathrm{d}s\right\}. \quad (15)$$

The next step is to rewrite the integral over trajectories $x(\tau)$ as the standard path integral representation for a free scalar field. For us it is important only that the m_0^2 term in (15) is becoming the standard mass term in the field theoretical language:

$$\mathcal{Z}=\sum_{M=0}^{\infty}\frac{1}{M!}Z^M$$
$$=\mathrm{const}\cdot\int D\phi\,\exp\Big\{-\int\mathrm{d}^4x\left[(\partial_\mu\phi)^2+m_0^2\phi^2+g_0V(x)\phi^2\right]\Big\}. \quad (16)$$

The whole machinery can be easily generalized to the case of charged particles (monopoles) with Coulomb-like interactions.

2.3 Monopole Condensation in Non-Abelian Case: Expectations

If we try to adjust the lessons from the compact $U(1)$ to the non-Abelian case then the good news is that, indeed, all the $U(1)$ subgroups of the color $SU(2)$ are compact. Moreover, dynamics of any subgroup of the $SU(2)$ is governed by the same running coupling $g^2(r)$. Thus, we could hope that the following simple picture might work: if the lattice spacing a is small we would not see monopoles because $g^2(a)$ falls below e^2_{crit}. However, going to a coarser lattice a la Wilson we come to the point where $g^2(a^2) \approx e^2_{crit}$. Then we apply the entropy-energy balance which works so well in case of the compact $U(1)$ and conclude that the monopoles of a critical size a_{crit} such that $g^2(a_{crit}) \sim 1$ condense in the QCD vacuum.

This simple picture is open, however, to painful questions. First, monopoles are defined topologically within a $U(1)$ subgroup [6]. However, it is only the $U(1)$ invariant action which has a non-vanishing minimum for a $U(1)$ topologically non-trivial object. There is no relation, generally speaking, between the full non-Abelian action and a $U(1)$-subgroup topology. For illustrations of this general rule see [3].

Therefore, there is no reason, at least at first sight, for the saturation of the functional integral at the classical solution with infinite action, see (4). This observation brings serious doubts on the validity of our simple dynamical picture.

3 Monopoles, as They Are Seen

3.1 Monopole Dominance

On the background of the theoretical turmoil, the data on the monopoles indicate a very simple and solid picture. We will constrain ourselves to the monopoles in the so called Maximal Abelian gauge and the related projection (MAP). We just mention some facts, a review and further references can be found, e.g., in [2].

Since the monopoles of the non-Abelian theory are expected to actually be $U(1)$ objects one first uses the gauge freedom to bring the non-Abelian fields as close to the Abelian ones as possible. The gauge is defined by maximization of a functional which in the continuum limit corresponds to $R(\hat{A})$ where

$$R(\hat{A}) \;=\; -\int d^4x\left[(A^1_\mu)^2+(A^2_\mu)^2\right] \tag{17}$$

where $1, 2$ are color indices.

[6] Note that a $SU(2)$-invariant definition of the monopoles is also possible [14]. However, their dynamical characteristics have not been measured yet and such monopoles are not considered here.

As the next step, one projects the non-Abelian fields generated on the lattice into their Abelian part, essentially, by putting $A^{1,2} \equiv 0$. In this Abelian projection one defines the monopole currents k_μ for each field configuration. Note that the original configurations which are used for a search of the monopoles are generated within the full non-Abelian theory. Upon performing the projection one can introduce also the corresponding Abelian, or projected action.

The relation of the monopoles to the confinement is revealed through evaluation of the Wilson loop for the quarks in the fundamental representation. Namely it turns out, first, that the string tension in the Abelian projection is close to the string tension in the original $SU(2)$ theory [15]:

$$\sigma_{U(1)} \approx \sigma_{SU(2)} . \tag{18}$$

Moreover, one can define also the string tension which arises due to the monopoles alone. To this end, one calculates the field created by a monopole current:

$$A^{mon}_\mu(x) = \frac{1}{2}\varepsilon_{\mu\nu\alpha\beta} \sum_y \Delta^{-1}(x-y)\, \partial_\nu m_{\alpha\beta}[y;k] , \tag{19}$$

where Δ^{-1} is the inverse Laplacian, and sums up (numerically) over the Dirac surface, $m[k]$, spanned on the monopole currents k. The resulting string tension is again close to that of the un-projected theory:

$$\sigma_{mon} \approx \sigma_{SU(2)} . \tag{20}$$

It might worth mentioning that these basic features remain also true upon inclusion of the dynamical fermions in $SU(3)$ case (full lattice QCD) [16].

3.2 Gauge-Invariant Properties of the Monopoles

Despite of the apparent gauge-dependence of the monopoles introduced within the MAP, they encode gauge-invariant information. In particular, we would mention two points: scaling of the monopole density and full non-Abelian action associated with the monopoles.

According to the measurements (see [17] and references therein) the monopole density ρ_{mon} in three-dimensional volume (that is, at any given time) is given in the physical units. In other words, the density scales according to the renormgroup as a quantity of dimension 3. Numerically:

$$\rho_{mon} = 0.65(2)\, (\sigma_{SU(2)})^{3/2} . \tag{21}$$

One important remark is in order here. While discussing the monopole density one should distinguish between what is sometimes called ultraviolet (UV) and infrared (IR) clusters [18]. The infrared, or percolating cluster fills in the whole lattice while the UV clusters are short. There is a spectrum of the UV clusters, as a function of their length, while the percolating cluster is in a single copy. The statement on the scaling (21) applies only to the IR cluster. We do not consider the UV clusters in this note.

Also, upon identification of the monopoles in the Abelian projection, one can measure the non-Abelian action associated with these monopoles. For practical reasons, the measurements refer to the plaquettes closest to the center of the cube containing the monopole. Since the self energy is UV divergent, it might be a reasonable approximation. The importance of such measurements is that we expect that it is the non-Abelian action which enters the energy-entropy balance for the monopoles.

The results of one of the latest measurements of this type are reproduced in Fig. 1 (see [5]).

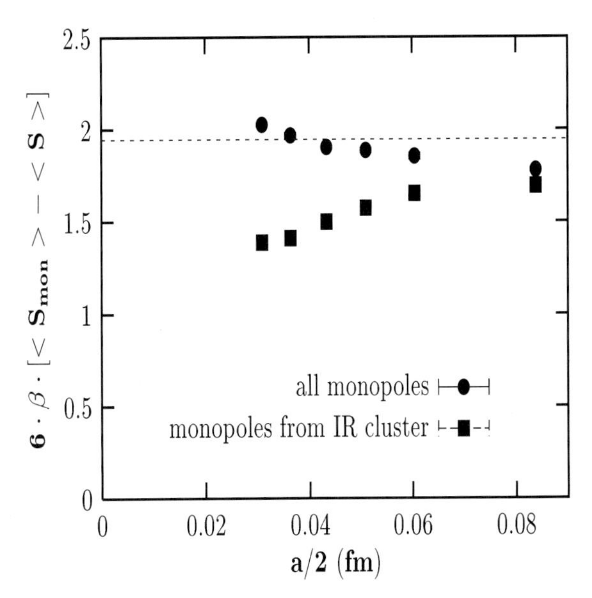

Fig. 1. The average excess of the full non-Abelian action on the plaquettes closest to the monopole, as a function of a half of the lattice spacing $a/2$. The data are reproduced from the first paper in [5]

What is plotted here is the average excess of the action on the plaquettes closest to the monopole (monopoles are positioned at centers of cubes). The action is the lattice units. In other words, the corresponding mass of the monopole $M_{mon}(a)$ of order $1/a$ if the action of order unit.

As is emphasized in [5], the IR and UV monopoles are distinguishable through their non-Abelian actions. For the UV monopoles the action is larger, in accordance with the fact that they do not percolate (condense). This is quite a dramatic confirmation that the condensation of the monopoles in the Maximal Abelian projection is driven by the full non-Abelian action, not by its projected counterpart.

4 Fine Tuning

Let us pause here to reiterate our strategy. We are assuming that the monopole condensation can be described within an effective Higgs-type theory like (1). In fact, even this broad assumption can be wrong but at this time it is difficult to suggest a framework alternative to the field theory. Next, we would like to fix the effective theory using results of the lattice measurements. Moreover we are interested first of all in interpreting data which can be expressed in gauge independent way. As the first step, we will argue in this section that the data on the monopole action [5] imply a fine tuning. By which we understand that

$$|M_{mon}(a) - \frac{ln7}{a}| \ll M_{mon}(a) \tag{22}$$

where $M_{mon}(a)$ is the monopole self energy [7] and ln7 is of pure geometrical origin (see (6)). Note that (22) looks similar to the fine tuning condition in the Standard Model.

4.1 Evidence

There are a few pieces of evidence in favor of the fine tuning (22):

a) Direct measurements indicate that the excess of the action is indeed related to the ln7, as is obvious from Fig. 1. Let us also emphasize that it is only the full non-Abelian action which "knows" about the ln7. The Abelian projected action is not related at all to the ln7 [5]. This illustrates once again that the dynamics of the monopoles in MAP is driven by the total $SU(2)$ action.

b) It is difficult to be more quantitative about the excess of the action basing on the direct data quoted above. In particular, we should have in mind that for finite a there are geometrical corrections to the equation (6). Indirect evidence could be more precise. In particular, it is rather obvious that the scaling of the monopole density (see (21)) implies:

$$|M_{mon}(a) - \frac{\ln 7}{\mathrm{a}}| \sim \Lambda_{QCD} \tag{23}$$

so that the action per unit length of the monopole trajectory does not depend on the lattice spacing a.

c) Also, independence on the lattice spacing of the temperature (3) of the phase transition suggests strongly validity of (23). Indeed, the measurements at the smallest a available, $a \sim 0.06 fm$, see Fig. 1, suggest

$$M_{mon} > 4\,GeV, \quad M_{mon} \gg T_{deconf}, \tag{24}$$

[7] We hope that the notations are not confusing: there are two monopole masses being discussed. One is the standard magnetic field energy (see (4)) and the other is what we call physical mass, m^2_{phys} and this mass determines propagation of a free monopole.

Moreover, it is well known that at the point of the phase transition the monopole trajectories change drastically. Such a sensitivity of the monopoles to the temperature is possible only if the effect of the self energy of the monopole is mainly canceled by the entropy factor, see (23).

Also, an analysis of the data in [19] suggests that

$$T_{deconf} \sim d_{mon}^{-1}, \tag{25}$$

where d_{mon} is the distance between the monopoles in the infrared cluster, $d_{mon} \sim 0.5 fm$ [5]. Thus the temperature is not sensitive to our ultraviolet parameter which is the size of the monopole.

d) Phenomenological fits suggest [11]:

$$M_{mon} \approx M_{mon}^{Coul}(a) + const, \quad const > 0\,, \tag{26}$$

where by M_{mon} we understand the action associated with the monopole. Note also that the Coulombic part of the mass, $M_{mon}^{Coul}(a)$ is of order $1/g^2a$.

Let us recall the reader that on the theoretical side our main concern was that there is no reason why $M_{mon}(a)$ cannot drop to zero. Now we see that our fears are not justified: the monopole self energy is even higher than it would be in the pure Coulomb-like case! As far as we concentrate on a single monopole there is no way to understand (26). But this is indeed numerically necessary for the fine tuning.

Thus, the fine tuning (22) seems to be granted by the data.

4.2 The Origin of the Huge Mass Scale

We are talking actually about small distances, by all the standards of QCD. The numerical value [5] of the size of the monopole (2) is much smaller than the inverse temperature of the phase transition.

The radius (2) is defined in terms of the derivative from the monopole action with respect to a, see [5]. What we would like to emphasize here is that the actual "physical size" of the monopole can be much smaller than (2). By the physical size R_{phys} we understand now the distances where the excess of the monopole action is parametrically smaller than the action associated with the zero-point fluctuations. It is the R_{phys} where the asymptotic freedom actually reigns, not R_{mon} quoted in (2).

No evidence exists at the moment that reaching R_{phys} is in sight, see Fig. 1. Indeed, in the lattice units used in Fig. 1 the excess of the action density of order Λ^4_{QCD} would look like having zero at $a = 0$ and approaching this zero as a^4. Having in mind the data showed in Fig 1 it is tempting to speculate that the onset of such a behavior is still far off from the presently available lattice spacings.

Moreover, as we will argue now it looks plausible that the R_{phys} is shifted to the scale

$$R_{phys} \sim (100\ GeV)^{-1}\,. \tag{27}$$

Before giving arguments in favor of (27) let us ask ourselves, why the estimate (27) is difficult to accept, at least at first sight so. The reason is obvious: one thinks usually about non-perturbative effects in quasi-classical terms, which work in the instanton case. Thus, one assumes that the probability to find non-perturbative effects is exponentially small at small $g^2(a)$, $exp(\; - c/g^2(a))$.

But the failure of such a logic in the monopole case is evident from the case of the compact $U(1)$, see above. Even the monopoles with infinite (Euclidean) action condense. Moreover, R_{phys} is naturally determined by the running of the coupling which is logarithmic and can result in huge factors in the linear scale.

Let us make simple estimates. Namely, the $U(1)$ critical coupling is well known, $e^2_{crit} \sim 1$. In the QCD case we can rewrite the condition (7) as a condition on the R_{phys}. In the realistic case we have at the LEP energies $E^2 \sim (100 \text{ GeV})^2$, $\alpha \approx 0.1$. Then

$$M_{phys} \; \sim \; \text{TeV} \tag{28}$$

and, remarkably enough, we are getting rather the weak interactions scale than $\sim \Lambda_{QCD}$.

Also, the $SU(2)$ lattice measurements typically refer to $\beta \sim 2.6$ while our guess about R_{phys} asks for measurements at $\beta \sim 4$ which are absolutely unrealistic at the moment.

Thus, we come to a paradoxical conclusion that the presently available β are too low to see dissolution of the monopoles at small distances. Moreover, because the running of the coupling is only logarithmic the scale of of the onset asymptotic freedom – which is defined now as the vanishing of the excess of the monopole action compared to the zero-point-fluctuations action– can be very far off.

It is amusing to notice [8] that in case of the $SU(3)$ gluodynamics on the lattice $g^2 = 1$, or $\beta = 6$ corresponds to the lattice spacing $a \approx 0.1 \;\; fm$ and the scale is:

$$R^{SU(3)}_{phys} \; \sim \; (2\text{GeV})^{-1} \, .$$

Thus, through dedicated studies of the monopoles in the $SU(3)$ case it is possible to clarify whether there is a crucial change in the monopole structure at the point $g^2(a) \approx 1$.

5 Conclusions

We have argued that data are emerging which indicate that QCD, when projected onto the scalar-field theory via monopoles corresponds to a fine tuned theory. Which is if course extremely interesting, if true, in view of the mystery of the fine tuning in the Standard Model. The monopoles which we considered are defined ("detected") through the Maximal Abelian projection. However, the mass scales which exhibit mass hierarchy are gauge independent. The scales are provided by the $SU(2)$ invariant action per unit length of the monopole trajectory, on one hand, and by the temperature of the phase transition, on the other.

[8] The observation is due to M.I. Polikarpov.

More generally, we have found that the polymer approach allows to get a new insight into the mechanism of the monopole condensation.

Acknowledgements

I am grateful to S. Caracciolo, M.N. Chernodub, F.V. Gubarev, R. Hofmann, K. Konishi, K. Langfeld, S. Narison, M.I. Polikarpov and L. Stodolsky for discussions. Special thanks are due to M.N. Chernodub for numerous communications and thorough discussions of the results.

I am grateful to the organizers of the Meeting and especially to Prof. J. Trampetic, for the invitation and hospitality.

References

1. Y. Nambu, Phys. Rev. D **10**, 3262 (1974); G. 't Hooft, in *High Energy Physics*, Editorici Compositori, Bologna, (1975); S. Mandelstam, Phys. Rep. C **23**, 516 (1976).
2. M.N. Chernodub, M.I. Polikarpov, in "Cambridge 1997, Confinement, duality, and nonperturbative aspects of QCD", p. 387; hep-th/9710205; T. Suzuki, Prog. Theor. Phys. Suppl. **131**, 633 (1998); A. Di Giacomo, Prog. Theor. Phys. Suppl. **131**, 161 (1998).
3. M.N. Chernodub, F.V. Gubarev, M.I. Polikarpov, V.I. Zakharov, Nucl. Phys. B **592**, 107 (2000); Nucl. Phys. B **600**, 163 (2001).
4. M.N. Chernodub, F.V. Gubarev, M.I. Polikarpov, V.I. Zakharov, Phys. Atom. Nucl. **64**, 561 (2001).
5. V.G. Bornyakov et al, hep-lat/0103032; V.A. Belavin, M.I. Polikarpov, A.I. Veselov, hep-lat/0110011.
6. K. Symanzik, in *Local Quantum Theory*, (1969) Varenna International School of Physics, Course XLV, p. 152.
7. M. Stone, P.R. Thomas, Phys. Rev. Lett. **41**, 351 (1978); S. Samuel, Nucl. Phys. B **154**, 62 (1979).
8. C. A. De Carvalhom, S. Caracciolo, J. Frohlich, Nucl. Phys. B **215**, 209 (1983).
9. A.M. Polyakov, Phys. Lett. B **59**, 82 (1975).
10. H. Shiba, T. Suzuki, Phys. Lett. B **343**, 315 (1995).
11. T. Suzuki, H. Shiba, Phys. Lett. B **351**, 519 (1995); S. Kato *et al*, Nucl. Phys. B **520**, 323 (1998); M.N. Chernodub et al, Phys. Rev. D **62**, 094506 (2000).
12. J. Ambjorn, B. Durhuus, Th. Johnsson, *Quantum Geometry*, Cambridge University Press (1997), Cambridge Monographs on Mathematical Physics.
13. G. Parisi, *Statistical Field Theory*, Addison-Wesley, (1988).
14. F.V. Gubarev, V. I. Zakharov, Nucl. Phys. Proc. Suppl. **106**, 622 (2002); Int. J. Mod. Phys. A **17**, 157 (2002).
15. T. Suzuki, I. Yotsuyanagi, Phys. Rev. D **42**, 4257 (1990); G.S. Bali, V. Bornyakov, M. Mueller-Preussker, K. Schilling, Phys. Rev. D **54**, 2863 (1996).
16. V. Bornyakov et al, hep-lat/0111042.
17. V. Bornyakov, M. Muller-Preussker, hep-lat/0110209.
18. T.L. Ivanenko, A.V. Pochinsky, M.I. Polikarpov, Phys. Lett. B **252**, 631 (1990); S. Kitahara, Y. Matsubara, T. Suzuki, Progr. Theor. Phys. **93**, 1 (1995); A. Hart, M. Teper, Phys. Rev. D **58**, 014504 (1998).
19. K. Ishiguro, Y. Nakatani, T. Suzuki, Prog. Theor. Phys. Suppl. **138**, 35 (2000).

Short Talks (on CD-ROM)

- Damir BOSNAR: Some Recent Results from the A1 Collaboration (Electron Scattering) at MAMI Mainz
- Maja BURIĆ: The Dyon Solutions in Born-Infeld-Yang-Mills Theory
- Martin BUYSSE: One-Loop Natural Relations in the Standard Model
- Herman J. M. CUESTA: Gravitational Waves from Neutrino Oscillations
- Goran DJORDJEVIĆ: Towards Adelic Noncommutative Quantum Mechanics
- Michael FAISST: Three-Loop Corrections to the rho-Parameter
- Svjetlana FAJFER: $\Delta S = 2$ Decays of B^- Meson
- Olga IGONKINA: Charmonium Hadro Production at HERA-B
- Larisa JONKE: Chern-Simons Matrix Model – Algebraic Approach
- Ritva KINNUNEN: Expectations for Charged Higgs in CMS
- Olga L. KODOLOVA: Heavy Ion Study in CMS
- Gordana MEDIN: Tracking in a High Rate Environment (HERA-B Experiment)
- Ivan MELO: Resonances from Strongly-Interacting Electroweak Symmetry Breaking Sector at Future e^+e^- Colliders
- Dieter MUELLER: Predictions for Deeply Virtual Compton Scattering
- Alexander NIKITENKO: Study of $H \to \tau$ Channels in CMS
- Hrvoje NIKOLIĆ: Self-gravitating Bosons at Finite Temperature
- Kornelija PASSEK: BLM Scale for the Pion Transition Form Factor
- Ivica PICEK: Selected Decays of Massive Neutrinos
- Michael PRASZALOWICZ: Pion Light Cone Wave Function in the Nonlocal NJL Model
- Predrag PRESTER: Black Hole Entropy from CFT at Horizon
- Tomislav PROKOPEC: Sources in Electroweak Baryogenesis
- Voja RADOVANOVIĆ: The Hawking Radiation of BTZ Black Hole and Its Quantum Corrections
- Gabriele SEGNERI: Squarks and Gluino Searches with CMS at LHC
- Igor SOLOVTSOV: Target Mass Effects and the Jost-Lehmann-Dyson Representation for Structure Functions
- Yasutaka TAKANISHI: Baryogenesis via Lepton Number Violation in Anti-GUT Model
- Gloria VUAGNIN: Observation of CP Violation in B Mesons with BaBar
- Jure ŽUPAN: The Rare Decay $D^0 \to \gamma\gamma$

About the Meeting

The 8th Adriatic Meeting was organized by:
Rudjer Bošković Institute, Zagreb; Universities of Split and Zagreb; Ludwig-Maximilians-Universität, Munich; Max-Planck-Institut, Munich and WIGV.

The Meeting was to large extent supported by VolksWagen Stiftung.
We would also like to thank for support:
Ministry of Science and Technology, Republic of Croatia; Croatian Academy of Sciences and Arts; Rudjer Bošković Institute; The Abdus Salam International Centre for Theoretical Physics

International Advisory Committee

D. Amati, K. Berkelman, M. Blagojević, S.J. Brodsky, M. Burić, G. Buschhorn, D. Chang, D. Denegri, N.G. Deshpande, G. Eilam, S. Fajfer, W.S. Hou, R. Jackiw, P. Jenni, K. Kleinknecht, L. Lederman, A. Ljubičić, D. Lüst, L. Maiani, H. Nicolai, J. Niederle, S. Pakvasa, R. Peccei, H. Pietschmann, J. Pišut, G. Pócsik, H. Rubinstein, A.I. Sanda, H. Satz, G. Senjanović, M. Shifman, D.V. Shirkov, K. Suruliz, D. Tadić

National Advisory Committee

M. Dželalija, M. Furić, B. Guberina, K. Kadija, M. Martinis, I. Picek, K. Pisk, N. Zovko

Organizing Committee

I. Andrić, Z. Antunović, N. Bilić, L. Jonke, L. Möller, S. Pallua, K. Passek, J. Trampetić – Chairman, J. Wess – Chairman

List of Participants

Guido ALTARELLI
CERN, Geneva
guido.altarelli@cern.ch

Gilvan A. ALVES
Centro Brasileiro de Pesquisas Fisicas, Rio de Janeiro
gilvan@cbpf.br

Ivan ANDRIĆ
Rudjer Bošković Institute, Zagreb
iandric@irb.hr

Željko ANTUNOVIĆ
University of Split
Zeljko.Antunovic@pmfst.hr

Ana BABIĆ
Rudjer Bošković Institute, Zagreb
ababic@thphys.irb.hr

Antun BALAŽ
Institute of Physics, Zemun
antun@phy.bg.ac.yu

Neven BILIĆ
Rudjer Bošković Institute, Zagreb
bilic@thphys.irb.hr

Ines BILJAN
Rudjer Bošković Institute, Zagreb
Ines.Biljan@hotmail.com

Loriano BONORA
SISSA, Trieste
bonora@he.sissa.it

Damir BOSNAR
University of Zagreb
bosnar@phy.hr

Wilfried BUCHMULLER
DESY, Hamburg
buchmuwi@mail.desy.de

Maja BURIĆ
University of Belgrade
majab@ff.bg.ac.yu

Gerd W. BUSCHHORN
Max-Planck-Institute, Munich
gwb@mppmu.mpg.de

Martin BUYSSE
Universite Catholique de Louvain
buysse@fyma.ucl.ac.be

Herman J. M. CUESTA
Centro Brasileiro de Pesquisas Fisicas, Rio de Janeiro
hermanjc@cbpf.br

Maro CVITAN
University of Zagreb
mcvitan@phy.hr

Ivan DADIĆ
Rudjer Bošković Institute, Zagreb
dadic@faust.irb.hr

Daniel DENEGRI
CERN, Geneva
denegri@cern.ch

Radovan DERMIŠEK
Charles University, Prague
dermisek@pacific.mps.
ohio-state.edu

N. G. DESHPANDE
University of Oregon, Eugene
desh@oregon.uoregon.edu

Marija DIMITRIJEVIĆ
University of Belgrade
dmarija@ff.bg.ac.yu

Goran DJORDJEVIĆ
University of Niš
gorandj@junis.ni.ac.yu

Goran DUPLANČIĆ
Rudjer Bošković Institute, Zagreb
gorand@thphys.irb.hr

John ELLIS
CERN, Geneva
john.ellis@cern.ch

Michael FAISST
University of Karlsruhe
mfaisst@particle.physik.
uni-karlsruhe.de

Svjetlana FAJFER
University of Ljubljana
svjetlana.fajfer@ijs.si

Szilard FARKAS
Eotvos Lorand University, Budapest
ifarkas@hu.ibm.com

Michael FINGER
Joint Institute for Nuclear Research, Dubna
michael.finger@cern.ch

Yitzhak FRISHMAN
Weizmann Institute, Rehovot
yitzhak.frishman@weizmann.ac.il

Nikola GODINOVIĆ
University of Split
Nikola.Godinovic@fesb.hr

Krzysztof GRACZYK
University of Wroclaw
kgraczyk@ift.uni.wroc.pl

Harald GROSSE
University of Vienna
grosse@doppler.thp.univie.ac.at

Branko GUBERINA
Rudjer Bošković Institute, Zagreb
guberina@thphys.irb.hr

Zsuzsanna GYORY
Eotvos Lorand University, Budapest
gyory@complex.elte.hu

Davor HORVATIĆ
University of Zagreb
davorh@phy.hr

Raul HORVAT
Rudjer Bošković Institute, Zagreb
horvat@lei3.irb.hr

Dario HRUPEC
University of Zagreb
dario.hrupec@fer.hr

Helmuth HUEFFEL
University of Vienna
helmuth.hueffel@univie.ac.at

Robert IENGO
SISSA, Trieste
iengo@he.sissa.it

Olga IGONKINA
ITEP, Moscow
olya@mail.desy.de

Roman JACKIW
MIT, Cambridge
jackiw@lns.mit.edu

Larisa JONKE
Rudjer Bošković Institute, Zagreb
larisa@irb.hr

Vladimir JURČIĆ
Institute of Physics, Zemun
juricic@phy.bg.ac.yu

Danijel JURMAN
Rudjer Bošković Institute, Zagreb
jurman@irb.hr

Harris KAGAN
Ohio State University, Columbus
KAGAN@mps.ohio-state.edu

Michael KATANAEV
Steklov Mathematical Institute, Moscow
katanaev@mi.ras.ru

Ritva KINNUNEN
Helsinki Institute of Physics
ritva.kinnunen@cern.ch

Dubravko KLABUČAR
University of Zagreb
klabucar@phy.hr

Konrad KLEINKNECHT
Johannes-Gutenberg Universitaet, Mainz
kleinknecht@dipmza.physik.uni-mainz.de

Olga L. KODOLOVA
Skobeltcyn Institute of Nuclear Physics, Moscow
Olga.Kodolova@cern.ch

Krešimir KUMERIČKI
University of Zagreb
kkumer@phy.hr

Duško LATAS
University of Belgrade
latas@ff.bg.ac.yu

Matt LILLEY
Heidelberg University
lilley@thphys.uni-heidelberg.de

Ante LJUBIČIĆ
Rudjer Bošković Institute, Zagreb
aljubic@irb.hr

Mladen MARTINIS
Rudjer Bošković Institute, Zagreb
martinis@irb.hr

Antonio MASIERO
SISSA, Trieste
masiero@he.sissa.it

Gordana MEDIN
University of Montenegro, Podgorica
medin@ifh.de

Blaženka MELIĆ
Rudjer Bošković Institute, Zagreb
melic@thphys.irb.hr

Stjepan MELJANAC
Rudjer Bošković Institute, Zagreb
meljanac@irb.hr

Ivan MELO
University of Zilina
melo@fel.utc.sk

Vesna MIKUTA-MARTINIS
Rudjer Bošković Institute, Zagreb
mikutama@irb.hr

Marijan MILEKOVIĆ
University of Zagreb
marijan@phy.hr

Lutz MOELLER
University of Munich
lmoeller@theorie.physik.
uni-muenchen.de

Dieter MUELLER
University of Wuppertal
dmueller@theorie.physik.
uni-wuppertal.de

Holger B. NIELSEN
Niels Bohr Institute, Copenhagen
hbech@nbivms.nbi.dk

Alexander NIKITENKO
ITEP, Moscow
Alexandre.Nikitenko@cern.ch

Hrvoje NIKOLIĆ
Rudjer Bošković Institute, Zagreb
hrvoje@thphys.irb.hr

Bene NIŽIĆ
Rudjer Bošković Institute, Zagreb
nizic@thphys.irb.hr

Sandip PAKVASA
University of Hawaii, Honolulu
pakvasa@phys.hawaii.edu

Silvio PALLUA
University of Zagreb
pallua@phy.hr

Kornelija PASSEK
Rudjer Bošković Institute, Zagreb
passek@thphys.irb.hr

Željko PASTUOVIĆ
Rudjer Bošković Institute, Zagreb
pastu@irb.hr

Matej PAVŠIČ
Jozef Stefan Institute, Ljubljana
matej.pavsic@ijs.si

Ivica PICEK
University of Zagreb
picek@phy.hr

Michael PRASZALOWICZ
Jagellonian University, Krakow
michal@th.if.uj.edu.pl

Predrag PRESTER
University of Zagreb
pprester@phy.hr

Tomislav PROKOPEC
Heidelberg University
T.Prokopec@thphys.
uni-heidelberg.de

Ivica PULJAK
University of Split
Ivica.Puljak@fesb.hr

Voja RADOVANOVIĆ
University of Belgrade
rvoja@ff.bg.ac.yu

Mitja ROSINA
University of Ljubljana
mitja.rosina@ijs.si

Yoram ROZEN
Technion, Haifa
rozen@phep7.technion.ac.il

Paolo SALUCCI
SISSA, Trieste
salucci@sissa.it

A. I. SANDA
Nagoya University
sanda@eken.phys.nagoya-u.ac.jp

Helmut SATZ
University of Bielefeld
satz@physik.uni-bielefeld.de

Peter SCHUPP
University of Munich
schupp@theorie.physik.
uni-muenchen.de

Gabriele SEGNERI
INFN, Pisa
gabriele.segneri@pi.infn.it

Dmitry SHIRKOV
Bogoliubov Lab at JINR, Dubna
shirkovd@thsun1.jinr.ru

Yu. A. SITENKO
Bogolyubov Institute for Theoretical Physics, Kiev
yusitenko@bitp.kiev.ua

Igor SOLOVTSOV
Bogoliubov Lab at JINR, Dubna
solovtso@thsun1.jinr.ru

Hrvoje ŠTEFANČIĆ
Rudjer Bošković Institute, Zagreb
shrvoje@thphys.irb.hr

Nikolaos G. STEFANIS
Ruhr-University Bochum
stefanis@tp2.ruhr-uni-bochum.de

Marko STOJIĆ
Rudjer Bošković Institute, Zagreb
mstojic@thphys.irb.hr

Hans STREMNITZER
University of Vienna
strem@ap.univie.ac.at

Dubravko TADIĆ
University of Zagreb
tadic@phy.hr

Yasutaka TAKANISHI
Niels Bohr Institute, Copenhagen
yasutaka@nbi.dk

Josip TRAMPETIĆ
Rudjer Bošković Institute, Zagreb
josip@thphys.irb.hr

Nikolai URALTSEV
INFN, Sezione di Milano
Nikolai.Uraltsev@mib.infn.it

Milivoj UROIĆ
University of Zagreb
muroic@phy.hr

Claudio VERZEGNASSI
University of Trieste
claudio@trieste.infn.it

Miroslav VESKOVIĆ
University of Novi Sad
veskovic@unsim.im.ns.ac.yu

Gloria VUAGNIN
INFN, Sezione di Trieste
vuagnin@SLAC.Stanford.EDU

Roland WALDI
University of Rostock
Roland.Waldi@physik.
uni-rostock.de

Julius WESS
University of Munich
julius.wess@physik.
uni-muenchen.de

Valentine ZAKHAROV
Max-Planck Institute, Munich
xxz@mppmu.mpg.de

Nikola ZOVKO
Rudjer Bošković Institute, Zagreb
zovko@thphys.irb.hr

Jure ŽUPAN
Jozef Stefan Institute, Ljubljana
jure.zupan@ijs.si

Druck: Strauss Offsetdruck, Mörlenbach
Verarbeitung: Schäffer, Grünstadt

Lecture Notes in Physics

For information about Vols. 1–589
please contact your bookseller or Springer-Verlag

Vol.590: D. Benest, C. Froeschlé (Eds.), Singularities in Gravitational Systems. Applications to Chaotic Transport in the Solar System.
link.springer.de/link/service/series/2669/tocs/t2590.htm

Vol.591: M. Beyer (Ed.), CP Violation in Particle, Nuclear and Astrophysics.
link.springer.de/link/service/series/2669/tocs/t2591.htm

Vol.592: S. Cotsakis, L. Papantonopoulos (Eds.), Cosmological Crossroads. An Advanced Course in Mathematical, Physical and String Cosmology.
link.springer.de/link/service/series/2669/tocs/t2592.htm

Vol.593: D. Shi, B. Aktaş, L. Pust, F. Mikhailov (Eds.), Nanostructured Magnetic Materials and Their Applications.
link.springer.de/link/service/series/2669/tocs/t2593.htm

Vol.594: S. Odenbach (Ed.),Ferrofluids. Magnetical Controllable Fluids and Their Applications.
link.springer.de/link/service/series/2669/tocs/t2594.htm

Vol.595: C. Berthier, L. P. Lévy, G. Martinez (Eds.), High Magnetic Fields. Applications in Condensed Matter Physics and Spectroscopy.
link.springer.de/link/service/series/2669/tocs/t2595.htm

Vol.596: F. Scheck, H. Upmeier, W. Werner (Eds.), Noncommutative Geometry and the Standard Model of Elememtary Particle Physics.
link.springer.de/link/service/series/2669/tocs/t2596.htm

Vol.597: P. Garbaczewski, R. Olkiewicz (Eds.), Dynamics of Dissipation.
link.springer.de/link/service/series/2669/tocs/t2597.htm

Vol.598: K. Weiler (Ed.), Supernovae and Gamma-Ray Bursters.
Online version forthcoming

Vol.599: J.P. Rozelot (Ed.), The Sun's Surface and Subsurface. Investigating Shape and Irradiance.
Online version forthcoming

Vol.600: K. Mecke, D. Stoyan (Eds.), Morphology of Condensed Matter. Physcis and Geometry of Spatial Complex Systems.
link.springer.de/link/service/series/2669/tocs/t2600.htm

Vol.601: F. Mezei, C. Pappas, T. Gutberlet (Eds.), Neutron Spin Echo Spectroscopy. Basics, Trends and Applications.
link.springer.de/link/service/series/2669/tocs/t2601.htm

Vol.602: T. Dauxois, S. Ruffo, E. Arimondo (Eds.), Dynamics and Thermodynamics of Systems with Long Range Interactions.
link.springer.de/link/service/series/2669/tocs/t2602.htm

Vol.603: C. Noce, A. Vecchione, M. Cuoco, A. Romano (Eds.), Ruthenate and Rutheno-Cuprate Materials. Superconductivity, Magnetism and Quantum Phase.
link.springer.de/link/service/series/2669/tocs/t2603.htm

Vol.604: J. Frauendiener, H. Friedrich (Eds.), The Conformal Structure of Space-Time: Geometry, Analysis, Numerics.
link.springer.de/link/service/series/2669/tocs/t2604.htm

Vol.605: G. Ciccotti, M. Mareschal, P. Nielaba (Eds.), Bridging Time Scales: Molecular Simulations for the Next Decade.
link.springer.de/link/service/series/2669/tocs/t2605.htm

Vol.606: J.-U. Sommer, G. Reiter (Eds.), Polymer Crystallization. Obervations, Concepts and Interpretations.
Online version forthcoming

Vol.607: R. Guzzi (Ed.), Exploring the Atmosphere by Remote Sensing Techniques.
link.springer.de/link/service/series/5304/tocs/t3607.htm

Vol.608: F. Courbin, D. Minniti (Eds.), Gravitational Lensing:An Astrophysical Tool.
link.springer.de/link/service/series/2669/tocs/t2608.htm

Vol.609: T. Henning (Ed.), Astromineralogy.
Online version forthcoming

Vol.610: M. Ristig, K. Gernoth (Eds.), Particle Scattering, X-Ray Diffraction, and Microstructure of Solids and Liquids.
link.springer.de/link/service/series/5304/tocs/t3610.htm

Vol.611: A. Buchleitner, K. Hornberger (Eds.), Coherent Evolution in Noisy Environments.
link.springer.de/link/service/series/2669/tocs/t2611.htm

Vol.612 L. Klein, (Ed.), Energy Conversion and Particle Acceleration in the Solar Corona.
Online version forthcoming

Vol.613 K. Porsezian, V.C. Kuriakose, (Eds.), Optical Solitons. Theoretical and Experimental Challenges.
link.springer.de/link/service/series/2669/tocs/t3613.htm

Vol.614 E. Falgarone, T. Passot (Eds.), Turbulence and Magnetic Fields in Astrophysics.
link.springer.de/link/service/series/5304/tocs/t3614.htm

Vol.615 J. Büchner, C.T. Dum, M. Scholer (Eds.), Space Plasma Simulation.
Online version forthcoming

Vol.616 J. Trampetic, J. Wess (Eds.), Particle Physics in the New Millenium.
Online version forthcoming

Vol.617 L. Fernández-Jambrina, L. M. González-Romero (Eds.), Current Trends in Relativistic Astrophysics, Theoretical, Numerical, Observational
Online version forthcoming

Vol.618 M.D. Esposti, S. Graffi (Eds.), The Mathematical Aspects of Quantum Maps
Online version forthcoming

Vol.619 H.M. Antia, A. Bhatnagar, P. Ulmschneider (Eds.), Lectures on Solar Physics
Online version forthcoming

Vol.620 C. Fiolhais, F. Nogueira, M. Marques (Eds.), A Primer in Density Functional Theory
Online version forthcoming

Monographs

For information about Vols. 1–30 please contact your bookseller or Springer-Verlag

Vol. m 31 (Corr. Second Printing): P. Busch, M. Grabowski, P.J. Lahti, Operational Quantum Physics. XII, 230 pages. 1997.

Vol. m 32: L. de Broglie, Diverses questions de mécanique et de thermodynamique classiques et relativistes. XII, 198 pages. 1995.

Vol. m 33: R. Alkofer, H. Reinhardt, Chiral Quark Dynamics. VIII, 115 pages. 1995.

Vol. m 34: R. Jost, Das Märchen vom Elfenbeinernen Turm. VIII, 286 pages. 1995.

Vol. m 35: E. Elizalde, Ten Physical Applications of Spectral Zeta Functions. XIV, 224 pages. 1995.

Vol. m 36: G. Dunne, Self-Dual Chern-Simons Theories. X, 217 pages. 1995.

Vol. m 37: S. Childress, A.D. Gilbert, Stretch, Twist, Fold: The Fast Dynamo. XI, 406 pages. 1995.

Vol. m 38: J. González, M. A. Martín-Delgado, G. Sierra, A. H. Vozmediano, Quantum Electron Liquids and High-Tc Superconductivity. X, 299 pages. 1995.

Vol. m 39: L. Pittner, Algebraic Foundations of Non-Com-mutative Differential Geometry and Quantum Groups. XII, 469 pages. 1996.

Vol. m 40: H.-J. Borchers, Translation Group and Particle Representations in Quantum Field Theory. VII, 131 pages. 1996.

Vol. m 41: B. K. Chakrabarti, A. Dutta, P. Sen, Quantum Ising Phases and Transitions in Transverse Ising Models. X, 204 pages. 1996.

Vol. m 42: P. Bouwknegt, J. McCarthy, K. Pilch, The W3 Algebra. Modules, Semi-infinite Cohomology and BV Algebras. XI, 204 pages. 1996.

Vol. m 43: M. Schottenloher, A Mathematical Introduction to Conformal Field Theory. VIII, 142 pages. 1997.

Vol. m 44: A. Bach, Indistinguishable Classical Particles. VIII, 157 pages. 1997.

Vol. m 45: M. Ferrari, V. T. Granik, A. Imam, J. C. Nadeau (Eds.), Advances in Doublet Mechanics. XVI, 214 pages. 1997.

Vol. m 46: M. Camenzind, Les noyaux actifs de galaxies. XVIII, 218 pages. 1997.

Vol. m 47: L. M. Zubov, Nonlinear Theory of Dislocations and Disclinations in Elastic Body. VI, 205 pages. 1997.

Vol. m 48: P. Kopietz, Bosonization of Interacting Fermions in Arbitrary Dimensions. XII, 259 pages. 1997.

Vol. m 49: M. Zak, J. B. Zbilut, R. E. Meyers, From Instability to Intelligence. Complexity and Predictability in Nonlinear Dynamics. XIV, 552 pages. 1997.

Vol. m 50: J. Ambjørn, M. Carfora, A. Marzuoli, The Geometry of Dynamical Triangulations. VI, 197 pages. 1997.

Vol. m 51: G. Landi, An Introduction to Noncommutative Spaces and Their Geometries. XI, 200 pages. 1997.

Vol. m 52: M. Hénon, Generating Families in the Restricted Three-Body Problem. XI, 278 pages. 1997.

Vol. m 53: M. Gad-el-Hak, A. Pollard, J.-P. Bonnet (Eds.), Flow Control. Fundamentals and Practices. XII, 527 pages. 1998.

Vol. m 54: Y. Suzuki, K. Varga, Stochastic Variational Approach to Quantum-Mechanical Few-Body Problems. XIV, 324 pages. 1998.

Vol. m 55: F. Busse, S. C. Müller, Evolution of Spontaneous Structures in Dissipative Continuous Systems. X, 559 pages. 1998.

Vol. m 56: R. Haussmann, Self-consistent Quantum Field Theory and Bosonization for Strongly Correlated Electron Systems. VIII, 173 pages. 1999.

Vol. m 57: G. Cicogna, G. Gaeta, Symmetry and Perturbation Theory in Nonlinear Dynamics. XI, 208 pages. 1999.

Vol. m 58: J. Daillant, A. Gibaud (Eds.), X-Ray and Neutron Reflectivity: Principles and Applications. XVIII, 331 pages. 1999.

Vol. m 59: M. Kriele, Spacetime. Foundations of General Relativity and Differential Geometry. XV, 432 pages. 1999.

Vol. m 60: J. T. Londergan, J. P. Carini, D. P. Murdock, Binding and Scattering in Two-Dimensional Systems. Applications to Quantum Wires, Waveguides and Photonic Crystals. X, 222 pages. 1999.

Vol. m 61: V. Perlick, Ray Optics, Fermat's Principle, and Applications to General Relativity. X, 220 pages. 2000.

Vol. m 62: J. Berger, J. Rubinstein, Connectivity and Superconductivity. XI, 246 pages. 2000.

Vol. m 63: R. J. Szabo, Ray Optics, Equivariant Cohomology and Localization of Path Integrals. XII, 315 pages. 2000.

Vol. m 64: I. G. Avramidi, Heat Kernel and Quantum Gravity. X, 143 pages. 2000.

Vol. m 65: M. Hénon, Generating Families in the Restricted Three-Body Problem. Quantitative Study of Bifurcations. XII, 301 pages. 2001.

Vol. m 66: F. Calogero, Classical Many-Body Problems Amenable to Exact Treatments. XIX, 749 pages. 2001.

Vol. m 67: A. S. Holevo, Statistical Structure of Quantum Theory. IX, 159 pages. 2001.

Vol. m 68: N. Polonsky, Supersymmetry: Structure and Phenomena. Extensions of the Standard Model. XV, 169 pages. 2001.

Vol. m 69: W. Staude, Laser-Strophometry. High-Resolution Techniques for Velocity Gradient Measurements in Fluid Flows. XV, 178 pages. 2001.

Vol. m 70: P. T. Chruściel, J. Jezierski, J. Kijowski, Hamiltonian Field Theory in the Radiating Regime. VI, 172 pages. 2002.

Vol. m 71: S. Odenbach, Magnetoviscous Effects in Ferrofluids. X, 151 pages. 2002.

Vol. m 72: J. G. Muga, R. Sala Mayato, I. L. Egusquiza (Eds.), Time in Quantum Mechanics. XII, 419 pages. 2002.

Vol. m 73: H. Emmerich, The Diffuse Interface Approach in Materials Science. VIII, 178 pages. 2003